Vertebrates

Structures and Functions

Biological Systems in Vertebrates

Series Editors

Hiran Dutta and Douglas Kline

Department of Biological Science

Kent State University

Kent, Ohio, USA

Books in this Series

J.N. Maina—Functional Morphology of the Vertebrate Respiratory Systems

Hans Ditrich—Renal Structure and Function in Vertebrates

Seth Kisia— Muscular System of Vertebrates

Maria Ogielska—Reproduction of Amphibians

Seth Kisia— Vertebrates: Structures and Functions

Vertebrates
Structures and Functions

Seth M. Kisia
Department of Veterinary Anatomy and Physiology
University of Nairobi
Nairobi
KENYA

CRC Press
Taylor & Francis Group
Boca Raton London New York

CRC Press is an imprint of the
Taylor & Francis Group, an **informa** business

CRC Press
Taylor & Francis Group
6000 Broken Sound Parkway NW, Suite 300
Boca Raton, FL 33487-2742

First issued in hardback 2017

ISBN-13: 978-1-5780-8606-1 (pbk)
ISBN-13: 978-1-1384-7590-8 (hbk)

Visit the Taylor & Francis Web site at
http://www.taylorandfrancis.com

and the CRC Press Web site
http://www.crcpress.com

Library of Congress Cataloging-in-Publication Data

Kisia, Seth M.
Vertebrates : structures and functions / Seth M. Kisia.
p. cm.--(Biological systems in vertebrates)
Includes bibliographical references and index.
ISBN 978-1-57808-606-1 (hardcover)
1. Vertebrates--Anatomy. 2. Vertebrates--Physiology. I. Title.
QL805.K67 2009
571.3'16--dc22

2009033178

Preface

Although vertebrates comprise of over 40,000 species that form about 2% of the total animal species on earth, they are the most conspicuous animals on the planet. Vertebrates are generally larger than other animal species. This group of chordates is quite diverse and ranges in weight from about 10mg to 150 tons while sharing many inherited features with the most remarkable one being possession of a backbone. Vertebrates have dominated other animal species due to the possession of complex body systems that are coordinated by relatively advanced nervous and endocrine systems. Vertebrates could have originated from a common ancestor in the sea about 500 million years ago before evolving into the current diverse groups that inhabit many niches including land, water and the air. Adaptive radiation was necessary in the diversification of the various vertebrate species together with their body systems that has resulted in the success of the group on earth. The evolutionary tree of vertebrates showing how members of the subphylum are related through phylogeny has been drawn based on fossil records and more recently by DNA analyses.

This volume is about the diversity of vertebrates and the unique features they possess that have enabled the group occupy most niches on earth. A comparative study of the body systems of various classes of vertebrates has been discussed. The book also looks at evolution of various vertebrate groups and the agents that have been responsible for these changes as well as how the subphylum has radiated into the many extant species. Evolution of vertebrates has been marked by major and minor extinctions with the latest event occurring during late Pleistocene (about 11,000 years ago) during which large mammals weighing more than 44kg and making up to about 70% of large North American mammalian species were lost. The current zoogeographic distribution of vertebrates has occurred partly due to local changes in environmental conditions and the drifting of continents as explained by the continental drift theory of Alfred Wagener (1880-1930).

The author wishes to acknowledge Amos T. Mwasela of the Department of Veterinary Anatomy and Physiology, University of Nairobi, Kenya, for assistance with technical knowledge on use of a computer programme that was quite useful in the illustrative work as well as artistic expertise that was required in improving many of the illustrations. The author also thanks Mr. Mwasela for drawing some of the sketches in the book. I also wish to thank Dr. Hiran M. Dutta, emeritus professor, Department of Biological Sciences, University of Yale, USA for editing the book and making various suggestions that helped improve the quality of the volume. I am also indebted to several colleagues and various friends who were a source of inspiration at the time of writing the book and the University of Nairobi for providing the necessary conducive environment during the period the book was being written.

Contents

Preface *v*

1. General Introduction to the Study of Vertebrates 1

2. Diversity, Distribution and Characteristics of Vertebrates 8

3. Organization of the Vertebrate Body 46

4. Skeletal System 81

5. Muscular System 119

6. Integument 153

7. Nutrition and Digestion 186

8. Respiratory System 218

9. Reproduction and Early Developmental Biology 248

10. Circulatory System 294

11. Nervous System and Endocrine Organs 335

12. Sense Organs 382

13. Excretion and Osmoregulation 423

14. Evolution of Vertebrates 449

Index *519*

Color Plate Section *543*

1

General Introduction to the Study of Vertebrates

Vertebrates are a major group of animals that are fairly well understood and have evolved complex body systems that enable them to compete effectively with and dominate other groups of animals. The complex action of the different vertebrate body systems is coordinated by relatively advanced nervous and endocrine systems. It is this complexity that makes vertebrates quite noticeable since they are able to perform rapid and complex movements, produce a variety of sounds and exploit the different niches in water, air and on land. It was recognized long back by the Greek physician and scientist Claudius Galen (130-200) that a living organism is not a summation of body parts but an indivisible unity.

Vertebrates range in size from the stout infantfish (*Schindleria brevipinguis*) of Australia's Great Barrier Reef and the Coral Sea to the blue whale (*Balaenoptera musculus*) (a mammal) that inhabits the world's oceans (Fig. 1.1). Female stout infantfish measure 8.9 mm whereas males are 8.6 mm in length and are smaller than the dwarf goby that was previously thought to be the world's smallest vertebrate. The infantfish is paedomorphic (adults retain larval characteristics) and males mature by 7.0 mm in length (Watson and Walker Jr., 2004). Another small fish, the paedormophic cyprinid (*Paedocypris progenetica*) has been discovered in acidic forest swamps of the Indonesian island of Sumutra (Kottelat et al., 2006; Lukas et al., 2007). Adult *Paedocypris progenetica* measure 7.9 mm long but are slightly broader than the stout infantfish. The blue whale can weigh over 150 tons and measures 25 m long.

The presence of **bony** or **cartilaginous vertebrae** instead of a notochord makes the backs of vertebrates more rigid than in other

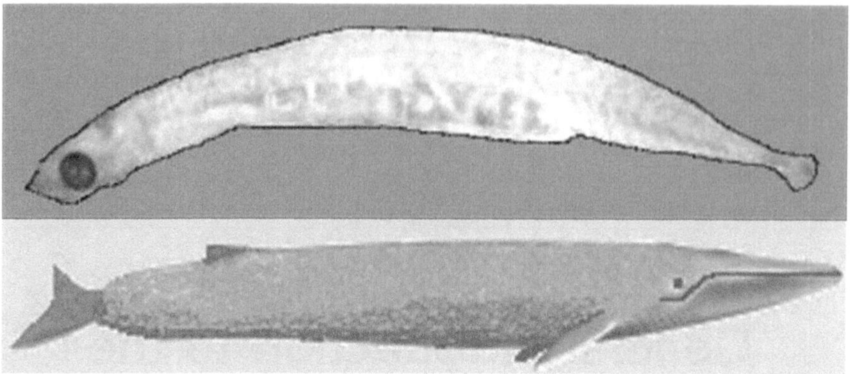

Fig. 1.1 The stout infantfish (above) is the world's smallest vertebrate and fish and the blue whale (below) is the largest animal in the world. Around 106 specimens of the infantfish are needed to just tip a kilogramme. The infantfish lacks teeth, scales and pigmentation except in its eyes. Blue whales are also the loudest animals on earth producing sounds of up to 188 decibels (human pain threshold is about 130 decibels). The heart of a blue whale weighs about 450 kg. Numbering about 1,300 to 2,000 individuals in the world's oceans, blue whales are considered endangered as a result of enlargement in fishing nets, pollution and illegal whaling.

animals and is one of the chief characteristics unique to vertebrates. The evolution of vertebrae in the early aquatic vertebrates was an early adaptive method that enabled vertebrates to perform effective waving back and forth movements that were vital for active swimming movements. This book compares the main characteristics, body systems, evolution and interaction of vertebrates with the environment and each other. Classification of some vertebrates continues to change as more information emerges from studies enhanced by new technology. Such information is important in understanding the current coexistence of vertebrates, the past history and prediction of the future state of this diverse group of animals.

BODY SYSTEMS OF VERTEBRATES

All vertebrates have 11 body systems that work together in a coordinated manner and show various degrees of complexity depending on the mode of life of the vertebrate and the environment in which it lives. Two systems viz. the nervous and endocrine have been considered together as they coordinate activities of the vertebrate body.

The **skeletal system** is the framework of the body and forms the passive part of the locomotor apparatus. The vertebrate skeleton shows variation in structure that has evolved due to various pressures on the system as a result of inhabiting different environments, feeding habits and types of movement that have been adopted. The general design of the skeletal system of vertebrates has many similarities that point to a likely past common ancestor and the changes that have occurred in the system are probably related to adaptation to various aquatic or terrestrial environments. The two types of supportive connective tissue of vertebrates are bone and cartilage and also the notochord that is present in part of the embryonic life of all vertebrates and persists to various degrees in many adults belonging to lower vertebrates. The skeletal system is an articulation of bones or cartilage that is supported by various types of soft tissue including ligaments, joint capsules and tendons of some muscles.

As the **muscular system** is closely related to the skeletal system in performing its functions, the two systems work together in bringing about movement in vertebrates. Since the muscular system forms a major proportion of the vertebrate body weight, its role in heat generation is vital. The presence of different types of muscle (viz. smooth, skeletal and cardiac muscles) has enabled vertebrates to perform complicated body movements when compared to other forms of life on Earth. Vertebrates are capable of performing various types of movement such as swimming, flying and various forms of terrestrial locomotion. Since muscle contraction brings about major body movements in vertebrates, the presence or absence of some major muscles or muscle groups and how similar muscles have undergone structural change can be used explain how certain specific movements occur in various vertebrates. The homology of some specific as well as groups of muscles is based on factors such as nerve supply, points of origin and insertion, appearance and location. Modified muscle organs such as the conducting tissue of the heart and electric organs of some fish have evolved to bring about coordinated contraction of the heart and defence or offence organs respectively.

The integument is a system that is in direct contact with the environment and is the most conspicuous body system. It comprises the skin and its derivatives and covers the entire vertebrate body except at its natural openings. The integument shows variation in structure and some of its functions in various vertebrates. The roles played by the integument such as mechanical protection, possession of sensory organs, thermoregulation mainly in birds and mammals and as an organ of osmoregulation and excretion are vital to the survival of vertebrates. The mammalian skin is an important system in homeostasis of body temperature due to the presence of structures such as sweat glands and its large surface area. Below the skin in birds and mammals is a hypodermis that is rich in fat and insulates the body from heat loss as they are homoeothermic and show little variation in their body temperature in relation to that of the environment. The derivatives of the skin such as nails, claws, horns and poisonous glands have been used as defence and offence weapons in some vertebrates whereas others such as scent glands produce secretions that are used in communication and species recognition. The mammary gland that is unique to mammals might have evolved from sweat glands and produces milk that is used to nourish the young. The presence of pigment producing cells (chromatophores) in the skin imparts various colors to vertebrates.

Nutrition and digestion in various vertebrates is quite diverse between and within major and even minor groups. The roles played by various major food groups, vitamins and certain minerals are of major importance to the survival of vertebrates. The digestive system and processes involved in procurement of food, its digestion and elimination of undigested material that has not been absorbed into the body system in vertebrate groups has been discussed in this chapter. The design of the vertebrate body and its appendages when present is related to methods of food acquisition. Various glands and hormones have evolved that enhance the role of the digestive system. Directly related to the system are teeth that vary depending on the diet of vertebrates.

The vertebrate **respiratory system** has evolved to meet the gaseous exchange requirements of vertebrates depending on whether they live on land, water or air. Since oxygen supply to body tissues is a major factor in determining vertebrate activity and most tissues in active vertebrates cannot last for long without oxygen supply, the design of the system and how it has evolved to meet metabolic requirements remains quite a challenging issue. Ectotherms have an advantage since they do not require as high a supply of oxygen as endotherms under normal circumstances. Endotherms have to maintain a relatively stable body temperature and high metabolic rates and their energy demands and activity is higher than that of ectotherms. The respiratory, cardiovascular and other related body systems have adapted to aquatic and aerial methods of respiration in both environments since there is great variation in the partial pressure of oxygen in these two environments that are inhabited by vertebrates. Birds have had to adapt to the high demands of energy and gas exchange on flight.

Reproduction and early developmental biology of vertebrates are vital processes that ensure the continued survival of a vertebrate species. An efficient reproductive system ensures the propagation of genes to the next generation and the survival of a particular species that is able to compete with other vertebrate species. Vertebrates display a variety of reproductive systems, environment in which they reproduce and nurture the young and the behavior they display at this time. The success of various vertebrates in colonizing different habitats depends to a large extent on their ability to reproduce and adapt to the environment. Vertebrates have evolved various methods of reproduction including oviparity, ovoviviparity and viviparity. Early developmental changes that occur in the zygote after fertilization vary in anamniotes and amniotes with the former requiring mainly an aquatic environment for the relatively shorter incubation period that leads to the production of young that have to start feeding on their own immediately after hatching and that sometimes do not resemble adults of their own species.

The **circulatory or cardiovascular system** plays a vital role in the overall activity of vertebrates as its function determines the efficiency of transporting oxygen and nutrients to all body systems and removal of metabolic wastes to sites of excretion. The simple single circuit system is found in nearly all fishes whereas the double circuit system of circulation is found in higher vertebrates. Vertebrates have various types of blood vessels that play specific roles in circulation. The lymphatic system is part of the overall system of circulation that drains extra-cellular fluid back to the cardiovascular system. Blood is the main transport medium of the vertebrate body and contains erythrocytes that play a role in the transport of oxygen and carbon dioxide and leucocytes that are important in the body's immune response. The liquid part of blood is known as plasma and contains mainly water and solutes that comprise proteins, glucose, amino acids, hormones, enzymes, urea, uric acid and gases. Exchange of material between blood capillaries and tissue cells occurs through the interstitial fluid whose composition determines the osmotic pressure and solute concentration that are responsible for movement of fluid and solutes in various body compartments respectively.

A **nervous system and endocrine organs** of the vertebrate body are different systems that are highly dependent on each other. Together, the two systems play the roles of communication, integration and control of other systems of the vertebrate body. The response that is brought about by the nervous system is much faster than that of hormonal action as hormones have to be transported in blood to target organs. A properly coordinated and integrated system is essential to the proper functioning of all body systems that ensure homeostasis and survival of the vertebrate. The level of development of the nervous and endocrine systems in various vertebrates differs depending on activities performed. The complexity that is displayed by these systems in higher vertebrates is responsible for the sophisticated behavior and intelligence in these animals. The various hormones produced in the vertebrate body by endocrine glands such as the pituitary, adrenal, thyroid and pineal glands act on specific target cells throughout the body. Related to hormonal action are prostaglandins that have an effect in tissues in which they are produced and have a short acting duration.

Sense organs of vertebrates can be grouped into two main categories: general and special sense organs. Vertebrates possess these organs to varying degrees. The numerous general sense organs or receptors such as senses for detecting touch, pain and temperature are distributed in various parts while special sense organs that are concerned with vision, sound, balance, smell or taste are located in specific areas of the vertebrate body. These sense organs have evolved to detect information about the environment and cause sensory transduction to convert the extra-cellular signals into electrical or energy signals. Vertebrates are then able to make the appropriate response according to the nature of signals.

Excretion and osmoregulation are necessary for the maintenance of constancy of the internal environment and involve various structures such as kidneys, gills and the digestive system that are vital and have evolved differently in vertebrates. The form in which the different wastes are excreted from the vertebrate body depends largely on the nature of the environment that is inhabited. Elimination of metabolic wastes from the body is closely related to osmoregulation in many vertebrates since most wastes, especially in many terrestrial vertebrates, are excreted in liquid form and there is loss of water. The conditions faced by freshwater and marine vertebrates in osmoregulation and mechanisms that have evolved to ensure their survival are quite different from those of terrestrial vertebrates.

Evolution of vertebrates has been a continuous process throughout vertebrate history. Evolution occurs as a result of various agents that bring about mutations. Vertebrates first evolved in the sea from a common ancestor about 500 million years ago. The first vertebrates to evolve were the jawless fish and the last ones to emerge were humans. The course of vertebrate history has been shaped by evolution. For vertebrates to continue to survive, changes in the environmental condition have to be accompanied by evolutionary changes that make the vertebrate adaptable. Many major vertebrate groups have emerged or become extinct in the course of evolution. The distribution of vertebrates throughout the world has to a great extent been influenced with the splitting of continents during the continental drift. Previous knowledge on evolution has relied on fossil remains and their distribution. Current knowledge on molecular biology gathered from nuclear and mitochondrial DNA analyses has been recently used to confirm or correct knowledge of the evolutionary tree.

Various **environmental conditions** have influenced the course of vertebrate evolution. Vertebrates have also influenced local environmental conditions to a certain extent. Past climatic conditions have had the maximum environmental impact on the history of vertebrates. Climatic changes in the form of rain, temperature, wind, snow and ocean currents have affected the dynamics of populations by disrupting life-supporting natural systems and processes. The interaction of vertebrates with each other has also affected various species of vertebrates. The human being has had the most effect on other vertebrate groups. Anthropogenic activities at present including human population pressure for space, pollution, hunting and environmental degradation have impacted negatively on vertebrates and threatened certain species. Such human activities have interfered with the natural evolution of many vertebrate species.

REFERENCES

Kottelat, M., Britz, R., Hui, T.H. and Witte, K.E. (2006). *Paedocypris*, a new genus of Southeast Asia cyprinid fish with a remarkable sexual dimorphism, comprises the world's smallest vertebrate. Proc. Biol. Sci., 273(1589):895-899.

Lukas, K., Maurice, T., Hui, T.H., Ng, P. and Ralf, B. (2007). Evolution of Miniaturization and the phylogenetic position of *Paedocypris*, comprising the world's smallest vertebrate. BMC Evolutionary Biology, 7(38):1-10.

Watson, W. and Walker Jr., H.J. (2004). The World's Smallest Vertebrate, *Schindleria brevipinguis*, A New Paedomorphic Species in the Family Schindleriidae (Perciformes: Gobioidei). Rec. Austral. Museum 56(2):139-142.

2

Diversity, Distribution and Characteristics of Vertebrates

WHAT ARE VERTEBRATES?

Vertebrates (which belong to the subphylum Vertebrata and phylum Chordata) have been in existence from late Cambrian about 500 million years ago. They represent a small group that comprises about 2% and over 40 thousand species of the widely distributed animals on Earth. One is quite familiar with vertebrates as they are relatively agile and large in size and human beings are vertebrates and interact with quite a number of them in their daily lives. The generally larger size of vertebrates in comparison to invertebrates is as a result of evolution of a versatile support system, mainly the skeletal system.

The classification of vertebrates has for a long time been based on **homology** (shared characteristics that have been inherited from a common ancestor). Anatomical structures and the pattern of embryonic development and more recently molecular biology techniques such as protein sequencing, DNA-DNA hybridization, chromosome painting and comparison of DNA sequences have been used in classifying vertebrates. The amino acids of proteins such as hemoglobin and cytochrome c and how they compare in various vertebrates have been useful in resolving some classification problems. Proteins that evolve fast such as fibrinogen have been used in drawing phylogenetic trees of recent evolutionary events such as mammalian relationships.

DNA analyzing techniques have been used in solving classification problems of some vertebrates. The mystery of the extinct New Zealand giant moa bird (*Dinornis*) was resolved recently when DNA analysis was applied to the bones of the bird. The size difference between the male and female moa birds is the largest of any known species of birds and terrestrial mammals with the largest females being 280% the weight and 150% the height of the largest males (Bunce et al., 2003). The two sets of bone remains belonging to male and female moas were previously thought to belong to two different species of birds. Bunce et al. (2003) found in a study that of the 11 recognized species of moa birds, 3 species were found throughout New Zealand and female birds ranged in height at the back from 1 to 2 m and varied in weight from 34 to 242 kg.

Some of the unique characteristics of vertebrates are derived from a multipotent population of neural crest cells that are present early during embryonic development and are found only in vertebrates. **Neural crest cells** are ectodermal in origin and originate from the most dorsal part of the neural tube at the border between the neural tube and the prospective epidermis (see Chapter 9). Neural crest cells are a classic example of embryonic induction. At the time of the neural tube closure, neural crest cells migrate from their original position along defined paths to other parts of the body and differentiate into or cause the formation of several structures such as part of the skeleton and connective tissue of the head, neurons and glial cells of sensory and autonomic nervous system, the adrenal medulla cells that produce epinephrine (adrenalin) and the pigment producing cells of the epidermis. For a review on neural crest induction, see Basch et al. (2004).

Somatic neural and neural crest stem cells are promising sources for cellular therapy of several neurodegenerative diseases (Widera et al., 2009). Neural crest-derived stem cells (NCSCs) can be found in deverse mammalian tissues including tissues that are known to be derived from the neural crest and tissues that are not specifically derived from the neural crest, such as bone marrow (Nagoshi et al., 2009).

Vertebrates as well as other chordates pass through a stage known as **pharyngula** (Fig. 2.1) during their embryonic development. This stage is characterized by the presence of the following features:

(a) A dorsal tubular nerve cord (tube)
(b) A flexible notochord
(c) A series of paired branchial (pharyngeal) grooves (gill pouches)
(d) A post-anal tail

The branchial grooves are matched to the inside by pharyngeal pouches. During further development in fishes, the grooves and pouches meet to form continuous paths of gill slits that allow water to pass from the pharynx, over the gills, to the outside of the body. The grooves and pouches disappear in other vertebrates. The mammalian eustachian tube and the auditory canal that are separated by the eardrum and connect the

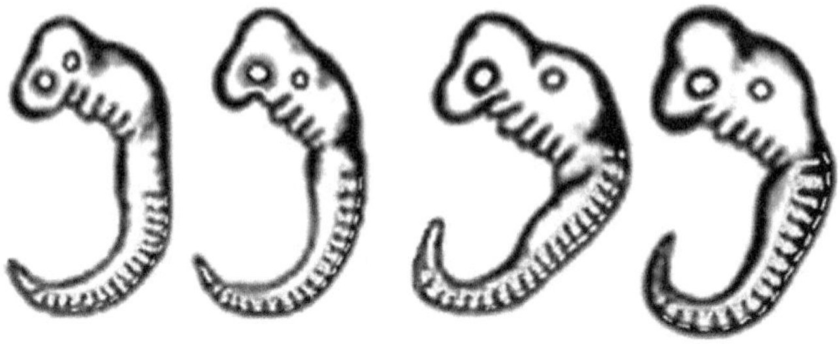

Fig. 2.1 A diagram of the pharyngula stage during embryonic vertebrate development belonging to fish (i), amphibian (ii), reptile (iii), bird (iv) and mammal (v). Note the branchial grooves on the ventral side, posterior to the head.

pharynx with the outside of the head are an example of remnants of gill slits.

A typical vertebrate feature is the presence of a bony or cartilaginous vertebral column that is made of articulating vertebrae and replaces the notochord long before attainment of maturity. The vertebral column is strong as well as flexible and serves as an anchorage for body limbs as well as protecting the spinal cord and its nerves. Vertebrates also possess a highly developed nervous system that enables them to react quickly to environmental changes, a chambered heart and blood vessels, a muscular mouth and pharynx and paired complex eyes. The presence of these features has enabled vertebrates to grow into large animals while still maintaining a relatively active life that gives them a competitive edge over other animals.

Although fish have paired fins (with the exception of agnathans that have only median fins), all other vertebrates possess either four limbs or are descendants of four limbed ancestors. The presence of four limbs (tetrapods) has enabled vertebrates to walk, run, swim and fly. Apart from a few cases of parthenogenesis (Gr. *parthenos*, virgin; *genesis*, descent or birth), vertebrates normally breed by sexual reproduction and either lay eggs or give birth to live young. The evolutionary diversity that characterizes vertebrates could have resulted from the quadrupling of their **homeotic (Hox) gene** clusters (on four separate chromosomes) that play a role in regionalization along the anteroposterior body axis.

Vertebrates inhabit different niches on Earth including tropical and polar zones, deep seas, high mountains and even air and comprise fishes, amphibians, reptiles, birds and mammals. The earliest vertebrates to evolve were fishes and the last group to appear was that of mammals. Many major groups of vertebrates that once dominated life on Earth have become extinct in the course of evolution.

MAJOR GROUPS OF VERTEBRATES AND THEIR CHARACTERISTICS

The world's geological history has experienced five massive extinctions and later a taxon specific extinction at the end of Pleistocene (the most recent ice age that lasted from about 1.8 million to 11 thousand years ago). Some of these events, among others, have shaped the course of vertebrate evolution. Extinction of species reduces diversity of life on Earth although it may lead to emergence of less known species that will radiate to occupy the space left by the extinct group. Current information about extinct and extanct vertebrates should help us understand forces that have shaped the course of evolution and should be useful in predicting the future of vertebrates on Earth.

FISH (PISCES)

Fish are the earliest (oldest) group of vertebrates and were the first to evolve about 500 million years back. Fish are the most numerous vertebrates with estimates of over 20,000 species, which is about half the number of all living vertebrate species. Fossil records show that many species of fish have become extinct in the course of vertebrate evolution.

Fish are ectothermic (cold-blooded) vertebrates that have **gills** and **fins** instead of pentadactyl limbs and primarily depend on water as their environment. Although most fish have a fusiform body that is ovoid in cross section, they display other shapes such as being laterally compressed, flattened (depressed), truncated and needle-like. The fins may be quite elaborate in some species but reduced or missing in others. A number of fish are highly colored with others closely matching their coloration with that of the environment they inhabit. Some fish are able to alter their coloration for purposes of camouflage or behavioral purposes such as breeding. Fish range in length from about 10 mm up to 20 m and in weight from about 1.5 g to several tons. The largest fish are the whale sharks and basking sharks.

Distribution of Fish

Fish inhabit most natural fresh and marine water bodies on Earth. Some fish live in thermal and alkaline springs that experience temperatures of slightly over 40°C and also cold Arctic Seas with temperatures below 0°C. Fish are not found in very hot ponds and extremely salty or alkaline lakes such as the Dead Sea of the Near East and the Great Salt Lake of Utah in the U.S.A. The diverse distribution of different fish in most waters on Earth is related to the geological history and development of the Earth as well as the ability of fish to undergo evolutionary change and adapt to changing habitats.

Major Groups of Fish

Living and extinct fish have been classified into four classes. The three classes of fish with living representatives are Agnatha, Chondrichthyes and Osteichthyes (Fig. 2.2). The only class of fish to become completely extinct is Placodermi.

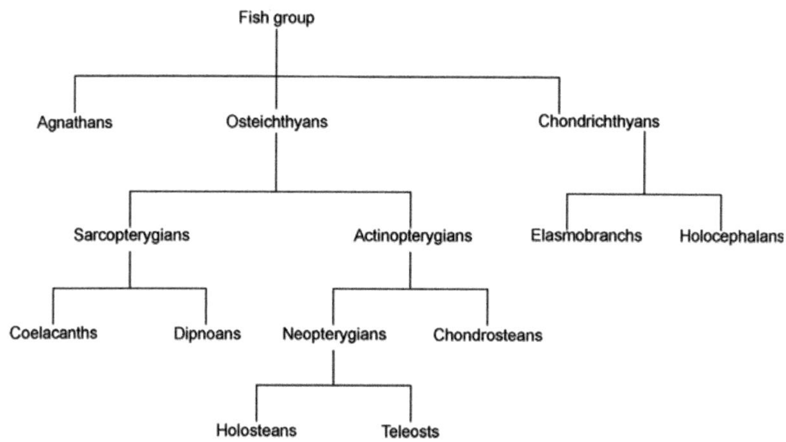

Fig. 2.2 A simple classification of the major groups of fish with living representatives. Many groups and individual species of fish have become extinct in the course of evolution.

Class Agnatha

Members of the class Agnatha comprise about 70 species of fish that lack jaws and are the most primitive vertebrates. Agnathans are represented by the extant **cyclostomes** (lampreys and hagfish) and the extinct ostracoderms. Extanct cyclostomes and most ostracoderms lack paired pectoral or pelvic appandages and have a suctoral or filter-feeding mouth. Agnathans are of importance from the evolutionary point of view as the group has the oldest known craniate fossils. Living agnathans also have many primitive characteristics.

Ostracoderms

Ostracoderms (Gr. *ostrakon,* shell; *derma,* skin) were the first vertebrates to evolve and lived from early Ordovician to late Devonian. They were mostly less than 20 cm long and had shell-like bony plates in their skin. These bony plates have been found in Cambrian and Ordovician marine or estuarine deposits. Remains belonging to about 200 species of this diverse group of vertebrates have been discovered. Ostracoderms were fish-like in appearance and had thick, flat bodies with a pair of side flaps they used for steering. Their internal skeleton was cartilaginous and has rarely been preserved. Ostracoderms became rapidly extinct near the end of Devonian. A type that lacked body armor survived and gave rise to the living lampreys and hagfish (Fig. 2.3).

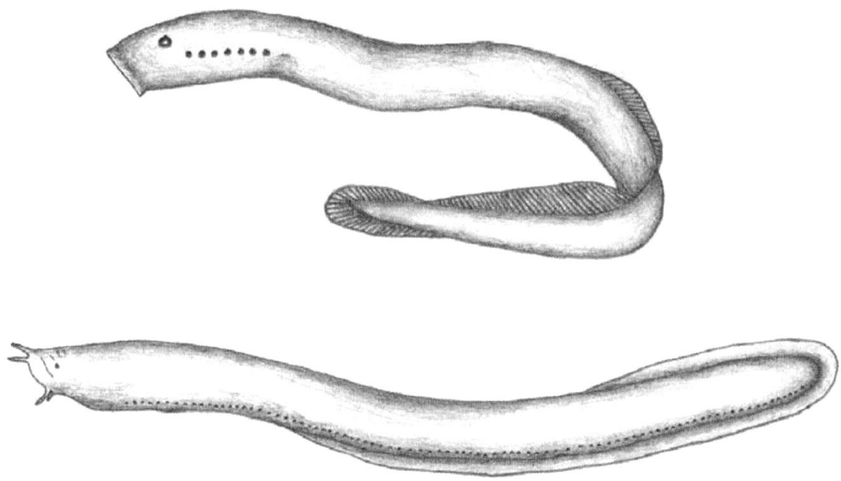

Fig. 2.3 Living cyclostomes. A lamprey (above) and a hagfish (below). Hagfish and lampreys are jawless fish that represent the oldest class of fish, Agnatha.

Lampreys

Lampreys, which comprise about 40 species of fish, are found in fresh and coastal waters of all continents except Africa. They are parasitic in nature and have been a major pest to fisheries in the Great Lakes of North America as they feed on tissues (blood and flesh) of living fish such as trout. The sea lamprey (*Petromyzon marinus*) is also parasitic to other fish such as sharks. The ability of lampreys to penetrate the dermal denticle armor of sharks, excrete a lot of urea and their great ability to deaminate amino acids is a case of an adaptation that has contributed to the evolutionary success of this ancient vertebrate (Wilkie et al., 2004). Lampreys are of little economic importance in terms of food to humans.

Lampreys have an eel-like body that lacks scales with well developed dorsal and caudal fins. They range in length from 15 to 100 cm long. Their notochord persists throughout life and is never completely replaced by a backbone. Like all invertebrates, the axons of lamprey neurons are unmyelinated. These fish have only an innate immune system. The mouth of a lamprey has a suctoral disk that bears horny teeth.

Hagfish

Hagfish are found in temperate and tropical marine waters. There are about 30 species of hagfish that range in length from 40 to 80 cm long. They are minor pests and feed on dead animals and soft-bodied invertebrates they detect by smell. Their body salt concentration is same as that of seawater as is the case with marine invertebrates. Hagfish have an almost cylindrical body that lacks scales and have a low caudal fin around the tail. They have comb-like horny teeth on the floor of the mouth. All hagfish have many large skin glands that secrete copious protective mucus.

Jawed Fish

Jawed fish (gnathostomes) have developed jaws that were derived from some of the hyoid arches (cartilaginous or bony) that support the pharyngeal region. The axons of neurons in jawed fish have myelin sheaths that permit a more rapid transmission of nerve impulses. These fish also have an adaptive immune system in addition to the innate type. Gnathostomes include jawed fish in the classes Chondrichthyes, Osteichthyes and the extinct Placodermi.

Class Chondrichthyes

Chondrichthyes (Gr. *chondros,* cartilage; *ichthys,* fish) comprise approximately 1000 species of living cartilaginous fish such as sharks, skates, rays and chimaeras (ratfish). The earliest cartilaginous fishes appeared late in the Devonian period and were much like sharks in appearance though their bodies were heavily armored and they were more sluggish. Chondrichthyans were the first vertebrates to develop internal fertilization as they live in a hyperosmotic environment and the ability to be in osmotic balance with marine water develops late in embryology in this group of fish. The internal skeleton of chondrichthyans remains cartilaginous throughout life and none of it is replaced by bone. The notochord is often reduced in cartilaginous fish and is partially replaced by cartilage.

The skin of cartilaginous fish is flexible and leathery and has minute dermal denticles or **placoid scales** that are remnants of the extensive dermal armor of their ancestors. The scales and teeth of these fish stop growing once fully developed but are replaced when worn out. The first gill pouch in cartilaginous fish is reduced to a **spiracle** or is lost and part of the pelvic fin of the male is modified into an intromittent organ for internal fertilization known as a **clasper**.

Cartilaginous fish tend to be heavy bodied and lack swim bladders and lungs that could render buoyancy to these fish. The presence of a less dense cartilaginous skeleton as opposed to bone, reduction of the ancestral body armor and presence of large quantities of lipids in their livers lightens their bodies and renders some buoyancy. Most cartilaginous fish have a heterocercal (assymetrical) tail.

Chondrichthyans are primarily marine fish and only rarely are these fish able to penetrate freshwater. Sharks, which are mainly confined to oceans, are the most numerous chondrichthyans. Chondrichthyans can be subdivided into two major groups (sub-classes): elasmobranchs and holocephalans.

Elasmobranchs

Elasmobranchs (Gr. *elasmos*, thin plate; *branchia*, gills) have **gill slits** (pouches) that open to the outside independently from the pharyngeal cavity (Fig 2.4). These slits are separated from each other by thin plates

Fig. 2.4 Some elasmobranchs. (i) tiger, (ii) hammerhead and (iii) whale sharks. Note the vertical gill slits that are found anterior to the pectoral fins. Gill slits are located on the ventral part of the head region in rays and skates.

of tissue that bear gills. Since the upper jaw of elasmobranchs is not fused to the cranium, this makes it possible for the upper and lower jaws of the fish to be raised and lowered relative to the cranium. The jaws can also be moved forwards and backwards. Elasmobranchs have an extensive network of electroreceptors. **Electroreceptors** are modified parts of the lateral line system that enable fish to detect action potentials generated by contraction of muscles of other organisms. Elasmobranchs also possess many small sensory organs known as **pit organs** that are located over the skin surface. These organs are free **neuromasts** whose role in these fish is not fully understood (Peach and Marshall, 2000). Most cartilaginous fish are elasmobranchs and examples include sharks, skates and rays. Whereas sharks generally hunt in open water, most skates and rays tend to be benthic.

Holocephalans

Holocephalans (Gr. *holos*, whole, *kephale*, head) (Fig. 2.5) have highly compressed and large heads with bodies that taper posteriorly into long tails and comprise about 30 species of fish also known as chimaeras (Gr. *chimaira*, monster) or ratfishes. The gill pouches on either side of the head are covered by a fleshly flap (operculum) leaving a single opening on either side of the head anterior to the pectoral fins. The upper jaw of holocephalans is firmly united to the cranium. Apart from possessing pelvic fin claspers, cephalic claspers are also present in some holocephalans. Dermal armor is present on the heads of many holocephalans.

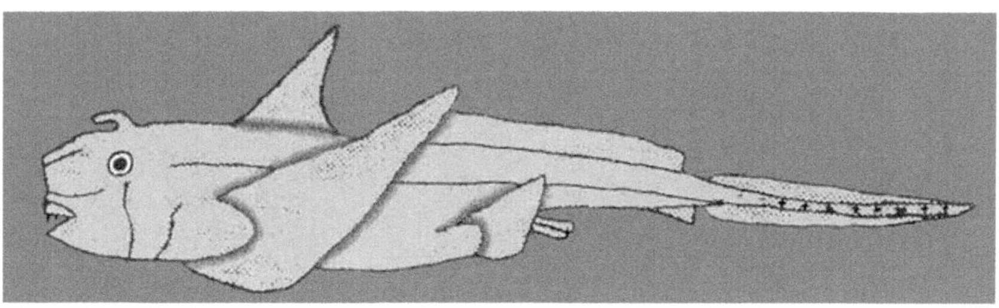

Fig. 2.5 A holocephalan, chimaera or ghost shark. Holocephalans belong to the order Chimaeriformes and first appeared during the Devonian but never radiated into many species. Chimaeras live at the bottom of most temperate oceans and can grow up to 2 m long. The skin of chimaeras is smooth and scaleless.

Economic Importance of Chondrichthyans

Chondrichthyans, especially sharks, have been a source of food and other products and also a danger to humans. Shark meat is eaten in most maritime countries and the dorsal fins of certain sharks have been used in making epicurean soup that has been responsible for the great exploitation of these sharks. Shark liver oil is also used as a lubricant and folk medicine against conditions such as rheumatism and coughs and also in tanning leather and preserving wood in certain parts of the world. Skin from sharks such as the tiger, sand, dusky and brown-sharks is used in making leather.

Shark cartilage products are widely used in the U.S.A for the treatment of various medical conditions such as cancer, arthritis, intestinal disorders and psoriasis. Shark cartilage contains angiogenesis inhibitors that have the ability to destroy and prevent the growth of new blood vessels that certain tumors need for nourishment and sustenance.

Certain species of sharks such as the white shark (man–eater), the hammerhead, tiger, blue and sand sharks have attacked people or boats. Other large sharks with big sharp triangular cutting teeth have also been implicated. Such sharks have very powerful bites as the upper jaw can also be moved relative to the braincase. Stingrays use their tails to lash out at enemies including man. Stingray lashes can be fatal. Rays have been known to be destructive to oyster and clam-beds.

Class Osteichthyes

The class osteichthyes (Gr. *osteon*, bone; *ichthys*, fish) is the most diverse of all fishes and vertebrates in general. The class comprises over 20,000 species of bony fishes. The internal skeleton of these fish has varying amounts of bone though most species have well ossified bones. Most bony fishes have thin bony scales embedded in their skin that represent the reduced heavy, bony surface ancestral armor. The gills of osteichthyans are covered by a bony operculum on either side of the head. Either lungs or swim bladders are present in bony fishes apart from some deepsea and bottom dwelling species where it has been secondarily lost. Bony fishes are divided into two groups (sub-classes): Actinopterygii and Sarcopterygii. Another major group, acanthodians, has become extinct.

Acanthodians

Acanthodians (Gr. *acanthodes*, spiny) were among the earliest jawed vertebrates to appear. Their early Ordovician and late Silurian fragmentary fossils comprising mainly of spines and pectoral bones were discovered in North America and China respectively. Acanthodians (spiny sharks) exhibited relatively little diversity. The earliest members of the group were mainly marine that later became freshwater after middle Devonian. Most acanthodians were less than 200 mm long, though some exceeded 1.0 m in length. Acanthodians had heterocercal tails and had stout bony spines that supported their fins except the tail fin while their bodies were covered by minute diamond-shaped scales.

Actinopterygii

Actinopterygii (Gr. *aktin*, ray; *pterygon*, wing or fin) are **ray-finned fishes** and include all bony fishes (about 20,000 species) except the lungfish and coelacanth. Actinophterygians have fins that are supported by bony rays that evolved from rows of minute bony scales. The scales of actinopterygians continue to grow throughout their lives. The group is divided into two infraclasses: Chondrostei and Neopterygii.

Chondrostei

Chondrosteans (Fig. 2.6) are the oldest group of actinopterygians and are represented by a few surviving species including freshwater bichirs (*Polypterus*), sturgeons (*Acipenser*), paddlefishes (*Polyodon*) and reedfish (*Erpetoichthys*). The internal skeleton of chondrosteans has a considerable amount of cartilage apart from bone. The base of the pectoral fin of chondrosteans is fleshy and has an endoskeleton with elements that are similar to both the fin rays of teleosts and the limb skeleton of tetrapods. Chondrosteans have simple scissor-like jaw actions that show limited lateral movement. Chondrosteans have a paired swim bladder that acts as a lung, a spiracle and an intestinal spiral valve. Bichirs and reedfishes have a long dorsal fin that is divided into many finlets. Sturgeons and paddlefishes have a braincase that is mainly cartilaginous with a few areas of isolated bone.

Bichirs comprise about 16 species and several subspecies and are mainly found in Africa. The snake-like bodies of these fish are covered with hard and rhomboid scales. The large pectoral fins of bichirs are used for bottom crawling in water and locomotion on land and the dorsal fin has 5 to 18 finlets. Bichirs periodically come to the surface of water to gulp in air to their lung-like swim bladder and will suffocate if prevented from doing so. The swim bladder in bichirs, unlike most fish species, is located ventral to the esophagus. The bichir has been shown to be the basal ray-finned fish (Kikugawa et al., 2004).

Sturgeons are found throughout the northern hemisphere in both fresh and seawater and produce the highly priced fish roe known as caviar. Most sturgeons are found in the Caspian Sea. The 20 species of sturgeons belong to the genus *Acipenser* and spend most of the year in the sea but occasionally ascend rivers to spawn. Only a few species of

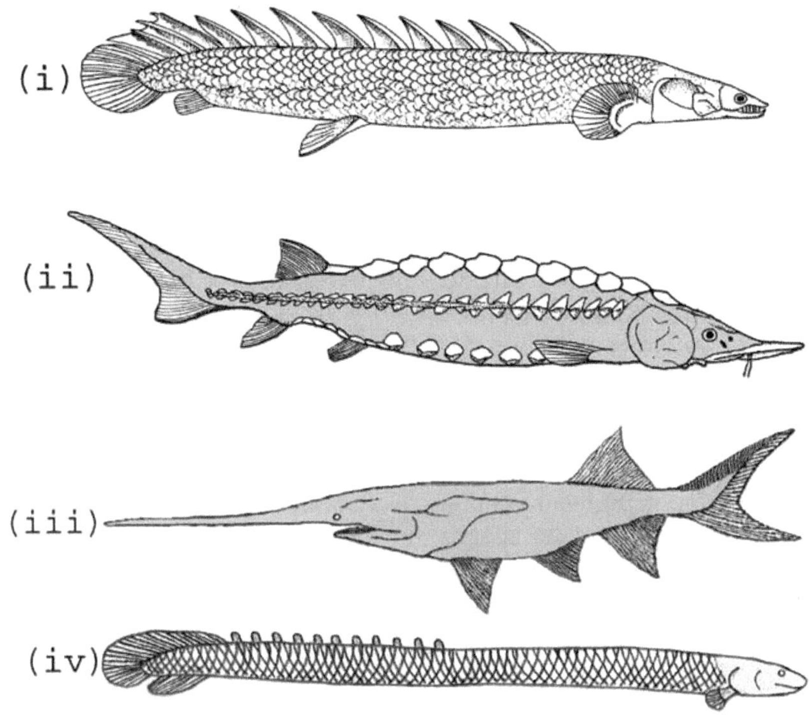

Fig. 2.6 Chondrosteans: bichir (i), sturgeon (ii), paddlefish (iii) and reedfish (iv). Chondrosteans are the oldest group of actinopterygians. Most chondrosteans lived during late Paleozoic. With the exception of the sturgeon, a source of food for human beings and caviar, chondrosteans are of little economic importance. Chondrosteans have left a poor fossil record since their internal skeleton is badly ossified.

sturgeons spend their entire life in seawater. Sturgeons possess several rows of bony plates on their body surfaces and have heterocercal tails. The bony plates are remnants of the ancestral body armor.

There are two living species of paddlefishes: the American species (*Polyodon spathula*) that is found in the slow-flowing Mississippi and Ohio Rivers and the Chinese paddlefish (*Psephurus gladius*) that inhabits in the River Yangtze. Paddlefishes have paddles or spatula-shaped snouts that measure about half the length of the fish. The heterocercal tail fins of paddlefishes are deeply forked. The American paddlefish can grow up to 1.5 m in length and weigh about 27 kg.

Reedfishes are represented by one species (*Erpetoichthys calabarius*) and are native to West Africa. The body surface of reedfishes is covered by thick rhomboid-shaped and shiny ganoin scales and their skulls are rigid. Reedfishes have epicercal tails in which the notochord extends upwards while fin rays extend back and down and have to periodically breathe in air to supplement oxygen uptake by the gills.

Neopterygii

Most living marine and freshwater fishes are neopterygians. They have thin and flexible scales (cycloid and ctenoid types) that enable their bodies to bend more rapidly and powerfully except gars. Their caudal fins are homocercal (superficially symmetrical). Neopterygians have shorter jaws than chondrosteans and the jaws can undergo lateral, forward and backward movements bringing about an increase in the volume of the oral cavity and enabling the fish to perform a variety of feeding movements. Neopterygians show a reduction in the number and increase in the level of ossification of the vertebrae. The external opening into the oral cavity is round and can be moved forwards resulting in a strong sucking action. The versatile feeding modes of neopterygians could have contributed to the success of this fish group. The swim bladder of neopterygians is fully differentiated and is a hydrostatic organ. Extanct neopterygians comprise teleosts, and holosteans.

Holostei

Holosteans were the dominant fishes in the middle of the Mesozoic era. There are fewer living holosteans than chondrosteans today and include the seven species of garpikes (gars) belonging to the genera *Lepisosteus* and *Atractosteus* of North and Central America and Cuba and bowfin (*Amia*) of eastern Canada (Fig. 2.7). Studies of cloned and sequenced

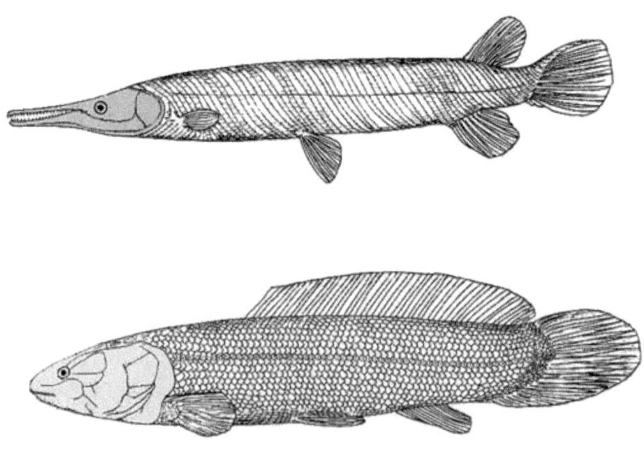

Fig. 2.7 Holosteans: garpike (above) and bowfin (below). Holosteans have remained almost unchanged throughout their evolutionary history. The bowfin has a long dorsal fin.

DNA-coded genes show that living holosteans are a monophyletic group that is the sister group to teleost fishes (Kikugawa et al., 2004). The jaws of holosteans are shorter and better supported than the jaws of chondrosteans. The posterior end of the maxilla is free, which enables it to move forwards. The mouth of holosteans is capable of producing powerful suction action.

Garpikes are primarily freshwater predators that feed on other small fish. The bodies of garpikes are covered by tough and rhombic or diamond-shaped ganoid scales that interlock and their beak-shaped snouts bear sharp teeth. The median fins of garpikes are located posteriorly on the body. The swim bladder of garpikes is used as an accessory air-breathing organ.

The only living representative of the bowfin family is *Amia calva*. Bowfins inhabit the sloughs and ponds of the River Mississippi and rivers of eastern North America and their bodies are covered with cycloid scales that make the body surface feel smooth and leathery. There is a large bony plate between the lower jaws of bowfins. Bowfins use the swim bladder for breathing and are primarily carnivorous.

Teleosts

Teleosts (infraclass Teleostei) (Fig. 2.8) include almost all the world's most important commercial and sport fishes such as herrings, salmons, tunas, cods, marlins and perches. Also known as the 'advanced bony fishes', the group comprises more than 20,000 species. Teleosts range in length from the smallest of all vertebrates, the stout infantfish (about 8.9 mm long) to the giant European catfish (5 m long). Large swordfishes and

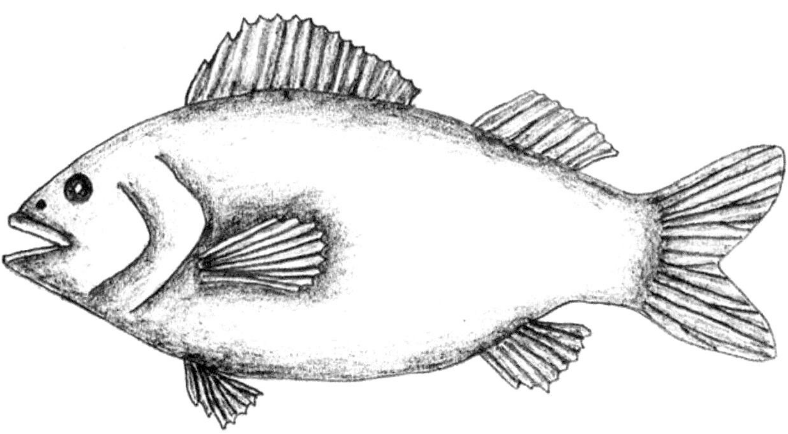

Fig. 2.8 A typical teleost. Teleosts are sometimes referred to as the 'advanced bony fishes' and comprise the world's main sport and commercial fishes.

marlins weigh as much as 500 kg. The ocean sunfish (*Mola*) can weigh up to 900 kg. About 10% of teleosts are less than 10 cm long whereas 80% are between 10 cm and 1.0 m in length. Teleosts became the dominant fishes of the sea and freshwater by the end of Cretaceous.

Teleosts (Fig. 2.8) have a free premaxilla (front upper jawbone) and a fully movable maxilla that make it possible for the front part of the mouth to protrude when the jaws are opened thus creating powerful suction action. The skeleton of teleosts is well calcified and is made of a scaffolding of bony struts making it lighter while enabling teleosts be active and rapid swimmers. The neural arches in the caudal vertebrae of teleosts are elongated into **uroneurals** that increase stiffness to the upper and lower parts of the caudal fin. Teleosts also show a reduction in the number of vertebrae and tail fin rays. Such changes have contributed to greater thrust and flexibility that is displayed by the teleostean body while at the same time decreasing the drag effect.

Economic Importance of Teleosts

Teleosts are an important source of food in most parts of the world. They form a major part of the sport fishing industry and many are pets in aquaria due to their esthetic beauty. Some teleosts such as wolf fishes are important in the leather industry and thread fishes and drums are a source of isinglass that is used in production of jellies.

Some teleosts are also a danger to life. The electric eel and electric catfish generate electricity that can stun other animals including humans. The South American piranhas and the African tiger fish have powerful jaws and sharp strong teeth they use for defense and offence. The barracuda, with sharp teeth, has been reported to attack swimmers in oceans.

A small Japanese freshwater fish, medaka (*Oryzias latipes*) and a small tropical aquarium fish, the zebrafish (*Danio rerio*) are important research animals. For a review on the history of medaka as a research organism and its genomic analyses, see Shima and Mitani (2004). Zebrafish are used as a model system for studying gene function during vertebrate developmental biology and the zebrafish have taken over from rats as the most popular laboratory animal.

Sarcopterygians

Sarcopterygians (Gr. *sarkodes*, fleshy; *pterygion*, wing or fin) are **fleshy finned** fish. They are thought to be the ancestrors of terrestrial vertebrates. Their paired fins have a skeletal axis surrounded by muscles. Sarcopterygians have electroreceptors. The subclass Sarcopterygii has the orders Dipnoi (lungfishes), Coelacanthini (coelacanths) and the extinct Rhipidistia. Coelacanths and rhipidistians are also known as **crossopterygians** (lobe finned fish). Sarcopterygians have been primarily freshwater fishes. Coelacanths adapted to a marine environment. Sarcopterygians were abundant during the Paleozoic era but most have become extinct.

Coelacanths

Coelacanths were thought to be extinct at the end of the Mesozoic era. In 1938, a living specimen of the fish was caught off the Comoro islands near Madagascar. South African ichthyologist J.L.B. Smith identified the fish and named it after Courteney Latimer, an associate who drew his attention to it, and the Chalumna river mouth where it was caught as *Latimeria chalumnae.* More than 200 specimens have been caught in the area since, including Indonesia (*Latimeria menadoensis*) and the East African coast (Kenya).

The coelacanth (Fig. 2.9) has paired fleshly fins and two dorsal fins. The anterior dorsal fin is ray-finned (actinopterygial) type and is not fleshy. The skull of the fish is divided into anterior and posterior units that are joined to each other at each side of the head. The base of the skull and vertebral column are incompletely ossified and the notochord persists in some areas. Large rough scales cover the body and the powerful tail fin is divided into three lobes (upper, middle and lower). A vestigial lung, mainly filled with fat, persists. The coelacanth is of interest from an evolutionary point of view.

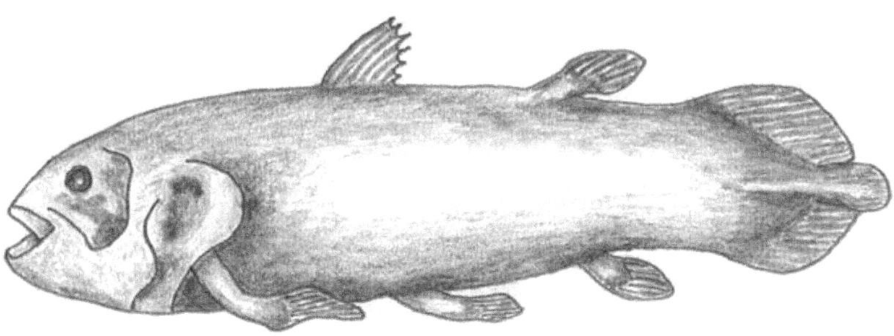

Fig. 2.9 The only living coelacanth of the genus *Latimeria.* This sluggish fish evolved early during the Devonian. Coelacanths were close relatives of rhipidistians that are believed to have been the ancestors of the first terrestrial vertebrates.

Lungfishes

Lungfishes have little ossification in their internal skeleton. The fish have lost pre-maxilla and maxilla and have well developed lungs and reduced gills. Only three forms of lungfishes survive today in rivers and ponds: the Australian lungfish (*Neoceratodus fosteri*), the South American lungfish (*Lepidosiren paradoxa*) and the African lungfish (*Protopterus*). The yellow marbled Ethiopian lungfish (*Protopterus aethiopicus*) is the most common and largest of the few surviving African species. The paired appendages of the South American and African types resemble tendrils whereas those of the Australian species are much broader (Fig. 2.10). The nostrils of lungfishes open internally and are used for breathing when the mouth is closed.

Lungfishes have two atria and the ventricle is partially divided. There is partial separation of oxygenated and deoxygenated blood. Although these fish excrete their nitrogenous wastes as ammonia when in water, the fish have an enzyme system that can convert ammonia into the less toxic urea during drought conditions. The enzymes system is highly developed in the African and South American lungfishes. These two forms burrow in mud and cocoon themselves in dry river and pond beds during drought leaving a small opening to the outside through which they breathe.

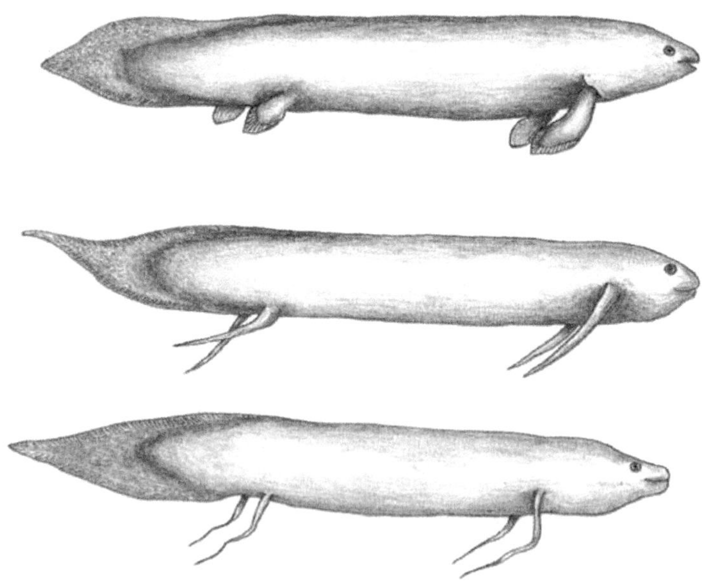

Fig. 2.10 The only surviving lungfishes. The Australian (*Neoceratodus*) (top), the South American (*Lepidosiren*) (middle) and the African (*Protopterus*) (bottom) lungfishes. The long tendril-like pectoral and pelvic fins of the African and South American lungfishes are in constant motion and their tips have a highly developed sense of touch.

DNA sequences on the three lungfish groups and the Indonesian coelacanth (*L. menadoensis*) show that lungfishes are the closest living relatives of land vertebrates (Brinkmann et al., 2004). This is supported by amino acid sequences deduced from nucleotide sequences of mitochondria cytchrome oxidase subunit 1 (col) genes that indicate a lungfish/tetrapod and coelacanth/lungfish clade (Yokobori et al., 1994).

Rhipidistians

The skeletal system of early rhipidistians (Gr. *rhipis,* fan) had a more ossified and stronger skeleton than that of lungfishes. A transverse joint is found in their cranium that allowed skull movements during feeding. A suture that represents this joint is found in the early amphibians. Rhipidistians became extinct during the Carboniferous period and could have given rise to amphibians.

Class Placodermi

This class of fish flourished during the Devonian period and became completely extinct during the Carboniferous. Placoderms (Gr. *plak,* plate; *derma,* skin) resembled ostracoderms in having extensive armor on the anterior part of the trunk and head. Bony scales covered the rest of the body. A joint existed between the head and thoracic parts of the fish. This made it possible for the skull to be raised. Pelvic and pectoral fins were present and the tail was heterocercal. The internal skeleton was cartilaginous and a notochord persisted throughout life.

Class Amphibia

Amphibians (Gr. *amphi,* both or double; *bios,* life) are semi-terrestrial and are the smallest group of tetrapods. They were the first group of vertebrates to live on land successfully. Transition from water to land occurred during the Devonian. With a few exceptions, amphibians spend part of their lives under water (breathing with gills) and the remainder on land where they use lungs for breathing. The only subclass of extanct amphibians is **Lissamphibia** and has three orders comprising about 4,000 species. **Anurans** (Gr. *a,* without; *oura,* tail) or **salientians** include frogs (Ranidae) and toads (Bufonidae), **urodeles** (Gr. *oura,* tail; *delos,* visible) or **caudatans** are newts and salamanders and **apodans** or **gymnophionans** are caecilians.

Frogs resemble toads though toads tend to be more terrestrial in their habits, have a drier skin and possess parotid glands behind the eyes that can produce poisonous substances. Frogs and toads are capable of jumping. Most species of salamanders are terrestrial as adults but some are aquatic such as the North American mudpuppy (*Necturus*) and the Mexican axolotl (*Ambystoma mexicanum*) that retain their larval gills (Fig. 2.11) and stay in water throughout their lifetime. Axolotl is threatened towards extinction in its natural habitat by pollution, exploitation and introduction of sport fish that are predators. Newts have well developed tails. Some urodeles have no lungs and depend entirely on cutaneous respiration. Caecilians are rare, limbless and worm-like amphibians that are adapted to burrowing.

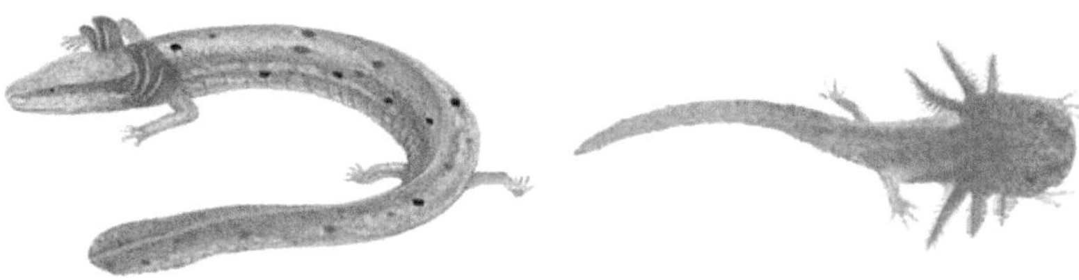

Fig. 2.11 The common mudpuppy *Necturus maculosus* (left) and the Mexican axolotl *Ambystoma mexicanum* (right). These aquatic salamanders that measure on average as adults about 20 to 30 cm long are neotenic or paedomorphic (mature as adults while retaining larval characteristics). The two species of salamanders can grow to over 40 cm in length.

Amphibians lay their eggs in water or a moist environment. These eggs are surrounded by a delicate membrane and contain limited supplies of yolk, so the larvae must feed soon after hatching. The larvae of amphibians such as anurans stay in water for some time before undergoing metamorphosis into the adult terrestrial amphibian. With the exception of one species, members of the South American frog genus *Eleutherodactylus* that has about 600 species lay eggs that undergo direct development in a terrestrial environment. The larval (tadpole) stage occurs within the egg on land that will hatch into a small frog that has sometimes bypassed stages present in the free-living larval aquatic stage (Hanken et al., 1997). It has been demonstrated in *E. coqui* that the dermal folds of this frog are homologous with the opercular folds of metamorphosing frogs (Callery and Elinson, 2000).

Features of Amphibians

Most amphibians have a soft, moist and glandular skin that lacks scales with the exception of a few species. The skin is well supplied with capillaries and serves as a respiratory membrane and also absorbs water. Amphibians thus have close contact with their environment that makes them susceptible to human activities such as pollution. Amphibians are among the first organisms to be influenced by activities such as pollution and climate change, thus can provide an early warning to environmental degradation.

Many amphibians (frogs and salamanders) have a strong skeleton and four-well developed limbs. The amphibian skull is flattened and broader than that of other tetrapods and the pelvis in most species is attached to the vertebral column by a single pair of sacral ribs and one sacral vertebra.

Since amphibians are ectothermic, those living on land are faced with more complex problems than fishes. Amphibians tend to avoid stressful habitats such as exposed areas that are directly exposed to solar radiation and great variability in temperatures.

Class Reptilia

Reptiles (L. *reptare,* to crawl or creep) were the dominant terrestrial vertebrates during the Mesozoic era. Reptiles lay leathery or brittle shelled eggs that are yolk-filled and more resistant to evaporation of fluids from inside when compared to an amphibian egg or give birth to live young. This has enabled reptiles to adapt to life on land more successfully than amphibians. The arrival such a **cleidoic** ('self contained') egg coincided with internal fertilization since the calcium rich shell is impervious to both sperm and water. The egg can also be retained in the oviduct and enables reptiles bypass the aquatic larval stage. The developing embryo produces four **extra-embryonic membranes** (chorion, allantois, yolk sac and amnion). Reptiles, birds and mammals are known as **amniotes** as their embryos produce these four membranes.

Reptiles have scales or modified scales that are keratinized and reduce loss of heat through evaporation of water. The lungs of reptiles generally have a larger surface area than those of amphibians. Reptiles have more control over their body temperature than amphibians. Reptiles can maintain a relatively high and constant body temperature during periods of activity.

The reptilian skeleton is stronger than that of amphibians. The reptilian skull is deeper and narrower than that of amphibians and the pelvic girdle is united to the vertebral column by at least two sacral vertebrae.

The major orders of reptiles include **Chelonia** (turtles), **Squamata** (snakes, lizards and amphisbaenians), **Thecodonta** or **Archosaura** (crocodiles and alligators) and **Sphenodonta** (tuataras or *Sphenodon*). The order Squamata is the largest reptilian group with about 6,300 of the reptilian species. Amphisbaenians or worm lizards (Fig. 2.12) comprise

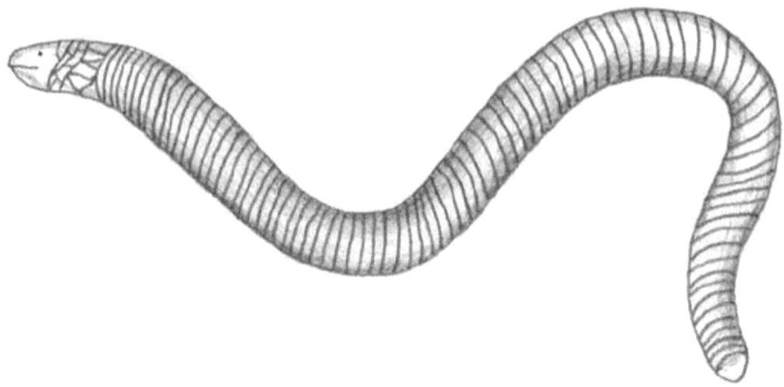

Fig. 2.12 Drawing of an amphisbaenian. Amphisbaenians are found in warm parts of the world including South America, southern North America and Africa. Most lack limbs and adults measure between 10 and 75 cm in length. The thick and bony skulls of amphisbaenians help these reptiles dig channels in the ground and their strong jaws are used for crushing insects and worms.

about 150 species of burrowing worm-like reptiles. Some 250 species of turtles survive today. Tuataras are represented by a single species that is protected on a few islands off New Zealand. The biggest living reptile is the estuarine crocodile that can measure over seven meters in length and the smallest reptile (which is a lizard) is the British Virgin Islands gecko that is only 18 mm in length.

Class Aves (Birds)

Aves (L. *aves*, birds; Gr. *ornithes*, birds) probably evolved from dinosaurs. The main differences between birds and their reptilian ancestors have arisen as a result of adaptations to flight. Birds are sometimes referred to as 'feathered reptiles' or 'hot lizards'. The main characteristic feature of birds is the possession of feathers that evolved from reptilian scales. Intrinsic integumentary muscles known as **arrectores plumarum** (mainly smooth muscles) are found in the dermis and attach to feather follicles. These muscles cause ruffling of feathers on contraction when it is cold or in response to danger. Diverticula of pulmonary (air) sacs enter the medullary spaces of bones through pneumatic foramens during early growth in birds. The bones of birds are thus hollow and contain air (are pneumatic) that makes them light but strong. The pectoral appendages are modified as wings and the sternum is broad and keeled in most birds. Such a large sternum provides a large surface area for origin of flight muscles. Several bones in the bird skeleton are fused. There is fusion of pelvic bones, sacral and lumbar vertebrae, thoracic vertebrae and bones of the hind limbs. Fusion of trunk vertebrae has resulted in birds having a short and rigid back.

Flight is physiologically expensive as a flying bird has a very high rate of metabolism that requires an efficient circulatory system. Birds have evolved a four- chambered heart and their lungs have evolved thin-walled air sacs that act as bellows that ventilate lungs during inspiration and expiration.

Birds lack teeth but have a horny bill instead. The function of teeth is taken over by the gizzard. During developmental stages, birds have two gonads. In the majority of birds, only the left ovary reaches full development.

The avian brain, in proportion to body weight, is larger than that of reptiles. The difference is found mainly in the cerebellum that is the center of site and muscular coordination.

Living birds comprise about 10,000 species that belong to two super orders: **Paleognathae** and **Neognathae**. Paleognathae (Gr. *palaios*, ancient; *gnathos*, jaw) are also known as terrestrial birds and have an archosaur-like bony palate that is solid. Most evolved in areas with low terrestrial predation and have reduced wings, a broad sternum that lacks a keel and strong legs (Fig. 2.13). Birds of the order Paleognathae include **ostriches** of Africa, **emus** of Australia, **rhea** of South America, **cassowary** of tropical Australia and New Guinea and **kiwi** of New Zealand (Fig. 2.14). The **dodo** (*Raphus cucullatus*) was another flightless forest dwelling bird that was found in Mauritius which is about 220 kilometers east of Madagascar in the Indian Ocean. This large and friendly bird that could weigh up to 23 kg became extinct about 1690 almost 30 years after its most recent sighting (Roberts and Solow, 2003). It has been proved by DNA analysis that the dodo belonged to the dove and pigeon family. The dodo became extinct due to destruction of the forest that cut off the dodo's food supply and destruction of the dodo nests by animals such as cats, rats, monkeys and pigs that were brought to the island by sailors who also ate some of the birds. Birds of the order Neognathae (Gr. *neos*, new) have a more mobile palate and include most of the birds.

Class Mammalia

Mammals are the most conspicuous vertebrates and comprise about 4500 species. Mammals nourish their young on milk from special secreting mammary glands (L. *mamma*, breast). Mammals have several unique features that differentiate them from other vertebrates and include:

Hair. Some mammals such as whales have secondarily lost hair except at the fetal stage. Associated with hair and its follicle is a muscle known as **arrector pili**. On contraction, this smooth muscle causes erection of

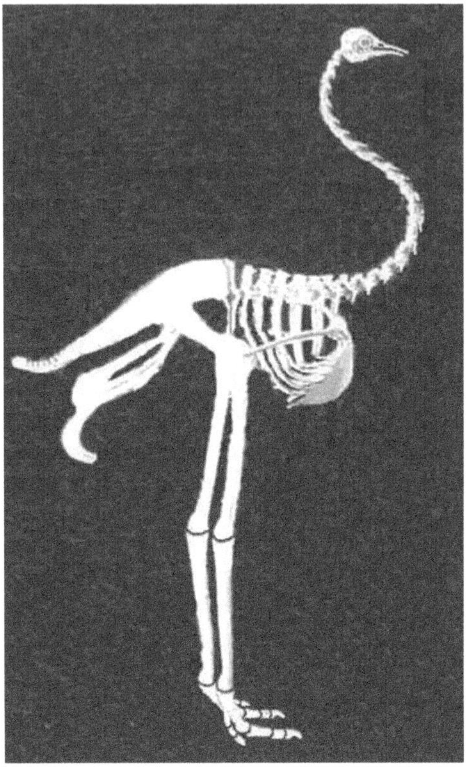

Fig. 2.13 A photograph of the skeleton of an ostrich showing the broad and unkeeled sternum. The upper limbs of the ostrich are quite reduced in size. The ostrich (*Struthio camelus*) is the largest living bird in the world and is found in the wild in grassland savannahs of certain parts of Africa. This ratite (flightless bird) is quite adapted to desert life and can reach a height of 2.6 m and weighs as much as 135 kg. The ostrich together with other ratites are believed to have evolved from flying ancestors in areas with low predators. The ostrich is the only bird with two toes on each foot.

individual hair that causes the hair to trap air thus increasing its insulating qualities. Hair follicles give rise to sebaceous glands that produce sebum, an oily substance that lubricates hair. Most mammals have three distinct types of hairs viz. guard, fur and vibrissae. Guard hairs protect the rest of hair, the underfur is primarily insulative and may differ in color from the guard hair and vibrissae or whiskers are stiff and elongate and function in tactile sensation.

Skeleton. The mammalian mandible (lower jaw) is hinged directly to the temporal bone of the skull forming the temporo-mandibular (mandibular) joint. The **quadrate bone** (Fig. 2.15) that is found between

Fig. 2.14 The living representatives of the order Ratitae that comprises flightless birds that belong to the subclass Paleognathae ('old jaw'). Paleognath birds were the first birds to evolve and include the largest of all living birds. A male ostrich (top left), emu (top right), cassowary (bottom left), rhea (bottom middle) and kiwi (bottom right). Some of the extinct ratites include the elephant bird of Madagascar and the moas of New Zealand that are thought to have been eliminated by human beings. Ratites rely on their running speed and strong kicks to defend themselves. The ostrich of Africa is the fastest bipedal runner and can attain speeds of up to 65 km/h while taking strides as long as 4.5 m at times. The other order of paleognath birds is Tinamiformes that includes the tinamous of South and Central America that have a keeled sternum and are weak fliers.

Color image of this figure appears in the color plate section at the end of the book.

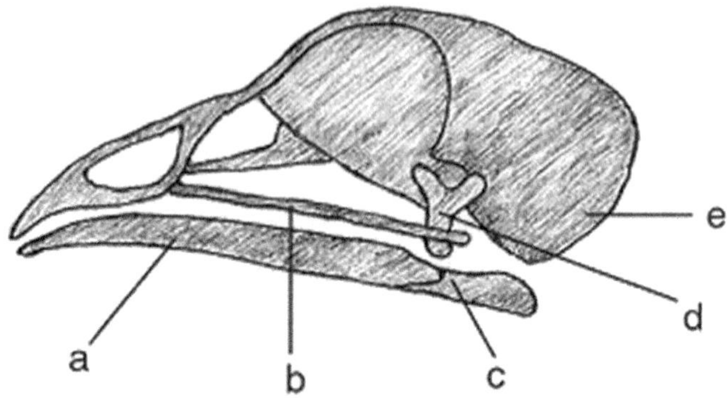

Fig. 2.15 The skull of a bird showing the location of the quadrate bone: (a) mandible, (b) zygomatic bone, (c) articular bone, (d) quadrate bone and (e) occipital bone. The articular and quadrate bones have evolved into the malleus and incus of the mammalian middle ear respectively.

the two bones forming a similar joint in other vertebrates is lacking in mammals. Mammals also have a chain of three tiny bones (ossicles) that transmit sound waves across the middle ear. These bones (from outside to inside) are the **malleus** (L. hammer), **incus** (L. anvil) and **stapes** (L. stirrup). The **cochlea** that detects sound in the inner ear is highly developed and gives mammals a keen sense of hearing. The separation of most of the mammalian oral and nasal cavities by a secondary palate enables them to feed and breathe at the same time. The nasal cavities of most mammals are large and their surface area is increased by the presence of the scroll-like **nasal turbinates** (**conchae**). The mammalian teeth are variously specialized for cutting, chewing and grinding.

Generally, the mammalian skeleton shows a number of advances over that of other vertebrates. The cartilages that will develop into long bones of lower vertebrates have a single (primary) center of ossification at the diaphysis (shaft). Replacement of cartilage by bone then spreads to the ends (epiphyses) that sometimes remain cartilaginous in adults. In addition to primary centers of ossification, mammals have secondary centers of ossification at the epiphyses of their long bones that make the bones of this class of vertebrates stronger at the articular surfaces before the skeleton is mature.

Bone sometimes develops within organs and non-bony connective tissue. Such bones are known as **splanchnic bones** and include the **os penis** or **baculum** that is found in the penis of carnivores, rodents, many bats, some insectivores and most primates excluding man; the **os cordis** of the heart in some large ungulates and **os rostrale** of the porcine snout. The os penis develops in the fibrous septum that lies between the corpora cavernosa above the urethra and is the most varied of all bones as far as the shape is concerned. The **os clitoridis** may be found in the clitoris of some female mammals. Bones found in some tendons of muscle insertion are known as **sesamoid bones** (Fig. 2.16). The skull of mammals is also more expanded than that of other vertebrates as a result of the expansion of the brain.

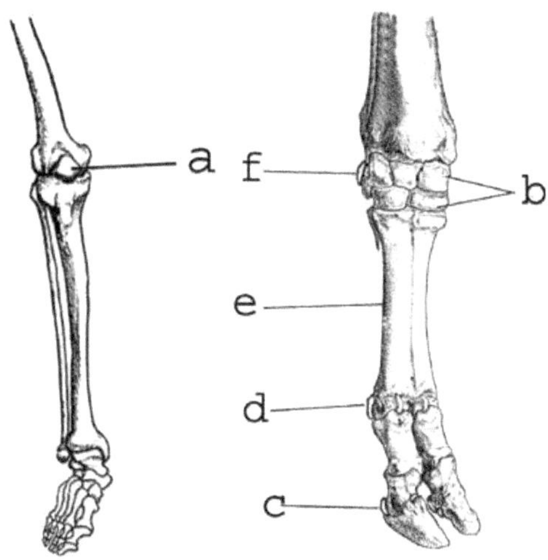

Fig. 2.16 Some sesamoid bones: (a) patella of the human being, (c) distal and (d) proximal sesamoid bones, (f) accessory carpal bone of an ungulate, and (b) carpal and (e) metacarpal bones.

Muscles. The flat muscular **diaphragm** (Gr. *dia*, through or across; *phragma*, a partition wall) separates the thorax from the abdominal cavity. The diaphragm is the chief muscle of inspiration and increases the efficiency and depth of breathing. **Cutaneous muscles (panniculus carnosus)** are sheaths of muscle found in fascia of most mammals and cause movement of the skin independent of deeper muscle masses. Such movement causes twitching of the skin to shake off insects and is also important in heat production during shivering. The cutaneous muscles of the face are well developed in carnivores and primates. They are responsible for facial expression of primates including the human.

Reproduction. Placental mammals carry their young in the uterus where they reach a relatively advanced stage of development before being born. In marsupials, the newborn is incompletely developed at birth and continues further development outside the uterus by attaching itself to the female's mammary glands located in a pouch or a groove. The embryos have to be reared in a warm environment before thermoregulatory mechanisms are fully developed.

Nervous system. The mammalian brain is greatly enlarged, especially in the cerebral hemispheres. Such an enlargement enables mammals to develop more complex instincts and undergo more advanced learning. Mammals are also able to adapt to short-term environmental changes by responding appropriately as a result of previous experiences. The superficial **gray matter** (**neopallium**) of the **cerebrum** forms the major portion of the cerebrum. It is the main center of neural function and is involved in intelligent response. In some mammals, the gray matter is quite expanded by the presence of folds or convolutions known as **gyri** (Gr. *gyros*, circle) that are located between deep grooves or furrows referred to as **sulci** (L. *sulcus*, groove).

Circulation. The mammalian aortic arch is unpaired and is derived from the fourth left aortic arch of the primitive vertebrate. The right arch persists in birds whereas both arches persist in fishes, amphibians and reptiles. The mature mammalian erythrocyte is biconcave in shape and lacks a nucleus whereas all other vertebrates have nucleated red blood cells. The right and left ventricles of mammals and birds are completely separate, so pulmonary and systemic blood is completely independent.

The ability to regulate their body temperatures and internal environment in hot and cold conditions has enabled mammals to exploit the different environments on Earth better than any other group of vertebrates. Several orders of mammals have become extinct in the last 185 million years of mammalian evolution. Mammals are divided into two major subclasses: prototheria and theria.

Prototheria

The only group of surviving prototherians is that of monotremes and belongs to the order **Monotremata** (Gr. *monos*, single/one; *trema*, hole). Monotremes are found in Australia and New Guinea and are represented by the duckbill platypus and two species of echidnas (spiny anteaters). They lay parchment-shelled eggs as their therapsid ancestors did. **Parchment-shelled eggs** have a porous eggshell and a bilaminar yolk sac membrane that permits uptake of uterine secretions during the intrauterine period. The eggs lose water very rapidly when exposed to air of lower water pressure. Monotremes have skeletal and other features that resemble those of reptiles such as presence of a cloaca. They are toothless and have a horny beak. Since monotremes lack organized nipples, milk flows from mammary ducts onto the abdominal fur.

Theria

Therians (Gr. *therion*, beast) are divided into metatheria (marsupials) and eutheria (placental mammals).

Metatheria

Marsupial embryos have a brief gestation period in the reproductive tract of the mother during which they receive nourishment from the yolk sac that grows into the uterus. The young are born in a relatively undeveloped state and attach themselves to nipples in a **pouch** (**marsupium**) or marsupial groove where they undergo further development. The female reproductive tract of marsupials is bifid (paired uterus and vagina). The penis of male animals is bifurcate and is posterior to the scrotum. Marsupials have a relatively smaller braincase than eutherians and their mandibles have an in turned angular process. An **epipubis** or **marsupial bone** (Fig. 2.17) that supports the pouch is present. The rod-shaped epipubis extends from the pubis into the ventral body wall of the abdomen. Only one marsupial, the opposum survives in North America. There are several other marsupials in South America and Australia that include kangaroos, wallabies, bandicoots, wombats, phalangers and koalas.

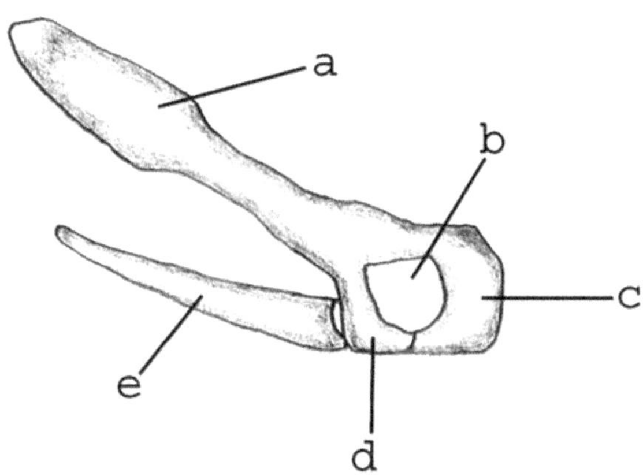

Fig. 2.17 The pelvic bones of a marsupial (a) ilium, (b) obturator foramen, (c) ischium, (d) pubis and (e) epipubic bone. The epipubic bone is also present in monotremes the extinct multituberculates.

Eutherian Mammals

Eutherian (Gr. *eu*, true or good) mammals retain embryos for a relatively longer time than marsupials so that they can undergo quite some development before birth. The eutherian egg contains little yolk. Extraembryonic membranes form an umbilical cord and placenta through which the embryo is able to receive nourishment and undergo gaseous exchange with the mother. Mammals have been divided into four clades that comprise closely related groups or orders based on analysis of DNA sequences of several nuclear and mitochondrial genes (Murphy et al., 2001 a and b, Reyes et al. 2004).

Mammalian Clades

A clade (Gr. *klados*, branch) forms a branch of a cladogram (phylogenetic tree) that is made up of a monophyletic group of organisms that share homologous features since they originated from a common ancestor. Clades are based on the type of data used in a cladistic analysis and may change with time when analysis using different data is applied. Cladistics involve the analysis of evolutionary relationships based on shared and derived similarities that seeks to establish points of separation of various lineages of organisms.

Afrotheria

The clade Afrotheria has the rank of cohort and includes six orders that are considered the oldest of the placental mammals. The clade exhibits extreme morphological diversity and large genome sizes (Zhao et al., 2009). Such physically disparate animals share morphological characteristics such as late eruption of permanent teeth, vertebral anomalies and testicondy (Asher and Lehmann, 2008). Afrotherians evolved on what later on became the continent of Africa during the breakup of **Gondwanaland** about 100 million years ago. The superorder includes **Tubulidentata** (aardvarks) of Africa and Eurasia, **Macroscelidea** (sengis or elephant shrews) of Africa, **Hydracoidae** (hyraxes) of Africa and South West Asia, **Sirenia** (manatees and dugongs or sea cows) found in the seas that neighbor Africa, Asia, North and South America and **Proboscidea** (elephants) of Africa and India. The order **Lipotyphyla** that comprises golden moles and tenrecs is polyphyletic according to molecular evidence and has also been placed under Afrotheria (Malia et al., 2002). The order **Afrosoricida** has been suggested by some scientists and includes tenrecs, golden moles and otter shrews.

Xenarthra

These are thought to have arisen in part of Gondwana that later became South America. Mammals of the clade Xenarthra belonged to the former order **Edentata** and include armadillos, giant anteaters and sloths that are found in North and South America. Phylogenetic analysis of conserved protein-coding sequences from several placental mammals has joined Xenarthra and Afrotheria on a common branch, **Atlantogenata** (Hallström et al., 2007).

Euarchontoglires

The orders included in this clade that is also known as **Supraprimates** are **Rodentia** (rodents), **Lagomorpha** (hares, rabbits and pikas), **Dermoptera** ('skin-winged') that comprise colugos or 'flying lemurs' that are found in the tropical rainforests of Southeast Asia, **Scandentia** and **Primates** (primates) of Eurasia, Africa, North and South America. A sister clade, Laurasiatheria, could have split from Euarchontoglires about 85 to 95 million years ago.

Laurasiatheria

The clade Laurasiatheria groups together several orders that were first found in the supercontinent of **Laurasia**. The orders are **Isectivora** (shrews, moles, hedgehogs and solenodon) found worldwide except in Australia, the polar region and many oceanic islands; **Chiroptera** (bats) found in tropical and temperate regions and **Pholidota** (pangolins or scaly anteaters) of Eurasia and Africa. Other orders include **Carnivora** (carnivores); **Perissodactyla** (titanotheres, chalicotheres, tapirs rhinoceroses and horses) of North America, Eurasia and Africa; **Artiodactyla** (pigs, cattle, sheep, goats, antelopes, pronghorns and giraffes) and **Cetacea** (whales). Artiodactyls (even-toed) and cetaceans are so closely related genetically that they are placed in the clade **cetartiodactyla**. Insectivores are the most diverse group in the clade Laurasiatheria.

The order Rodentia (with approximately 1,700 species) is the largest mammalian group. Another large group is Chiroptera with 1,000 species. The orders Dermoptera and Proboscidea are the smallest and are represented by two and three living species respectively. The two living species of lemurs are *Galeopterus variegatus* that is found in Southeast Asia including Java and Borneo and the Philippine species (*Cynocephalus volans*). The Indian elephant has one species (*Elephas maximus*) whereas the African elephant has two species, the African bush elephant (*Loxodonta africana*) and the African forest elephant (*Loxodonta cyclotis*).

ADAPTIVE RADIATION OF VERTEBRATES

Adaptive radiation entails the diversification of a species as it adapts to different ecological niches. Species that are able to adapt successfully become specialized for surviving in the new environment by natural selection. Throughout their evolutionary history, vertebrates have undergone adaptive radiation that has involved an increase in the number and diversification of the group with each species exploiting various habitats differently. It has been recognized that an increase in the number of Hox genes may have been a major event in adaptive radiation of the group. Invertebrates have one Hox cluster whereas vertebrates have at least four. Constraints on vertebrate Hox cluster structure lead to an association between the retention of duplicated Hox clusters and adaptive radiations (Wagner et al., 2003). Amphioxus has one Hox gene cluster with a similar genomic organization to the four Hox mammalian clusters and is a living representative of an important intermediate stage in the gene cluster evolution (Garcia-Fernandez and Holland, 1994).

Fish

The earliest vertebrates evolved in the ocean from invertebrate ancestors that possessed gill slits which they mainly used for filter feeding and exchanged gases with the environment through the skin. The first vertebrates to evolve were fish. The fish like **ostracoderms** are considered to be the first vertebrates. They lived from early Ordovician to late Devonian. They were abundant mainly in seawater although some groups are thought to have been freshwater. Some 200 species of this diverse group have been discovered. They are considered to have been bottom swimmers as is evident from the abrasions on the ventral surface of their heads. After true fish appeared about 400 million years ago, ostracoderms rapidly become extinct. An armorless type survived and gave rise to modern cyclostomes (lampreys and hagfish). Lampreys and hagfish are thought to be monophyletic (Gr. *phyle,* tribe) according to DNA studies (Kuraku et al., 1999; Delarbre et al., 2002; Furlong and Holland, 2002). At Devonian, most of the fish had occupied freshwater that later on became their habitat for subsequent radiations.

Placoderms were primarily freshwater fish but most become secondarily sea fish. They comprised about six diverse orders that radiated into several body shapes and ecological niches. The earliest placoderms appeared in early Silurian. Placoderms reached their greatest diversity in the Devonian ('Age of Fishes'). Only two of these orders (Arthrodira and Antiarchi) are well known. Placoderms lasted about 50 million years and all became extinct quite suddenly during early Carboniferous for unknown causes. Some preserved specimens of placoderms are found in Cleveland deposits.

The earliest **cartilaginous fish** appeared late in the Devonian and originated in the sea. They resembled the living sharks but were more sluggish and were heavily armored. They were the first vertebrates to develop internal fertilization as they lived in a hypertonic environment since the ability to be in osmotic balance with seawater develops later in embryology. One line of the early cartilaginous fish led to the freshwater pleuracanths that flourished for millions of years but became extinct during the Triassic. Living sharks and rays appeared later in the Jurassic.

The earliest **bony fish** appear in freshwater deposits of the Devonian period. They possessed gills and lungs that were pouched outgrowths from the pharynx. The lungs assisted the gills in gaseous exchange when water was stagnant or too warm to contain enough oxygen. Bony fish diversified through the Devonian period and some migrated to the sea. Their lungs were transformed into the swim bladder (with the exception of lungfish and other fish such as bichirs). The kidneys were also transformed to adapt them to a hypertonic environment. **Chondrosteans** flourished during the later part of the Paleozoic and were succeeded by holosteans in the Permian. From the Jurassic onwards, **holosteans** dominated the world's oceans but diminished in the Cretaceous and are now largely extinct. **Teleosts** first appeared in the Middle Triassic and did not change much till Late Jurassic. By the end of Cretaceous, teleosts had become the dominant fish in both sea and freshwater. Teleosts showed their greatest radiation after Eocene.

Amphibians

Amphibians first appeared during the Devonian period and were so successful during the Carboniferous (Mississipian and Pennsylvanian periods) that this period is known as the 'Age of Amphibians'. The earliest amphibians were icthyostegalians (of subclass Labyrinthodontia) that had a mixture of piscine and terrestrial characteristics such as possession of fishlike caudal fins and four legs.

During Permian, the Earth became colder and dryer. The number of amphibians began to decline with the extinction of labyrinthodonts. The only amphibian subclass with living representatives is Lissamphibia that has the orders Urodela, Anura and Apoda. Labyrinthodonts could have given rise to the first terrestrial vertebrates.

Reptiles

Reptiles evolved from amphibians during the Carboniferous. They were adapted to survive the cold and dry Permian because of their development of a cleidoic egg that could be laid on land without the danger of dessication and evolution of internal fertilization. Reptiles underwent remarkable adaptive radiation during the Mesozoic era ('Age of Reptiles'). They were the dominant terrestrial vertebrates and produced several lines of descent that include synapsids and sauropsids. **Synapsids** had their legs under, other than the sides of their bodies and could run fast. A line of active terrestrial therapsids evolved from synapsids and later on gave rise to mammals.

The line **Sauropsida** gave rise to all the other reptiles that included turtles, plesiosaurs, ichthyosaurs and diapsids. Plesiosaurs and ichthyosaurs were marine reptiles that became extinct by the end of the Mesozoic era. Diapsids developed the ability to convert nitrogenous wastes into the almost insoluble uric acid that does not require much water for excretion. This method of excretion has reduced water intake by diapsids and their ancestors. Diapsid evolution gave rise to squamates (lizards, snakes and amphisbaemians), tuataras (*Sphenodon*) and archosaurs. Archosaurs were able to run fast by rising up on their larger hind legs while using their long tails for balance. Archosaurs include crocodiles and alligators and the extinct dinosaurs that were the dominant reptiles during the Mesozoic and from some of which evolved today's birds.

Birds

Birds resemble archosaurs from which they evolved during the Jurassic period. The closest living relative of the bird is the crocodile. Avian flight developed early in avian evolution as is seen in the 135 million year old sparrow sized *Sinornis* that was discovered in China (Sereno and Chenggang, 1992). The early birds had teeth as is also seen in the Jurassic bird *Archaeopteryx*. The rest of the birds lack teeth and belong to the superorders Paleognathae and Neognathae. Radiation of neognathous birds has given rise to about 23 monophyletic lineages that have adapted to different foods, methods of locomotion and habitat.

Mammals

Mammals first appeared in the early part of the Mesozoic era after branching off the now extinct reptilian order Therapsida. Mammals co-existed less prominently with reptiles for about 150 to 200 million years. Prototherians and therians first appeared about 185 million years ago. Therians later diverged into Methatheria and Eutheria about 125 years ago. Adaptive radiation of mammals began about 65 million years ago thus the Cenozoic era is known as the 'age of mammals'. The extinction of many reptiles at the end of the Mesozoic about 65 million years ago could have created habitats for mammals, for example ungulates replacing the herbivorous ornithischian dinosaurs and cetaceans taking over from plesiosaurs and icthyosaurs.

Several mammals have become extinct during the course of evolution. The latest mammalian extinction occurred during late Pleistocene (about 11,000 years ago) when a variety of animals mainly mammals become extinct across North America. The animals were larger than 40 kg in weight and include the mammoths, mastodons, saber-toothed cats, giant ground sloths, longhorned bisons and the native horses and camels. The great teratorn birds also became extinct at the same period. Mammoths, mastodons and modern elephants belong to the order Proboscidae.

ECTOTHERMY AND ENDOTHERMY

Ectothermy (Gr. *ektos*, outside; *thermos*, heat) is a condition where an animal's body temperature is influenced by the temperature of the environment. The body temperature of an ectotherm depends on the rate at which it generates heat internally and the rate at which the heat is lost or gained from the environment. An ectotherm lacks the ability to maintain a stable body temperature. Vertebrate ectotherms include fishes, amphibians and reptiles. Their body temperatures and metabolic rates vary greatly depending on environmental temperatures. An increase in body temperatures by 10°C can double the body metabolism and vice versa. Ectotherms found in the tropics generally have higher body temperatures and metabolic rates throughout the year than those found in temperate regions. Some ectothermic vertebrates can attain body temperatures that are very different from their surroundings. Fast swimming fishes with representatives from tunas, billfish and sharks generate a lot of heat by action of their muscles during rapid swimming. Some of this energy can be conserved for some time thus enabling the fish to conserve higher body temperatures in relation to their surroundings. These fish share various specialized anatomical features such as structure of their swimming muscles and some form of regional endothermy known as **heterothermy** (Katz, 2002). Heterothermy is kept by maintenance of elevated temperatures in swimming muscles or presence of muscle-derived tissues specialized for delivering warm blood to the head.

Ectothermy has some advantages in energy costs. Most fish, amphibians and reptiles do not have to spend a lot of energy to sustain higher stable body temperatures. Lower environmental temperatures slow down the body activities of ectotherms and reduce their energy costs. Water has a high thermal stability and requires much energy before its temperature can change and so aquatic ectotherms are not exposed to as much diurnal temperature fluctuations as ectotherms living on land.

Endothermy (Gr. *endon*, within) is seen in vertebrates that generate their body heat from within their bodies. Such vertebrates are able to efficiently retain the heat and maintain high metabolic rate due to presence of insulation such as hair, feathers and subcutaneous fat and also presence of mechanisms that will lead to a generation of more heat when environmental temperatures fall. It has been suggested that the base of chemical thermoregulation in birds and mammals is a thermoregulatory muscle tonus that remains unknown (Dol'nik, 2003). Birds and mammals are endotherms and most maintain their body temperatures between 35°C and 40°C. The amount of blood that flows to the skin can also be reduced to conserve heat and vice versa. When heat needs to be lost, many birds and mammals will pant and some mammals will sweat thus losing water by evaporation in the process.

Endothermy develops later during embryogenesis. In birds, endothermy is closely related to the maturational state of a hatchling. Newly hatched chicks of different birds show varying degrees of development. **Altrical chicks** (Fig. 2.18) hatch with little or no feathers and little motor activity and have to depend on their parents for feeding and warmth. Altricials are not yet endothermic and are still exothermic in heat regulation. Altricial birds include songbirds, hummingbirds, woodpeckers, pelicans, swifts and kingfishers. The more mature **precocial chicks** have downy coats and more developed nervous and muscular functions at the time of hatching. Precocial chicks can walk from the nest, feed themselves and are also able to regulate their body temperature soon after hatching. Chickens, ducks, geese, ostriches, quails, turkeys, swans and pheasants are precocial birds.

Birds and mammals evolved endothermy independently as they originated from different groups of reptiles. As they evolved alongside mammals, dinosaurs had to compete effectively with the endotherms for the 170 million years when the two groups coexisted. The first mammals are thought to have lived like the burrowing and nocturnal tenrec of Madagascar that maintains a body temperature of 28°C to 30°C. Dinosaurs are thought to have been endotherms (Bakker, 1971). Therapsids (mammal-like reptiles) are known to have had increased activity, a change that was necessary before endothermy. For more information on origins of endothermy, see McNab (1978) and Bennet et al. (2000).

Fig. 2.18 American Robin altricial chicks with eyes that are yet to open and a poor cover of feathers (left) and precocial ducklings (right).

Endothermy has its advantages and disadvantages. Endotherms are able to maintain high levels of metabolism and activity even when ambient temperatures are low. Endothermy also enables birds and mammals to live in environments that are extreme such as deserts and the Arctic. Endothermy is expensive in energy costs especially when food is scarce. Birds and mammals generally consume more food than ectotherms of equivalent weight.

In hot and arid environments, the camel (Fig. 2.19) has to adapt to conditions that vary and at times provide limited water for thermoregulation. The body temperature of a healthy camel that has adequate supplies of water varies by about 2°C from about 36°C in the morning to 38°C in the afternoon. Body temperature fluctuations increase when water supplies are limited to as high as 6°C to 7°C. Fluctuations in body temperature save on body water since such changes reduce the amount of water required for evaporative cooling of the body and favor heat gain by the animal from the surrounding environment. Camels lose heat at night resulting in a fall in body temperature that is below normal. Such great lose of body heat at night ensures that during daytime it will take a much longer time before the top body temperature is attained. The thick fur of the camel also plays an important role in thermoregulation since it has good insulating properties against heat. The Arctic fox (Fig. 2.19) has adapted to the cold polar environment by developing a wide thermoneutral zone that can tolerate temperatures as low as -40°C. The fur also provides some insulation against the low temperatures. Fluffing of the fur traps unstirred air that retains body heat.

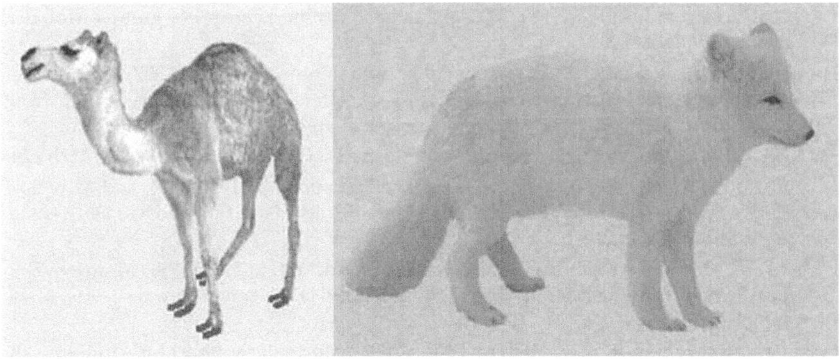

Fig. 2.19 A dromedary or Arabian camel (*Camelus dromedarius*) (left) and the Arctic fox (*Alopex lagopus*) (right). Camels have the ability to tolerate dehydration and can survive a water loss of up to 40% of their body weight during which they produce urine that is as thick as syrup and contains twice as much salt as seawater. Dromedary camels weigh between 300 to 690 kg whereas the Arctic fox weighs between 2.5 and 4.5 kg.

REFERENCES

Asher, R.J. and Lehmann, T. (2008). Dental eruption in afrotherian mammals. BMC Biol., 6:14.

Bakker, R.T. (1971). Dinosaur physiology and the origin of mammals. Evolution, 25:636-658.

Basch, M.L., Garcia-Castro, M.I. and Bronner-Fraser, M. (2004). Molecular mechanisms of neural crest induction. Birth Defects Res. Part C Embryo Today, 72(2):109-123.

Bennet, A.F., Hicks, J.W. and Cullum, A.J. (2000). An experimental test of the thermoregulatory hypothesis for the evolution of endothermy. Evolution, 54(5):1768-1773.

Brinkmann, H., Venkatesh, B., Brenner, S. and Meyer, A. (2004). Nuclear protein-coding genes support lungfish and not the coelacanth as the closest living relatives of land vertebrates. Proc. Natl. Acad. Sci. U.S.A., 101(14):4900-4905.

Bunce, M., Worthy, T.H., Ford, T., Hoppit, W., Willerslev, E., Drummond, A. and Cooper, A.C. (2003). Extreme reversed sexual size dimorphism in the extinct New Zealand moa *Dinornis*. Nature, 425(6954):172-175.

Callery, E.M. and Elinson, R.P. (2000). Opercular development and ontogenic re-organization in a direct-developing frog. Dev. Genes Evol., 210(7):377-381.

Delarbre, C., Gallut, C., Barriel, V., Janvier, P. and Gachelin, G. (2002). A complete mitochondrial DNA of the hagfish *Eptatretus burgeri*: the comparative analysis of mitochondrial DNA sequences strongly supports the cyclostome monophyly. Mol. Phylongenet. Evol., 22(2):184-192.

Dol'nik, V.P. (2003). The origin of homoiothermy-unresolved problem. Zh. Obshch. Biol., 64(6):451-462.

Furlong, R.F. and Holland, P.W. (2002). Bayesian phylogenetic analysis supports monophyly of ambulacraria and of cyclostomes. Zoolog. Sci., 19(5):593-599.

Garcia-Fernandez, J. and Holland, P.W. (1994). Archetypal organization of the amphioxus Hox gene cluster. Nature, 370(6490):504-505.

Hallström, B.M., Kullberg, M., Nilsson, M.A. and Janke, A. (2007). Phylogenomic data analyses provide evidence that Xenarthra and Afrotheria are sister groups. Mol. Biol. Evol., 24(9):2059-2068.

Hanken, J., Klymkowsky, M.W., Alley, K.E. and Jennings, D.H. (1997). Jaw muscle development as evidence from embryonic repatterning in direct-developing frogs. Proc. Roy. Soc. Lond. Series B. Biol. Sci., 264(1386):1349-1354.

Katz, S.L. (2002). Design of heterothermic muscle in fish. J. Exp. Biol., 205 (Pt. 15):2251-2266.

Kikugawa, K., Katoh, K., Kuraku, S., Sakurai, H., Ishinda, O., Iwabe, N. and Miyata, T. (2004). Basal jawed vertebrate phylogeny inferred from multiple nuclear DNA-coded genes. BMC. Biol., 2(1):3.

Kuraku, S., Hoshiyama, D., Katoh, K., Suga, H. and Miyata, T. (1999). Monophyly of lampreys and hagfishes supported by musclear DNA-coded genes. J. Mol. Evol., 49(6):729-735.

Malia, M.J. Jr., Adkins, R.M. and Allard, M.W. (2002). Molecular support for Afrotheria and the polyphyly Lipotyphala based on analyses of the growth hormone receptor gene. Mol. Phylogenet Evol., 24(1):91-101.

McNab, B.K. (1978). The evolution of endothermy in the phylogeny of mammals. American Naturalist, 112:1-21.

Murphy, W.J., Eizirik, E., Johnson, W.E., Zhang, Y.P., Ryder, O.A. and O'Brien, S.J. (2001a). Molecular phylogenetics and the origins of placental mammals. Nature, 409 (6820):614-618.

Murphy, W.J., Eizirik, E., O'Brien, S.J., Madsen, O., Scally, M., Douady, C.J., Teeling, E., Ryder, A.O., Stanhope, M.J., de Jong, W.W. and Springer, M.S. (2001b). Resolution of the early placental mammal radiation using Bayesian phylogenetics. Science, 294(5550):2348-2351.

Nagoshi, N., Shibata, S., Nakamura, M., Matsuzaki, Y., Toyama, Y. and Okano, H. (2009). Neural crest-derived stem cells display a variety of characteristics. J. Cell Biochem., 107(6):1046-1052.

Peach, M.B. and Marshall, N.J. (2000). The pit organs of elasmobranches: a review. Philos. Trans. R. Soc. Lond. B. Biol. Sci., 355(1401):1131-1134.

Reyes, A., Gissi, C., Catzeflis, F., Nevo, E., Pesole, G. and Saccone, C. (2004). Congruent mammalian trees from mitochondrial and nuclear genes using Bayesian methods. Mol. Biol. Evol., 21(2):397-403.

Roberts, D.L. and Solow, A.R. (2003). Flightless birds: when did the dodo become extinct? Nature, 426(6964):245.

Sereno, P.C. and Chenggang, R. (1992). Early evolution of avian flight and perching: New evidence from the lower Cretaceous of China. Science, 255:845-848.

Shima, A. and Mitani, H. (2004). Medaka as a research organism: past, present and future. Mech. Dev., 121(7-8):599-604.

Wagner, G.P., Amemiya, C. and Ruddle, F. (2003). Hox cluster duplications and the opportunity for evolutionary novelties. Proc. Natl. Acad. Sci. U.S.A., 100(25):14603-14606.

Widera, D., Zander, C., Heidbreder, M., Kasperek, Y., Noll, T., Seitz, O., Saldamli, B., Sudhoff, H., Sader, R., Kaltschmidt, C. and Kalschmidt, B. (2009). Adult palatum as a Novel Source of Neural-Crest Related Stem Cells. Stem Cells, 27(8):1899-1910.

Wilkie, M.P., Turnbull, S., Bird, J., Wang, Y.S., Claude, J.F. and Youson, J.H. (2004). Lamprey parasitism of sharks and teleosts: high capacity urea excretion in an extanct vertebrate relic. Comp. Biochem. Physiol.. A Mol. Integr. Physiol., 138(4):485-492.

Yokobori, S., Hasegawa, M., Ueda, T., Okada, N., Nishikawa, K. and Watanabe, K. (1994). Relationship among coelacanths, lungfishes and tetrapods: a phylogenetic analysis based on mitochondrial cytochrome oxidase I gene sequences. J. Mol. Evol., 38(6): 602-609.

Zhao, F., Qi, J. and Schuster, S.C. (2009). Tracking the past: Interspersed repeats in an extinct Afrotherian mammal, *Mammuthus primigenius*. Genome Res., 19(8):1384-1392.

3

Organization of the Vertebrate Body

The different parts of the vertebrate body are organized to fit together and function effectively as a unit. There are several levels of organization that function harmoniously with each other for the survival of vertebrates. Starting from the lowest to the highest levels are chemical, organelle, cellular, tissue, organ and system levels of organization.

Chemical level. This is the level at which organization of the vertebrate body begins. More than 100 atoms combine to form molecules that will in turn combine to form even larger molecules known as macromolecules. All these structures have unique relationships and form a semi-fluid material known as cytoplasm. The cytoplasm of different vertebrate cells has its unique characteristics that are related to its functions.

The different chemicals can be organized into larger structures that perform specific functions in cells known as **organelles**. Organelles are incapable of surviving on their own outside a cell. More than 20 organelles found in vertebrate cells are necessary for the survival of cells. Examples of organelles include mitochondria, endoplasmic reticula, Golgi apparatus and lysosomes.

Cellular level. Cells represent the basic level of organization in living matter that is capable of independent survival as is exhibited by unicellular organisms. The cells of vertebrates have a common basic structure but differ in structural details and shapes as they perform various specialized functions and are also located in different parts of the body that are subjected to various forces and strains. Each cell has a limiting membrane, a nucleus that contains the genetic material, a cytoplasm, organelles and cellular inclusions.

A **tissue** is a group of similar cells that are specialized to perform a particular function. Tissue cells make contact with each other with varying amounts of non-living intercellular substances (matrix) between them. As cells are able to sense changes in their mechanical environment and promote alterations and adaptations in tissue structure and function, mechanical strain regulates important processes such as cell division and differentiation that determine tissue form (Benjamin and Hillen, 2003). There are four basic (major) tissues that make up the vertebrate body and include epithelial, connective, nervous and muscular tissues. The basic tissues are able to meet all the structural and functional needs of the vertebrate body. Tissues work together in the vertebrate body.

An **organ** consists of several different tissues that work together as a unit to perform a specialized function. There are several organs in the vertebrate body, for example the liver, kidney, heart and pancreas. Organs differ in shape, location, tissue composition and function and play specific roles in the vertebrate body.

System level. Several organs can be part of a more complex structure that is specialized to perform a unique and major function in the body. There are eleven major systems in the vertebrate body and include the integumentary, muscular, skeletal, circulatory, nervous, endocrine, lymphatic/immune, digestive, respiratory, urinary and reproductive systems. Although these systems may differ structurally in many vertebrates, they perform similar general functions in these animals. The overall functions of all body systems are vital to the survival of the entire organism.

BASIC TISSUES OF VERTEBRATES

All tissues of the body perform various unique functions that together maintain homeostasis that ensures the survival of the entire vertebrate body. The cells that make up body tissues are held together by intercellular material that is also known as the matrix. The matrix varies in quantity and composition greatly depending on the function played by the tissue in the body. A mineralized matrix will render rigidity whereas presence of fibers will render flexibility to the connective tissue. The four basic tissues are epithelial, connective, muscle and nervous tissues and each has various specialized subtypes.

Epithelial Tissue

Epithelial tissue (Gr. *epi*, upon; *thele*, delicate skin) covers the body surface, the inner surface of hollow structures and also forms many glands. Material into and out of the inner body cells has to cross this epithelium. The epithelium is found in many parts of the body and serves protective (skin), secretive (glandular epithelium) absorptive, excretive (as in kidney tubules) and sensory (skin, nose and ear) functions. To be able to carry out these varied functions, epithelial tissue has cells that are arranged differently or have undergone specialization. The eight types of covering epithelial tissue are made of differently shaped cells that are arranged in single or several layers depending on the function performed (Fig. 3.1).

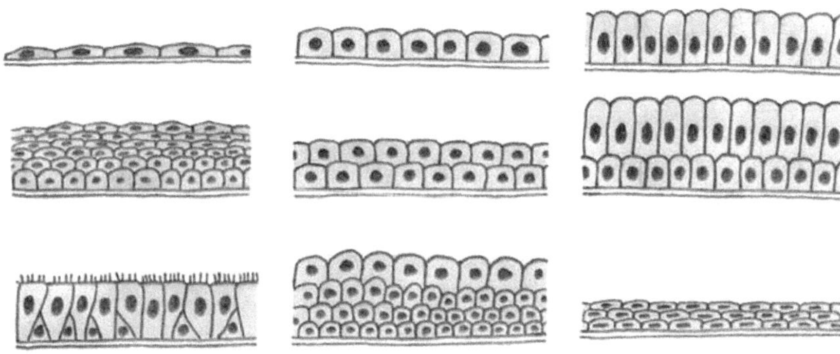

Fig. 3.1 Various types of epithelia found in vertebrates. **Top row**: simple squamous (left), simple cuboidal (middle) and simple columnar (right). **Middle row**: stratified squamous (left), stratified cuboidal (middle) and stratified columnar (right). **Bottom row**: pseudostratified columnar (left), relaxed (middle) and stretched (right) transitional epithelia.

Covering epithelia are a continuous layer or layers of cells with small quantities of intercellular substance and cover an external or internal surface of the body and its organs. An epithelium comprises a free border that is adjacent to the external environment or a lumen, cells and a basal layer that supports the cells. Epithelia are classified according to the arrangement or shape of cells that form these tissues. A **simple epithelium** is one cell thick whereas a **stratified epithelium** is several cells thick. **Pseudostratified epithelia** are simple epithelia that comprise a layer of epithelial cells that are quite variable in shape. All cells in a pseudostratified epithelium rest on a basal lamina but not all cells reach the free border. Since nuclei of a pseudostratified epithelium are located at various levels because they occupy the widest parts of the cells, the epithelium appears as if it is made up of several layers of cells. Pseudostratified epithelia occur in the respiratory system, epididymis and excretory ducts of some glands such as those of the parotid and male urethra. The pseudostratified epithelia of epididymis and respiratory system contain long microvilli and cilia respectively. The olfactory epithelium is also of a ciliated pseudostratified type. Cells that are arranged in close apposition to each other in a manner that resembles an epithelium but do not have a free surface or basal lamina are known as **epithelioid cells**. Examples of epithelioid cells include synovial cells that line the synovial cavity but lack a basal lamina, Leydig cells of the testis and luteal cells of the ovary that lack a free border and some cells associated with pathological cases such as cancerous cells.

According to the shape of cells, which are squamous, cuboidal, columnar and transitional epithelia. **Squamous epithelia** (L. *squama*, scale) are made up of thin cells in cross-section that appear like paving stones on the surface whereas **cuboidal epithelia** have cells that are isodiametric in shape. A **columnar epithelium** has cells that are higher than being wide. **Simple squamous epithelia** are made up of a single layer of flat plate-like cells that contain little cytoplasm with each cell covering a large surface area. The endothelium of blood and lymphatic vessels, excretory ducts of many exocrine glands and Bowman's or glomerular capsule of the kidney are made up of simple squamous epithelia. Serous cavities such as the peritoneum, pleural membrane and pericardium are lined with simple squamous epithelia that cover large areas known as **mesothelia**. An **endothelium** is a type of epithelium that forms the inner lining of blood and lymph vessels.

Cells of a **simple cuboidal epithelium** have approximately equal width and height and may change their shape according to their physiological state. Simple cuboidal epithelia are found in excretory ducts of exocrine glands, some kidney tubules, the pigmented epithelium of the retina, thyroid follicles of the thyroid gland, surface of the ovary and the anterior part of the lens of the eye. Some cuboidal cells have surface specializations such as microvilli that are found in the thyroid gland. A **simple columnar epithelium** is a layer of cells that are taller than being wide and the cells may change their shapes depending on their various functions. Simple columnar epithelia are found in structures including excretory ducts of exocrine glands, intestines, ventricular ependymal lining of the central nervous system, bronchioles and the gall bladder. Simple columnar epithelia that contain cilia and are found in structures such as small bronchi, uterus, oviducts, central canal of the spinal cord and paranasal sinuses.

A **stratified squamous epithelium** is important in providing mechanical protection to body organs. The cells of a stratified squamous epithelium vary in shape from the basal membrane to the free border or surface. Cells that rest on the basal membrane are thick and thin out towards the free surface. Stratified squamous epithelia can be keratinized or non-keratinized. **Keratinized stratified squamous epithelia** contain deposits of keratin in the superficial cells of the epithelia. The cytoplasm of the superficial cells of keratinized stratified squamous epithelial cells lack nuclei and contain mainly keratin making the cells dry and scalelike structures such as the integument and the initial portion of the digestive system of some vertebrates. The superficial cells of a **non-keratinized stratified squamous epithelium** are viable and contain nuclei and such an epithelium is found in the skin of fish, initial and terminal parts of the digestive system, the vestibule of the nose and the female genital system.

All the cells in a **stratified cuboidal epithelium** that normally comprises two or three layers are of uniform appearance. Stratified cuboidal epithelia are found in the excretory ducts of some exocrine glands such as salivary, mammary and sweat glands and the pancreas. A **stratified columnar epithelium** is rare and has at least two layers of cells. The lower basal cells are lower in height but the superficial cells are columnar in shape. Stratified columnar epithelia are found in some large excretory ducts of exocrine glands, vas deferens, anorectal junction, transitional areas between stratified squamous and pseudostratified epithelia and fornix of conjuctiva. A **transitional epithelium** or **urothelium** was previously thought to be an intermediate form between stratified squamous and stratified columnar epithelia. A transitional epithelium changes its shape greatly depending on the physical forces it is subjected to. In the relaxed state (contracted condition), the epithelium is several layers of cells thick while in the stretched or expanded state the cells rearrange themselves into layers that can be as low as two cells thick. Transitional epithelia are found in parts of urinary passages that include the renal calyx, pelvis, ureter, urinary bladder and urethra.

Glandular epithelia are epithelia that are found in glands and are mainly composed of specialized secretory cells. **Exocrine glands** have ducts and their secretory products pass through these openings to the surface of the epithelia. The ducts may actively reabsorb water from the secretory product thus concentrating it in the process. Secretory products of exocrine glands include enzymes and mucin. Hormones are produced by **endocrine glands**. Endocrine glands lack ducts and secrete their products into blood vessels (capillaries). Hormones act on target organs that are located some distance away from the endocrine gland and have to be transported there by the circulatory system. Some of the endocrine glands include the pituitary, adrenal, thyroid and parathyroid glands. Organs that perform other functions but also produce hormones include the pancreas, gonads, gastrointestinal mucosa, placenta and the heart.

All glands originate from the epithelium (Fig. 3.2). Certain epithelial cells start to proliferate into areas underlying the epithelium. In exocrine glands, a tubule forms in the middle of the cord of ingrowing epithelial cells as a result of degeneration of some cells. The cord of cells continues to multiply deeper into the the wall of the tubular structure with the tubule increasing in length until finally a gland with a distal secretory

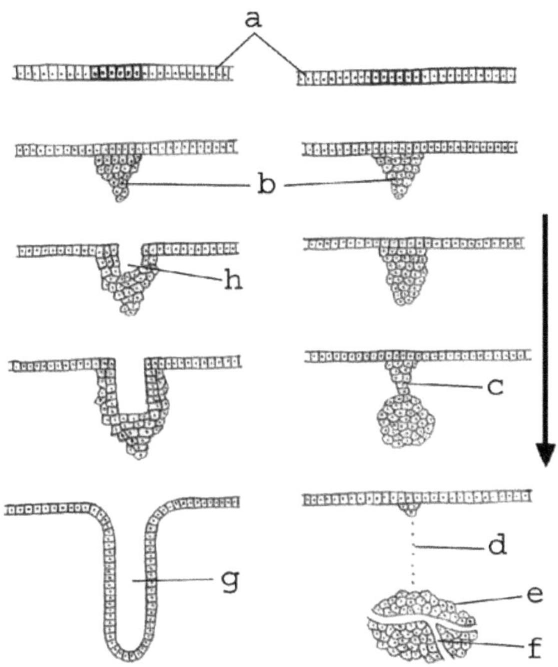

Fig. 3.2 Development of exocrine (left) and endocrine (right) glands. (a) epithelium, (b) cord of ingrowing epithelial cells, (c) degenerating ingrowing cells, (d) lost epithelial connection, (e) endocrine gland cell, (f) capillary, (g) duct of exocrine gland and (h) developing tubule. The secretory part of an exocrine gland is distal to the duct. Arrow indicates the various stages in development of the glands.

region and a proximal duct that is continuous with the epithelium is formed. Development of endocrine glands is also accompanied by growth of epithelial cells deeper into the wall of a tubular organ. The cord of cells nearest the epithelium then degenerate so that the clump of cells that are distal from the surface lose their epithelial connection. The distal cells then develop into secretory cells as they are invaded by blood vessels.

Exocrine glands are classified according to shape of the secretory region and the nature of the ducts (Fig. 3.3). A gland is **tubular** when the secretory part is cylindrical in shape and **alveolar** or **saccular** when this region is sac-like in appearance. The gland is referred to as **simple** when the duct is not branched and **compound** when the duct is branched. A **simple tubular gland** has an unbranched duct and a non-expanded secretory part such as the crypt of Lieberkühn in the small intestine. A simple tubular gland that has a coiled secretory region is referred to as

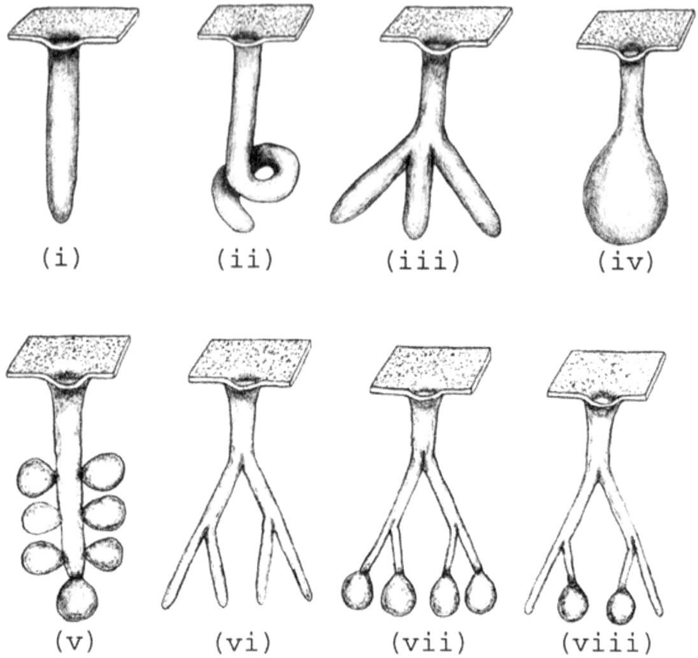

Fig. 3.3 Various types of exocrine glands. Simple tubular (i), simple coiled tubular (ii), simple branched tubular (iii), simple alveolar or saccular (iv), simple branched alveolar (v), compound tubular (vi), compound alveolar (vii) and compound tubuloalveolar (viii).

a **simple coiled tubular gland**. Sweat glands are simple coiled tubular glands. Fundic glands of the stomach are **simple branched tubular glands** since their ducts are not branched but their secretory part is tubular and branched. **Simple alveolar glands** such as the mucous and poison glands in the skin of frogs have an expanded secretory region. The secretory part of **simple branched alveolar glands** such as Meibomian (tarsal) glands of the eyelid is expanded and branched. **Compound tubular glands** such as Brunner's (submucosal) glands of the small intestines have branched ducts and branched tubular secretory regions. Brunner's glands are sometimes classified as compound tubuloalveolar glands since some of the secretory units are expanded. In **compound alveolar glands**, the expanded secretory units are branched as well as the ducts. Lactating mammary glands are compound alveolar glands. Glands such as submaxillary glands are **compound tubuloalveolar** in nature because their tubular and alveolar secretory units are branched.

Secretions from glands are produced by merocrine, apocrine and holocrine modes of secretion (Fig. 3.4). In **merocrine secretion**, membrane bound secretory vesicles are formed and accumulate below the free surface of the cell. The vesicles coalesce with the cellular membrane on the apical surface and release their products by exocytosis. There is no loss of cytoplasm in merocrine secretion. Most glands including goblet cells, eccrine sweat glands and the pancreas release their products by merocrine secretion. **Apocrine secretion** is accompanied by loss of cytoplasm since secretion accumulates below the free surface of the cell before it is released together with the apical part of the cell that breaks away. Mammary glands and apocrine sweat glands of mammals release their products by apocrine secretion. Complete breakdown of secretory cells in order to release secretory products as occurs in sebaceous glands is known as **holocrine secretion**.

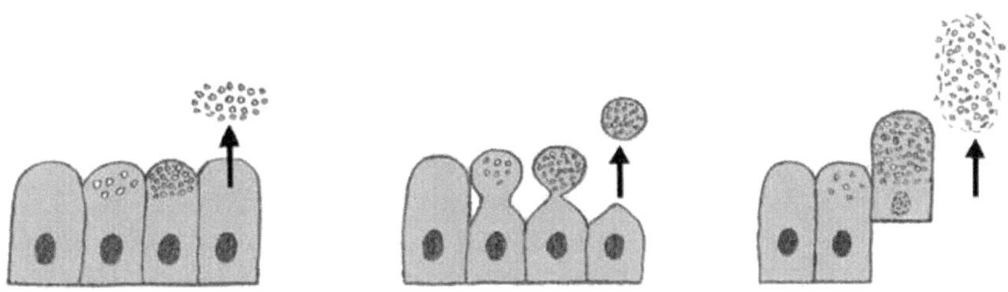

Fig. 3.4 Merocrine (left), apocrine (middle) and holocrine (right) modes of glandular secretion.

Connective Tissue

The connective tissue performs several functions in the vertebrates body that include supporting the various body parts and connecting them together, protection of the body from foreign material and transport of substances throughout the body. All connective tissues except blood have varying quantities of connective tissue cells, a relatively large matrix consisting of connective tissue fibers (collagen, reticular and elastic) and the ground substance. The ground substance is a viscous gel-like solution that surrounds the cells and fibers. The nature of the matrix together with the fibers is responsible for the consistency of connective tissues such as rigidity, flexibility, toughness and softness. Most of the major types of vertebrate connective tissues are classified as **fibrous**, **bone**, **cartilage** and **blood**.

Fibrous Connective Tissues

The predominant feature of fibrous connective tissues is extracellular fibers. The type of fibers present and how they are arranged differs in the various types of fibrous connective tissues. Fibrous connective tissues include loose, adipose, reticular and dense fibrous connective tissues.

Loose Connective Tissue

Loose or areolar connective tissue (Fig. 3.5) is one of the most common connective tissues in the vertebrate body since it connects many neighboring structures. The tissue is stretchable since its fibers are wavy, loosely arranged and running in various directions with amorphous ground substance filling the spaces between the fibers and cells. During dissection, bubbles appear as loose connective tissue is pulled apart thus the name areolar that means 'like a small space'. Loose connective tissue is found in many organs and around blood vessels and nerves.

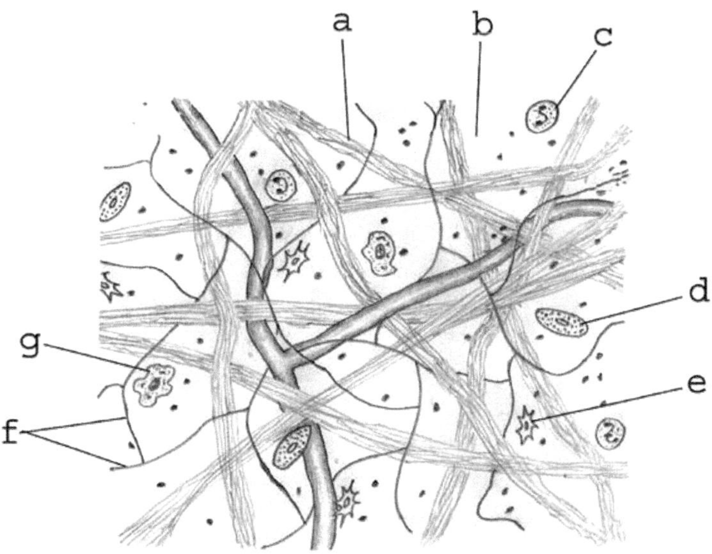

Fig. 3.5 Loose or areolar connective tissue. (a) collagen fibers, (b) amorphous ground substance, (c) white blood cell, (d) mast cell, (e) fibroblast, (f) elastic fibers and (g) macrophage.

Adipose Tissue

The main cell type of adipose tissue (Fig. 3.6) is the fat cell or adipocyte that acts as a store for fat. Adipose tissues also act as protective pads around some organs. In birds and mammals, adipose tissue forms a thermal insulating layer of fat under the skin. Adipose tissue is highly vascularized and and has little ground substance, fibroblasts, mast cells and macrophages. A dense network of reticular fibers surrounds the fat cells.

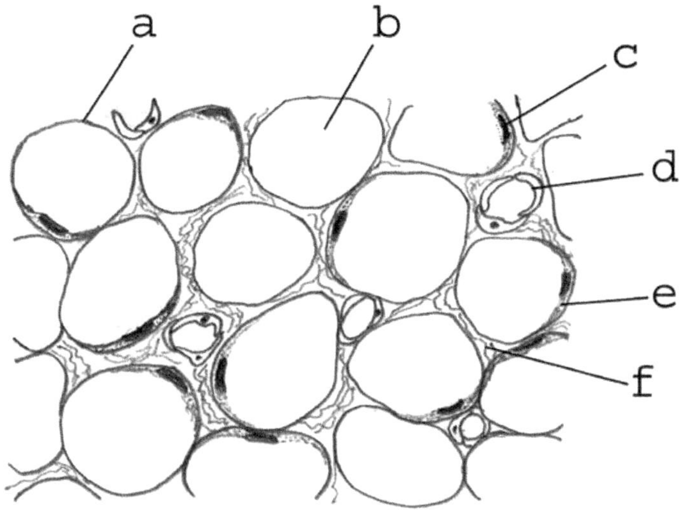

Fig. 3.6 Adipose tissue. (a) fat cell, (b) accumulated fat, (c) nucleus of adipose cell, (d) capillary, (e) cytoplasm and (f) reticular fibers in the ground substance. In prepared slides, adipose cells are normally empty since fat is soluble in most of the solvents that are used in tissue preparation.

Reticular Tissue

Reticular connective tissue (Fig. 3.7) has large reticular cells with many branching processes that form a three-dimensional web and are closely associated with reticular fibers. Between the reticular cells are also other cells including macrophages, histiocytes, leucocytes and erythrocytes. Reticular tissue is the basic tissue of lymphoid organs such as lymph

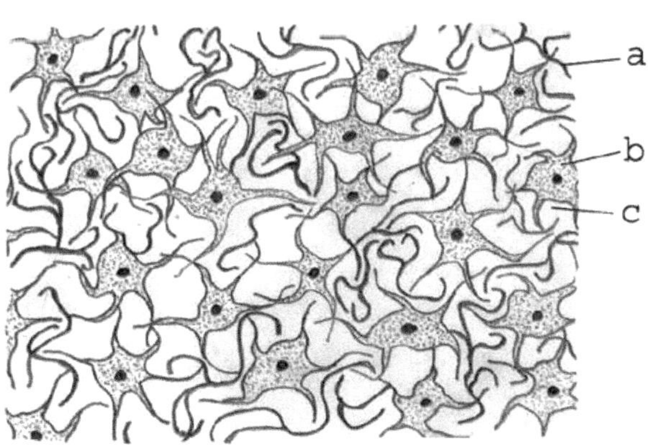

Fig. 3.7 Reticular connective tissue. (a) reticular fiber, (b) reticular cell and (c) amorphous ground substance.

nodes, thymus, spleen, tonsils, bone marrow as well as the liver and is part of the reticuloendothelial system that plays a role in the body's defense system. The reticular web filters foreign material from lymph and blood while the phagocytic reticular cells engulf the material. Reticular cells also synthesize reticular fibers.

Dense Fibrous Tissue

Dense fibrous connective tissue (Fig. 3.8) is characterized by the predominance of connective tissue fibers that are mainly collagen over other connective tissue material Elastic fibers may also be present. The fibers impart great tensile strength to this tissue. The bundles of collagen fibers are elongated and arranged in a wavy and regular manner with little ground substance (matrix) between the fibers in **dense regular connective tissue**. Cells such as fibroblasts are arranged in between the fibers. Dense regular connective tissue is found in tendons, ligaments and some organ capsules. Ligaments form strong bands that unite bones at joints and tendons act as points of muscle origin and insertion. In **dense irregular connective tissue**, the fibers run in various directions and intertwine with each other to form a strong mat that can withstand stress

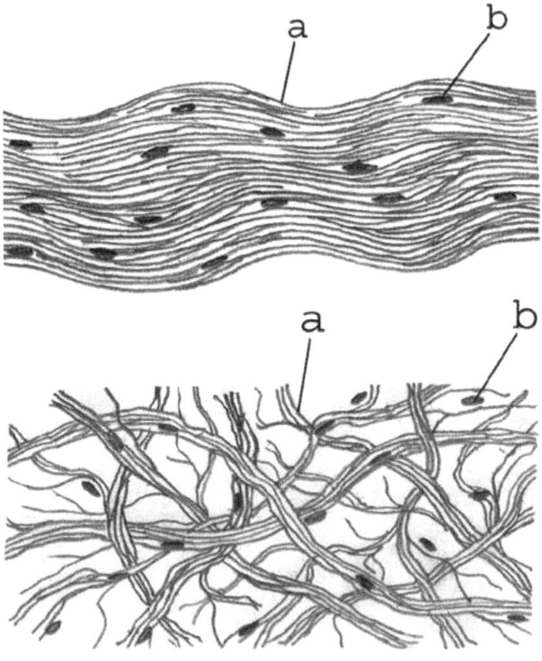

Fig. 3.8 Dense regular (top) and irregular (bottom) connective tissues. (a) collagen fibers and (b) fibroblasts.

from any direction while allowing for adaptation to changes in size of an organ. Dense irregular tissue is found in the dermis, fascia, lining of the digestive system, aponeurosis, joint capsule, pericardium and capsules of various organs including the spleen, kidney, liver and testis.

Cartilage

A cartilage is a semi-rigid form of connective tissue that is composed of cells known as **chondrocytes** and a large amount of fibers and gel-like ground substance that are located between the cells. Characteristics of cartilage arise from the nature and predominance of intercellular components over cells. A cartilage has considerable tensile strength due to the presence of collagen and elastic fibers and can bear considerable weight as a result of its pliability. A cartilage is pliable due to the nature of its ground substance that is a glycoprotein and has a high percentage of sulphated polysaccharide units.

Distribution of cartilage in mature higher vertebrates such as birds and mammals is limited. It is more extensive in immature vertebrates of such groups since it forms a template for most of the developing bone and has the capacity for rapid growth while maintaining a considerable degree of stiffness. The internal skeleton of some vertebrates such as cartilaginous fish is entirely cartilage whereas others such as chondrosteans and lungfishes still retain a considerable amount of cartilage in their skeleton.

The three types of cartilage (Fig. 3.9) differ slightly from each other. **Hyaline cartilage** is the most common type and forms the skeleton of the embryo. It is also found in articulating surfaces of bone, nasal septum, larynx, trachea and bronchi. The collagen fibers have the same index of refraction as the ground substance and as a result are not visible in normal histological preparations of hyaline cartilage. **Elastic cartilage** is found in areas where there is need for elasticity and rigidity such as the mammalian external ear, epiglottis, auditory canal and eustachian tube. Elastic cartilage differs in structure from hyaline cartilage because it contains elastic fibers that branch and anastomose and are visible in ordinary histological sections. **Fibrocartilage** is the least common type of cartilage and is the strongest of the three types. It is a transitional form between cartilage and dense connective tissue. It provides great tensile strength to the cartilage due to the presence of numerous collagen fibers that are arranged in a plane that is parallel to the direction of stress. There is little ground substance in fibrocartilage and chondrocytes are scattered between the collagen fibers. Fibrocartilage is found where some tendons attach to the bone, intervertebral disks, ligaments and some articular cartilages.

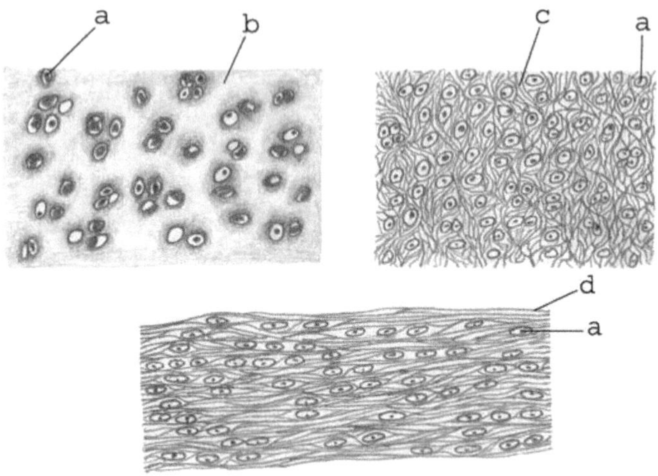

Fig. 3.9 The three types of cartilage. Hyaline (top left), elastic (top right) and fibrocartilage (bottom). (a) chondrocytes in lacunae, (b) matrix containing collagen fibers and ground substance, (c) elastic fibers and (d) collagen fibers. Fibers are not visible in histological sections of the hyaline cartilage as the type of collagen present has the same index of refraction as the ground substance.

Sometimes calcium salts are deposited in certain parts of cartilage leading to calcified cartilage. Calcification of cartilage is different from ossification that is concerned with bone formation and calcified cartilage differs from the bone structurally. Calcification is seen in some parts of the vertebrae in many cartilaginous fish and is a process that strengthens cartilage.

A cartilage displays interstitial and appositional growth (addition of new layers of cartilage to the surface). Interstitial growth is brought about by differentiation of mesenchymal cells into cartilage precursor cells known as **chondroblasts** at the inner part of cartilage. These cells divide by mitosis to give rise to several chondroblasts that will synthesize ground substance and fibrous material. Appositional growth occurs at the periphery of cartilage where there is a zone of condensed connective tissue called **perichondrium**. The perichondrium has an inner chondrogenic layer comprising chondroblasts with cartilage forming potential and an outer fibrous layer that consists of collagen fibers and fibroblasts. A cartilage lacks blood vessels, a lymphatic system and nerves so nutrients are transported to it by diffusion.

Cartilage Homeostasis

The abundant extracellular matrix of a cartilage plays a major role in the regulation of chondrocyte function and is maintained by a limited number of chondrocytes. Soluble and insoluble matrix-derived factors play an important role in cartilage homeostasis. Molecular signaling via soluble mediators has been shown to be crucial to cartilage homeostasis. The macromolecules of the matrix such as collagen and hyaluronate and some growth factors that are released under specific conditions communicate with chondrocytes via specific membrane receptors and affect their behavior thus maintaining a close interaction between the extracellular and intracellular media (van der Kraan et al., 2002). Mechanical stress can also affect the regulation of chondrocyte biosynthetic and catabolic activity. Reactive oxygen species such as nitric oxide, peroxynitrite and superoxide that increase during joint diseases such as osteoarthritis and rheumatoid arthritis also play a major role in regulation of metabolic activities of chondrocytes (Henrotin et al., 2003). The hormone calcitonin has a direct effect on chondrocytes and synthesis of the matrix and inhibits degredation of cartilage (Karsdal et al., 2006).

Bone

Bone provides the greatest strength with the least amount of weight of any substance within the vertebrate body. Bone consists of cells referred to as **osteocytes** and collegen fibers that are embedded in a mineralized hard substance (glycoprotein) that is rigid and strong. The minerals include calcium salts such as phosphates, carbonates and flouorides. Bone gives internal support to the entire vertebrate body and provides points of attachment of muscles and tendons. It also protects the brain, organs of the body and bone marrow. Bone is dynamic and is renewed and remodeled throughout life in vertebrates. It has been shown that bone produces at least two hormones, fibroblast growth factor 23 (FGF 23) and osteocalcin (Fukumoto and Martin, 2009). FGF 23 is produced by osteocytes and inhibits 1 alpha-hydroxylation of vitamin D and promotes excretion of phosphorus in the kidney. Osteocalcin is produced by osteoblasts and acts on the pancreatic beta cells leading to an increase in glucose metabolism due to increased insulin sensitivity.

The mineralized part of bone forms the inorganic component of bone. Mineralization makes bone hard and rigid. The level of mineral deposition in bone depends on the use the bone is put to in the vertebrate body. In load-bearing bones such as the mammalian femur and humerus, the inorganic part of bone makes up to two-thirds of completely developed bone. This level of mineralization is lower in bones that are put to lighter use. Collagen forms the main part of the organic framework of bone and gives the bone toughness and resilience.

Bone has a canalicular system whereby osteocytes are trapped in spaces known as calcification of the intercellular substance. Osteocytes

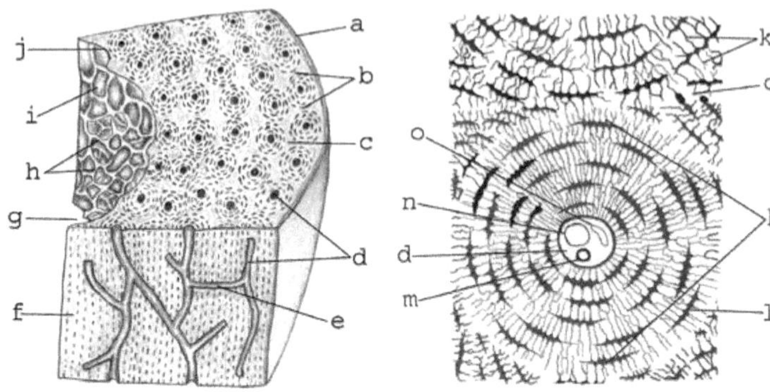

Fig. 3.10 Structure of bone. Longitudinal and cross-sections of bone showing both compact and cancellous bone (left) and a higher magnification of compact bone (right). (a) periosteum, (b) Haversian system or osteon that forms structural cylindrical units in compact bone, (c) interstitial bone, (d) Haversian and (e) Volkmann's canals that contain blood vessels and nerves, (f) compact bone, (g) medullary marrow cavity, (h) trabeculae, (i) cancellous (spongy) bone, (j) endosteum, (k) canaliculi, (l) lacuna containing osteocyte, (m) artery, (n) vein and (o) lymphatic vessel.

are mainly arranged in circular (concentric) layers around **Haversian canals** forming cylindrical structural units known as **osteons** or **Haversian systems** (Fig. 3.10). The osteocytes have long cytoplasmic processes that are encased in tiny canals known as **canaliculi** that extend from one lacuna to another. The canaliculi permeate the bony matrix and provide a system for the nourishment of osteocytes.

Bone is covered with a fibrous layer known as **periosteum** that has an inner part that contains many cells such as osteoblasts (bone forming cells), nerves and blood vessels. Small vessels branch from these blood vessels and enter Volkmann's canals through the many microscopic pores of bone. Volkmann's canals run at right angles to the long axis of bone. Vessels of these canals eventually supply those of Haversian canals and the neighboring osteocytes (through the canalicular system). Larger blood vessels penetrate the compact layer of bone directly through nutrient foramina to supply the spongy bone and medullary cavity.

Incineration of bone destroys the organic matrix, leaving bone ash (inorganic part) that is brittle and is easy to break. Dissolution of the inorganic compounds of bone using acid leaves an elastic and plastic mass of mainly collagenous material (organic part) that can be bent or cut. In both treatments, the characteristic shape of bone is maintained, showing how the two structural elements relate to each other.

In ordinary histological sections, bone has to be decalcified by agents such formic and nitric acid before sectioning. These agents soften bone by causing decalcification while retaining the cells (osteocytes) and the organic intercellular substance.

BONE GROWTH

Bone formation is known as **ossification** or **osteogenesis** and starts from an already existing connective tissue. There are two types of ossification that depend on specific cells differentiating into bone within two different cellular environments viz. intramembranous and intracartilaginous (endochondral) ossification. The two types of bone formed by these processes are structurally similar. **Heterotopic** or **heteroplastic bone** develops in tissue that has fully differentiated. Such splanchnic bones include the baculum or os penis, sesamoid bones, os cordis and os rostrale.

INTRAMEMBRANOUS OSSIFICATION

Intramembranous ossification is a process of bone formation directly from **mesenchyme** (Fig. 3.11) and leads to formation of **membrane** or **dermal bone**. Intramembranous ossification is widespread in fishes but confined mainly to the skull in tetrapods. This process of bone formation gives rise to some bones of the skull, the lower jaw, some pectoral girdle bones, vertebrae in some teleosts, urodeles and apodans, dentin and bone that develops in the skin. Sites of intramembranous ossification have mesenchymal cells, fibroblasts, amorphous ground substance, collegen fibrils and many small blood vessels. Mesenchymal cells will then differentiate into **osteoblasts** that will later form into clusters that will become centers of ossification. The osteoblasts secrete a matrix that comprises collagen fibers and other intercellular substances that together are referred to as **osteoid** or **prebone**. During the process of calcification, osteoblasts deposit calcium salts into osteoid resulting in formation of immature woven bone that consists of small rods or bony spicules that radiate in different directions in line with the random organization of collagen fibers.

Gradually, the bony spicules will enlarge to become trabeculae (L. *trabecule*, small beam) as more bone is laid down. Osteobasts then become trapped in the bony matrix and lose their ability to form the bone to become osteocytes. Trabeculae surround blood vessels and connective tissue. Many trabeculae will give rise to trabecular, spongy or cancellous bone. With further development, layers of bone are added to the surface of the trabeculae that form the outer surface of bone until all spaces between the trabeculae are filled with bone to form a compact bone. Spaces between trabeculae of spongy bone that were occupied by blood vessels will become filled with red bone marrow. As membrane bone matures, the connective tissue that surrounds the bone changes into periosteum. Membrane bone attains its adult size and shape through

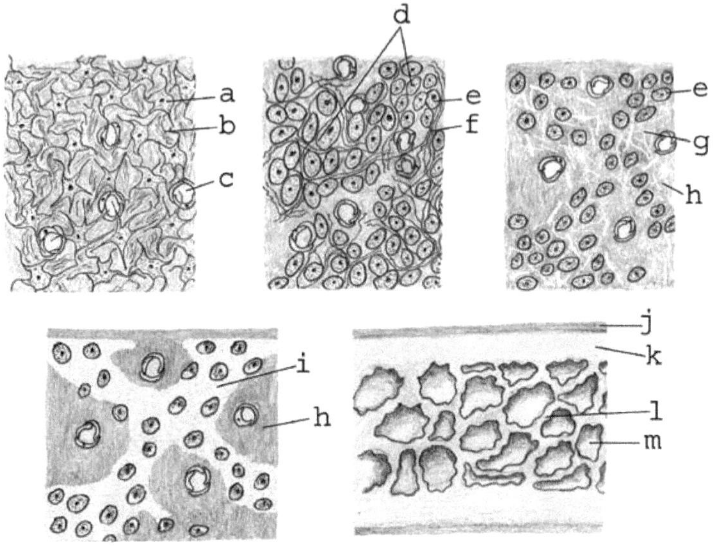

Fig. 3.11 Intramembranous bone ossification. Mesenchymal cells in well vascularized tissue (top left), osteoblasts that have been formed from mesenchymal cells arrange themselves into centers of ossification (top middle), formation of bone spicules (top right), formation of trabeculae (bottom left) and mature membrane or dermal bone (bottom right). (a) mesenchymal cell, (b) collagen fibrils, (c) blood vessel, (d) ossification centers, (e) osteoblasts, (f) collagen fibers, (g) bone spicules, (h) connective tissue with blood vessels, (i) immature trabecular bone, (j) periosteum, (k) compact bone, (l) trabecula in cancellous (spongy) bone and (m) spaces in cancellous bone. The spaces in cancellous bone that formed the internal vascular tissue become bone marrow.

destruction and reconstruction of the newly formed bone. Bone increases in size by appositional growth. Some surfaces of the trabecula have osteoblasts that are involved in bone formation while other surfaces are occupied by multinucleated **osteoclasts** that erode bone. Osteoclasts result from the union of osteogenic cells.

ENDOCHONDRAL OSSIFICATION

Endochondral ossification occurs around and within hyaline cartilage (Fig. 3.12). Bone formed from such cartilage is known as cartilage replacement bone or simply replacement bone. Examples of replacement bone include bones of extremities, vertebral column, pelvis and base of skull. The design of replacement bone is more complex than that of dermal bone as replacement bone has to be subjected to more complex stresses in life. Details of ossification vary in different vertebrates (Shapiro, 1992).

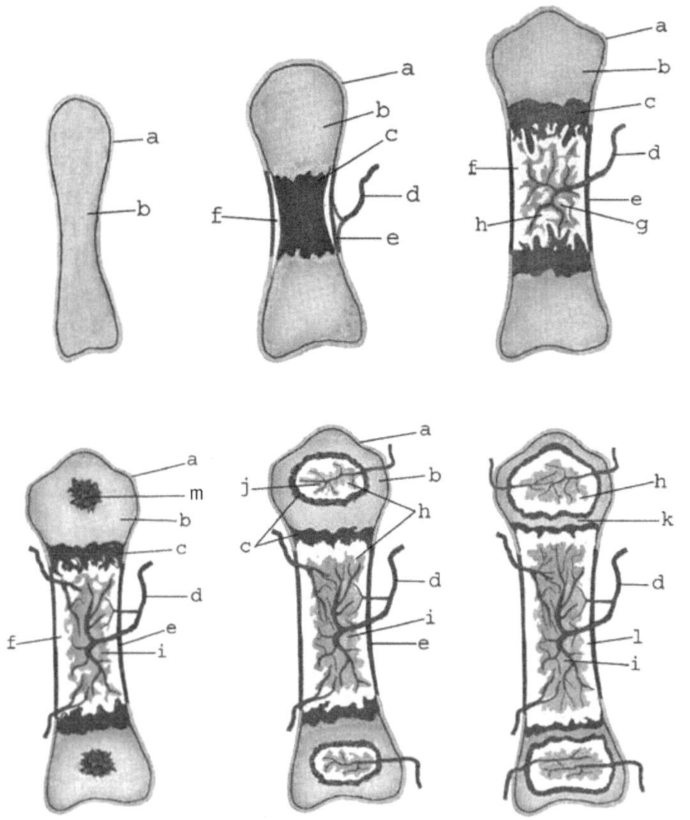

Fig. 3.12 Various stages during endochondral ossification in a long bone. Hyaline cartilage (top left), beginning of calcification and bone collar formation (top middle), formation of primary ossification center and penetration of blood vessels into bone (top right), lengthening of bone collar and formation of a medullary cavity (bottom left), formation of secondary ossification centers at the epiphyses (bottom middle) and formation of epiphyseal cartilages at either end of bone with enlargement of secondary ossification centers (bottom right). (a) perichondrium, (b) hyaline cartilage, (c) calcified cartilage, (d) blood vessel, (e) periosteum, (f) bone collar, (g) open space forming in bone, (h) cancellous bone, (i) medullary cavity, (j) space in secondary ossification center, (k) epiphyseal cartilage or plate, (l) compact bone, and (m) secondary ossification center.

During endochondral ossification in mammals, chondrocytes enlarge (hypertrophy) so that the intercellular space becomes thin. The chondrocytes start secreting alkaline phosphatases. Calcification of intercellular spaces also occurs during this time. Chondrocytes are prevented from receiving nutrients due to calcification and die as a result, leaving large honey-combed cavities. The perichondrium is invaded with numerous capillaries at the same time. Undifferentiated cells of the perichondrium produce osteoblasts (instead of chondroblasts) and start laying down bone collar around the middle of the shaft of the developing bone. **Bone collar** is the bone that forms around hyaline cartilage and holds together the bone shaft that has been weakened by the formation of large cavities. The perichondrium changes to periosteum with formation of the bone collar. Endochondral ossification starts in the middle of cartilage. Vascular connective tissue invades the interior of cartilage that contains the honeycomb cavities. Some of the undifferentiated cells in the invading tissue differentiate into osteoblasts that line up along the remaining intercellur spaces and start the primary center of ossification. Calcification leads to formation of cancellous (immature) bone that consists of interlacing bony trabeculae. Spaces between the trabeculae are filled with bone marrow. In long bones, cancellous bone in the center is resorbed by osteoclasts leading to formation of a marrow cavity.

Endochondral ossification proceeds from the center of a long bone to either end. Secondary centers of ossification are seen in bones with more than one center of ossification. Some cartilage is not ossified such as the epiphyseal plate (disk) of long bones in growing vertebrates and articular cartilage. The epiphyseal cartilage is a point of increase in length of long bones. Growth in thickness of bone occurs by the continued formation of the perichondral bone at the peripheral part of the cartilage that is part of the periosteum.

The perichondrium plays a vital role in the normal invasion of blood vessels into the hypertrophic cartilage and both the perichondrium and blood supply are necessary for endochondral ossification (Colnot et al., 2004). Osteoblasts vary their synthetic activity in response to signals that determine cell differentiation and bone matrix production and skeletal morphology depends on the local regulation of bone formation (Wurtz and Berdal, 2003). It has been shown that mesenchymal stem cells can be induced to produce bone using bone morphogenetic protein-2 (Noel et al., 2004) outside the normal location whose environmental factors are important in determining the differentiation capacities of stem cells in vivo.

Bone Homeostasis

The continuous activities of osteoblasts and osteoclasts in the periosteum and endosteum of bones of vertebrates are in response to factors that affect calcium homeostasis and mechanical forces of gravity and exercise. The process of bone resorption by the large and multinucleated **osteoclasts** and formation by **osteoblasts** are in balance in the adult skeleton thus maintaining a constant homeostatically controlled amount of bone. Bone is quite dynamic. Old bone matrix that could result from fractures is gradually resorbed and replaced by new bone matrix. During fractures, a substance within bone is believed to send out signals to the circulatory system that leads to dissolution of the damaged tissue that is followed by invasion of the site by undifferentiated cells. The undifferentiated cells will transform into chondrocytes on receiving the right signals. A process similar to endochondral ossification follows leading to repair of the fracture by production of bone. Using a negative feedback mechanism that regulates mineral absorption from the digestive system, mineral excretion in the kidneys and mineral deposition and dissolution in the skeletal system, the bone is able to maintain an internal equilibrium of bone matrix density and blood calcium and phosphate levels.

The endocrine system controls mineral homeostasis in the body. Parathyroid hormone is produced by the parathyroid glands whenever calcium levels fall below a certain level. The hormone is the major hormonal regulator of calcium homeostasis. The hormone stimulates the kidney to produce adenylate cyclase that leads to an increase in the excretion of phosphate in urine. This action lowers the level of phosphate in blood. Parathyroid hormone also stimulates the production of 1,25-dihydroxyvitamin D3 in the body. Van Driel et al. (2004) have reviewed the role of vitamin D in regulation of mineral homeostasis and bone metabolism. The vitamin and its derivatives promote absorption of calcium in the intestine and also stimulate production of calcium from bone. When calcium levels in blood are high, calcitonin is produced by the thyroid gland and decreases the blood calcium levels by inhibiting the loss of calcium from bone to blood.

Bone is a reservoir of calcium and phosphate in the body mainly in the form of calcium phosphate. Calcium can be drawn upon during periods of high demand such as pregnancy and lactation. Calcium is also important in various physiological functions including muscle contraction. Phosphate is the key to energy production in the form of ATP, formation of the cell structural phospholipid and the nucleic acids DNA and RNA.

Mechanical forces such as gravity and increased muscular activity stimulate deposition of additional bone. Low gravity and lack of exercise stimulate bone resorption leading to osteoporosis. Osteoblasts are sensitive to cyclic pressure stimuli and respond by proliferating.

Nervous Tissue

As connective tissue that is adapted to transmission of impulses and integration of the body's activities, nervous tissue is the most complex connective tissue in the vertebrate body. Nervous tissue has developed more excitability and conductivity characteristics than any other tissue in the body. Nervous tissue is found in the nervous system that consists of the brain, spinal cord and nerves. Nerve tissue consists of the transmitting cells known as neurons and supportive cells known as neuroglia.

NEURONS

A neuron (Gr. *neuron*, nerve) consists of a cell body, dendrites, an axon and a terminal transmission segment (Fig. 3.13). The **cell body**, also known as **soma** or **perikaryon**, contains the nucleus and is the site of synthesis of the neurotransmitter substances that carry signals between cells. The cell body receives impulses from several tree-like **dendrites**. The **axon** is a long slender process that extends from the cell body and is also known as a nerve fiber. Axons conduct impulses away from the cell body towards the central nervous system or from the central nervous system to effectors and branch frequently. The distal ends of axons form branches that are the terminal transmission segments known as

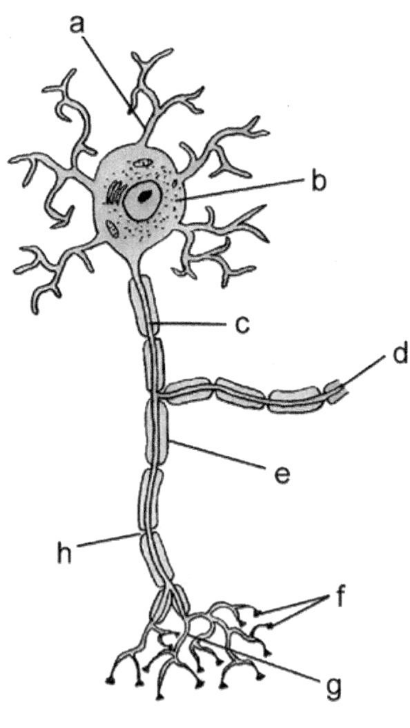

Fig. 3.13 Structure of a neuron. (a) dendrite, (b) cell body or perikaryon containing a nucleus and other organelles, (c) axon, (d) axon branch, (e) Schwann cell, (f) synaptic knobs, (g) telodendron and (h) node of Ranvier.

telodendria that end in **synaptic knobs**. Synaptic knobs are rich in mitochondria and vesicles that contain neurotransmitter substances.

Axons vary considerably in length and diameter. Some axons are only a few millimeters in length whereas others are more than a meter long. Generally, the wider the diameter of an axon, the higher the speed with which it is able to transmit impulses. Many nerves in the vertebrate nervous system contain myelin sheath. Myelin sheath is formed when Schwann cells wrap themselves around an axon several times in a spiral manner forming a sheath-like structure. Myelinated neurons transmit impulses at a higher speed than non-myelinated neurons.

Neurogenesis is the development of nervous tissue and occurs mainly during embryologic development. Adult neurogenesis from stem cells occurs in many vertebrate species including the human being and is important in brain repair (Doetsch and Scharff, 2001).

Types of Neurons

Neurons are classified according to their structure and function. Structural classification takes into consideration the number of extensions from the cell body. There are three types of neurons according to this classification and include multipolar, bipolar and unipolar neurons (Fig. 3.14). **Multipolar neurons** have several dendrites and one axon and form most of the neurons in the brain and spinal cord. **Bipolar neurons** have one axon and one dendrite that bears many branches and their distribution in the body is limited to the inner ear, olfactory neurons and the retina of the eye. Neurons with a single process arising from the

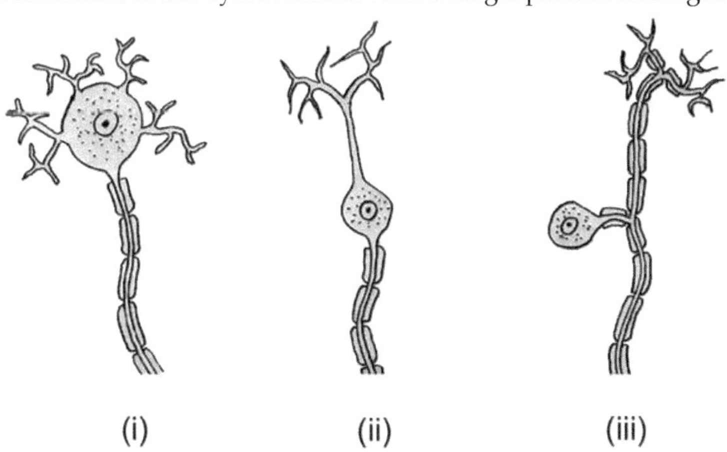

Fig. 3.14 Types of neurons. (i) multipolar, (ii) bipolar and (iii) unipolar or pseudounipolar neurons. The cell body is the circular structure with a nucleus. In a unipolar neuron, only one process initially arises from the cell body before dividing into two processes.

cell body are known as **unipolar** or **pseudounipolar neurons** and are sensory neurons. The single process arising from the cells body branches into two processes that form one axon which runs towards the dendrites and the other that leads to the central nervous system.

The functional classification of neurons is based on the direction in which impulses are conducted and includes affarent, efferent and interneurons (Fig. 3.15). **Afferent** or **sensory neurons** transmit nerve impulses from various parts of the body to the brain or spinal cord. **Efferent** or **motor neurons** transmit impulses from the central nervous system to effectors mainly muscles and glands. **Interneurons** make a link between afferent and efferent neurons and are located in the central nervous system. Afferent and efferent neurons mainly run in the same peripheral nerves apart from the **dorsal** and **ventral nerve roots** that lie close to the spinal cord. Afferent neurons enter the spinal cord via dorsal roots whereas efferent neurons exit through ventral roots. Such an arrangement whereby impulses are conducted to the central nervous system by sensory neurons and away from it through motor neurons is known as a **reflex arc**.

After embryological development, the nervous system has attained its general form although still undergoing various changes that will lead to the maturity of the system on attainment of adult life. In early life, several synapses are made and broken as the correct nervous system pathways are established. Sensory stimulation from the environment plays a role in directing the establishment of synapses in the nervous

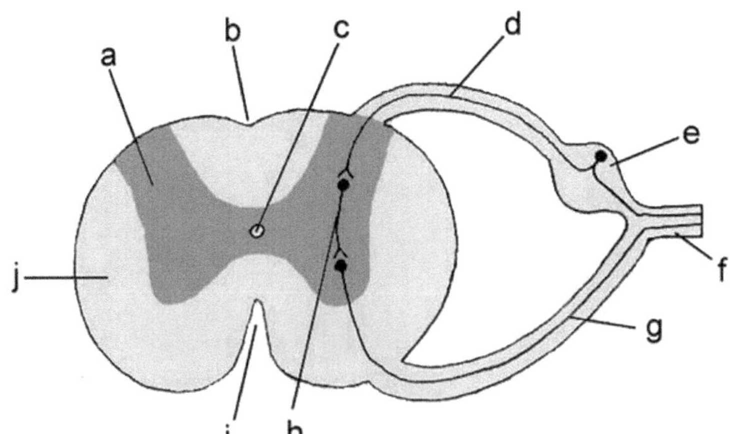

Fig. 3.15 The neurons of a reflex arc and a cross section through the spinal cord. (a) gray matter, (b) dorsomedian or posterior median sulcus, (c) central canal, (d) afferent or sensory neuron that is unipolar with a cell body in (e) a dorsal root ganglion, (f) spinal nerve, (g) efferent or motor neuron in a ventral nerve root, (h) interneuron, (i) ventromedian fissure or anterior median fissure and (j) the lateral column that forms white matter.

system and this contributes to early learning and memory. With old age in some vertebrates, there is degeneration of neurons, glia and blood vessels that supply these tissues resulting in destruction of nervous tissue and loss of memory and coordination.

Neuroglia

The non-nervous cells of the nervous system are known as neuroglia or glia (Gr. *glia*, glue) and are found in the central nervous system. These cells, unlike neurons, are capable of cell division throughout adult life. Neuroglia are approximately 10 to 50 times as numerous as neurons. Glial cells provide structural support and functional integrity to neurons. The main glial cells in the vertebrate central nervous system include astrocytes, oligodendrocytes, microglia and ependymal cells (Fig. 3.16) while Schwann cells (Fig. 3.17) are associated with the peripheral nervous system.

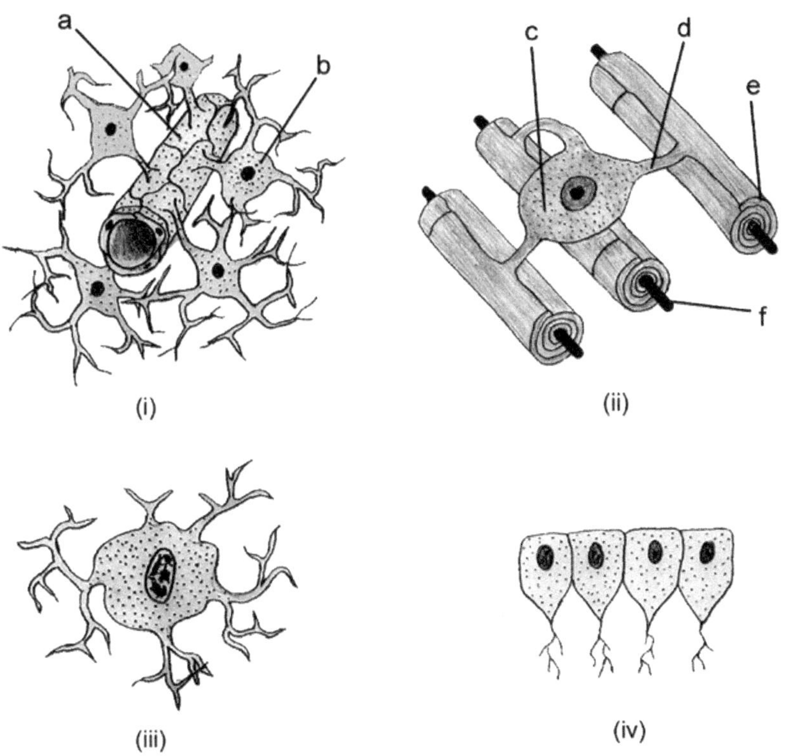

Fig. 3.16 Neuroglia or glia of the central nervous system. (i) astrocytes (b) surrounding a capillary with their processes, (ii) oligodendrocyte (c) with its processes forming myelin sheaths around neurons, (iii) microglia and (iv) ependymal cells. (a) astrocyte process, (d) oligodendrocyte process, (e) myelin sheath and (f) axon of a nerve.

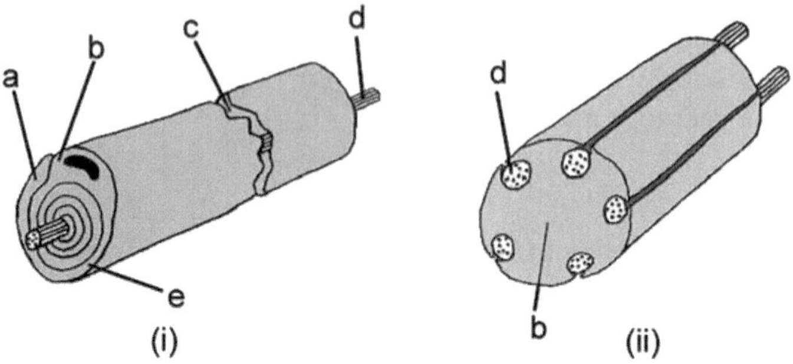

Fig. 3.17 (i) Myelinated and (ii) non-myelinated neurons. (a) myelin sheath, (b) Schwann cell, (c) node of Ranvier, (d) axon of a nerve and (e) neurilemma or sheath of Schwann.

Astrocytes (Gr. *astron,* star) are star-shaped cells that possess long processes that make contact with blood vessels and form tight sheaths around brain capillaries. Astrocytes are the most numerous and largest neuroglia and are found only in the brain and spinal cord. Astrocytes are part of the blood-brain barrier. The endothelial cells of brain capillaries form the other barrier with their tight junctions with each other. Astrocytes also play a role in regulating the composition of ions and electric balance of extra-cellular fluid as well as degrading substances.

Oligodendrocytes (Gr. *oligos,* few; *dendron,* tree; *kytos,* cell or hollow vessel) have fewer processes and are smaller than astrocytes. The processes wrap around nerve fibers in the central nervous system forming fatty myelin sheaths. Myelination of nerve fibers by oligodendrocytes continues after birth. Oligodendrocytes also hold nerve fibers together. Myelination of axons gives regions of the central nervous system a whitish color in unstained sections and is known as the **white matter**. The **gray matter** contains cell bodies with their dendrites and non-myelinated axons. Although oligodendrocytes are widely distributed in the mature central nervous system, they originate from very restricted regions of the embryonic neural tube and might have as much in common with ventral neurons (mainly motor neurons) as is the case with other types of glia (Richardson et al., 2000). Oligodendrocytes and astrocytes are sometimes referred to as **macroglia** (Gr. *macros,* large).

The smaller **microglia** (Gr. *micros,* small) are normally stationary but when activated enlarge and move about performing the role of phagocytes by engulfing degenerating brain tissue and foreign material. Microglia are thus important in repair of brain tissue and inflammation. Microglia are developmentally unrelated to other nervous system cells as they are of mesodermal origin whereas macroglia develop from the neural tube ectoderm. There is evidence to indicate that microglia may participate in the regulation of non specific inflammation as well as adaptive immune responses (Aloisi, 2001). Microglia engulf invading neutrophils after traumatic injury thus maintaining the immunological integrity of the healthy brain (Neumann et al., 2008).

Ependymal cells line the fluid-filled cavities of the brain and spinal cord thus playing the role of endothelial cells. Some ependymal cells are ciliated and play the role of propelling cerebrospinal fluid in the central nervous cavities whereas others produce some of the fluid.

The first glial cells to appear during embryonic development are **radial glial cells**. These cells possess radial processes that extend from the cells to the walls of neural tubes and act as guiding cables during the radial migration of neurons mainly during embryonic development. Radial glial cells are present in the central nervous system of adult sub-mammalian vertebrates but change into astrocytes and ependymal cells in mammals after birth. Radial glial cells also produce neurons in regenerating adult central nervous system of vertebrates. Most dividing progenitor cells in the embryonic human ventricular zone express radial glial proteins (Weissman et al., 2003). The significance in guiding migrating neurons by neural glial cells has increased with the evolutionary expansion of the mammalian neocortex and reaches a peak in the gyrencephalic human forebrain (Rakic, 2003).

Schwann cells surround axons of the peripheral nerves (Fig. 3.17) along their length forming white fibers and are of neural crest origin. Since Schwann cells support nerve fibers and sometimes form myelin sheath around axons, they are the functional equivalent of oligodendrocytes of the central nervous system. Schwann cells and Schwann-cell precursors are an important source of developmental signals in embryonic and neonatal nerves (Jessen and Mirsky, 1999). The signals regulate the survival and differentiation of early nerves. Schwann cells are also necessary for successful nerve regeneration following injury. Schwann cells can wrap themselves around an axon several times forming several layers of membrane containing the white fatty myelin (Gr. *myelos*, marrow) known as myelin sheath. There are gaps between neighboring myelin sheaths known as **nodes of Ranvier**. The nucleus and cytoplasm of a Schwann cell is pushed to the periphery forming a **neurilemma** or **sheath of Schwann** that plays a role in regeneration of damaged axons. In non-myelinated nerves or gray fibers, a Schwann cell surrounds axons of several fibers without forming several layers around the fibers and perform a supportive role. Such nerve fibers are referred to as non-myelinated as layers of the Schwann cell plasma membrane cannot be visualized under light microscopy. Transmission of impulses is faster in myelinated nerves when compared to non-myelinated fibers as impulses jump from one node of Ranvier to the other. The nodes of

Ranvier are the only parts of an axon that have a plasma membrane separating the interior of an axon from extra-cellular fluid.

Muscular Tissue

Muscular tissue is specialized for contractility due to the presence of contractile units (proteins) in the muscle fibers (cells). Muscular tissue is thus specialized for movement of the vertebrate body and its parts. The three types of muscular tissues are skeletal, smooth and cardiac (Fig. 3.18). **Skeletal muscle** forms the bulk of muscle tissue and is able to bring about movement of vertebrates or their appendages as it is attached to the skeletal system. **Smooth muscle** (visceral muscle) tissue is found in the walls of hollow organs and brings about changes in the diameter of their lumina. **Cardiac muscle** tissue is found in the wall of the heart and is responsible for the pumping action of this organ.

Smooth (Unstriated) Muscle

Of the three types of muscle found in vertebrates, smooth muscle is the simplest and was the first to evolve. Smooth muscle fibers are fusiform

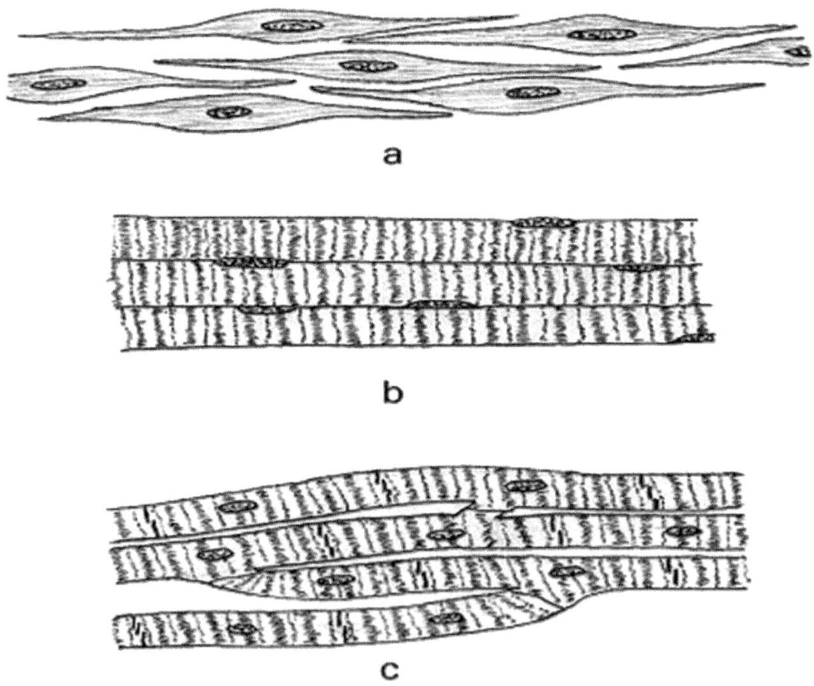

Fig. 3.18 Types of muscle fibers in vertebrates. (a) smooth, (b) skeletal and (c) cardiac muscle fibers.

in shape and fairly circular in cross section. Smooth muscle fibers often branch especially at the end of each cell and each fiber has a centrally located nucleus. Smooth muscle fibers are offset with respect to each other. Thick central portions of smooth muscle fibers are juxtaposed to thin ends of adjacent fibers giving rise to a dovetailed arrangement that is similar to the way a tenon fits into a mortise. As a result of such an arrangement, smooth muscle appears as a mosaic of rounded or irregularly arranged polygonal profiles of various sizes in cross section.

Smooth muscle fibers vary in length from about 20 micrometers in the wall of small blood vessels to about 500 micrometers in the wall of a pregnant uterus. An average length in smooth muscle fibers of about 200 micrometers is found in the wall of intestines. The structural lattice of smooth muscle is composed of the 'cytoskeleton' that pervades the cytoplasm and consists of longitudinally arranged actin filaments, and a network of desmin intermediate filaments as well as the membrane skeleton at the cell surface that provides anchorage for the cytoskeleton and contractile apparatus. Intermediate filaments play an important role in development of normal contractile force and not just cell shape structure (Small and Gimona, 1998). The proteins actin, myosin and tropomyosin of smooth muscle are not arranged in a regular manner as is seen in cardiac and skeletal muscle so smooth muscle is not striated.

Contractions of smooth muscle are slower when compared to those of other muscle types. Smooth muscle is capable of sustaining contractions for a long time with relatively little expenditure of energy. Contractions are initiated in smooth muscle by nerve impulses, hormonal stimulation or local changes within the muscle such as stretching of the muscle fibers. There are two types of smooth muscle fibers depending on their physiological properties. **Unitary (visceral)** smooth muscle contracts spontaneously and slowly and is found in the wall of the gastrointestinal tract, the uterus and the urinary ducts. Unitary smooth muscle contractions are usually initiated by the stretching of muscle fibers and are thus myogenic. Nerve fibers modulate the rate and force of contraction of unitary smooth muscle fibers. The action potential of some fibers spreads slowly to other neighboring fibers. **Multiunit** smooth muscle fibers contract on stimulation by a nerve (neurogenic) or a hormone. Many multiunit fibers usually contract slowly but simultaneously. Multiunit fibers are found in the iris of the eye and walls of many blood vessels and sperm ducts. Such contractions are more carefully regulated than those of the unitary smooth muscle contractions. Some smooth muscle fibers are capable of undergoing hyperplasia and hypertrophy such as uterine smooth muscles during pregnancy.

Skeletal Muscle

Most of the vertebrate muscle is skeletal and is almost exclusively attached to the skeleton. Skeletal muscle is responsible for the movement of the skeleton and the whole individual. Cutaneous muscle is skeletal muscle found below the skin in fascia of mammals in some body parts and lacks skeletal attachment. Skeletal muscle movements are mainly voluntary although several contractions are controlled by the nervous system subconsciously.

Each skeletal muscle fiber is a long cylindrical cell that contains several hundred nuclei i.e. a **syncytium**. The syncytium results from the end-to-end fusion of several myoblasts during embryonic development. The nuclei are located peripherally whereas the sarcoplasm is densely packed with longitudinally running myofibrils that are parallel to each other. Skeletal muscle fibers may run the whole length of a muscle. In most muscles however, the fibers are shorter than the entire muscle and end in connective tissue within the muscle. Mammalian skeletal muscle fibers usually have a diameter of 10-100 micrometers. The myofibrils in a skeletal muscle fiber are responsible for the striations as they are in register with each other, so the striations appear to extend across the whole width of the fiber. The sarcoplasm of the fibers corresponds to the cytoplasm of other cells of the body and contains organelles and inclusions that are typical of such cells.

Each skeletal muscle fiber is covered with a limiting membrane known as **sarcolemma**, a basal lamina and reticular fibers. Muscle fibers are separated from each other by collagen fibers that make up the

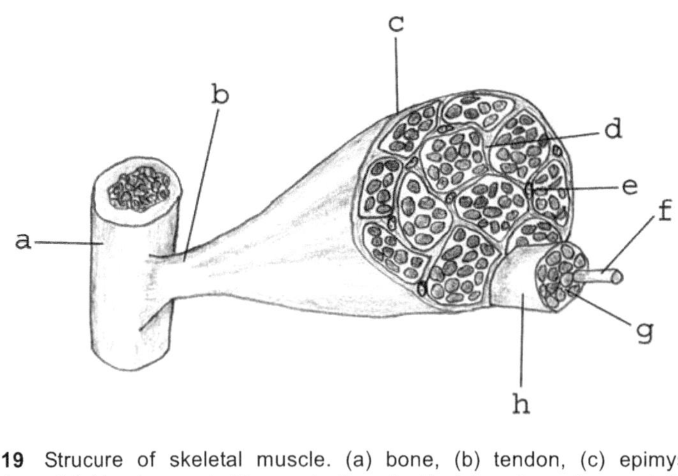

Fig. 3.19 Strucure of skeletal muscle. (a) bone, (b) tendon, (c) epimysium, (d) perimysium, (e) blood vessel, (f) muscle fiber (cell), (g) endomysium and (h) muscle fascicle.

endomysium (Fig. 3.19). Several muscle fibers are surrounded by connective tissue fibers known as **perimysium** and such a group of fibers constitutes a muscle **fascicle** (L. *fasikyulus,* small bundle). Many fascicles are grouped together to form an entire muscle such as the triceps. An entire muscle is covered with a connective tissue membrane known as **epimysium**.

Cardiac Muscle

Cardiac muscle is striated and is found in the heart. Cardiac muscle fibers are separate cellular units that are about 80 micrometers long and are joined end to end by specialized junctions known as **intercalated disks** that run transversely in relation to the long axis of fibers. Cardiac muscle fibers frequently branch and connect with adjacent fibers to form a three dimensional complex. The nuclei of cardiac muscle fibers are located deeply in the interior of the cells. At intercalated disks are patterns of ridges and papillary projections on each cell that fit into corresponding grooves of neighboring cells thus forming interdigitations at such junctions.

Contractions of cardiac muscle are involuntary (myogenic) and originate from the conducting tissue of the heart although the contractions are regulated by the sympathetic and parasympathetic (autonomic) nervous systems. Cardiac muscle does not fatigue and the arrangment of myofilaments ensures maximal overlap bringing about stronger and faster contractions than is the case with smooth muscle.

Conducting Tissue of the Heart

The conducting tissue of the heart (Fig. 3.20) is modified cardiac muscle that is specialized in generation of stimuli as well as transmission of impulses to the rest of the cardiac muscle thus initiating contraction of the heart. The action of the conducting tissue of the heart is able to bring about coordinated contraction of the atria and ventricles in organized succession. Such action ensures that the heart is able to function as an effective pump. The conducting system of the heart consists of the sinoatrial node, atrioventricular node and atrioventricular bundle.

Sinoatrial node ('pacemaker') is located under the epicardium at the junction of the anterior (superior) vena cava and right atrium in mammals. The cells of the node are fusiform, smaller than those of cardiac muscle and are embedded in dense connective tissue. The nodal fibers make contact with those of the atria. Electrical impulses that bring about cardiac muscle contraction originate at the sinoatrial node and are modulated by the sympathetic and parasympathetic nervous systems that innervate the node. Impulses will spread from the node to the rest of

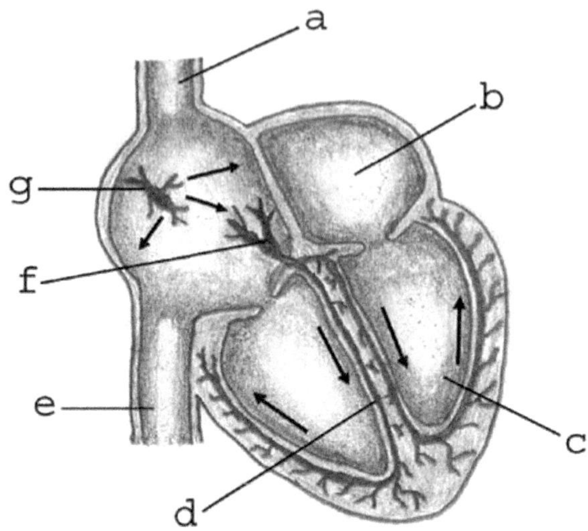

Fig. 3.20 Conducting tissue of the mammalian heart. (a) anterior or superior vena cava, (b) left atrium, (c) left ventricle, (d) atrioventricular bundle, (e) posterior or inferior vena cava, (f) atrioventricular node and (g) sinoatrial node. Arrows show how the electrical impulses that originate from the sinoatrial node spread out to other parts of the myocardium.

the atrial fibers and atrioventricular node.

The **atrioventricular node** lies beneath the endocardium in the interatrial septum. The fibers of the atrioventricular node are similar to those of the sinoatrial node.

Atrioventricular bundle originates from the anterior part of the atrioventricular node and proceeds in the fibrous part of the interventricular septum before dividing into right and left bundles that will eventually spread into the myocardium of the right and left ventricles. The fibers of the bundle are similar at their origin to those of other conducting tissues of the heart but become larger than those of cardiac muscle further in the right and left bundles. At this level they are known as **Purkinje fibers** that have relatively sparse myofibrils which are not as consistent in their orientation as the cardiac muscle fibers. Eventually the Purkinje fibers lose their cytological features and are continuous with cardiac muscle fibers.

FINE STRUCTURE OF MUSCLE

Smooth muscle fibers have dense areas on their sarcolemma at points of contact with other fibers. Such points are also areas of myofilament attachment. In visceral smooth muscle, there are some areas between adjacent fibers that lack intercellular substance and where membranes of neighboring cells are in close contact. These areas are gap junctions (nexi) and are areas of low electrical resistance that permit free movement of ions that enable spread of impulses throughout smooth muscle fibers. When smooth muscle contracts, the dense areas on the sarcolemma that act as points of myofilament attachment will exhibit depressions.

Smooth muscle fibers lack striations as the proteins in the filaments are normally assembled when required for muscle contraction but are degraded with muscle relaxation. The myofilaments in smooth muscle are not in register with each other (Fig. 3.21) as is the case with skeletal and cardiac muscle. Smooth muscle is able to exhibit a greater degree of movement and can contract to shorter lengths than the other two muscle types. Since sarcoplasmic reticulum is not developed much in smooth muscle fibers, calcium required for contraction is stored outside the cell.

The endoplasmic reticulum of striated muscle fibers is modified into a system of membranous tubules known as the **sarcoplasmic reticulum** (Fig. 3.22) that stores calcium ions. The sarcoplasm is packed with myofibrils that are made up of finer thick and thin filaments. Each myofibril consists of several **sarcomeres** that constitute the contractile units. Sarcomeres are found between two successive **Z-lines**. Sarcomeres are made of several zones that bear different names. The **A-bands** are relatively wide and form the dark stripes that are visible under a microscope. In between the A-bands of neighboring sarcomeres are the lighter stripes known as the **I-bands**. Skeletal and cardial muscle fibers have **transverse (T) tubules** that are formed by the inward invagination of the sarcolemma. T-tubules enable electrical signals to spread from the sarcolemma into the fibers. Next to a T-tubule are two **tubular sacs** of the sarcoplasmic reticulum in skeletal muscle fibers. The three tubules are known as the **triad** and enable electrical impulses travelling along the T-tubule spread to the membrane of sarcoplasmic reticular sacs.

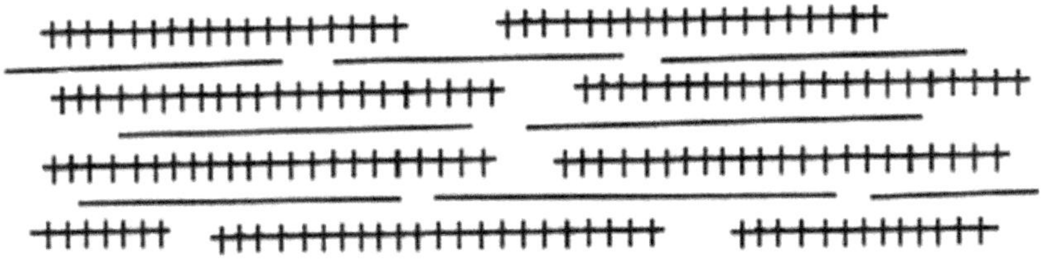

Fig. 3.21 Arrangement of myofilaments in smooth muscle. Smooth muscle lacks cross striations since the myofilaments are not in register with each other. Smooth muscle is capable of shortening to a greater extent than either skeletal or cardiac muscle as a result of this arrangement of myofilaments.

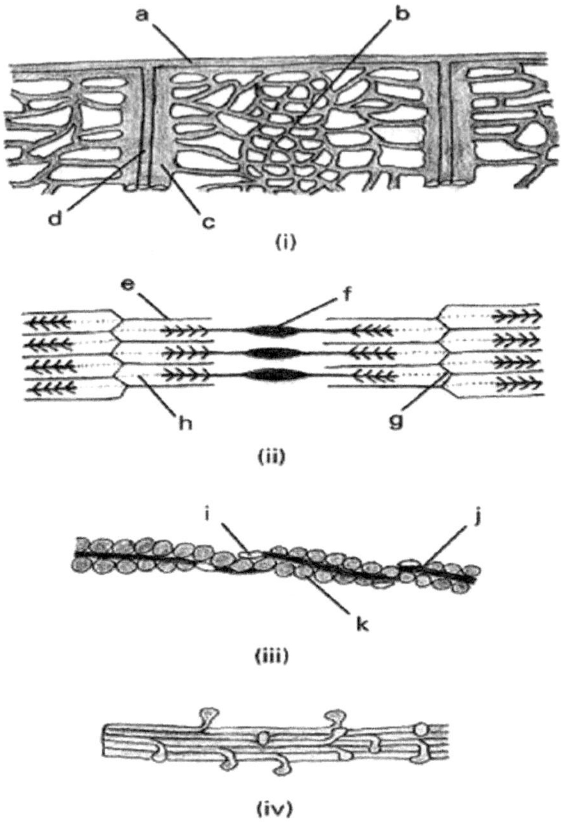

Fig. 3.22 Fine structure of striated muscle. (i) part of a single fiber, (ii) arrangement of thick and thin filaments, (iii) thin filament and (iv) the golf club shaped myosin molecules of a thick filament. (a) sarcolemma, (b) sarcoplasmic reticulum, (c) tubular sacs of sarcoplasmic reticulum or terminal cisternae, (d) transverse tubule, (e) and (f) thin and thick myofilaments respectively, (g) Z-line, (h) elastic filament, (i) troponin, (j) tropomyosin and (k) globular actin molecule. The area between two neighboring Z-lines constitutes a sarcomere (the contractile unit of striated muscle).

Myofilaments

In each myofiber are about a thousand parallel subumits known as myofibrils. Each myofibril is about 1 mm thick and contains thousands of thin and thick myofilaments that lie side by side. The thin filaments are made of **actin, troponin** and **tropomyosin**. The globular actin molecules form the bulk of the thin filament. When not involved in muscle contraction, the active sites on actin molecules are covered with the long tropomyosin molecules. The thick filaments are made of mainly **myosin** molecules. Myosin molecules resemble golf clubs with their long shafts stacked together whereas their heads (cross bridges) stick out towards the thin filaments. There are nearly 20 classes of myosin that are classified into two groups according to their head structure as double- or single-headed myosins. The double-headed myosins have been shown to induce sliding movements among neighboring actin filaments and for generating tension or elongation in actin bundles (Tanaka-Takiguchi et al., 2004).

Blood is a unique connective tissue as it is liquid in form and neither contains fibers nor ground substance. Generally blood has two parts: the liquid part known as plasma and the cellular part. Plasma contains dissolved solutes and proteins that are vital for the nutrition, excretion, protection and maintenance of the correct blood pressure. The cellular component consists of red blood cells that transport oxygen, the white blood cells that are involved in the body's defense system and platelets that play a role in clotting of blood. Blood is discussed separately in Chapter 10.

REFERENCES

Aloisi, F. (2001). Immune function of microglia. Glia, 36(2):165-179.

Benjamin, M. and Hillen, B. (2003) Mechanical influences on cells, tissues and organs—'Mechanical Morphogenesis'. Eur. J. Morphol. 41(1):3-7.

Colnot, C., Lu, C., Hu, D. and Helms, J.A. (2004). Distinguishing the contributions of the perichondrium, cartilage and vascular endotherlium to skeletal development. Dev. Biol., 269(1):55-69.

Doetsch, F. and Scharff, C. (2001). Challenges for brain repair: Insights from Adult Neurogenesis in Birds and Mammals. Brain Behav. Evol., 58(5):306-322.

Fukumoto, S. and Martin, T.J. (2009). Bone as an endocrine organ. Trends Endocrinol. Metab., 20(5):230-236.

Henrotin, Y.E., Bruckner, P. and Pujol, J.P. (2003). The role of reactive oxygen species in homeostasis and degradation of cartilage. Osteoarthritis Cartilage, 11(10):747-755.

Jessen, K.R. and Mirsky, R. (1999). Schwann cells and their precursors emerge as major regulators of nerve development. Trends Neurosci., 22(9):402-410.

Karsdal, M.A., Tanko, L.B., Riis, B.J., Sondergard, B.C., Henriksen, K., Altman, R.D., Qvist, P. and Christiansen, C. (2006). Calcitonin is involved in cartilage homeostasis: is calcitonin a treatment for OA? Osteoarthritis Cartilage, 14(7):617-624.

Neumann, J., Sauerzweig, S., Rönicke, R., Gunzer, F., Dinkel, K., Ullrich, O., Gunzer, M. and Reymann, K.G. (2008). Microglia cells protect neurons by direct engulfment of invading neutrophil granulocytes: a new mechanism of CNS immune privilege. J. Neurosci., 28(23):5965-5975.

Noel, D., Gazit, D., Bouquet, C., Apparailly, F., Bony, C., Plence, P., Millet, V., Turgeman, G., Perricaudet, M., Sany, J. and Jorgensen, C. (2004). Short-term BMP-2 expression is sufficient for in vivo osteochondral differentiation of mesenchymal stem cells. Stem Cells, 22(1):74-85.

Rakic, P. (2003). Elusive radial glial cells: historical and evolutionary perspective. Glia, 43(1):19-32.

Richardson, W.D., Smith, H.K., Sun, T., Pringle, N.P., Hall, A. and Woodruff, R. (2000). Oligodendrocyte lineage and the motor neuron connection. Glia, 29(2):136-142.

Shapiro, F. (1992). Vertebral development in the chick embryo during days 3-9 of incubation. J. Morphol., 213:317-333.

Small, J.V. and Gimona, M. (1998). The cytoskeleton of the vertebrate smooth muscle cell. Acta Physiol. Scand., 164(4):341-348.

Tanaka-Takiguchi, Y., Kakei, T., Tanimura, A., Takagi, A., Honda, M., Hotani, H. and Takiguchi, K. (2004). The elongation and contraction of the actin bundles are induced by double-headed myosins in a motor concentration-dependent manner. J. Mol. Biol., 341(2):467-476.

Van der Kraan, P.M., Buma, P., van Kuppevelt, T. and van den Berg, W.B. (2002). Interaction of chondrocytes, extracellular matix and growth factors: relevance for articular cartilage tissue engineering. Osteoarthritis Cartilage, 10(8):631-637.

van Driel, M., Pols, H.A. and van Leeuwen, J.P. (2004). Osteoblast differentiation and control by vitamin D and vitamin D metabolites. Curr. Pharm. Des., 10(21):2535-2555.

Weissman, T., Noctor, S.C., Clinton, B.K., Honig, L.S. and Kriegstein, A.R. (2003). Neurogenic Radial Glial cells in Reptile, Rodent and Human: from mitosis to migration. Cerebr. Cortex. 13(6):550-559.

Wurtz, T. and Berdal, A. (2003). Osteoblast precursors at different anatomic sites. Crit. Rev. Eukaryot Gene Expr., 13(2-4):147-161.

4

Skeletal System

The vertebrate skeletal system or passive locomotor apparatus is composed of a bony or cartilaginous framework that is able to bear body weight in these animals. Whereas the endoskeleton of vertebrates is capable of gradual growth along with the rest of the body, the exoskeleton of insects is shed off periodically as the insect increases in size. During early embryological development, the vertebral column of vertebrates passes through the notochord stage that is not yet bone or cartilaginous tissue. The skeletal and muscular systems are also known as the locomotor apparatus and form a complete structural unit. The action of the skeletal system is also complemented by connective tissue elements such as ligaments that are made of dense regular connective tissue. The presence of joints at points of skeletal contact makes it possible for movement to occur.

The skeletal system plays several other roles in vertebrates. Several body cavities are surrounded by the skeletal system thereby providing protection to the viscera. The skeleton is able to offer firmness and shape to vertebrates due to its structural strength that results from its chemical composition. The three types of skeleton present in the vertebrates are the notochord, cartilage and bone. Dense regular connective tissues such as ligaments and tendons offer structural support.

THE NOTOCHORD

The notochord (L. *notos*, back; *chorda*, string, cord) is a flexible but firm hydroskeleton that is located in the back area below the neural tube (Fig. 4.1) of early embryos in chordates. In vertebrates, the notochord is replaced later on in life by a bony or cartilaginous vertebral column

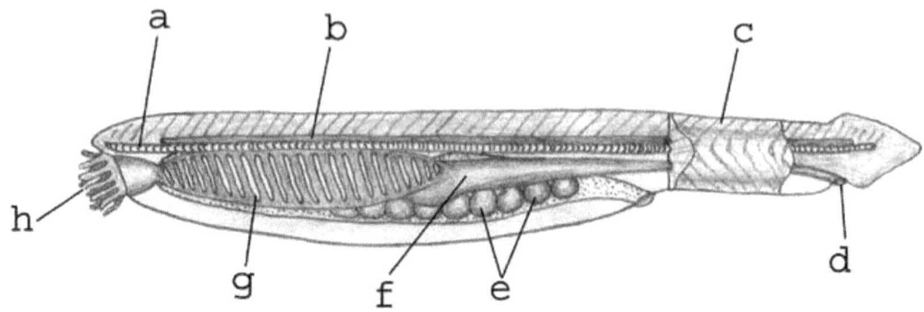

Fig. 4.1 A cephalochordate (Gr. *kephale*, head) showing the location of the notochord. (a) notochord, (b) neural tube, (c) myomere or muscle segment, (d) anal opening, (e) gonads, (f) gut, (g) gill slit and (h) mouth. The notochord of cephalochordates extends into their heads, leaving little room for the brain.

except in the hagfish and some early tetrapods. Varying amounts of notochord still persist in adult stages of some lower vertebrates. In mature higher vertebrates such as mammals, the notochord is represented by jelly-like remnants known as **nuclei pulposi** (L. *nucleus pulposus,* pulpy nucleus) of the **intervertebral disks** (Fig. 4.2). The notochord runs the entire length of the future vertebral column and ensures that the embryo does not shorten when it contracts but will bend from side to side. The cells of the notochord are liquid filled and are surrounded by a tough connective tissue sheath rendering them turgid (Fig. 4.3).

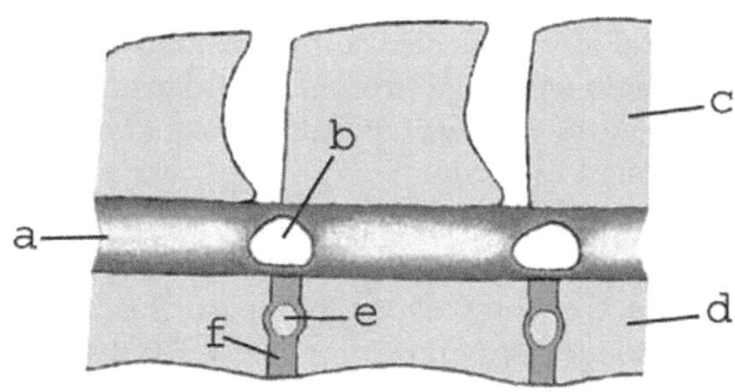

Fig. 4.2 A longitudinal section through some vertebrae. (a) neural canal that contains the spinal cord, (b) intervertebral canal through which spinal nerves leave the neural canal, (c) neural spine (spinous process), (d) vertebral centra, (e) nucleus pulposus and (f) annulus fibrosus that surrounds (e). (e) and (f) form the intervertebral disk.

BLOOD SUPPLY TO MAMMALIAN BONE

Blood vessels from the periosteum penetrate microscopic pores of bone and lead to Volkmann's canals that run at right angles to the long axis of bone. The blood vessels then join the Haversian systems to supply

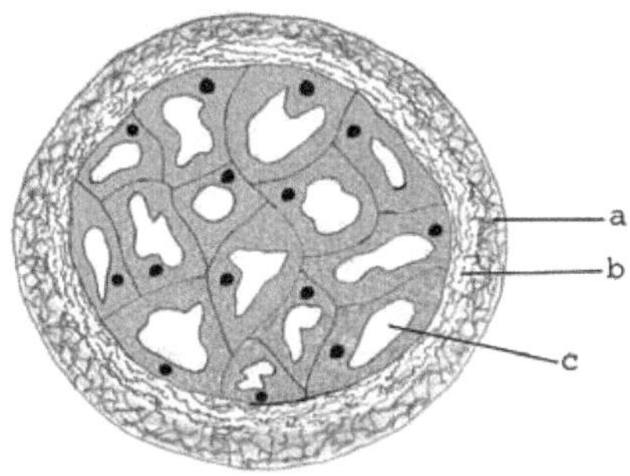

Fig. 4.3 A cross-section of the notochord. (a) connective and (b) collagenous sheaths and (c) vacuole in a notochord cell. The fibers of the sheaths are fibrous and elastic in nature and surround the large and vacuolated cells of the notochord. The vacuoles are surrounded by a dense cytoplasmic mesh of keratin-like intermediate filaments whose thickness varies in different vertebrate groups. Notochord cells exert pressure against the sheaths due to the presence of vacuoles. The arrangement of layers that form the sheath vary in structure and thickness in different vertebrates.

osteocytes. Larger blood vessels penetrate the compact bone through nutrient foramina and lead to the medullary cavity where they supply the bone marrow. The bone marrow has sinusoidal capillaries with gaps between endothelial cells and a basement membrane that is incomplete to permit passage of a lot of material through the gaps. The bone marrow consists of reticular cells that form a three-dimensional network and are supported by reticular fibers. The spaces between reticular cells are occupied by blood cells in various stages of maturation and a network of blood vessels.

TYPES OF SKELETON

Vertebrates have a dermal or integumentary skeleton and an endoskeleton. The endoskeleton comprises visceral and somatic skeletal systems. The dermal skeleton develops directly from mesenchyme while the endoskeleton forms in cartilage although it may remain cartilaginous throughout. Although these skeletal systems show varying degrees of development in vertebrates, the somatic skeleton of cartilage or cartilage replacement bone is the one that forms the major part of vertebrate skeleton.

INTEGUMENTARY (DERMAL) SKELETON

The integumentery skeleton consists of bony plates and scales that are embedded in the skin or are found beneath it. Early vertebrates such as ostracoderms had so much dermal bone that they were referred to as 'armored fishes'. This armor was reduced with the evolution of later fish.

Most living vertebrates lack dermal skeleton save for some elements in the head region. The dermal bone of fishes includes the scales that represent the remnants of the early ancestral armor. Dermal elements of vertebrates are retained in the skull (save for bones at its base) and the pectoral girdle.

VISCERAL SKELETON

The visceral skeleton is also known as the **splanchnocranium** (Gr. *splanchnon*, gut or visceral) and consists of the skeleton of the jaws and gill arches and their skeletal element derivatives such as jaws and ear ossicles. The visceral skeletal system is of neural crest origin. The ventral foregut endoderm exerts a regionalized patterning activity on the cephalic neural crest to build up the primary facial and visceral skeleton in jaws and neck (Ruhin et al., 2003). The cranial neural crest plays a major role in the development of the skeleton of the vertebrate head as well as cranial muscles and this might be a primitive feature of cranial development in vertebrates (Ericsson et al., 2004).

FISH VISCERAL SKELETON

The fish visceral skeleton consists of seven pairs of cartilages or bone in seven visceral arches and several serial ventral cartilages in the pharyngeal floor (Fig. 4.4). The visceral skeleton of bony fishes resembles that of sharks except for the presence of bone in the former. The caudal ends of the cartilaginous dorsal **palatoquadrate** or **pterygoquadrate** undergo ossification to become the quadrate bones whereas the remaining part becomes the palantine and pterygoid bones in bony fishes. The posterior end of mandibular (Meckel's) cartilage becomes the articular bone.

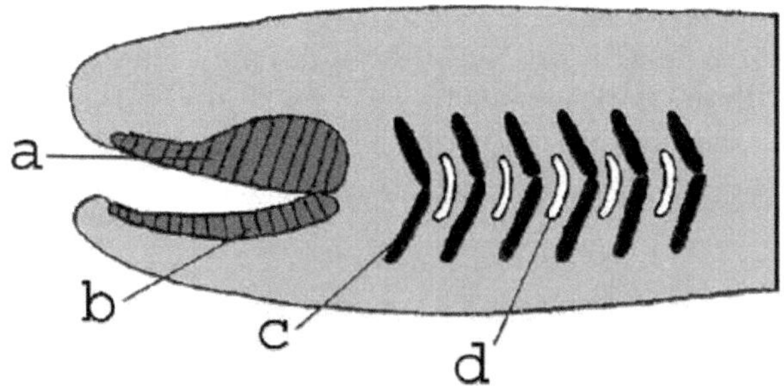

Fig. 4.4 The visceral arches of vertebrates. (a) palatoquadrate, (b) mandibular cartilage, (c) hyoid arch and (d) gill slit. (a) and (b) are derived from the mandibular or first arch that eventually develop into jaws in gnathostomes. The upper part of the hyoid arch in jawed fishes is the **hyomandibular cartilage** or **bone** that connects the lower jaw with the skull.

Feeding movements in many bony fishes are brought about by **cranial kinesis** (intracranial mobility) that involves relative motion between the upper jaw and bones of the cranium. This movement is found in many other vertebrates but is lost in mammals. Cranial kinesis optimizes biting and rapid feeding and allows a quick change in the shape and size of the mouth.

The lamprey visceral skeleton consists of a **latticework** or **branchial basket** that consists of seven visceral arches that support the pharynx and gill pouches (Fig. 4.5). In hagfish, the branchial latticework that is present in lampreys is represented by a single ring-cartilage that is found between the oral opening and esophagus.

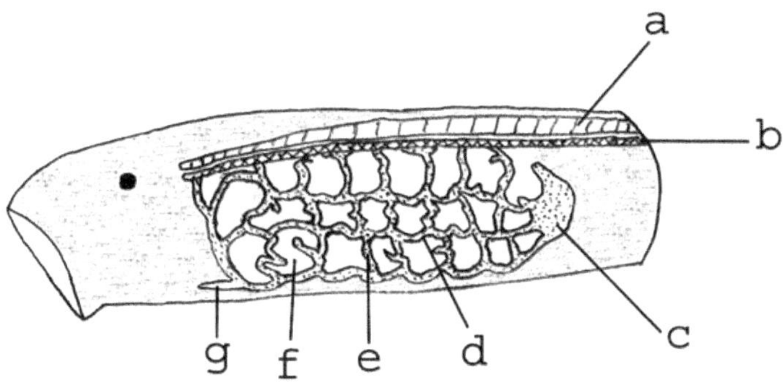

Fig. 4.5 The branchial basket of a lamprey. There are seven branchial bars in a branchial basket. (a) spinal cord, (b) notochord, (c) pericardial cartilage, (d) horizontal bar, (e) branchial arch, (f) gill slit and (g) styloid process.

JAW SUSPENSION

The upper jaws of fishes are supported from the neurocranium by amphistylic, autostylic and hyostylic methods of suspension (Fig. 4.6). In **amphistylic suspension** (Gr. *amphi*, both; *stylos*, pillar), the hyomandibular (part of the hyoid arch) supports the posterior part of the palatoquadrate from the cranium and the palatoquadrate has one or more movable articulations with the cranium. This type of suspension was found in the extinct cartilaginous and bony fishes. **Autostylic suspension** (Gr. *autos*, self) is found in holocephalans, lungfishes and tetrapods and in these vertebrates the palatoquadrate is fused to or forms a tight joint with the cranium. The hyomandibular does not play a role in supporting jaws from the braincase in this type of suspension and is incorporated into the middle ear as the stapes in tetrapods. **Hyostylic suspension** is where the palatoquadrate has lost connection with the cranium but is still supported by the hyomandibular. Hyostylic jaws are more flexible and allow for protrusion of jaws forwards and downwards as is seen in living cartilaginous and bony fishes.

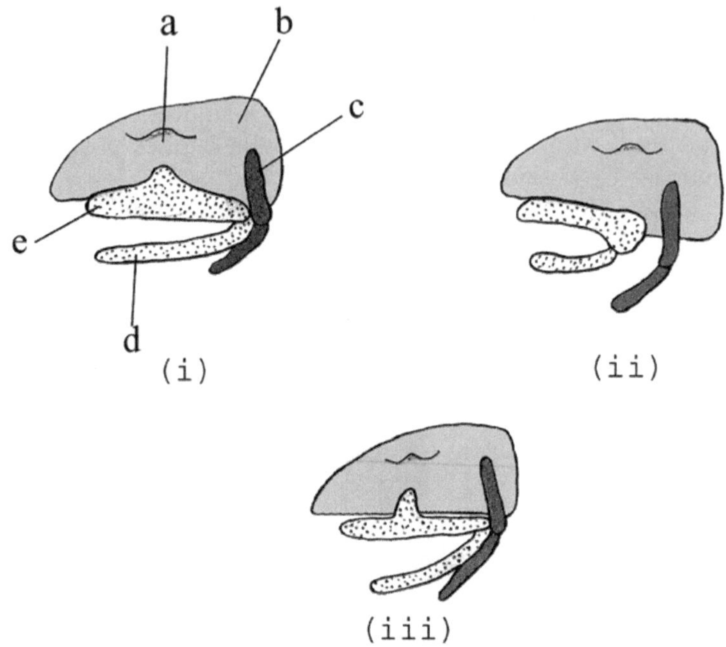

Fig. 4.6 Suspension of the palatoquadrate cartilage in vertebrates. (i) Amphistylic, (ii) autostylic and (iii) hyostylic suspension. (a) orbit, (b) cranium, (c) hyomandibular cartilage (part of the hyoid arch), (d) mandibular cartilage and (e) palatoquadrate.

TETRAPOD VISCERAL SKELETON

The transition from aquatic to terrestrial forms of life was accompanied by major changes in the visceral skeleton. The visceral skeleton lost some of its structures while others were modified to cope with life on land. The posterier part of palatoquadrate cartilage undergoes ossification to form the quadrate (Fig. 2.15) in amphibians, reptiles and birds. The quadrate articulates with both the braincase and the articular bone of the lower jaw. In mammals, the quadrate detaches from the palatoquadrate to become the incus of the middle ear, leaving the dentary (lower jaw) to articulate directly with the temporal bone of the skull forming the temporomandibular joint.

The mandibular (Meckel's) cartilage in reptiles is surrounded with dermal bones whereas that of birds and mammals shows few or no remnants of the cartilage. In mammals, the articular bone of the mandibular cartilage is taken up by the middle ear becoming the malleus.

The hyomandibular cartilage of sharks is found between the quadrate and the otic capsule. The cartilage loses articultion with the quadrate in tetrapods and ossifies to become part of the stapes of the middle ear. Branchial arches III, IV and V become part of the hyoid apparatus of tetrapods while arches VI and VII are lacking. The mandible (lower jaw) of cartilaginous fish may have originated as part of the mandibular cartilage. In bony vertebrates, the cartilage is strengthened and replaced by several dermal bones.

DERMATOCRANIUM

The dermatocranium forms most of the skull. Dermal bones first appeared in ostracoderms and other early fish as large dermal plates that covered the head. They attached to the splanchnocranium and the chondrocranium. Dermal bones include bones that form most of the roof of the skull and palate of most vertebrates. Living cyclostomes and cartilaginous fish lack a dermal skeleton. Higher vertebrates show a greater degree of fusion of dermal bones. The major dermal bones are grouped into several series for purposes of description: orbital, cheek, temporal, facial, vault, palatal and mandibular series. The orbital series consists of the jugal, lacrimal, prefrontal, postfrontal and postorbital (Fig. 4.7). The cheek series comprises the squamosal and quadrotojugal. Intertemporal, supratemporal and tabular bones form the temporal series. The facial series has the teeth bearing bones premaxilla and maxilla as well as the nasals. The frontals and parietals that are quite enlarged in mammals and the post-parietal form the vault series whereas

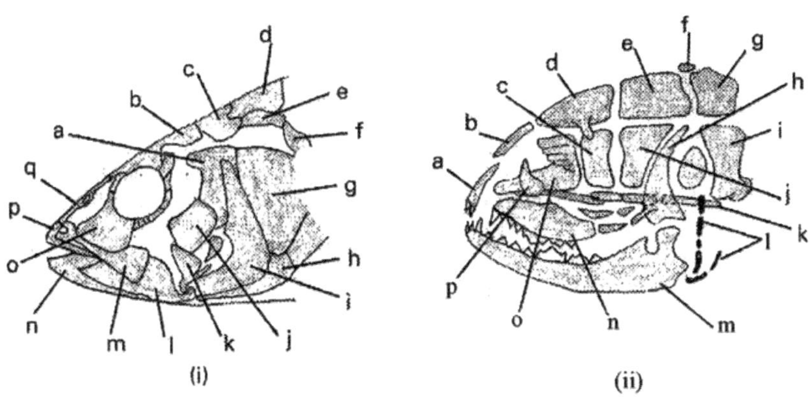

Fig. 4.7 Some skull bones in a teleost (i) and a young mammal (ii). **(i)** (a) hyomandibular, (b) frontal, (c) parietal, (d) supraoccipital, (e) posttemporal, (f) supracleithrum, (g) opercle, (h) interopercle, (i) preopercle, (j) metapterygoid (k) quadrate, (l) articular, (m) maxilla, (n) dentary, (o) lachrymal, (p) premaxilla and (q) nasal. **(ii)** (a) incisive, (b) nasal, (c) presphenoid, (d) frontal with a process to (c), (e) parietal, (f) interparietal, (g), (i) and (k) occipital, (h) temporal, (j) basisphenoid, (l) hyoid, (m) mandible, (n) maxilla, (o) ethmoid and (p) lachrymal.

the squamosal and quadratojugal form the cheek series. The palatal series includes the vomer, palatine, ectopterygoid, pterygoid and parasphenoid.

Mammals possess a secondary palate that consists of bone that completely separates the rostral part of nasal and oral cavities. The presence of a secondary palate could be a way of strengthening the rostrum in mammals (Thomason and Russell, 1986). The palate enables mammals feed while breathing at the same time. The lower jaw forms the mandibular series that includes the dentary that bear teeth, splenials, angular, surangular, prearticular and coronoid. The bones of the mandibular series are represented in mammals by the mandible.

Calcification of jaws, gill arches, vertebral arches and fins has also been demonstrated in sharks (Kemp and Westrin, 1979). Multiple layers of calcification in the jaws of some large sharks of the families Lamnidae and Carcharhinidae has been associated with strengthening of jaws in these groups for biting purposes (Dingerkus et al., 1991)

TEETH

It has been stated that teeth originated from separate skin denticles and their origin is correlated to the origin of jaws. It has also been argued that fossil jawless fish possessed pharyngeal denticles with relevant developmental controls prior to the evolution of jaws (Johanson and Smith, 2003). Many toothed non-mammalian vertebrates show continuous tooth replacement and it has been proposed that epithelial stem cells are involved in the process (Huysseune and Thesleff, 2004). Teeth will be covered under the section on nutrition and digestion (Chapter 7).

SOMATIC SKELETON

The **somatic skeleton** (Gr. *somatikos*, bodily) (Fig. 4.8) forms the bulk of vertebrate skeletal system and consists of most of the axial and appendicular skeletal systems. The somatic skeleton is of mesodermal origin. The neck vertebrae, trunk and appendicular skeletal systems and caudal or coccygeal vertebrae are also known as the **post-cranial skeleton**. Some of the somatic skeleton forms part of the posterior or lower skull and lower jaws.

AXIAL SKELETON

The axial skeletal system has a characteristic feature of segmentation and consists of the head, vertebral column median fins, the ribs and the sternum.

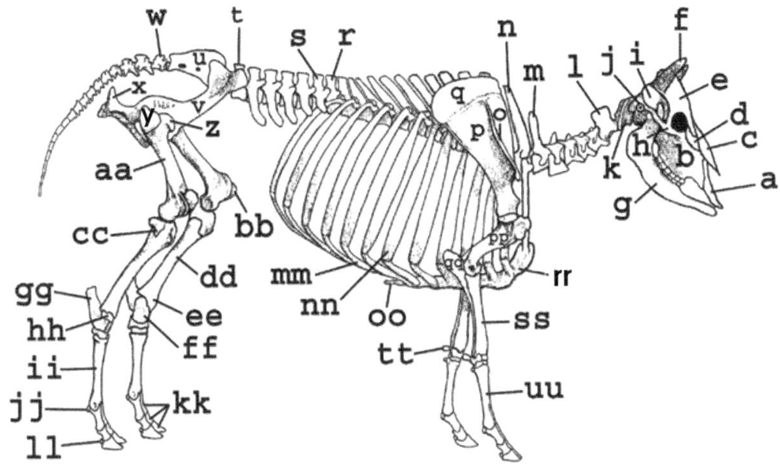

Fig. 4.8 A bovine (ox) skeleton. (a) incisive bone (premaxilla), (b) maxilla, (c) nasal, (d) lachrymal and (e) frontal bones, (f) cornual process, (g) mandible, (h) zygomatic bone, (i) temporal fossa, (j) external auditory meatus, (k) atlas, (l) axis, (m) last cervical vertebra, (n) first thoracic vertebra, (o) spinous process of (p) scapula, (q) scapular cartilage, (r) last thoracic vertebra, (s) first lumbar vertebra, (t) last lumbar vertebra, (u) sacrum, (v) ilium, (w) first caudal vertebra, (x) ischium, (y) greater trochanter and (z) head of femur, (aa) femur, (bb) patella, (cc) proximal rudiment of fibula, (dd) tibia, (ee) medial malleolus, (ff) talus, (gg) calcaneus, (hh) malleolar bone, (ii) metatarsal bone, (jj) proximal sesamoid bone, (kk) digits, (ll) distal sesamoid bone, (mm) costal cartilage, (nn) rib, (oo) xiphoid cartilage, (pp) humerus, (qq) olecranon process of ulna, (rr) sternum, (ss) radius, (tt) accessory carpal bone and (uu) metarcarpal bone.

NEUROCRANIUM

The neurocranium is the ossified chondrocranium. It forms the base of the cranium and bones that surround the (nasal, optic and otic) sensory organs. In cartilaginous fish, the chondrocrocranium consists of olfactory capsule, optic capsule and the trabecular, parachordal and occipital cartilages. Bony vertebrates have a neurocranium that is represented by the sphenoid, optic, otic and nasal capsules (including its turbinates in mammals) and the occipital bone.

VERTEBRAL COLUMN

The vertebral column consists of segmented units known as vertebrae. In fishes, the vertebral column plays a major role in movement as it prevents the body from shortening during muscle contraction and so brings about lateral undulations while swimming whereas in terrestrial vertebrates the column acts as a beam that supports axial body weight by transferring it to the girdles. The vertebral column plays a role in locomotion in some terrestrial vertebrates. A typical vertebra (Fig. 4.9) consists of a **body** or **centrum**, one or two **arches** and several **processes**. The vertebral body forms the major point of contact between neighboring vertebrae and shows differences at these points in various vertebrates (Fig. 4.10). The bodies are concave at either ends in **amphicoelous vertebrae** (Gr. *amphi*, both or double; *koilios*, hollow) of most fish, some salamanders and caecilians. **Opisthocoelous vertebrae** (Gr. *opisthen*, behind) are convex in front and concave at the back and such vertebrae are found in most salamanders and ungulates such as cattle. Anurans and living reptiles have **procoelous vertebrae** (Gr. *pro*, before) that are concave at the front and convex at the back. Vertebrae with saddle shaped ends are **heterocoelous** in nature and are found in the necks of turtles and birds whereas most mammals have **acoelous (platycoelous) vertebrae** (Gr. *a*, without) that have flat ends. Studies on vertebral patterning in zebrafish show that vertebral bodies arise by secretion of bone matrix from the notochord rather than somites (Fleming et al., 2004). Centra do not form via an intermediate cartilage stage and lack osteoblasts. Vertebrae from various vertebrates differ greatly between species and also body regions in individual species and this could have resulted from the demands that have been placed on these skeletal segments.

Each vertebra has a **neural** or **vertebral arch** that extends from the vertebral body to surround the neural canal that is occupied by the spinal cord. At the base of neural arches on either side of neighboring vertebrae are **interverbral foramina** through which spinal nerves from the spinal cord pass. **Hemal arches** or **chevrons** extend ventrally from caudal vertebrae in the tail region of fishes and some tetrapods and protect the caudal blood vessels they surround. Hemal arches also act as points of muscle attachment. The vertebral bodies replace the notochord in

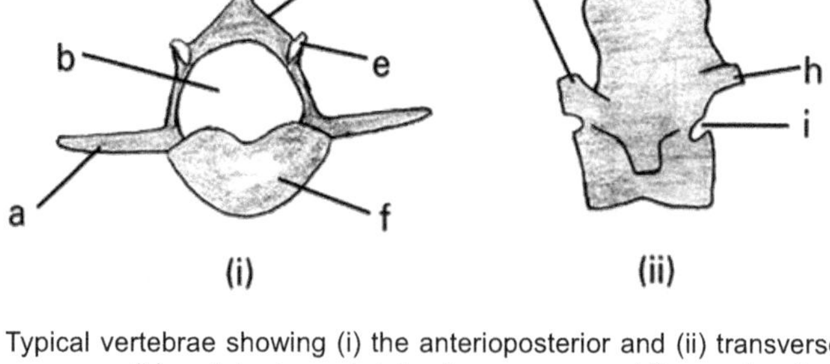

Fig. 4.9 Typical vertebrae showing (i) the anterioposterior and (ii) transverse views. (a) transverse process, (b) vertebral canal, (c) spinous process, (d) neural arch, (e) articular process (zygapophysis), (f) centrum (body), (g) and (h) anterior and posterior articular processes (pre- and postzygapophyses) and (i) part of intervertebral canal.

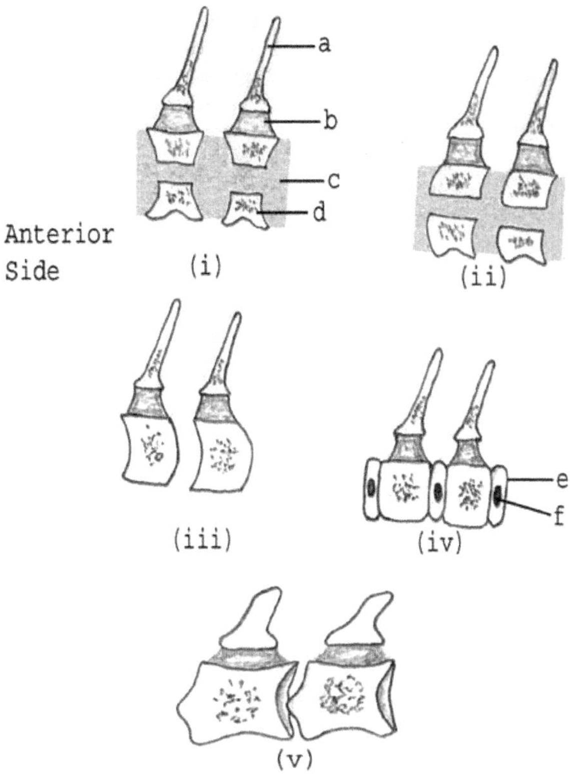

Fig. 4.10 Various shapes of vertebral bodies at their points of contact. Amphicoelous (top left), opisthocoelous (top right), procoelous (middle left), acoelous (middle right) and heterocoelous vertebrae. In heterocoelous vertebrae, one of the surfaces of neighboring vertebrae is saddle-shaped in the transverse plane while the other surface it makes contact with is saddle-shaped in the vertical plane. Heterocoelous vertebrae do not allow for rotation of the spine although permit most movement.

tetrapods and many fish though it persists in adult early groups such as lungfishes, coelacanth and chondrosteans to varying degrees. Between the vertebral centra of neighboring vertebrae are **intervertebral disks** or **pads** that are made of remnants of the notochord known as **nuclens pulposus** to the inside and connective tissue or fibrocartilage to the outside referred to as **annulus fibrosus**. There is normally one vertebral centrum for every body segment. Some fishes such as sharks have two vertebral centra per body segment thus displaying diplospondyly. **Diplospondyly** (Gr. *diplos*, double; *spondylos*, vertebra) increases flexibility of the back.

Most terrestrial vertebrates and some teleosts have a pair of articular processes at either end of their neural arches known as **pre-** and

postzygapophyses (Gr. *zygon*, yoke; *apo*, away from; *physis*, growth) that form synovial joints between successive vertebrae. These processes further strengthen the joints formed between vertebrae and can also restrict movement in certain directions such as flexion and torsion of the vertebral column.

The vertebral processes project from the neural arches and vertebral centra. They are sites of muscle attachment and some articulate with ribs. The processes together with the soft tissue between them also increase rigidity of the vertebral column. Spinous processes extend from the arches dorsally whereas transverse processes extend from the base of the arches or centrum laterally. Transverse processes are the more common processes and divide expaxial from hypaxial muscles. **Diapophyses** and **parapophyses** that are located on vertebral centra and transverse processes respectively articulate with ribs.

Vertebral Column of Fishes

Since water offers buoyancy to fish, the vertebral column does not have to support body weight a lot as is the case in tertrapods. Except for the first one or two vertebrae that have been modified for joining the vertebral column to the skull, the other vertebrae are basically similar. Fish are incapable of moving their heads independent of their bodies. Most vertebrae differ slightly depending on the species and the amount of undulations that occur in the different body regions. Generally, the fish vertebrae articulate with each other by contact of their vertebral bodies since zygapophyses are not present.

Hagfish lack a vertebral column and a notochord persists in its place throughout life. Chondrosteans such as sturgeons and paddlefishes, lungfishes and the coelacanth have incomplete centra with a notochord. Cartilage is found in the notochord sheath of these fishes and provides structural support. The only parts of vertebral column found in lampreys are paired lateral neural cartilages. Cartilaginous fishes have vertebrae with **cartilaginous dorsal plates** or **dorsal intercalary plates** that represent neural arches and lie between neighboring arches (Fig. 4.11). The vertebrae of teleosts are well ossified with amphicelous centra that contain a constricted notochord. A neural arch is found in each teleostean vertebra whereas the caudal vertebrae possess hemal (ventral) arches (Fig. 4.12).

VERTEBRAL COLUMN OF TETRAPODS

Tetrapods possess vertebrae that are specialized depending on their location along the column and the functions they perform. Starting from

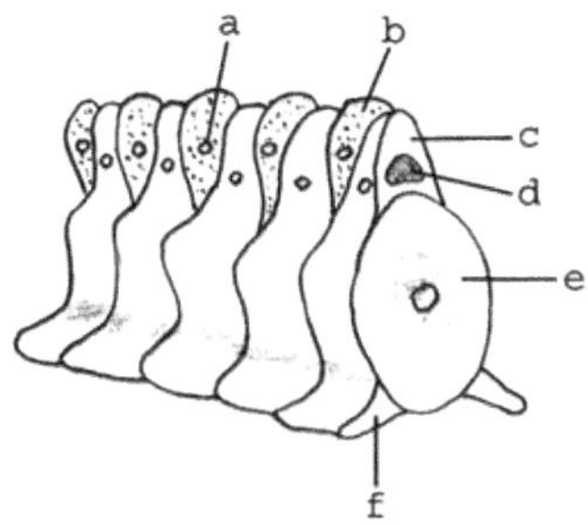

Fig. 4.11 Arrangement of vertebrae in cartilaginous fish. (a) spinal nerve foramen, (b) dorsal intercalary plate, (c) neural arch, (d) neural canal, (e) vertebral centrum and (f) transverse process. Cartilaginous fish lack typical fish vertebral columns. Dorsal intercalary plates are located between arches of neighboring vertebrae. Small **ventral intercalary plates** may be found between the bases of hemal arches in cartilaginous fish.

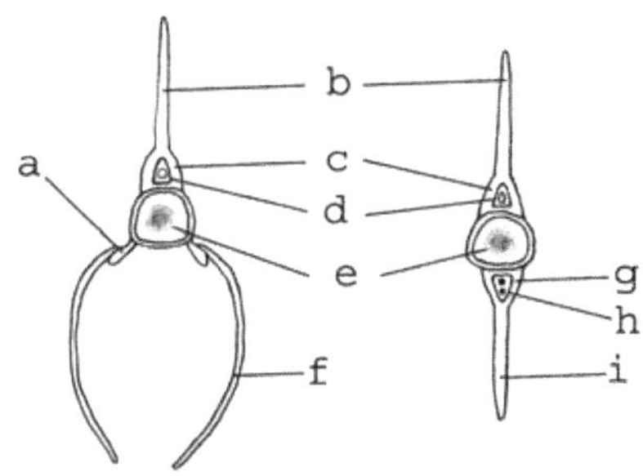

Fig. 4.12 Teleostean body trunk vertebra with ribs (left) and caudal vertebra (right). (a) transverse process, (b) neural spines, (c) neural arches, (d) neural canals with spinal cords, (e) centra or bodies, (f) ventral rib, (g) hemal arch, (h) hemal canal containing caudal vein and artery and (i) hemal spine.

the anterior to posterior regions of the body are cervical, thoracic, lumbar, sacral and caudal or coccygeal vertebrae (depending on the species).

CERVICAL VERTEBRAE

The cervical (neck) vertebrae show variation in number and structure in tetrapods. With evolution, several vertebrae that were anterior to the present most cranial cervical vertebra were incorporated into the chondrocranium. Amphibians have a short neck with a single cervical vertebra. The two convex occipital condyles at the back of the head form a joint with a pair of articular facets on the anterior part of the cervical vertebra making it possible for amphibians to move their heads up and down. There are about seven cervical vertebrae in reptiles enabling this group of vertebrates to exhibit greater flexibility of the neck. The number of cervical vertebrae in the 'S' shaped cervical vertebral column of birds varies greatly. The domestic fowl and duck have 14, the goose has 17 and the swan has as many as 25. Mammals, despite the great variation in the length of the neck, have seven cervical vertebrae except the manatee (a sea cow) and two-toed sloth that have six and the three-toed sloth that has either eight or nine cervicals.

The first two cervical vertebrae (Fig. 4.13) of reptiles, birds and mammals bring about free movement of the head and due to their specialized function differ in structure from other vertebrae in the neck region. The first cervical vertebra is known as the **atlas** and lacks most of the centrum. In reptiles and birds, the atlas articulates with a single occipital condyle forming a ball and socket joint while the craniovertebral joint is formed between two occipital condyles and two articular fovea of the atlas in mammals. Articulation of the atlas with the skull brings about flexion and extension movements of the head in relation to the neck ('nodding') in mammals. The second cervical vertebra, the **axis**, has a tooth-like structure known as the **dens** or **odontoid process** that protrudes into the vertebral canal of the atlas. There are no zygapophyses between the atlas and axis. The joint between the atlas and axis permits

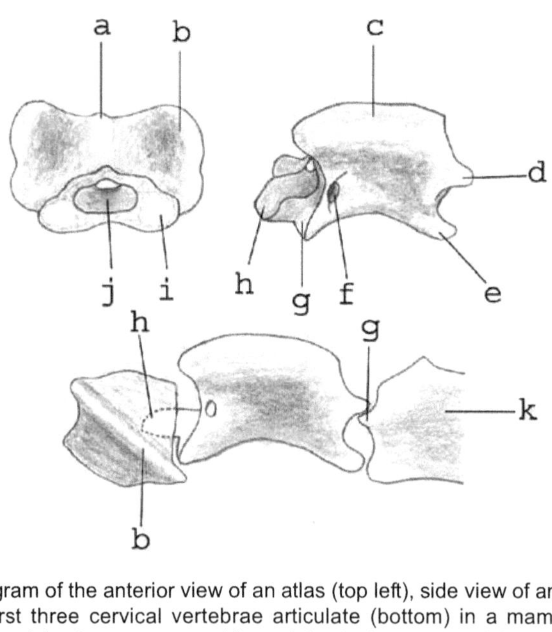

Fig. 4.13 Diagram of the anterior view of an atlas (top left), side view of an axis (top right) and how the first three cervical vertebrae articulate (bottom) in a mammal. (a) dorsal tubercle, (b) wing, (c) spinous process, (d) caudal articular process, (e) transverse process, (f) lateral vertebral foramen, (g) cranial articular process, (h) dens or odontoid process, (i) cranial articular fovea, (j) spinal canal and (k) third cervical vertebra. The condyloid joint between the occipital condyles and the cranial articular fovea of the atlas make it possible to perform the 'nodding' movement whereas the pivot joint between the atlas and axis enable a mammal to perform the side to side movements of the head.

sideways movements of the head. In reptiles, movements between the atlas and the head and between the atlas and axis are restricted due to the presence of a **proatlas** (Fig. 4.14) that is located dorsally between the atlas and skull. The transverse processes of birds and mammals have the transverse foramen in which lies the vertebral artery and vein.

DORSAL REGION

Dorsals are vertebrae that are found between the cervicals and sacrals and articulate with ribs. Dorsals are found in fish, amphibians and snakes. Crocodiles and some other reptiles have several pairs of abdominal ribs that are embedded in the ventral abdominal wall and unite ventrally. Ribs are confined to the anterior part of the trunk in lizards, birds and mammals. The equivalent dorsal region in these species is divided into thoracic and lumbar vertebrae in these vertebrates and their thoracic vertebrae articulate with ribs. There is considerable fusion of thoracic vertebrae in birds forming the **notarium**. In mammals, the embryonic ribs of the lumbar vertebrae that are present during the

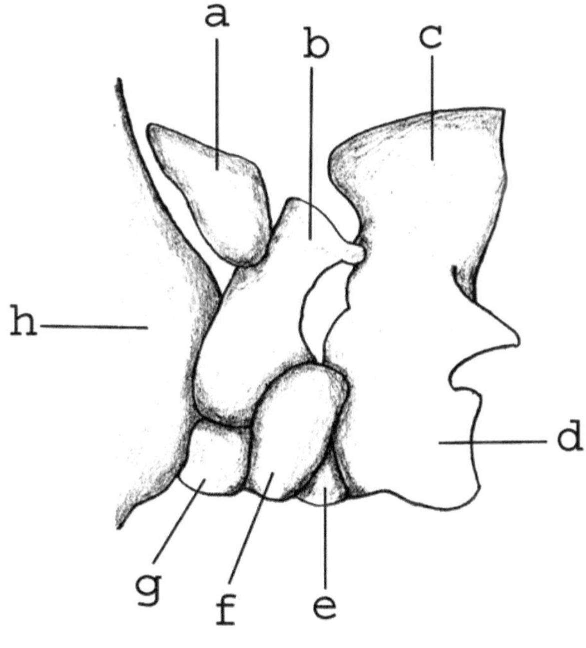

Fig. 4.14 The first two cervical vertebrae in some reptiles. (a) proatlas, (b) neural arch of atlas, (c) neural arch of axis, (d) centrum and (e) intercentrum of axis, (f) centrum and (g) intercentrum of atlas and (h) occipital bone. The proatlas is a vestigial neural arch that earlier in life belonged to a vertebra that was located in front of the atlas. The proatlas articulates with a large facet on the anterior part of the atlas and is found in some extinct amphibians, some reptiles such as crocodiles and embryos of many tetrapods.

embryonic stage fuse to the sides of the vertebrae forming **pleurapophyses**. Although most mammals have 12 to 15 thoracic and 4 to 7 lumbar vertebrae, cetaceans have about 9 thoracic and 20 lumbar vertebrae. The high numbers of lumbar vertebrae that also lack zygapophyses in cetaceans enable these marine mammals to perform dorsoventral swimming movements effectively.

SACRUM

Sacral vertebrae are posterior to the lumbar and anterior to the caudal vertebrae and articulate with the pelvic girdle. Amphibians have single sacral vertebra, reptiles and most birds have two and most mammals have three to five such vertebrae. The sacral vertebrae are fused into a sacrum in species with more than one such vertebra. There is fusion of the last thoracic, all lumbar, all sacrals and first few caudal vertebrae forming a **synsacrum** in birds (Fig. 4.15). The synsacrum provides strong support for onward transmission of forces during bipedal locomotion.

CAUDAL REGION

The last in the series of the vertebral column are the caudal vertebrae (L. *cauda*, tail). The number of caudal vertebrae is reduced in living

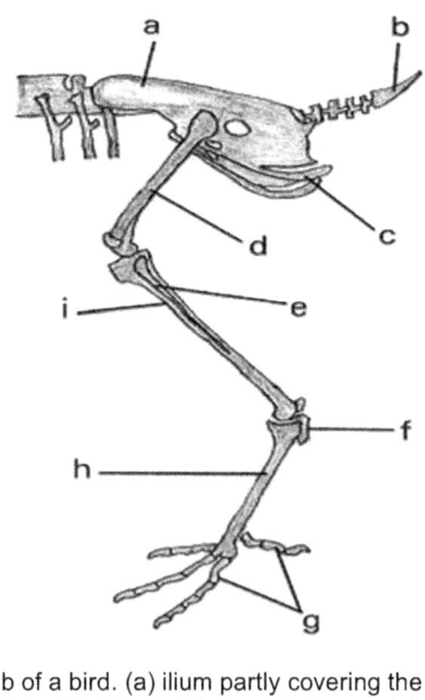

Fig. 4.15 The pelvic limb of a bird. (a) ilium partly covering the synsacrum, (b) pygostyle, (c) pubis, (d) femur, (e) fibula, (f) ridges of tarsometatarsus, (g) digits, (h) tarsometatarsus and (i) tibiotarsus.

tetrapods when compared to their ancestral representatives. Distally, the centra and processes are reduced till the last caudal vertebrae consist of only cylindrical rods of centra. Anurans have a portion of unsegmented vertebral column in the caudal series known as the **urostyle** (Gr. *oura*, tail; *stylos*, pillar). The last few caudal vertebrae (about four to seven) in birds are fused to form the **pygostyle** (Gr. *pyge*. rump) that bears the 'steering' feathers. Between the synsacrum and pygostyle are unfused caudal vertebrae that make it possible for the bird to change the tail position. The number of caudal vertebrae in mammals varies considerably as are functions of the tail. Some apes, including humans, have coccygeal vertebrae (Gr. *kokkyx*, cuckoo) that are a series of three to five fused caudal vertebrae.

EVOLUTION OF VERTEBRAE

The evolution of vertebrae in various vertebrates and regional differences seen in the same vertebrates have been brought about by forces placed on the vertebrate by the environment inhabited such as aquatic or terrestrial and the type of movement displayed. At first, vertebrae evolved as small structures of cartilage or bone that rested on the notochord. Such structures initially acted as points of muscle attachment and did not necessarily support the body. The notochord is still a prominent structure in the vertebral column of some early fishes represented by agnathans, lungfishes, the coelacanths and chondrosteans (Fig. 4.16). These fish are not fast swimmers so the backbone need not be very strong.

Vertebrae have developed from various centers of ossification. These centers have increased in size in higher vertebrates while the size of the notochord has decreased. The early crossopterygian vertebrae had a **hypocentrum** or **intercentrum** (ventral arch base) below the notochord

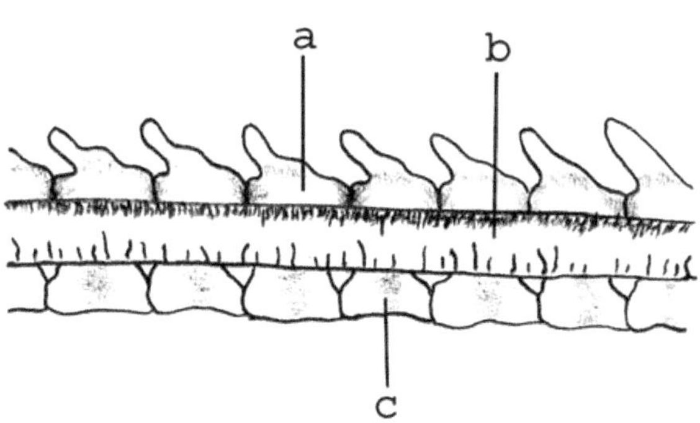

Fig. 4.16 Lateral view of the vertebral column in which a notochord persists throughout life. (a) neural arch, (b) notochord and (c) ventral arch. Fish with such a notochord have simple vertebrae.

and two small **pleurocentra** on the dorsolateral surface of the notochord (Fig. 4.17). Evolution of vertebrae led to an increase in the size of the pleurocentra and a decrease in the size of the hypocentrum with the increasing supportive role played by the vertebral column. The centrum is formed mainly from the pleurocentrum in amniotes while the intercentrum forms part of the interverebral disk whose other part comes from remnants of the notochord in mammals. The centra of cartilaginous fish are frequently strengthened by deposition of calcium salts. Peignoux-Deville et al. (1981, 1982) have demonstrated the presence of lamellar bone tissue containing osteoblasts and osteocytes in the neural arches of dogfish vertebrae and discuss the presence of osseus tissue in elasmobranch endoskeleton in relation to the evolution of the gnathostome skeleton and the endocrinological control of calcium metabolism.

RIBS

The nature of vertebrate ribs, their location and points of attachment vary in vertebrates. Ribs are found in the myosepta of the body trunk and caudal regions. They may be bony or cartilaginous and may be long or short. Ribs normally articulate with vertebrae and may or may not articulate with the sternum. Agnathans lack ribs. Some teleosts have two types of ribs known as dorsal and ventral ribs. **Dorsal** or **intermuscular ribs** develop in myosepta at its points of intersection with the horizontal septum whereas **ventral** or **subperitoneal ribs** are found at points of intersection of myosepta with the peritoneum. Ventral ribs normally encircle the body cavity and only attach to vertebrae. After studying the development of ribs in polypterids (a taxon that plays a crucial role in

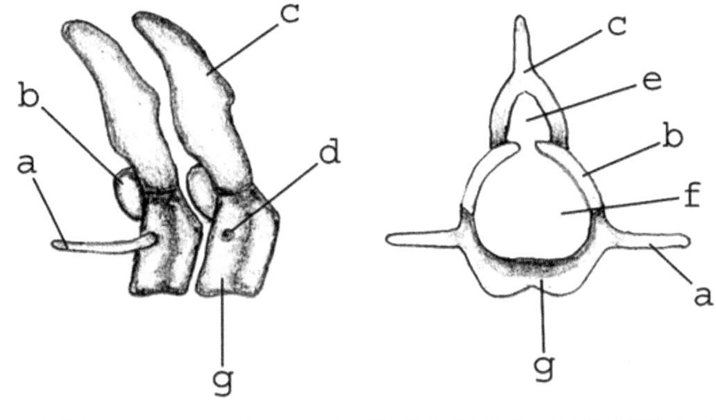

Fig. 4.17 Vertebrae of an early crossopterygian such as a rhipidistian. (a) rib, (b) pleurocentrum, (c) neural arch, (d) parapophysis, (e) vertebral canal, (f) notochordal canal and (g) intercentum.

discussions on rib homology), Britz and Bartsch (2003) argue that dorsal ribs do not exist and ribs of gnathostomes are ventral ribs. Most teleosts have ventral ribs only. Cartilaginous fish have short dorsal ribs that attach to the centra of vertebrae and are found in myosepta.

Tetrapods have ribs that normally articulate with vertebrae in movable joints. Apart from acting as points of muscle attachment, tetrapod ribs play a major role in respiration in most species. Each tetrapod rib has two points of articulation with vertebrae (Fig. 4.18). The **head** or **capitulum** of a rib forms a joint with the **parapophysis** on the centra of two neighboring vertebrae and their intervertebral disk while the **tubercle** or **tuberculum** articulates with the transverse process or **diapophysis**. Early tetrapods had ribs that articulated with vertebrae from the atlas to the end of the body trunk. Later tetrapods have ribs that are restricted to the body trunk or its anterior part. All tetrapod trunk vertebrae with the exception of the most caudal are associated with a pair of ribs each during embryonic development. Later on, some of these may be fused to the caudal trunk vertebrae (in species that lack them in this

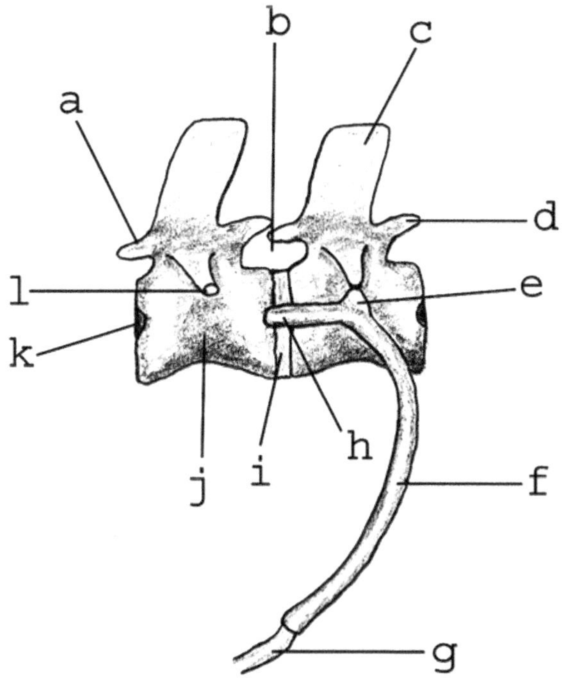

Fig. 4.18 A tetrapod rib and how it articulates with vertebrae. (a) anterior zygapophysis or articular process, (b) intervertebral canal, (c) spinous process, (d) posterior zygapophysis, (e) costal or transverse tubercle or tuberculum, (f) shaft or vertebral rib, (g) costal cartilage or sternal rib, (h) head or capitulum, (i) intervertebral disk or body, (j) centrum, (k) part of parapophysis and (l) facet for costal tubercle on diapophysis.

region) forming enlarged transverse processes known as **pleurapophysis** (Gr. side, rib). Salamanders, caecilians and early frogs have short ribs. In relative terms, ribs do not play a major role in small amphibians.

Tetrapod ribs have several parts depending on their points of articulation. **Vertebral ribs** articulate with vertebrae and **sternal ribs** that are found in birds and mammals articulate with the sternum. **Floating ribs** do not articulate with a sternum even though one is present. Sternal ribs are ossified in birds but remain cartilaginous in mammals. **Uncinate processes** are found on the vertebral ribs of birds. They project from the posterier aspect of ribs and lie against the lateral surface of the next caudal rib thus stabilizing the trunk skeleton further.

STERNUM

Fish and snakes lack a sternum and so were the early amphibians. The sternum shows various degrees of development in tetrapods. Frogs and most salamanders have a small sternum. In amniotes, the sternum is composed of a plate-like structure that is made of cartilage and replacement bone. The sternum of birds, bats and the extinct pterosaurs has a keel.

MEDIAN FINS

Median fins are a group of fins located in the sagittal plane of fish and include **dorsal**, **caudal** and **anal fins** (4.19). Also known as

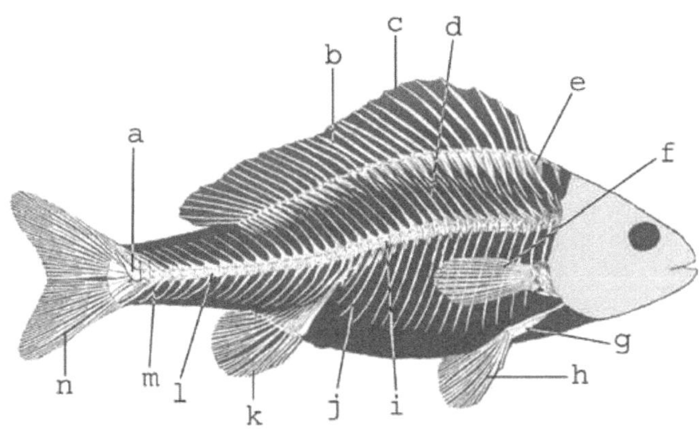

Fig. 4.19 Diagram showing the fins of fish and related structures. (a) urostyle, (b) lepidotrichia, (c) dorsal fin, (d) neural spine, (e) pterygiophore, (f) pectoral fin, (g) basipterygium, (h) pelvic fin, (i) body trunk vertebra, (j) ventral (pleural) rib, (k) anal fin, (l) caudal vertebra, (m) hemal spine, (n) caudal (tail) fin. Caudal vertebrae start from the caudal end of the body cavity.

perissopterygium (Gr. *perissos*, odd; *pteryg*, fin or wing), the median fins are unpaired. The **adipose fin** is also a median fleshy fin without rays that is located behind the dorsal and in front of the caudal fin in primitive teleosts such as salmonids, lizardfishes and characids. Medium fins first appeared in early vertebrates before the origin of the paired fins as is evident from the fossil records. A 530-million-year-old vertebrate that is fish-shaped and characterized by single dorsal, ventral and caudal fins *Haikouichthys ercaicunensis* was found in Chengjiang fauna of China (Zhang and Hou, 2004) and might provide more insight into the early evolution of vertebrates. It has been demonstrated that in all fishes, the exoskeletal fin rays differentiate in the same direction as the endoskeletal support structures and this indicates complete developmental intergration (Mabee et al., 2002).

Median fins are found in all fish groups. Some fish have two dorsal fins and many have an anal fin. Median fins are supported externally by spines or fin rays known as ceratotrichia or lepidotrichia. **Ceratotrichia** (Gr. *kerat*, horn; *trich*, hair) are unsegmented and are found in cartilaginous fish. **Lepidotrichia** (Gr. *lepis*, scale) are found in the median fins of bony fishes and are segmented and often branch distally. In dorsal and anal fins, lepidotrichia attach to rods of cartilage or bone that are located deeper and are known as **pterygiophores** (Gr. *phoros*, bearing) and as a result pterygophores are usually closely associated with neural or hemal spines of the vertebrae.

There are several types of caudal fins in fish (Fig. 4.20). **Homocercal fins** (Gr. *homos*, same; *kerkos*, tail) are superficially symmetrical but the

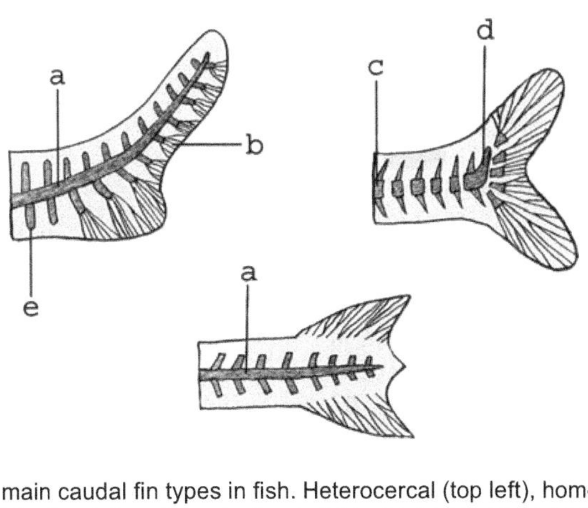

Fig. 4.20 The main caudal fin types in fish. Heterocercal (top left), homocercal (top right) and diphycercal (bottom) fins. (a) notochord, (b) lepidotrichia or fin rays, (c) vertebra (d) urostyle and (e) hemal spine.

skeleton of the vertebral column is slightly tilted upwards at its caudal end as is seen in teleosts. Homocercal fins are supported mainly by enlarged hemal spines that are located behind and below the urostyle known as **hypural bones** (Gr. *hypo,* under; *oura,* tail) while a small part of the fins is supported by a few neural spines referred to as **epiural bones** (Gr. *epi,* upon). Epiural bones are found in front of the urostyle. A **heterocercal fin** (Gr. *heteros,* other) is characterized by superficial assymetry with an enlarged dorsal lobe that results from a considerable upturn of the vertebral axis at its rear end. Heterocercal fins are found in sharks. **Diphycercal fins** (Gr. *diphyes,* two fold) are superficially and internally symmetrical. The vertebral column of a diphycercal fin runs to its posterior end in a straight manner thus dividing the fin into two symmetrical upper and lower lobes as in the case with lungfishes and the coelacanth caudal fins.

Median fins play a role in stabilizing the fish and preventing it from rolling from side to side as well as generating locomotor forces necessary for propulsion through water during steady swimming and maneuvering. The orientation and magnitude of locomotor forces, the mechanisms by which they are generated and how the median fin forces are modulated has been examined in some ray-finned fish clades using Digital Particle Image Velocimetry (Lauder et al., 2002).

APPENDICULAR SKELETON

The appendiculer skeleton (L. *appendo,* to hang something on) consists of the pectoral and pelvic girdles together with the skeleton of the paired fins and limbs. Some vertebrates such as agnathans, apodans, snakes and some lizards lack an appendicular skeleton that is much reduced or modified in others such as cetaceans and seals.

PECTORAL LIMB

Evidence from fossil records of early vertebrates shows that median fins appeared before the paired fins. The origin of the genetic program for fin development could have originated in the midline before being co-opted during the evolution of the paired fins.

Body support was not a major problem to the early and living aquatic vertebrates so their paired appendages were mainly fins. Transition from water to land was accompanied by several changes in the paired fins. The first transition was the evolution of the larger lobed fins that had a skeletal axis and stronger musculature. These fins were further modified into short but powerful legs in the first terrestrial vertebrates. The pectoral and pelvic girdles were also modified and strengthened to support body weight on land. The pectoral girdle that was attached to the skull via the scapula as is seen in crossoptergians which underwent several changes that led to its caudal migration and attachment to the rib cage by muscles in early amphibians. The skull was able to move freely after this shift by the pectoral girdle. Despite the pectoral girdle and skull being dissociated in higher vertebrates, the influence of the cranial neural crest cells in organizing skeletomuscular connectivity has been shown to extend beyond the head into the trunk (McGonnell et al., 2001).

PECTORAL GIRDLE

The pectoral girdle (L. *pectus*, breastbone) is a brace for pectoral (anterior) appendages of vertebrates. The girdle is made up of membrane and replacement bones in bony vertebrates. McGonnell (2001) has reviewed the changes in the pectoral girdle across different vertebrate taxa and indicated, where known, the developmental mechanisms underlying these changes.

FISH PECTORAL GIRDLE

The pectoral girdle of early fishes consisted of the coracoid, scapula and suprascapula (replacement bones) and the dermal bones that included the clavicle, cleithrum, supracleithrum and posttemporal (Fig. 4.21). There was a reduction in the number of these bones in later fish.

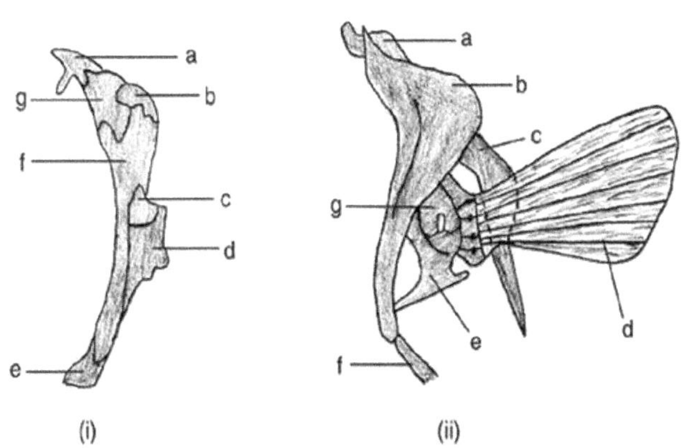

Fig. 4.21 The pectoral girdles of bony fishes. (i) shows the pectoral girdle of a primitive actinopterygian such as *Polypterus*: (a) posttemporal, (b) postcleithrum, (c) scapula, (d) coracoid, (e) clavicle, (f) cleithrum and (g) supracleithrum; (ii) represents a girdle of higher actinopterygians: (a) supracleithrum, (b) cleithrum, (c) postcleithrum, (d) pectoral fin, (e) coracoid, (f) basipterygium of pelvic girdle and (g) scapula.

Although the pectoral girdles of living bony fishes have reduced the sizes of their coracoid and scapula, the size of the cleithrum and supracleithrum has increased. The supracleithrum is connected to the skull by the post-temporal bone. The pectoral girdle of cartilaginous fish (Fig. 4.22) lacks dermal bones.

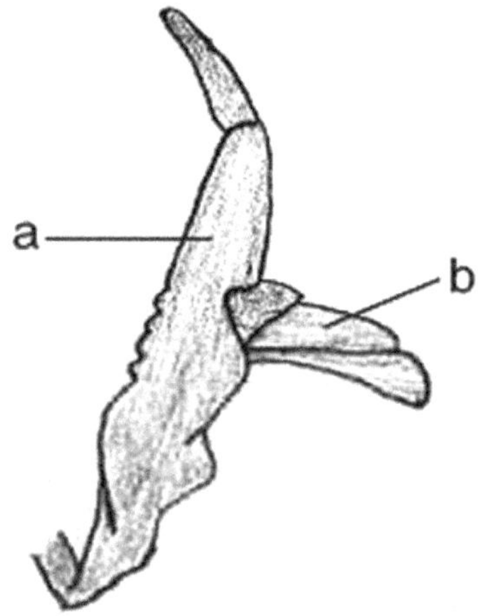

Fig. 4.22 The pectoral girdle of a shark. (a) scapulocoracoid cartilage and (b) a basal pterigophore.

TETRAPOD PECTORAL GIRDLE

The early tetrapods possessed a pectoral girdle that resembled that of early bony fishes in many ways but was lacking the post-temporal bone. The **interclavicle** is a remnant of a large dermal bone and is found in several amniotes such as crocodiles, alligators, birds and monotremes (Fig. 4.23). The bone has replaced the post-temporal bone of early tetrapods. The pectoral girdle of living tetrapods consists of the scapula, clavicle and coracoid (Fig. 4.24). These three bones show varied development in vertebrates depending on the use the pectoral limb has been put to such as gripping, climbing or walking. In vertebrates performing several of these functions, the shoulder requires to be anchored more firmly to the trunk. These bones are thus fully developed in birds and monotremes. The scapula is present in all tetrapods including those with vestigial pectoral limbs. The coracoid and clavicle are reduced or missing in several tetrapods.

PECTORAL LIMBS

Tetrapods normally have five segments in their pectoral (anterior limbs). These segments have been modified by forces of evolution to serve

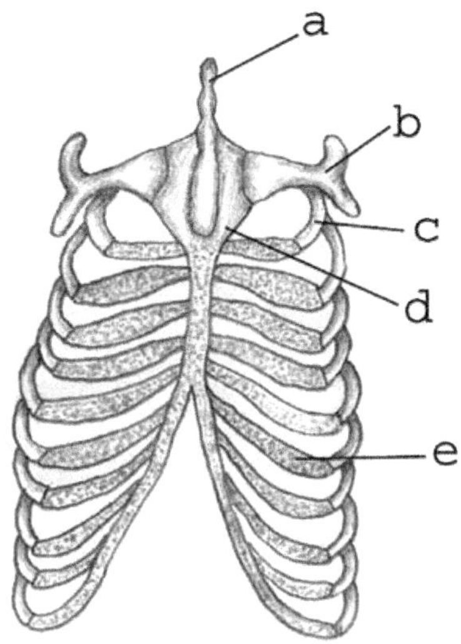

Fig. 4.23 Ventral view of the body trunk skeleton of a crocodile. (a) interclavicle, (b) coracoid, (c) 1st trunk rib, (d) sternum and (e) coastal cartilage of rib.

various functions such as wings for flying, flippers for swimming or arms for performing several functions. Starting from the proximal to the distal end of tetrapod limbs are the following segments: brachium (upper arm) consisting of the humerus, antebrachium (forearm) that includes the ulna

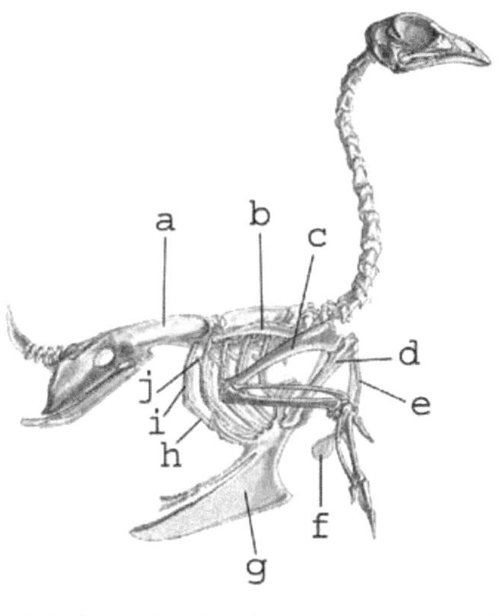

Fig. 4.24 Part of a bird skeleton showing the pectoral girdle. (a) ilium, (b) scapula, (c) humerus, (d) coracoid, (e) clavicle, (f) furcula, (g) sternum, (h) sternal rib, (i) vertebral rib and (j) uncinate process.

and radius, carpus (wrist) made of carpal bones, metacarpus (palm) comprising metacarpals and digits.

The bones of the forelimb possess various structures or processes for muscle attachment that differ in appearance depending on the type of functions that are performed and the demands that are placed on the limb. The typical limb of tetrapods is the **pentadactyl limb** (Gr. *penta*, five) that has five toes. Tetrapods evolved from an ancestor that had limbs with five toes. The pentadactyl limb is an example of a homologous structure of vertebrates. Dinosaurs possessed pentadactyl limbs for the millions of years they lived. Even though the number of digits in different vertebrates may vary from five, vertebrates develop from an embryonic five-digit stage. The pentadactyl limb reaches the highest level of diversity in mammals.

Early tetrapods had short limbs with the first proximal limb segment extending straight from the body in a horizontal manner in relation to the ground. Some living tetrapods such as crocodiles still have such a posture. In higher tetrapods such as birds and mammals, the appendages have undergone rotation leading to the long axis of the humerus running nearly parallel to the vertebral column (Fig. 4.25). The skeleton of the pectoral limb in aquatic mammals has been modified into flippers (Fig. 4.26).

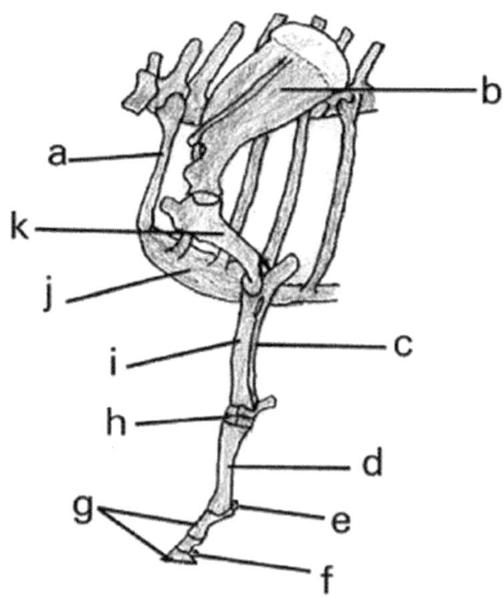

Fig. 4.25 Pectoral limb of a mammal. (a) first rib, (b) scapula, (c) ulna, (d) metacarpus, (e) and (f) proximal and distal sesamoid bones respectively, (g) phalanges (digits), (h) carpal bones, (i) radius, (j) sternum and (k) humerus.

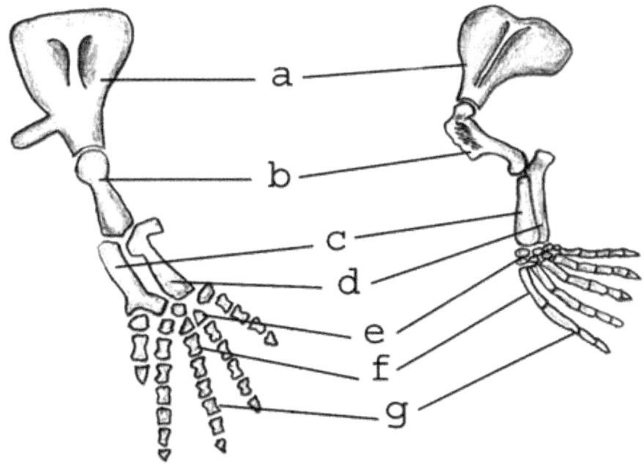

Fig. 4.26 Bones of the pectoral limb in a whale (left) and seal (right). (a) scapula, (b) humerus, (c) radius, (d) ulna, (e) carpals, (f) metacarpals and (g) digits.

WINGS

The forelimb in birds, bats and extinct pterosaurs underwent modifications as an adaptation to flight (Fig. 4.27). The shoulder girdle of birds is well developed and consists of a scapula, a powerful coracoid that forms a movable joint with the sternum and paired clavicles that fuse to form the **furcula** (L. *furcula*, small fork) or the **wishbone**. There is loss of some bones and also fusion of others in the bird forelimb. Fusion of the distal row of carpals with the metacarpals to form the **carpometacarpal bones** and the fusion of carpal bones with each other has left two carpal bones. Carpometacarpal bones have also fused with each other leaving one major carpometacarpal bone and a thin one (third carpometacarpal bone) that is fused proximally and distally to the main carpometacarpal bone. Wings of birds and pterosaurs are supported with limb bones. Bats show some adaptations that are similar to those of birds and pterosaurs. Bat bones tend to be light and slender and the bat radius is shortened and quite thin. Bats also have five digits, four of which are highly elongated (II, III, IV and V) and support the **patagium**. The first digit of the bat wing is small and clawed and is used to climb or walk on the ground.

PELVIC GIRDLE

The pelvic girdle (L. *pelvis*, basin) lacks dermal bones and forms the role of bracing the posterior limbs. The girdle is absent or totally lacking in some vertebrates. Cetaceans have vestigial elements of the girdle embedded in the body wall whereas the girdle is lacking in some salamanders, snakes, manatees and dugongs.

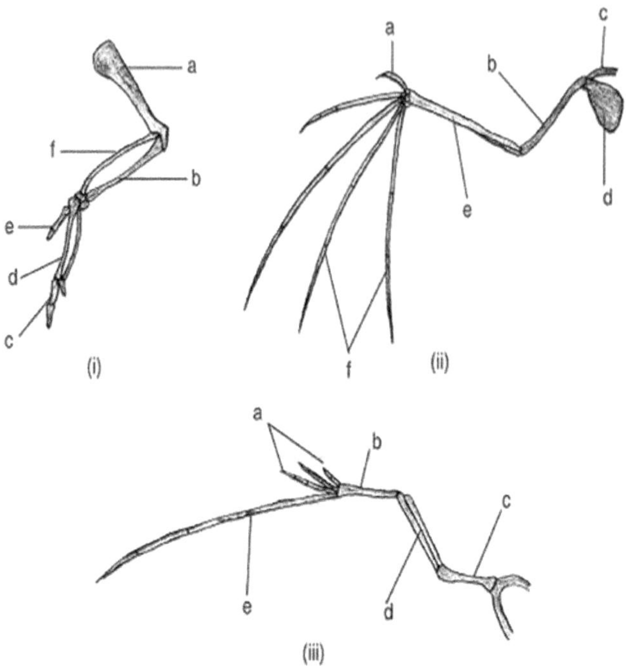

Fig. 4.27 The wing skeleton of (i) a bird, (ii) bat and (iii) the extinct pterosaur. **(i)** (a) humerus, (b) ulna, (c) phalanges, (d) carpometacarpus, (e) digit and (f) radius. **(ii)** (a) first digit, (b) humerus, (c) clavicle, (d) scapula, (e) radius and (f) phalanges that articulate with metacarpals proximally. **(iii)** (a) first three digits, (b) fourth metacarpal, (c) humerus, (d) radius and ulna and (e) phalanges of the fourth digit.

PELVIC FINS

Jawed fishes, with the exception of eels, have pelvic fins. **Agnathans** lack pelvic fins. The location of the fins in fish and their structure varies greatly. The fins are supported by a pelvic girdle that is embedded in the body musculature of fish. The fins are located in the abdominal region in most soft-rayed fishes such as salmonids, cyprinoids and clupeoids or may be located below the pectrorals anteriorly as is the case with spiny-rayed species (acanthopterygians). In some fish such as blennies, the fins are located in a more anterior position below the pharyngeal cavity. When located in an anterior position, the pelvics are connected to the pectorals by a ligament (Fig 4.28). Pelvic fins are modified into **claspers** (Fig. 4.29) that are intromittent organs in cartilaginous fish. Pelvic fins are primarily used for steering fish in a similar manner to rudders in ships. Sarcopterygians have lobed fins that have a skeletal axis that is covered by muscle. Other fish have fin folds that contain basal cartilages and several rows of radials.

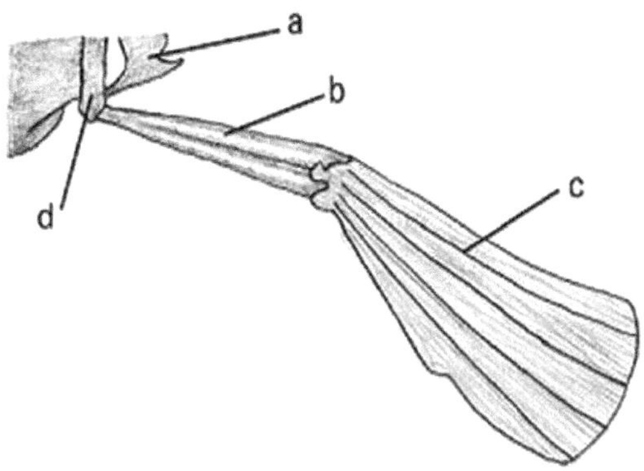

Fig. 4.28 Pelvic girdle of a teleost. (a) coracoid, (b) basipterygium, (c) pelvic fin and (d) supracleithrum.

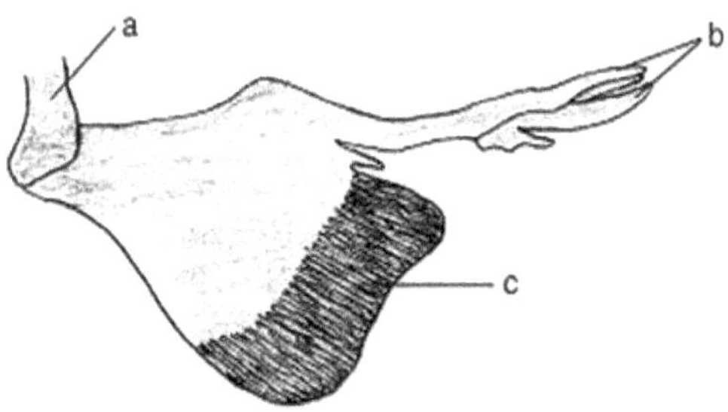

Fig. 4.29 A shark clasper. (a) ischiopubic cartilage, (b) clasper and (c) pelvic fin.

PELVIC LIMBS OF TETRAPODS

The tetrapod pelvic girdle comprises the ilium, pubis and ischium (Fig. 4.30). In amphibians, reptiles and mammals, the pubis and ischium of opposite sides are united at the midventral pelvic symphysis. At the point where the three pelvic bones unite is a socket known as the **acetabulum** (L. vinegar cup) that articulates with the head of the femur.

Birds are bipedal as their pectoral girdles are modified into wings. The pelvic girdle and pelvic appendages in birds are very strong as they support all the weight of the bird and act as shock absorbers during landing. The pelvic girdle of birds is large as a result of their bipedal standing posture. The birds also require a solid connection between the pelvis and the vertebral column and a large surface area for attachment of muscles that have to support the body weight. All the pelvic bones are firmly united in birds. The ilium is elongated and is fused with the synsacrum. The pubis is a thin rib-like bone on the ventral border of the

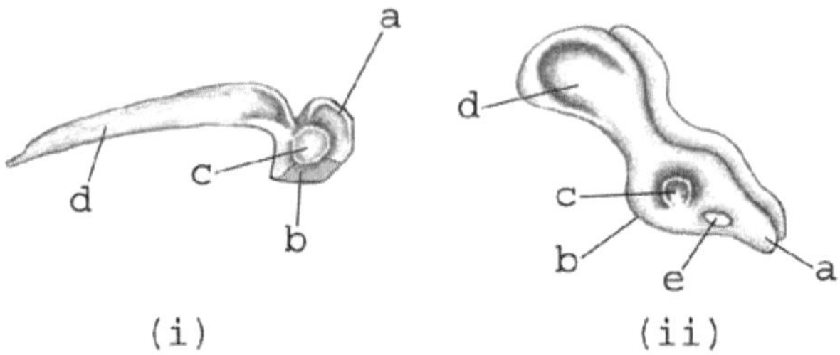

Fig. 4.30 Pelvic girdle of a frog (i) and a mammal (ii) (a) ischium, (b) pubis, (c) acetabulum, (d) ilium and (e) obturator foramen. The pubis of the frog is more cartilaginous than osseus.

ischium. The lack of a pelvic symphysis as a result of non-union of the pubis and ischia from opposite sides has resulted in a pelvic canal that is relatively much wider than is found in other terrestrial vertebrates. The slanting position of the body trunk and the wide pelvic canal allow for the caudal displacement of viscera in birds thus bringing the center of gravity over the hind legs.

The pelvic limbs of **tetrapods** consist of the femur, tibia and fibula, tarsals, metatarsals and digits (Fig. 4.31). The distal end of the tibia has fused with the proximal row of tarsal bones forming **tibiotarsus** in birds.

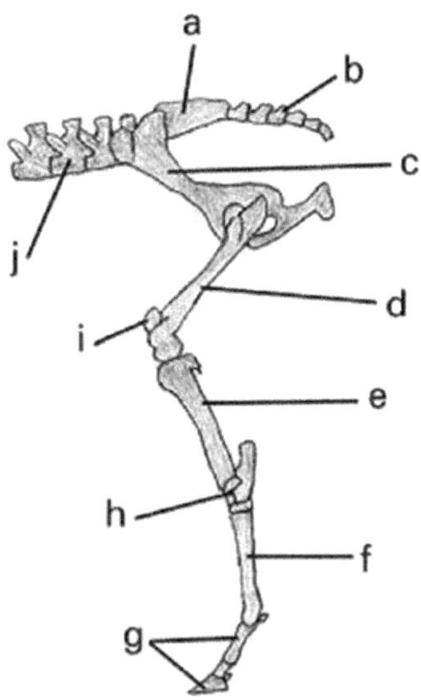

Fig. 4.31 Pelvic limb of a mammal. (a) sacrum, (b) caudal vertebrae, (c) pelvic girdle, (d) femur, (e) tibia, (f) metatarsus, (g) phalanges, (h) tarsal bones, (i) patella and (j) lumbar vertebrae.

The metatarsal bones of birds have undergone fusion with each other and also with the distal row of tarsal bones forming the **tarsometatarsal bone**. The distal bones of the pelvic limbs of tetrapods are comparable to those of the pectoral limbs. In birds and mammals, the long axis of the femur runs in an almost parallel manner to the vertebral column. The number of digits on each pelvic limb of a tetrapod varies from one to five depending on species.

The **digits** of some vertebrates have undergone changes that enable them to perform various functions. The primate hands are capable of grasping due to the presence of an opposable thumb that is at a wider angle from the index finger. The presence of an opposing thumb to fingers is possible due to the presence of a saddle joint between the first metacarpal or metacarpal of the thumb and the trapezius (a carpal bone). The saddle joint also enables the metacarpal perform flexion and extension in one plane as well as abduction and adduction in another plane. The giant panda *Ailuropoda melanoleuca* and the lesser panda *Ailurus fulgens* (Fig. 4.32) are rare and endangered mammals that are able to grab bamboo shoots due to the presence of a modified carpal sesamoid bone or the **radial sesamoid bone** beneath the thumb pad that functions as a sixth digit and is opposable to the other five (Fig. 4.33). The radial sesamoid bone supports the thumb pad or 'false thumb' above it that is more developed in the giant panda.

The fossil remains of a rhipidistian, *Eusthenopteron* show presence of bones in their paired fins that are very similar to the bones of tetrapod limbs. The limbs have a single bone proximally that is similar to the femur or humerus and articulate with paired bones that are similar to the

Fig. 4.32 A giant panda or panda bear (left) and the lesser or red panda (right). Pandas are found mainly in parts of South China, Tibet and Nepal. Male giant pandas weigh up to 115 kg while the lesser panda is slightly larger than the domestic cat and can measure up to 60 cm in length. Pandas are close to both bears and raccoons behaviorally and anatomically. There is much debate among scientists as to where to classify pandas.

Color image of this figure appears in the color plate section at the end of the book.

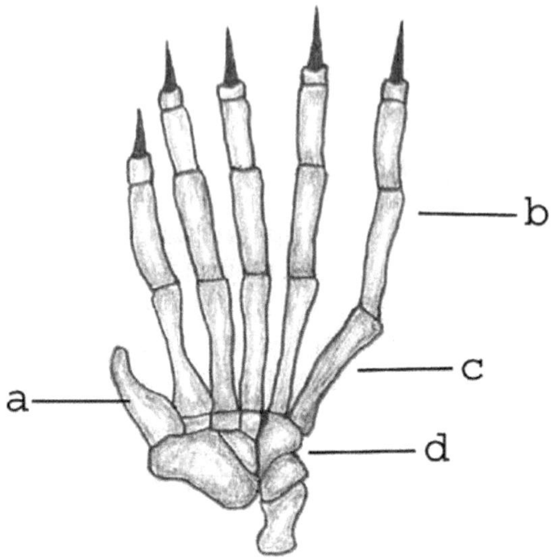

Fig. 4.33 Part of the distal forelimb skeleton of a panda. (a) radial sesamoid bone, (b) digits or phalanges with claws, (c) metacarpal bones and (d) carpal bones. The radial sesamoid bone is found on the medial side and is opposable to the first two digits thus making it possible for pandas to grasp bamboo. To the lateral side is the **accessory carpal bone** that forms a pincer-like structure together with the radial sesamoid bone. The two bones enable the panda to grasp material quite skillfully. The radial sesamoid bone is much smaller in bears.

radius and ulna or tibia and fibula of tetrapods. *Eusthenopteron* could have used such limbs to walk on the sea, lakes or riverbeds. *Eusthenopteron* lacked digits but had fin rays instead.

In a **plantigrade stance** (L. *planta*, sole of foot; *gradus*, step), the whole hand or foot is in contact with the ground as is seen in amphibians, many reptiles and some mammals such as insectivores, bears and apes (Fig. 4.34). The **digitigrade foot** (L. *digitus*, digit) is where the manus (L. hand) and pes (L. foot) are raised leaving toes to make contact with the ground. The first digit is normally lost or reduced. Digitigrades include rabbits, rodents and many carnivores. **Unguligrades** (L. *ungula*, hoof) have reduced the number of digits present in their limbs. The unguligrade foot makes contact with the ground at the tips of the most distal digits and hooves have replaced claws. Unguligrades include ungulates such as ruminants, horses and pigs.

VERTEBRATE JOINTS

Joints (articulations) are points of contact between cartilage or bones in the vertebrate body. During early embryonic development, the

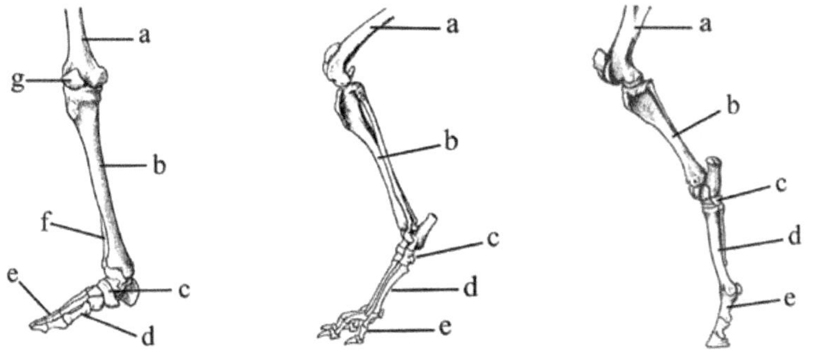

Fig. 4.34 Various stances adopted by mammals as they make contact with the ground in their anatomical position. Plantigrade (left), digitigrade (middle) and unguligrade (right). (a) femur, (b) tibia, (c) tarsal bones, (d) metatarsals, (e) digits, (f) fibula and (g) patella (knee cap). The tips of the distal digits make contact with the ground in ungulates.

cartilaginous or bony parts of the skeleton are united by interstitial tissue. The composition of this tissue determines the degree of movement. Many joints in the vertebrate body permit considerable movement whereas others are completely immovable or permit only limited motion.

Joints are normally classified according to their function or structure. Functionally, joints are divided into three classes according to the degree of movement permitted: **synarthroses** (Gr. *syn*, together; *arthrosis*, articulation) are immovable joints; **amphiarthroses** are slightly movable joints while **diarthroses** are freely movable joints. The structural classification of joints is based on the nature of connective tissue that unites the bones together.

STRUCTURAL CLASSIFICATION OF JOINTS

According to the nature of the tissue present, joints are classified as fibrous, cartilaginous or synovial. In **fibrous joints** (Fig. 4.35), connective tissue unites the articulating bones and permits limited movement in some of the joints while it does not allow movement in most joints. Fibrous joints include **sutures** of the skull, **syndesmoses** (between the ulna and radius and between the tibia and fibula) and **gomphoses** that unite the roots of teeth to bone. **Cartilaginous joints** articulate with each other through hyaline cartilage (synchondroses) or fibrocartilage (sympyses). **Synchondroses** are seen between the epiphyses and diaphyses of long bones of growing vertebrates and between the ribs and the sternum. **Symphyses** are found in areas such as between the centra of the vertebral column that are united by intervertebral disks and between the halves of the pelvic bones (pelvic symphyses). Synovial

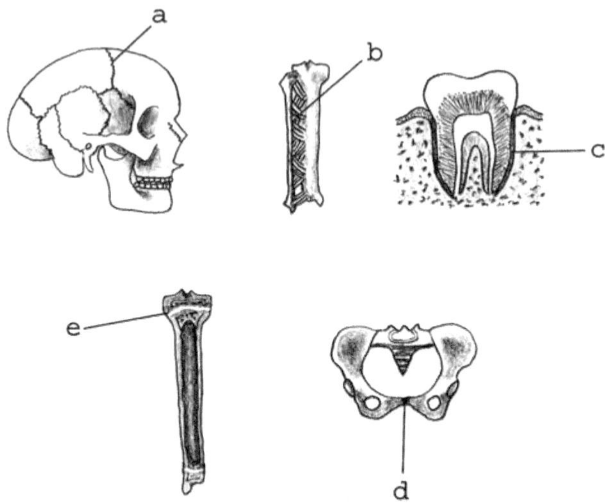

Fig. 4.35 Examples of fibrous and cartilaginous joints. (a) coronal suture between the frontal and parietal bones, (b) syndesmosis between the tibia and fibula, (c) periodontal membrane that anchors a tooth to the bony socket, (d) pubic symphysis and (e) a synchondrosis that is a point of increament in the length of a long bone in immature animals.

joints are the most movable joints in the body. Most joints of tetrapods are synovial joints.

SYNOVIAL JOINTS

Synovial joints (L. *synovia*, joint oil) are freely movable articulations that are anatomically the most complex and most numerous in the vertebrate body. Most joints of the appendicular skeleton are synovial joints. Synovial joints (Fig. 4.36) are characterized by the presence of the following structures: articular cartilage, joint capsule, fibrocartilaginous disks, joint ligaments, joint cavity and bursae.

Articular cartilage is a thin layer of hyaline cartilage that covers and cushions the articular surfaces of bones. The cartilage imparts elasticity and a high degree of adaptability to the articular surfaces and provides the surfaces with the ability to absorb shock. The **joint capsule** is the extension of the periosteam over a **joint cavity** and covers the cavity completely. The capsule consists of an outer fibrous layer and an inner synovial layer that produces the lubricating **synovial fluid**.

Sometimes **fibrocartilaginous disks** are found in synovial joints that are not congruent. These pads of fibrocartilage are known as **menisci** if they partially separate the cavity and are sickle-shaped or **articular**

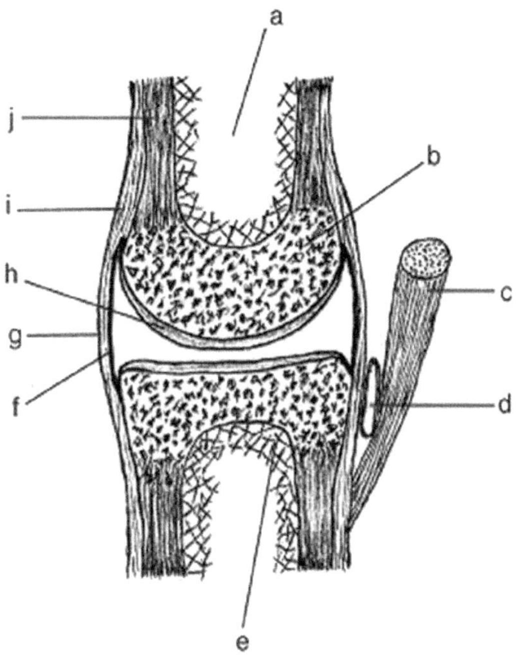

Fig. 4.36 A synovial joint. (a) medullary cavity of a long bone, (b) epiphysis, (c) tendon of muscle insertion, (d) synovial bursa, (e) spongy bone, (f) synovial membrane, (g) outer fibrous membrane, (h) articular cartilage, (i) periosteum and (j) compact bone.

cartilage when they completely divide the joint cavity into two. They also act as shock absorbers and are thought to be remnants of the original interstitial tissue that fills the joint cleft during embryonic development. **Ligaments** are strong cords of dense regular connective that bind bones of joints together. They can be independent or form part of the outer fibrous joint capsule.

Bursae are associated with some synovial joints. They are closed pillow-like structures that contain synovial fluid. Bursae are found between bony prominences and soft tissue such as tendons of muscle insertion. They cushion muscle tendons and joints and also facilitate movement of these structures.

TYPES OF SYNOVIAL JOINTS

The various types of synovial joints present in the vertebrate body are classified according to the shape of their articular surfaces and the main type of movements performed (Fig. 4.37). Mammals have quite a variety of synovial joints that include uniaxial, biaxial and multiaxial joints.

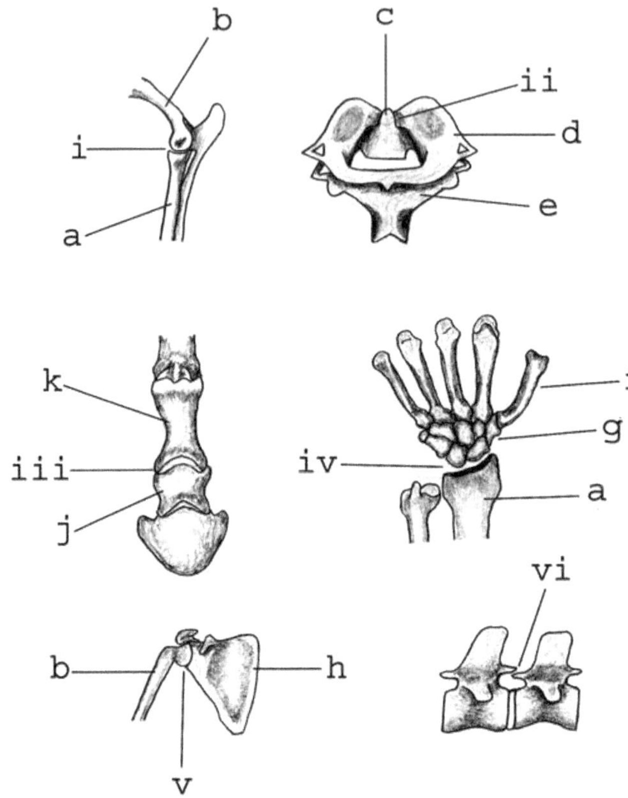

Fig. 4.37 The various types of synovial joints: (i) hinge, (ii) pivot, (iii) saddle, (iv) condyloid, (v) ball and socket and (vi) gliding joints. (a) radius, (b) humerus, (c) dens or odontoid process, (d) atlas, (e) axis, (f) metacarpal and (g) carpal bones, (h) scapula and (j) second and (k) first phalanges.

UNIAXIAL JOINTS

These are synovial joints that perform movements around one axis and one plane. The two types of uniaxial joints present in mammals are hinge and pivot joints. **Hinge** or **ginglymus joints** are those in which the articulating surfaces of the bones form a hinge-shaped unit similar to a common door hinge. Hinge joints permit only flexion and extension movements as is found in the articulating end of the humerus and ulna, the knee and the interphalangeal joints. **Pivot joints** are characterized by a projection of one bone articulating with a ring or notch of another bone. An example of a pivot joint is found between the dens of the axis and the ring shaped vertebral canal of the atlas. This joint permits the sideways movement of the head.

BIAXIAL JOINTS

Biaxial joints permit movements around two perpendicular axes in two perpendicular planes and are represented by the saddle and condyloid joints. In **saddle joints**, the articulating ends of bones resemble miniature saddles that are reciprocally arranged. The surfaces of saddle joints are sagittaly concave but transversely convex and vice versa thus performing flexion and extension as well as sideways movements when collateral ligaments donot impose restrictions. Saddle joints are found between the interphalangeal joints of many mammals and the first metacarpal and the trapezius carpal bone of primates. **Condyloid** or **ellipsoidal joints** are characterized by the presence of a condyle (cylindrical end of some bones) that fits into an elliptical socket. The condyles of the occipital bone form a condyloid joint with elliptical depressions in the atlas, permitting the back and forth movements of the head.

MULTIAXIAL JOINTS

Multiaxial joints show movement around three or more axes and in three or more planes. Examples of synovial joints include the ball and socket and gliding joints. **Ball and socket joints** involve articulation between a ball-shaped head and a concave depression and are the most movable joints in the vertebrate body. The shoulder and hip joints are ball and socket joints. **Gliding joints** are characterized by relatively flat articulating surfaces that allow limited gliding movements along various axes. Gliding joints are found between the zygapophyses of successive vertebrae. As a group, gliding joints are the least movable synovial joints.

REFERENCES

Britz, R. and Bartsch, P. (2003). The myth of dorsal ribs in gnathostome vertebrates. Proc. R. Soc. Lond. B. Biol. Sci., 270 Suppl. 1:51-54.

Dingerkus, G., Seret, B. and Guilbert, E. (1991). Multiple prismatic calcium phosphate layers in the jaws of the present-day sharks (Chondrichthyes: Selachii). Experientia, 47(1):38-40.

Ericsson, R., Cerny, R., Falck, P. and Olsson, L. (2004). Role of cranial neural crest cells in visceral arch muscle position and morphogenesis in the Mexican exolotl, *Ambystoma mexicanum*. Dev. Dyn., 231(2):237-247.

Fleming, A., Keynes, R. and Tannahill, D. (2004). A central role for the notochord in vertebral patterning. Development, 131(4):873-880.

Huysseune, A. and Thesleff, I. (2004). Continuous tooth replacement: the possible involvement of epithelial stem cells. Bioessays, 26(6):665-671.

Johanson, Z. and Smith, M.M. (2003). Placoderm fishes, pharyngeal denticles and the vertebrate dentition. J. Morphol., 257(3):289-307.

Kemp, N.E. and Westrin, S.K. (1979). Ultrastructure of calcified cartilage in the endoskeletal tesserae of sharks. J. Morphol., 160(1):75-109.

Lauder, G.V., Nauen, J.C. and Drucker, E.G. (2002). Experimental hydrodynamics and evolution: Function of median fins in ray-finned fishes. Integr. Comp. Biol., 42:1009-1017.

Mabee, P.M., Crotwell, P.L., Bird, N.C. and Burke, A. C. (2002). Evolution of median fin modules in the axial skeleton of fishes. J. Exper. Zool. (Mol. Dev. Evol.)., 294:77-90.

McGonnell, I.M. (2001). The evolution of the pectoral girdle. J. Anat., 199 (pt. 1-2):189-194.

McGonnell, I.M., Mckay, I.J. and Graham, A. (2001). A population of caudally migrating cranial neural crest cells: functional and evolutionary implications. Dev. Biol., 236(2): 354-363.

Peignoux-Deville, J., Lallier, F. and Vidal, B. (1981). Demonstration of bone tissue in the axial skeleton of a chondrichthian: the dogfish (*Scyliorhinus canicula*). R. Seances Acad. Sci III., 292(1):73-78.

Peignoux-Deville, J., Lallier, F. and Vidal, B. (1982). Evidence for the presence of osseus tissue in dogfish vertebrae. Cell Tissue Res., 222(3):605-614.

Ruhin, B., Creuzet, S., Vincent, C., Benouaiche, L., Le Douarin, N.M. and Couly, G. (2003). Patterning of the hyoid cartilage depends upon signals arising from the ventral foregut endoderm. Dev. Dyn., 228(2):239-246.

Thomason. J.J. and Russell, A.P. (1986). Mechanical factors in the evolution of the mammalian secondary palate: a theoretical analysis. J. Morphol., 189(2):199-213.

Zhang, X.G. and Hou X.G. (2004). Evidence for a single median fin-fold and tail in the Lower Cambrian vertebrate *Haikouichthys ercaicunensis*. J. Evol. Biol. 17(5):1162-1166.

5

Muscular System

The muscular system forms the active part of the locomotor apparatus. There are several hundred muscles present in higher vertebrates. These muscles have evolved to perform general and specific functions that make various vertebrates unique. Having the ability to contract, enables the muscle to perform several functions in the vertebrate body such as movement in the environment and within body systems, heat generation and maintenance of body posture. Muscle also plays a major role in support of body weight by bracing the bones across joints and also in the general appearance of a vertebrate. The muscular activities are coordinated in their actions by the nervous and endocrine systems. Muscles also perform their activities together with accessory structures that include fasciae, bursae and tendon sheaths. Muscle fibers (cells) or myocytes develop from mesenchyme. The fibers have the ability to contract due to the presence of contractile proteins chiefly actin and myosin. The three types of muscle tissue found in vertebrates are smooth, skeletal and cardiac muscles that are associated with body organs, the skeleton and heart respectively.

PARTS OF A SKELETAL MUSCLE

A typical skeletal muscle consists of an active part known as a **belly** and a passive part known as a **tendon** (Fig. 5.1). Tendons serve as points of muscle origin or insertion. The way a muscle attaches to tendon is important in terms of power generation and speed of contraction. Muscles generally attach to bone at an acute angle (tangentially). At the point of attachment to the bone, there is intertwining of the fibers of a tendon with those of the periosteum or the tendon can penetrate bone

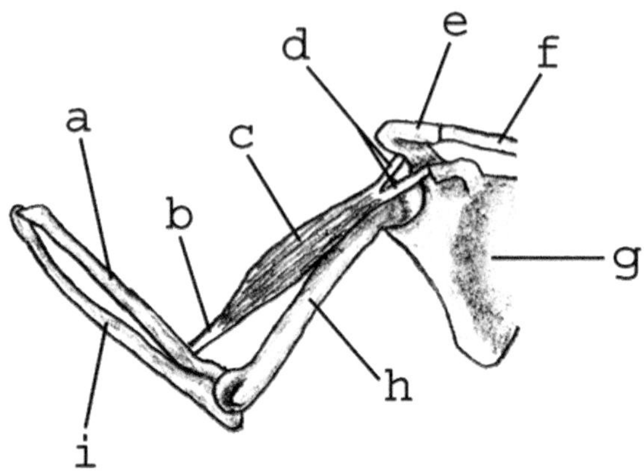

Fig. 5.1 Points of attachment in a skeletal muscle. (a) radius, (b) tendon of insertion of biceps brachii muscle, (c) muscle belly or body, (d) tendons of origin, (e) acromion process, (f) clavicle, (g) costal surface of scapula, (h) humerus and (i) ulna. The origin of a muscle is the point that remains stationary when a muscle contracts whereas the insertion is the part that moves with muscle contraction. Many muscles in the body have more than one point of origin and insertion.

through the periosteum in the form of fibers. A muscle will thus have a head that is the point of origin, a belly that is in the middle and is normally fleshy and a tail that acts as the point of insertion. The point of muscle origin is the proximal end that lies nearer the center of the body axis whereas the point of insertion is the more distal end.

A muscle belly can be more or less spindle shaped or flat. It is surrounded by connective tissue fibers that are mainly composed of collagen and form the epimysium. The collagen fibers are arranged in spiral coils. Muscle bellies are separated from each other by intermuscular connective tissue that facilitates free movement between adjacent muscles. An individual muscle fiber is surrounded with a network of reticular fibers. Perimysial fibers are arranged in spiral coils and surround groups of muscle fibers that form bundles or fascicles. Such an arrangement enables muscle fibers to increase in diameter during muscle contraction. The perimysium has some space that facilitates movement during muscle contraction and is occupied by blood vessels and nerves.

Tendons are normally white and shiny and are attached to ends of muscle fibers. Tendons act as passive tension bands that enable the transmission of muscle contractions to parts they are attached to such as the skeleton and connective tissue elements. Spindle shaped tendons are known as **tendinous cords** and are attached to spindle shaped muscles whereas flat tendons are known as **aponeurosis** and attach to flat muscle bellies. A muscle may have intersecting bands of tendinous tissue within the belly known as **tendinous inscriptions** as is the case with the **rectus abdominis** of mammals. Bundles of tendon fibers (chiefly collagen) are directed along the lines of tension. The tendon fibers are arranged in a parallel manner in short tendons and steep, spiral coils in long tendons. Tendons have great tensile strength and contraction of muscles is not able to break them. Tendons have low elasticity and can be stretched up to 5% of their original length. Tendons show low resistance to compression. There is normally deposition of fibrocartilage or bone known as **sesamoid bone** in parts of tendon likely to encounter a lot of pressure or was once subjected to a lot of tension in the course of evolution.

MUSCLE SHAPES

Muscles have been described according to their various shapes. A tendon normally broadens out, when close to a muscle belly, and covers the belly from outside or invades the muscle fibers during which it may split into several sheets. In **strap-shaped muscles** and **fusiform muscles** (Fig. 5.2a), some of the muscle fibers run in a parallel manner to the long axis of the muscle and to each other and may be as long as the muscles. Strap-shaped muscles have relatively broad points of attachment to tendons whereas fusiform muscle fibers attach to narrow tendons at either end thus concentrating the force of contraction to a smaller area.

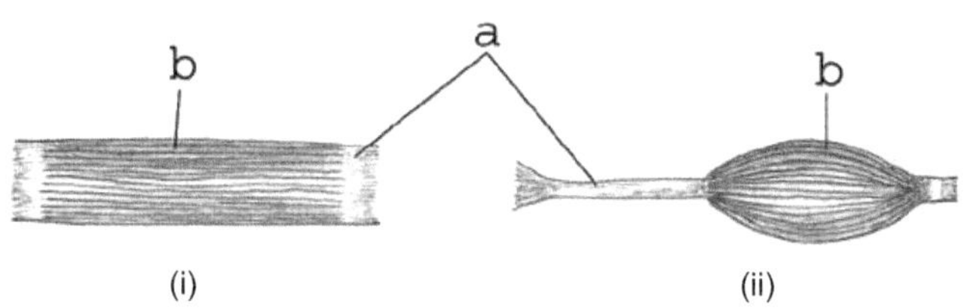

Fig. 5.2a Strap-shaped (i) and fusiform muscles (ii) (a) tendons and (b) muscle belly.

In most cases, muscle fibers are diagonally arranged and attach to tendons at acute angles in a similar manner to the way the barbs of a feather attach to a shaft (Fig. 5.2b). Such an arrangement is known as pennation. In a **unipennate** muscle, there are two opposing tendon sheets that cover a muscle. **Multipennate** muscles are characterized by the presence of more than one tendon sheet at either end for muscle attachment. Examples of multipennate muscles include **bipennate** and **quadripennate** muscles.

An increase in pennation shortens muscle fibers and increases muscular power since there is an increase in physiological diameter of a muscle that results from an increase in the number of muscle fibers

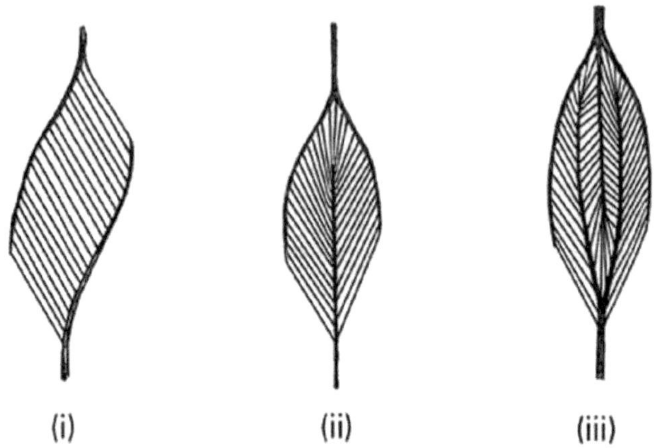

Fig. 5.2b Pennation in skeletal muscle. (i) unipennate, (ii) bipennate and (iii) quadripennate. The thicker lines represent tendon while the thinner ones show the course of muscle fibers.

without enlarging the volume of the muscle. Physiological diameter is measured when all muscle fibers are cut at right angles whereas anatomical diameter is a measure of the greatest transverse diameter of an entire muscle. Greater pennation also increases the duration a muscle can remain in a contracted state though the lifting height is shortened due to the shortening of muscle fibers. A decrease in pennation is accompanied by an increase in the length of muscle fibers. Such muscles shorten greatly during contraction and show greater movement though they fatigue rapidly as they expend a lot of energy.

INNERVATION OF MUSCLE

Nerves that supply skeletal muscle contain motor and sensory fibers as well as autonomic fibers. **Motor nerves** originate from the central nervous system and carry stimuli for muscle contraction. Motor nerves terminate at the **motor end plates** on the sarcolemma of muscle fibers. **Sensory fibers** originate on tendon organs and the intrafusal fibers of muscle spindles and transmit impulses to the central nervous system. **Golgi tendon organs** or tendon organs (Fig. 5.3) are a group of tendon fibers that are encapsulated by nerve endings and are found in mammalian tendons. The nerves entwine the tendon fibers and are arranged in series in relation to the fibers. Tendon organs act as mechanoreceptors and are sensitive to mechanical distortion such as tension changes brought about by muscle contraction and can detect the active muscles and magnitude of the forces generated in the process. Tendon organs are also able to sense passive tension changes in muscle

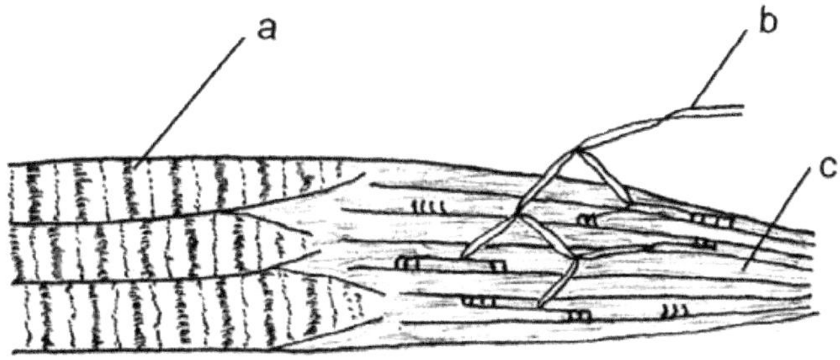

Fig. 5.3 Golgi tendon organs. (a) skeletal muscle fiber, (b) sensory neuron and (c) tendon fibril. These proprioceptors are mechanoreceptors that provide information about limb position and movement.

that could result from muscle damage due to eccentric contractions (Gregory et al., 2003).

Muscle spindles (Fig. 5.4) play a proprioceptive role such as sensing limb position as the passive stretching of their **intrafusal fibers** results from muscle stretch and is detected by their sensory nerve endings. The information is transmitted to the spinal cord that will initiate nerve reflexes that will increase motor impulse stimulation of the **extrafusal**

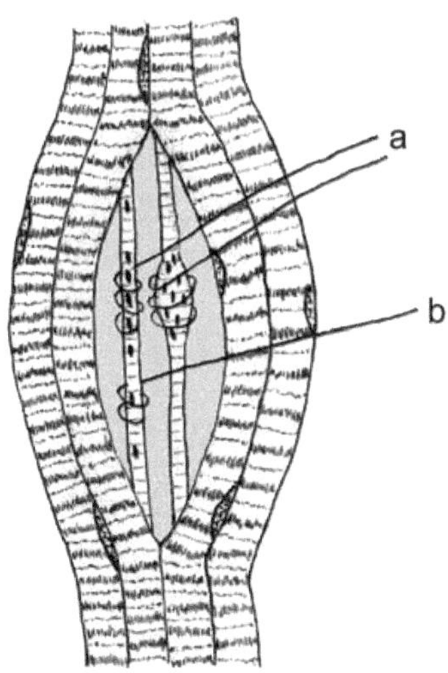

Fig. 5.4 A muscle spindle is located between muscle fibers and consists of 2 to 10 specialized intrafusal muscle fibers of two types: nuclear bag and nuclear chain. Nuclear bag fibers are thicker in the middle than the rest of the fibers. The annulospiral primary sensory endings (a) originate from the central (equatorial) region of both nuclear bag and chain fibers whereas the secondary sensory endings (b) do so from nuclear chain fibers. Motor nerves supply muscle spindle fibers at either side of the central region and lead to contraction of these fibers on stimulation. A muscle spindle is surrounded with a connective tissue envelope.

fibers. The extrafusal fibers contract, shortening the muscle and compensating for the stretch. Autonomic nerves originate from the central nervous system and innervate the blood vessels thus regulating blood supply to muscle.

MOTOR UNIT

A motor unit (Fig. 5.5) comprises all muscle fibers that are innervated by branches of the same motor neuron. When a motor neuron conducts an action potential, all fibers of a motor unit will contract simultaneously as they have similar characteristics. As most vertebrate muscles are activated in an 'all-or-none' manner whereby there is no part activation, increase in muscle contractile force results from recruitment of more motor units. The number of fibers supplied by a motor neuron varies depending on type of movements performed. Few fibers are normally supplied by a motor neuron if they perform precise movements such as those of the extraocular muscles of the eye and the hand muscles of primates. In muscles that are adapted to performing powerful movements such as the large hind limb and abdominal muscles that do not produce any precise movements apart from performing course control mainly, a motor neuron may supply several hundred muscle fibers.

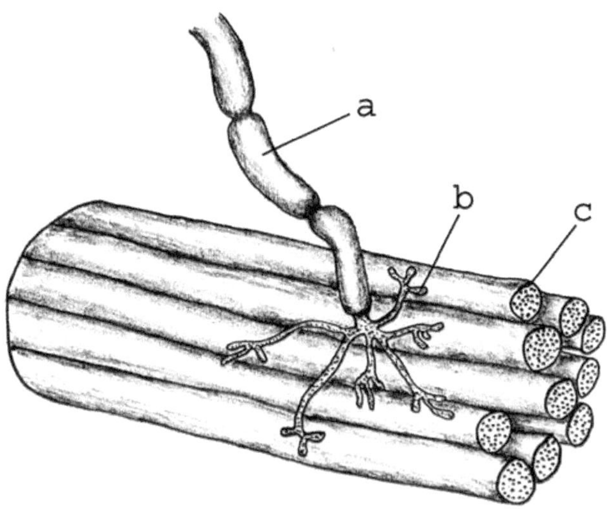

Fig. 5.5 A motor unit. (a) motor neuron, (b) neuromuscular junction and (c) muscle fiber. All the muscle fibers and the motor neuron that supplies them form a motor unit.

BLOOD SUPPLY TO MUSCLE

Striated muscle has a rich blood supply and in cardiac muscle the ratio of capillaries can be higher than that of muscle fibers. Slow phasic (red) fibers have a much higher blood supply than fast phasic (white) fibers. Krogh (1919) described the arrangement of blood vessels in skeletal muscle. Small arteries generally run parallel to the course of muscle fibers. The arteries eventually give rise to arterioles more or less at right angles and generally at a constant rate. Arterioles penetrate between the fibers and each divides into two capillaries that run along muscle fibers in opposite directions (Fig. 5.6). Two capillaries from opposite directions meet to give rise to a venule. The venules run in a perpendicular manner to the muscle fibers and will eventually join small veins. The small veins run together with corresponding small arteries along the muscle fibers. Blood tends to flow in the same direction in adjacent capillaries (Eriksson and Myrhage, 1972). Although blood capillaries tend to be mainly parallel to the course of muscle fibers, some muscles have tortuous blood vessels. The cat soleus muscle shows one of the highest tortuosity levels of 12% (Mathieu et al., 1983).

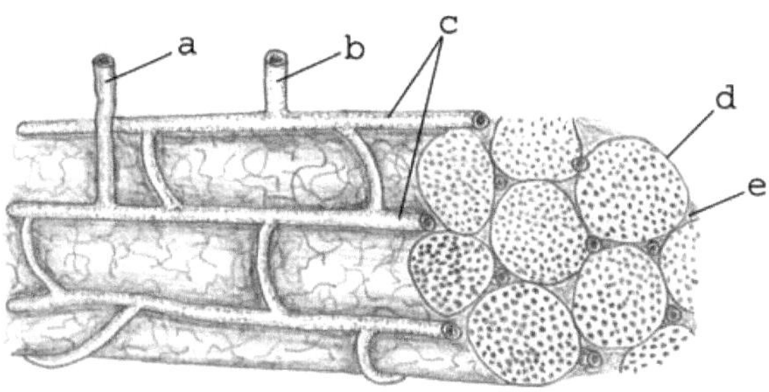

Fig. 5.6 Blood supply to skeletal muscle fibers. (a) venule, (b) arteriole, (c) capillaries, (d) muscle fiber and (e) endomysium. In the human being, 20% of the cardiac output goes to skeletal muscle where the muscle accounts for about 40% of the body weight.

During muscle contraction, capillaries tend to be more tortuous in relation to muscle fibers and such an arrangement increases the ratio of capillaries to fibers (Mathieu-Costello, 1987, Mathieu-Costello et al., 1988). An increase in tortuosity during muscle contraction increases capillary length and surface densities in relation to muscle fiber volume. There is dilation of capillaries and an increase in flow of blood during skeletal muscle contraction (Krogh, 1922). Muscle capillaries remain patent during contraction despite the compressive stress in a direction that is perpendicular to the course of muscle fibers and capillaries (Fung, 1977).

ACCESSORY STRUCTURES ASSOCIATED WITH MUSCLES

Accessory structures are vital to the proper functioning of some skeletal muscles as they support and facilitate the function of such muscles and their respective tendons. Accessory structures associated with muscle include fasciae, synovial bursae and tendon sheaths.

Fasciae are tough and thin connective membranes (sheets) that surround muscles, bones and joints and are mainly composed of collagen and elastic fibrils (dense connective tissue). Fasciae not only support and provide protection but also give structure to the body. The color of fascia depends on its fiber composition. Fasciae run mainly obliquely to the course of muscle fibers. The body surface is covered by superficial and deep fasciae. **Superficial fascia** lies directly under the hypodermis and contains fat and water. In some areas of the mammalian body, superficial fascia contains thin striated muscle known as **cutaneous muscle** that causes skin movement in local parts of the body. **Deep fascia** is located below superficial fascia and assists in muscle movement since it can surround a muscle or a group of muscles or may divide into several layers that give rise to inter-muscular septae. This fascia is normally thick and tougher and is a site of muscle attachment in certain parts of the body. In some areas of the body, deep fascia is rich in elastic fibers such as the yellow abdominal tunic of some mammals. Body cavities are surrounded by **subserous fascia** that is located between deep fascia and membranes that surround cavities of the body. Spaces are found between deep and subserous fasciae. Such spaces allow for ease of movement of internal body organs.

Synovial bursae (Fig. 5.7) are small and enclosed fluid-filled structures in loose connective tissue. They contain synovial-like fluid and are located between the bone and the overlying movable or tightly stretched muscle, tendon, ligament, fasciae or skin. Friction between the moving surfaces is reduced due to the presence of bursae. Hundreds of bursae are found in various parts of the body. The wall of a bursa comprises a tough outer fibrous membrane and an inner synovial membrane that often contains villi. The presence of bursae in various parts of the body depends on usage, age and nutritional status.

Tendon sheaths (Fig. 5.8) are tube-like structures that surround tendons in areas they are under high tension such as lying over bone. Tendon sheaths are filled with synovial fluid. Their walls have fibrous and synovial layers. The presence of synovial fluid enables the smooth gliding movements of tendons. Tendon sheaths leave a small area where they do not surround tendon known as **mesotendon**. Mesotendon is a double layer of tissue that has blood vessels and nerves.

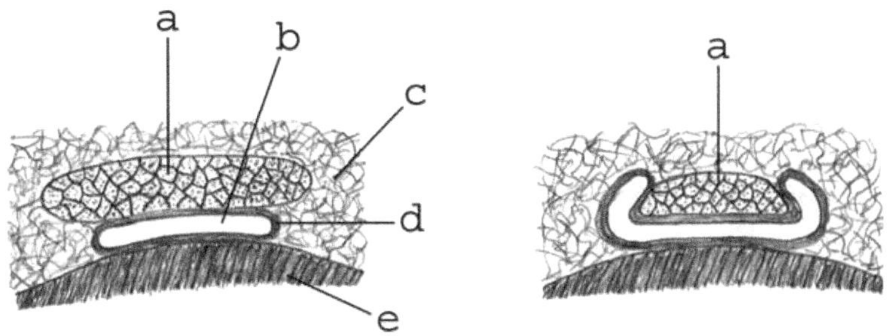

Fig. 5.7 Diagram of a cross section through subtendinous bursa (left) and enveloping bursa (right). (a) tendon, (b) space filled with synovial-like fluid, (c) loose connective tissue, (d) fibrous layer and (e) supporting structure. The space in a bursa is surrounded by a synovial membrane.

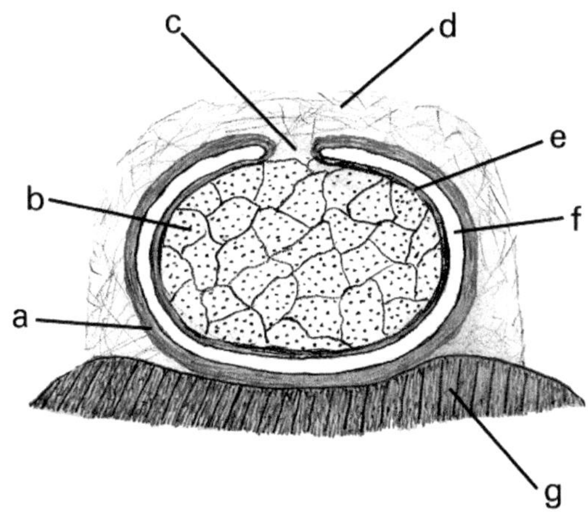

Fig. 5.8 Tendon sheath. (a) parietal fibrous layer, (b) tendon fiber bundle that is surrounded by peritendineum, (c) mesotendineum that contains blood vessels and nerves, (d) loose connective tissue, (e) epitendineum that surrounds the entire tendon, (f) cavity of tendon sheath with synovial fluid and (g) hard structure such as bone.

NAMING OF MUSCLES

Skeletal muscles are normally named according to the mode of action and appearance as well as their points of origin and insertion. Muscles that move a distal limb segment toward a more proximal position (flex joints) are known as **flexor** muscles. **Extensor** muscles bring about action that is antagonistic to that of flexor muscles i.e. their action will widen the angle between two limb segments. Contraction of flexors and extensors simultaneously will fix a joint. Contraction of **adductors** moves a body part nearer the median plane whereas **abductors** will perform an opposite movement. **Rotators** cause movement of a bone around the longitudinal axis. Muscles that cause a limb to rotate forwards and inwards are known as **pronators** while their antagonistic counterparts are referred to as **supinators**. Muscles that support each other to perform a similar action are known as **synergists** (Gr. *syn*, together, *ergon*, work) whereas those that oppose each other in action are **antagonistic**. The action performed by **dilators, levators, depressors** and **tensors** is self-explanatory.

Muscles that have two, three and four heads are referred to as **biceps, triceps** and **quadriceps** respectively. A **digastric** muscle has an intervening tendon that divides it into two bellies. **Sphincters** are circular or ring-shaped muscles that are found at natural body orifices. Muscles that lack a skeletal attachment such as the striated muscles of internal organs are known as **intrinsic** muscles.

MUSCLE CONTRACTION

Stimulation of a skeletal muscle fiber by a motor neuron will result in a series of events that will bring about contraction. Impulses reaching the **motor end plates** (Fig. 5.9) that constitute the terminal ends of motor neurons on the sarcolemma will cause release of the neurotransmitter substance acetylcholine that will diffuse across the synaptic cleft to the sarcolemma of the adjacent muscle fiber where it will stimulate acetylcholine receptors. This action will initiate an electrical impulse in the sarcolemma that will spread to the T-tubules and sarcoplasmic reticular sacs. The impulse will initiate release of calcium ions from the sarcoplasmic reticulum into the sarcoplasm. Calcium ions will bind to the

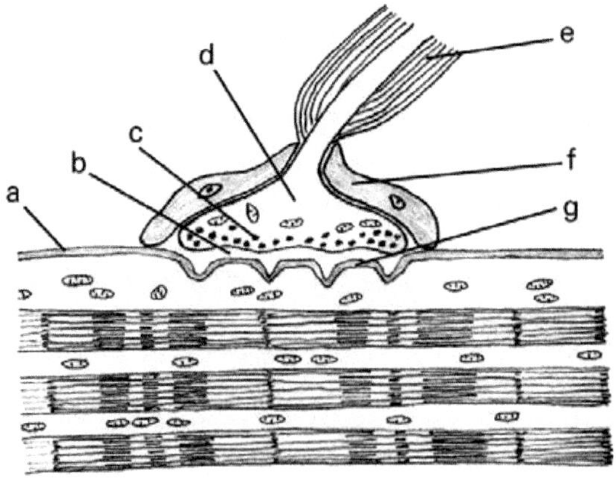

Fig. 5.9 Neuromuscular junction. (a) sarcolemma, (b) synaptic cleft, (c) synaptic vesicle, (d) motor neuron fiber, (e) myelin sheath, (f) Schwann cell and (g) motor endplate.

troponin molecules in the thin filaments. Troponin normally holds tropomysin strands in a position that shields the chemically active sites of actin. Binding of troponin by calcium ions leads to a shift in the position of tropomysin thereby exposing the active sites on actin molecules. Myosin cross-bridges of the thick filament will bind to the active sites. Myosin cross bridges will bend with great force, pulling the thin filaments and Z-lines longitudinally towards the center of a sarcomere. Each head will free itself and attach to the next active site before bending again. The eventual result is the shortening of the entire myofibril and muscle fiber in this sliding filament theory.

During **muscle relaxation**, calcium ions are actively pumped back into the sarcoplasmic reticulum in a process that takes a few milliseconds. The active transport carriers of the sarcoplasmic reticulum have greater affinity for calcium ions than troponin molecules. Without calcium bound troponin, tropomyosin is free to cover the active sites on the thin filaments. Forces external from muscle will then return muscle to its longer relaxing length.

The energy required during muscle contraction is provided by the breakdown of the energy rich bonds of adenosine triphosphate (ATP). Breakdown of glucose provides the energy that is used in the synthesis of ATP. During aerobic (oxygen requiring) respiration, glucose is broken down to carbon dioxide and water. Aerobic respiration ensures the production of the maximal amount of energy from a glucose molecule. Anaerobic respiration is a rapid process that takes place in the absence of oxygen. During the process, glucose is broken down to lactic acid. Anaerobic respiration does not lead to production of as much energy as aerobic respiration per molecule of glucose since a lot of energy is still stored in lactic acid bonds. Anaerobic respiration is an important catabolic process in muscle fibers that have a low blood supply and also those that generate great force rapidly.

RIGOR MORTIS

During rigor mortis (L., stiffness of death), there is stiffness of some postural skeletal muscles. Rigor mortis is observed sometimes shortly after death and is due to the presence of intact myosin and actin cross-bridges. After death, there is depletion of ATP that is necessary for the proper functioning of the calcium ion pump in the sarcoplasmic reticulum that pumps the ions back to this organelle from the sarcoplasm. The calcium ion concentration of the sarcoplasm remains high thus saturating the binding sites on troponin molecules. Since the active sites on the thin filaments are exposed as a result of a shift in the position of tropomyosin, there will be sustained binding of many actin sites by myosin cross-bridges. Skeletal muscle will exhibit sustained contraction in the process. Actin and myosin will remain bound during rigor mortis until muscle decomposition occurs as a result of tissue decay and digestion by lysosomal enzymes.

TYPES OF MUSCLE CONTRACTION

Vertebrate skeletal muscle shows different types of contraction depending on the fiber types present and also the type of movement performed. The amount of energy consumed also varies depending on the type of contraction. The three main types of muscle contraction are isometric, isotonic and negative work contractions.

Isometric contractions (Gr. *isos,* equal; *metron,* measure) are characterized by an increase in tension of muscles although there is no change in muscle length. Such contractions do not expend as much energy as isotonic contractions (Abbot, 1951). Initiation of an isometric contraction expends more energy than is later on needed for maintenance of the contraction. The slow and tonic muscle fibers show isometric contractions and are slow contracting in a process that saves a lot of energy as the cross-bridge cycle is longer when compared to that of fast fiber types.

During **isotonic contractions** (Gr. *tonos,* tension), there is development of force and shortening of muscles. The vertebrate species and type of muscle involved during isotonic contraction will determine contractile efficiency. Slow muscles perform more work than fast muscles per molecule of ATP degraded so they have a higher maximum efficiency. When the optimum velocity of contraction is exceeded, energy utilization decreases as the thin filaments are moving too fast for enough cross-bridges to make contact and complete their cycle.

Negative work contractions occur during muscle stretching while at the same time force or tension is developed. Negative work contractions occur when an animal walks downhill during which muscles will develop more tension than would be the case by some muscles during an isometric contraction.

Striated vertebrate muscle can shorten by up to 60% of its stretched length. Longer muscles have greater contraction distances. Shortening comes to an end when the myosin filaments make contact with the Z-lines. Smooth muscle is capable of much more shortening in comparison to the other muscle types since the thin filaments can form cross-bridges with several myosin filaments in succession.

SKELETAL MUSCLE FIBER TYPES

Skeletal muscle fibers have been classified according to functional and structural characteristics. The types present in vertebrates include tonic, slow phasic (red), fast phasic glycolytic (white) and the intermediate fast phasic oxidative (pink). These fiber types have distinct locations in fish while they occur in different proportions in muscles of other vertebrates depending on the type of contractions and movements that are performed.

Tonic muscle fibers are very slow contracting fibers that perform a continuous and partial contraction in a muscle. As they do not produce a significant response when stimulated with a single stimulus, tonic fibers do not bring about noticeable contractions and movement, as only a few fibers contract and cause tautness. Tonic fibers multiply innervated and show a graded response to stimulation of different frequencies that bring about contraction in relays by groups of fibers in a muscle. Tonic fibers are important in maintenance of posture as they can sustain isometric contractions at a low energy cost due to their long cross-bridge cycles. Muscle tone is maintained by negative feedback mechanisms centered in the spinal cord.

Slow phasic (red) fibers contain a high content of the oxygen carrying pigment in muscle known as myoglobin. Slow fibers have thick myofilaments that are made of a type of myosin that reacts at a slow rate. Due to their slow contraction, the fibers are able to synthesize ATP at a rate that meets the energy demands of myosin and avoid fatigue. Production of ATP is enhanced by the presence of a high density of mitochondria, high content of myoglobin and high density of capillaries. There is complete breakdown of glucose to carbon dioxide, water and energy in slow fibers. These fibers are suited for contractions in postural muscles.

Fast phasic (white) fibers have a faster type of myosin and a sarcoplasmic reticulum and T-tubule system that is efficient at delivering calcium ions to the sarcoplasm. The fibers contain little myoglobin and due to their high rate of contraction deplete ATP rapidly. Although the fibers are rich in glycogen, they have to depend on anaerobic respiration for regeneration of ATP since their mitochondrial volume density is low. Anaerobic respiration yields about 5% of the energy stored in the chemical bonds of glucose molecules. Fast fibers cannot sustain a contraction for long as a result of the low energy (ATP) available and accumulation of lactic acid resulting from anerobic respiration. Fast muscle fibers are abundant in muscles that generate great force quickly but not for long such as the bulk of fish musculature and muscles that move fingers in primates.

Intermediate phasic (pink) fibers possess characteristics between those of fast and slow fibers. They are similar to fast phasic fibers save for the high density of mitochondria. Intermediate fibers are more fatigue resistant than fast fibers and can contract with greater force more rapidly than the slow fibers. These fibers are predominant in muscles that perform postural support and can also generate rapid and powerful contractions such as the soleus of mammals that supports the hind limb and can also be used in walking, running and jumping.

DEVELOPMENT OF VERTEBRATE MUSCLES

Apart from the small muscles within the iris that develop from the same ectodermal tissue (the optic cup) that gives rise to the iris and the retina, all muscular tissue in vertebrates develops from mesenchyme of the mesoderm. The visceral muscles contribute to most of the musculature of organs including the heart and blood vessels. Most of the somatic muscles are found in the body wall and the appendages and develop from somites.

All visceral muscles are supplied by the autonomic nervous system and apart from cardiac muscle are all smooth muscles. These muscles are derived from the lateral plate mesoderm that extends laterally and ventrally between the archenteron and the surface ectoderm (see Chapter 9).

The somatic muscles are normally divided into the centrally occurring axial and the laterally located appendicular muscles. Axial musculature is divided according to the region of the body it arises from into extrinsic ocular, branchiomeric, epibranchial, hypobranchial, epaxial and hypaxial muscles.

Rostral to the ears of amniotes is condensed mesoderm known as somitomeres that is divided into several groups by grooves and lies on each side of the neural tube. The first three and the fifth masses give rise to the **extraocular** (extrinsic) muscles of the eye. **Branchiomeric** muscles comprising mandibular, hyoid and branchial muscles develop from the fourth and the other postotic mesodermal masses. In all these muscle groups, muscles that develop from each of the masses of condensed mesoderm is supplied by a specific cranial nerve. The more cranial myotomes give rise to the **epibranchial** muscles located above the gill region in fish and the **hypobranchial** muscles below the same region in vertebrates. The rest of the myotomes give rise to the muscles of the body trunk and tail (epaxial and hypaxial musculature).

Body trunk myotomes of gnathostomes are divided by a longitudinal **horizontal septum** in the region where the transverse processes of the vertebrae will develop. It extends from the vertebral column to the body below the skin. The septum divides the trunk musculature into **epaxial** muscles above and **hypaxial** muscles below the septum. Epaxial muscles are innervated by dorsal rami (L. *rami*, branch) whereas hypaxial muscles are innervated by the ventral rami of spinal nerves. During muscle development, each myotome increases in size and divides into the smaller group of epimeres and the larger group of hypomeres. Each spinal nerve grows into its respective divisions. These divisions of nerves establish permanent connections with their respective column of myotomes. Epaxial myotomes then divide further into deep and superficial portions. The deep portions generally retain their segmented nature and will give rise to the short intervertebral muscles though some will fuse over a few consecutive segments. The superficial portion of epaxial somites undergoes fusion over several segments and by splitting longitudinally will give rise to the long extensor muscles of the neck and trunk in higher vertebrates. Regional hypaxial somites will spread ventrally and laterally in the body wall to develop into muscles of the shoulder girdle, scalenus, the wide muscles, the thorax and abdomen, the rectus abdominis and muscles of the pelvic girdle.

Appendicular musculature in fishes develops from migration of hypaxial myotomal masses into the relevant locations. In higher vertebrates such as birds and mammals, limb muscles develop from pre-existing local mesoderm that is located within the developing limb buds. The mesoderm thickens then fuses. Degeneration of mesodermal masses occurs in some areas during which the masses are molded into certain units. Branches of spinal nerves extend into these premuscle masses and retain these early connections as the limbs elongate and the muscles differentiate. The muscles then develop into dorsal limb extensors followed by ventral limb flexors. Anterior and posterior branches of spinal nerves supply the extensor and flexor groups of muscles respectively. Development of forelimb muscles normally precedes that of the hindlimb.

During formatin of **skeletal muscles fibers**, myoblasts proliferate and start synthesizing myosin, actin and other muscle proteins. Myoblasts also fuse into multinucleated cells known myotubules that continue synthesizing contractile proteins of muscle and their assembly into myofibrils. At the beginning, myofibrils have very few myofilaments but more are added at the periphery of growing myofibrils until the mature state is attained. The myotubules give rise to the skeletal muscle fibers.

Post-embryonic increase in size of skeletal muscles in lower vertebrates is by hyperplastic and hypertrophic development whereas this process is by hypertrophy in higher groups. Hypertrophy mainly involves an increase in the density of myofilaments.

Most **smooth** muscle develops from the splanchnic mesoderm that migrates to the wall of different hollow organs. Mesenchymal cells present undergo elongation and aggregate. **Cardiac** muscle also develops from the splanchnic mesoderm that surrounds the primitive cardiac tube. The mesoderm thickens later to form a wall that will then transform into the myocardium. Myofibrils in cardiac muscle fibers develop first at the periphery before spreading to other parts of the cells.

It has been shown that **hematopoietic stem cells** from the adult bone marrow have the ability to differentiate into mature cells of various tissues including skeletal muscle (Corbel et al., 2003; Heike and Nakahata, 2004). Such cells could be useful in regeneration of damaged skeletal myofibers.

Normal muscular activity requires the correct level of calcium ions in the blood and tissue fluids otherwise its activity will be compromised. Bone is a calcium reserve in the body so calcium is normally mobilized from it for use by muscle and other body systems. The glycogen stored in the muscle fibers as well as glucose in blood are the main sources of energy needed for muscular activity. Muscle contributes to **heat homeostasis** as heat is produced by muscular activity and helps maintain the body temperature. Excess heat can result from vigorous muscular activity and has to be lost from the body to prevent a harmful rise in body temperature. Muscle repair follows injury that results from the wear and tear as a result of normal muscular activity as well as major accidental injuries. Muscle healing requires normal nerve contact, angiogenesis (development of blood vessels) and formation of a connective tissue matrix. Cytokines (growth factors produced mainly dy leucocytes) as well as other proteins such as fibroblast growth factor and platelet-derived growth factor may play a role in processes occurring in repair of skeletal muscle following injury, aging as well as muscle homeostasis in general (Cannon, 1995).

Muscle homology in the major groups of vertebrates and within specific vertebrates can be established from various structures including nerve supply, similarity in points of attachment and similarity in muscle functions performed. A similar nerve supply is normally retained in homologous muscles. In the course of evolution, some of the muscles in vertebrates have undergone various changes including fusion with each other, splitting into separate muscles, increased or reduced prominence and changes in their points of origin and insertion. Such changes in muscles are accompanied by altered roles that can enhance or decrease performance or totally change the role played by the muscle.

CRANIAL MUSCLES

Extraocular (external ocular) muscles (Fig. 5.10) attach to the surface of the eye and are responsible for moving the eye within the orbit. The six muscles are the most rostral somatic muscles and develop from the head myotomes. Extraocular muscles are found in jawed fishes and tetrapods unless their eyes have degenerated into vestigial organs as seen in the South American and African lungfishes. Some tetrapods also have a seventh muscle, **retractor bulbi** that attaches to the back of the eyeball and is responsible for pulling the eyeball deeper into the socket for the **nictitating membrane** to be drawn across its surface rapidly. This muscle and membrane are lacking in humans. The six extraocular muscles include **dorsal (superior)** and **ventral (inferior) oblique**; **dorsal (superior), ventral (inferior), medial** and **lateral recti** muscles. Since the lateral rectus muscle and retractor bulbi develop from the fifth somitomere, they are innervated by the **abducens nerve (VI)**. The dorsal oblique develops from the third somitomere and is innervated by the **trachlear nerve (IV)**. The other extraocular muscles are innervated by the **oculomotor nerve (III)** as they develop from the first and second somitomeres. The intrinsic muscles of the eye develop in situ from the head mesenchyme.

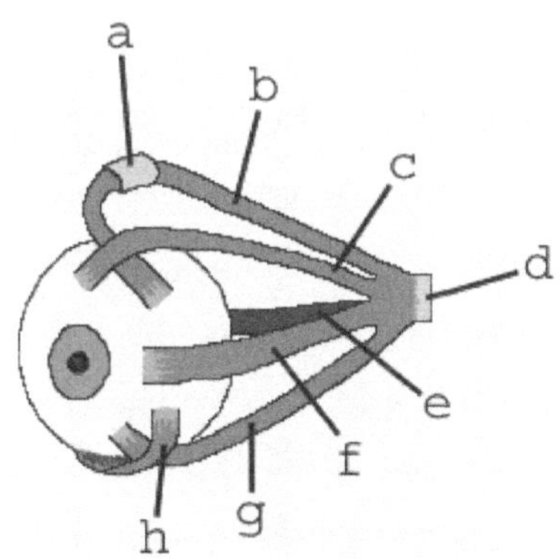

Fig. 5.10 Lateral view of extraocular or extrinsic muscles of the left eye. (a) trochlea, (b) superior oblique, (c) superior rectus, (d) annulus of Zinn, (e) medial rectus, (f) lateral rectus, (g) inferior rectus and (h) inferior oblique.

Color image of this figure appears in the color plate section at the end of the book.

Branchiomeric muscles develop from myotomes that are located caudal to those that develop into extraocular muscles. These muscles are striated and thus voluntary. Branchiomeric muscles are several in fish and move the visceral skeleton. They play a major role in breathing and feeding in this group of vertebrates. The muscles are named according to roles played such as **adductor mandibulae, intermandibularis, constrictors** and **levators** (Fig. 5.11). As gills were lost with life of tetrapods on land, there were changes in the visceral skeleton that resulted in loss or transformation of many branchiomeric muscles. The transformed branchiomeric muscles in tetrapods are those that are associated with the visceral arch derived structures such as jaws, auditory ossicles and larynx and include **masseter, temporalis, pterygoids, digastric, mylohyoid, platysma, sternomastoid** and **cleidomastoid** (Fig. 5.12).

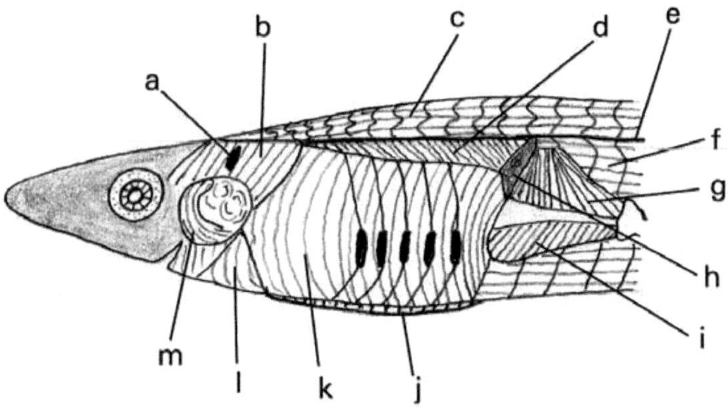

Fig. 5.11 Superficial branchiomeric muscles of a shark. (a) spiracle, (b) levator hyomandibulae, (c) epibranchial musculature, (d) cucullaris (tapezius), (e) horizontal septum, (f) hypaxial myomeres, (g) adductor, (h) scapula (i) abductor, (j) hypobranchial muscles, (k) ventral hyoid constrictors, (l) intermandibularis and (m) adductor mandibulae.

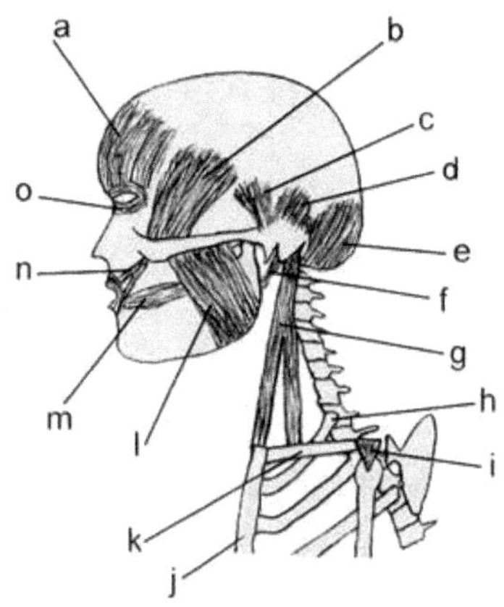

Fig. 5.12 Some of the visceral arch derived muscles of a human being. (a) frontal portion of occipitofrontalis, (b) temporalis, (c) anterior and (d) posterior auricular, (e) occipital portion of occipitofrontalis, (f) digastricus, (g) sternocleidomastoideus, (h) first rib, (i) deltoideus, (j) sternum, (k) clavicle, (l) masseter, (m) buccinator, (n) zygomaticus and (o) orbicularis oculi.

Epibranchial and Hypobranchial Muscles

These muscles are located dorsally and ventrally in the head and anterior trunk region. Epibranchial muscles in tetrapods support the body against gravity on land and show considerable fusion of myomeres that has resulted in longer muscles unlike in the case of fish and some amphibians such as salamanders that move by lateral undulations. Hypobranchial muscles extend from the visceral arches to the pectoral girdle and run in a longitudinal manner. Hypobranchial muscles of teleosts show some degree of complexity when compared to sharks as a result of the expansion of the buccopharyngeal cavity during feeding in teleosts. Terrestrial vertebrates have evolved several hypobranchial muscles as a result of adopting complex food gathering methods in the absence of an aquatic environment that offers some degree of buoyancy to food.

Between the hyoid bone and mandibles is a pair of **coracomandibularis** in sharks. This muscle has evolved into the **geniohyoid** (Gr. *geneion*, chin) of tetrapods that also have a muscular tongue. The geniohyoid extends from the mandibles to the tongue and has some fibers that run into the tongue. Since the muscular tongue of tetrapods plays a great role in food and liquid intake, the hypobranchial muscles of the group are quite complex and consist of **intrinsic** and **extrinsic** tongue muscles. The intrinsic **lingualis** consists of fibers that run in a longitudinal, transverse and perpendicular manner and forms the bulk of the tongue. These bundles of fibers change the shape of the tongue through contraction and relaxation. The extrinsic group of lingual muscles includes the **genioglossus**, **hyoglossus** and **styloglossus**. These muscles insert on the tongue and originate from neighboring structures and are responsible for moving the tongue about.

The hollow **ballistic** or **superconducting tongue** of the chameleon (Fig. 5.13) covers a forward pointing and long tapering cartilage known as the **entoglossal process** or **hyoid horn** that is attached to the center of a U-shaped hyoid bone. The entoglossal process runs to the tip of the tongue. The **accelerator** or **ring shaped muscles** of this tongue are sphincter- or ring-like in arrangement and contract at high speed against the entoglossal process thus creating a squeezed forward thrust that ejects the tongue off the process and out of the mouth at high speed and with a strong force. Lubricated cavities are present between the muscles and the entoglossal process and reduce friction during muscle contraction so that the tongue can shoot out at high speed. The tongue still retains a constant volume and changes in the normal resting dimensions are compensated for by elongation of the tongue. The chameleon uses such a mechanism together with the sticky **tongue pad**

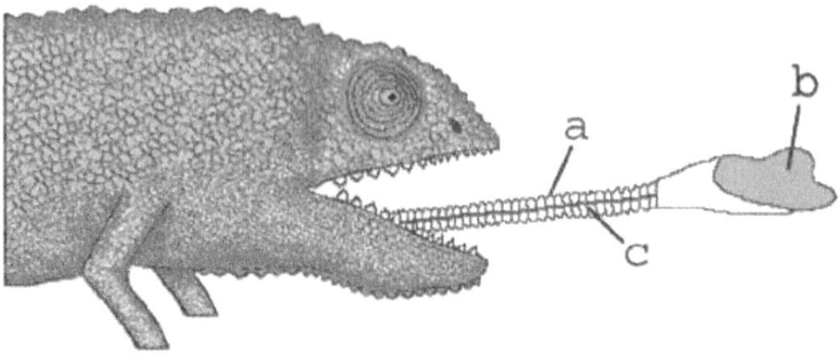

Fig. 5.13 The ballistic tongue of a chameleon and associated structures. (a) tongue, (b) tongue pad and (c) accelerator muscle. In the resting state, the tongue is anchored to the back of the mouth by the U-shaped hyoid bone. The tongue can be as long as one to one and a half times the length of a chameleon's body length when fully extended and attains this state in about one-sixth of a second. After a strike, the elastic and stretched hyoglossal muscles that contract at a slower speed than the accelerator muscles retract the rest of the tongue to the mouth.

that contains numerous epithelial glands as well as several abrasive papillae that can stick into irregular surfaces of the victim to capture prey such as insects at a distance up to one and a half times the chameleon's body length. To the posterior part of the club-shaped pad has a flap of skin that partly wraps around the prey after a strike.

Chameleons have also modified the primitive iguanian system by including a suction component that is generated by two modified intrinsic tongue muscles that pull the tongue pad inwards in their prehension mechanism (Herrel et al., 2000). The super-contracting muscles of the chameleon tongue show a large myofilament overlap at maximal tongue extension that results in retractor force production that is almost constant for a wide range of projection distances (Herrel et al., 2001). Terrestrial amphibians also capture their prey using their tongues. Some salamanders can project their tongues to capture prey that is up to 80% of their body length away from the tip of the mouth.

The straplike hyoid muscles are located ventrally along the neck and include **sternohyoideus**, **sternothyroideus** and **omohyoideus**. Contraction of hyoid muscles pulls the hyoid bone and larynx caudally and also causes the dilation of the pharynx.

Axial (Trunk) Muscles

The axial muscles associated with the trunk perform a major role in locomotion and breathing. In fish as well as other aquatic vertebrates, axial muscles form most of the musculature and are the main muscles of propulsion in fish. Changes in lives of vertebrates from water to land have been accompanied by enlargement of appendicular as well as a reduction in axial musculature. Amphibians represent a group to vertebrates with axial musclature that is transitional between that of fishes and reptiles. Axial musculature shows a higher degree of segmentation in lower when compared to higher vertebrates (Fig. 5.14). The muscle segments or **myomeres** are separated by **myoseptae** and represent myotomal segments that have not fused. Axial musculature is also divided into the upper epaxial and lower hypaxial muscles by the **horizontal septum** that runs at the level of the transverse processes of vertebrae. Alternate contractions of myomere segments in fish on either side bring about the undulating motion that is responsible for swimming movements in water.

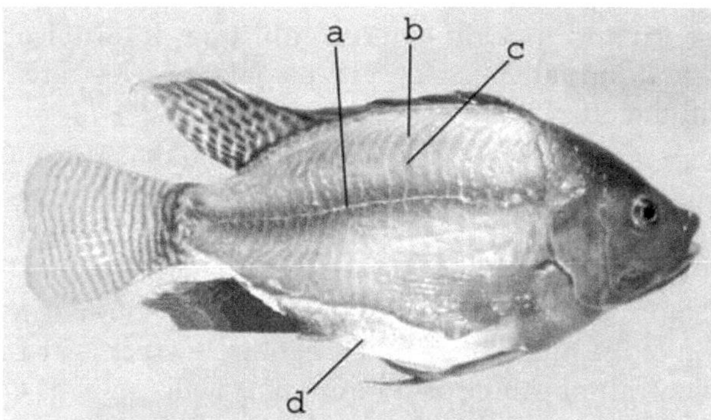

Fig. 5.14 Fish body trunk muscle. (a) horizontal septum, (b) myosepta, (c) myomere and (d) skin. The horizontal septum divides body trunk musculature into epaxial (above) and hypaxial (below) muscles. The red muscle is the dark longitudinal strip that is divided into lower and upper parts by the horizontal septum and is located superficially in most fish.

Epaxial Muscles

Epaxial muscles are located dorsal to the horizontal septum and lie above the vertebrae and their transverse processes. In fish and urodeles such as salamanders that have weak limbs and have to depend on lateral undulations for movement, epaxial muscles are massive and segmented and are considered a single dorsolateral trunk muscle known as **dorsalis trunci**. Dorsalis trunci is much reduced in relative volume in tetrapods since limbs have taken over the role of propulsion from the axial muscles. Epaxial muscles of tetrapods play a major role in the dorsoventral bending of the spine and movement of the head. Such movements are not possible in fish. In higher vertebrates such as mammals, the complex head and trunk movements have been accompanied by evolution of several specific short and long muscles. The long muscles of the back and loins originate from the sacrum and pelvis and extend up to the head.

Starting laterally and moving inwards, the long muscles include the **iliocostalis**, **longissimus**, **spinalis** and **multifidi**. The spinalis and multifidi show segmentation in mammals whereas the same group is partly segmented in reptiles that rely on undulations for movement.

Hypaxial Muscles

Hypaxial musculature normally lies below the vertebrae and their transverse processes. In amphibians, hypaxial muscles include a subvertebral group that lies below the transverse processes of vertebrae known as the **rectus abdominis** that runs between the pectoral and pelvic girdles. The lateral group consists of a broad group of muscles that cover the body cavity and include **external oblique**, **internal oblique** and **transverse abdominal** muscles. Hypaxial muscles are not much developed in the amphibian tail.

Hypaxial muscles of other tetrapods include the abdominal muscles as well as muscles of respiration. The abdominal muscles line the ventral or anterior and lateral walls of the abdomen and support abdominal viscera that can be quite substantial in large herbivores. Although the abdominal muscles are continuous flat sheets, evidence of original segmentation of the muscles is still evident in the **rectus abdominis** that lines the ventral or anterior abdominal wall and has **tendinous inscriptions**. The mammalian thorax has undergone structural changes as its wall possesses ribs that play a major role in respiration. The external and internal oblique muscles in the thoracic wall region of mammals have evolved into **external** and **internal intercostal** muscles that are responsible for inspiratory and expiratory movements on contraction respectively. The intercostal muscles extend more posteriorly in tetrapods with ribs in the abdominal region. Other muscles of respiration such as the **dorsal serrate** muscles probably separated from superficial hypaxial muscles. The **diaphragm** is the principal muscle of inspiration in mammals and develops from the ventral part of cervical myotomes and is supplied by the phrenic nerve that also originates from the cervical region.

Mammals have a sub-vertebral group of muscles that are located below the thoracic and lumbar vertebrae. The muscles are more developed in the lumbar region and attach to the pelvic girdle. The group consists of the **psoas minor**, **iliopsoas** and **quadratus lumborum** and on contraction cause the arching of the back. Arching of the back is quite marked in mammals with relatively longer lumbar regions such as carnivores and cats.

Appendicular Muscles

The fish appendicular muscles move the fins whose role is mainly to maintain stability, brake and maneuver. Such movements are much more limited when compared to those of tetrapod limbs. The appendicular muscles of fish fins comprise ventral and dorsal muscles. **Ventral (abductor)** muscles are found on the cranioventral aspect of fins and move them ventrally and cranially on contraction. The **dorsal (adductor)** muscles are located caudodorsal to the fins and move fins caudally and dorsally.

The complex movements and supportive role performed by tetrapod limbs have resulted in the evolution of complex appendicular muscles that form the bulk of tetrapod body musculature. The roles played by ventral and dorsal muscles of fish have been reversed in tetrapod muscle groups. The ventral muscles of tetrapods are involved in bringing about adduction or flexion whereas the dorsal muscle group abduct or extend limbs. The arrangement, size and presence or absence of certain muscles varies in tetrapods depending on the type of movement performed.

Pectoral Girdle Muscles

Despite having less musculature (in terms of weight) than hindlimbs, forelimbs are the main weight supporting limbs of the tetrapod trunk and support up to 60% of the trunk weight in many mammals. The pectoral limb is attached to the trunk by **serratus ventralis**, **pectoral**, **trapezius**, **rhomboideus** and **latissimus dorsi** muscles (Figs. 5.15 and 5.16). The pectoralis of amphibians and reptiles plays a major role in adduction and raising their bodies from the ground. The **supracoracoideus** is also present in amphibians and reptiles and plays a role in attaching the forelimb to the body trunk. The pectoralis and supracoracoideus are quite large in birds and are the main flight muscles. The pectoralis can be so large in birds that their weight can exceed that of all other bird skeletal muscles when the power of flight is fully developed. Supracoracoideus appears in mammalian embryos (Chen, 1955) but grows into the **supraspinatus** and **infraspinatus** with further development of the individual. **Coracobrachialis** is present in birds and is quite large in amphibians and reptiles that have a large coracoid. Apart from bats where it causes wing adduction, coracobrachialis is reduced or lacking in mammals that have small or no coracoids. Most mammals have a reduced or missing clavicle and simplification of their pectoral girdle has resulted in union of the **cleidomastoid** belly and the **cleidobrachialis**, as is the case in man, resulting in a single muscle known as **brachiocephalicus**.

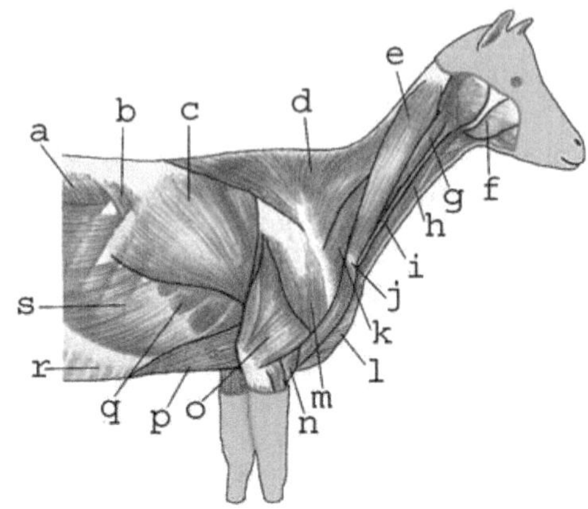

Fig. 5.15 Superficial muscles of the anterior body trunk of a goat. (a) internal oblique abdominal muscle, (b) serratus dorsalis caudalis, (c) latissimus dorsi, (d) trapezius, (e) cleidooccipitalis of the brachiocephalicus, (f) masseter, (g) cleidomastoideus, (h) sternomandibularis, (i) external jugular vein, (j) clavicular tendon, (k) omotransversarius, (l) superficial pectoral, (m) deltoideus, (n) brachialis, (o) triceps brachii, (p) deep pectoral muscle, (q) serratus ventralis, (r) rectus abdominis and (s) external oblique abdominal muscle.

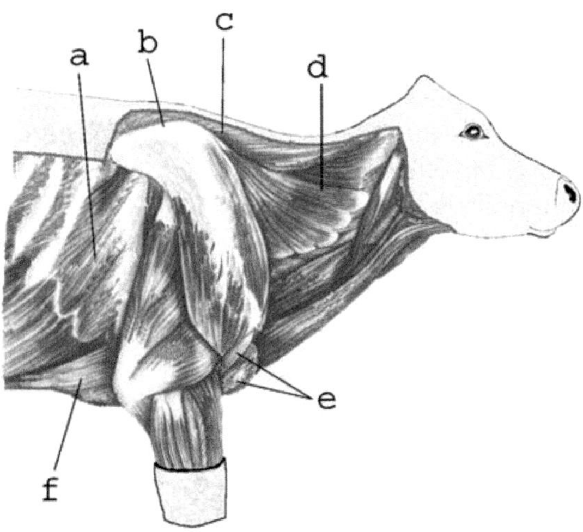

Fig. 5.16 Anterior body trunk muscles showing some of the main muscles that attach the pectoral limb to the body trunk in an ox. (a) thoracic part of serratus ventralis, (b) scapular cartilage, (c) rhomboideus, (d) cervical part of serratus ventralis, (e) superficial pectoral and (f) deep pectoral muscles.

The **deltoid** muscle originates partly from the scapula spine in mammals whereas it originates from the anterior border of the scapula and clavicle in amphibians and reptiles. Such an arrangement in the origin of the deltoid muscle shows that the spine might be the original anterior border of the scapula. The **teres major** that occupies the caudal border of the scapula separated from the **latissimus dorsi** with evolution. The **subscapular** muscle that lies on the medial surface of the scapula in mammals arose as a result of the expansion of the **subcoracoscapularis** of amphibians and reptiles. The **cutaneous muscle** of the trunk (cutaneous trunci) could have originated from the appendicular muscles as it attaches to the base of the latissimus dorsi and the pectoral muscles. This thin muscle lies in superficial fascia and is capable of contracting independent of deep muscles.

The distal muscles of the pectoral limb vary in vertebrates and show different degrees of tendinous components that impart strength to the limbs. Although individual muscles of the forelimb are weaker than those of the hindlimb, they are more tendinous in nature. In large herbivores, the distal parts of the pectoral limb are quite tendinous as they contain tendons of muscle insertion.

PELVIC GIRDLE MUSCLES

The pelvic girdle muscles of mammals are not as numerous and as active as the pectoral girdle muscles since the mammalian pelvis is firmly united to the vertebral column by a tight joint together with its ligaments. Since the pelvic limb is involved in propelling the body forwards, it has evolved intrinsic muscles that are stronger and more complex in structure and arrangement than those of the forelimb. The bellies of some of these muscles can be quite large and lie in the proximal part of the limb whereas the distal part is mainly tendinous. Several of these muscles are long and act on several joints (Figs. 5.17 and 5.18). The muscles of the pelvic limb in birds are the second largest and strongest group of muscles in this vertebrate class after the pectoral group as they perform several functions such as walking, climbing and perching. The muscle bellies of the pelvic limbs are located proximally with the rest of the legs consisting of smaller muscles and flexor and extensor tendons of toes.

Large ungulates have evolved mechanisms that minimize energy expenditure when in a standing position. The support column of the hind limb requires greater stability than the forelimb in a standing position due partly to the greater angling of its joints. Ungulates have evolved mechanisms that stabilize the stifle (knee) and hock (ankle) (Fig. 5.19) thus minimizing energy expenditure. The stifle is stabilized by the **quadriceps femoris** and **patellar mechanism** whereby the patella is shifted medially by extension of the quadriceps thus preventing further flexion of the stifle while the hock is extended by the contraction of the **gastrocenemius** and the **superficial digital flexor muscle**. The joints of

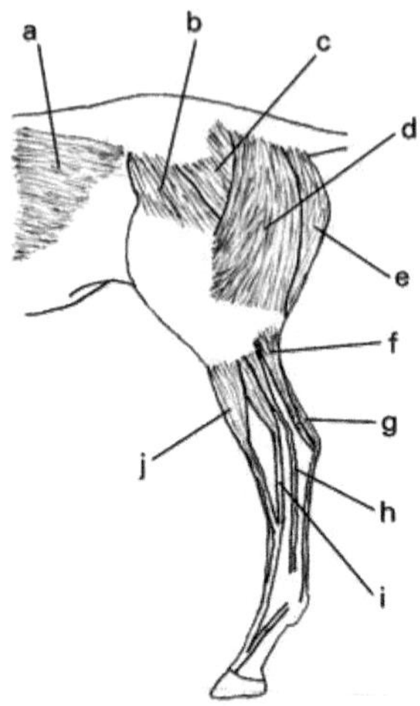

Fig. 5.17 Lateral superficial muscles of the pelvic limb of a horse. (a) external oblique abdominal, (b) tensor fasciae latae, (c) superficial gluteal, (d) biceps femoris, (e) semitendinosus, (f) gastrocnemius, (g) superficial digital flexor, (h) flexor hallucis longus, (i) lateral digital extensor and (j) long digital extensor.

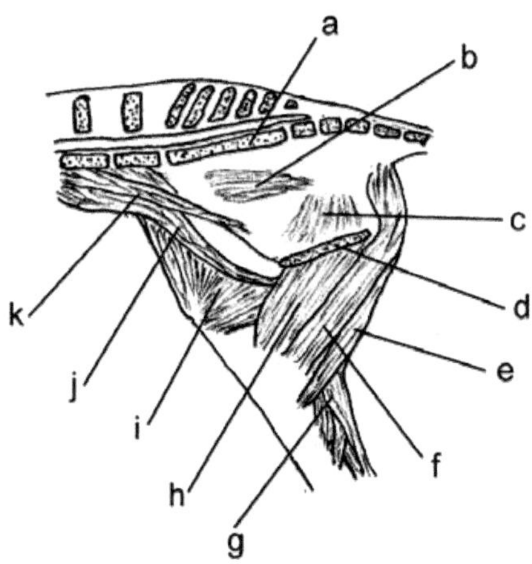

Fig. 5.18 Some of the medial muscles of a horse pelvic limb. (a) body of sacrum, (b) iliac and (c) pubo-ischiatic portions of the internal obturator muscle, (d) median section through the pelvic symphysis, (e) semitendinosus, (f) semimembranosus, (g) gastrocnemius, (h), adductor, (i) quadriceps femoris, (j) iliopsoas (psoas major and iliacus) and (k) psoas minor.

the digits receive support partly from the tendons of the **long flexor muscles**. Tendons have the ability to store energy as elastic strain energy since they can be stretched up to 5% of their original length at one stage of their locomotory cycle. Such storage is achieved when tendons change their lengths and not that of muscles.

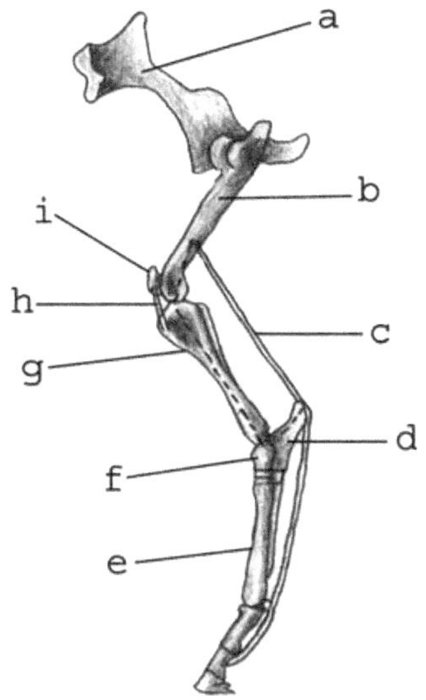

Fig. 5.19 Diagram showing structures that stabilize the hind limb of a large ungulate. (a) pelvis, (b) femur, (c) superficial digital flexor muscle, (d) calcaneus, (e) metatarsal bone, (f) talus or astragalus, (g) tibia, (h) patella ligament and (i) patella. The dotted lines together with the superficial digital flexor muscle and the gastrocnemius that lies between the femur and the talus demarcate a parallelogram that forms part of the locking mechanism. The quadriceps femoris is found anterior to the femur.

At times of inactivity, vertebrates with less tendinous material in their limb muscles reduce the weight they put on limbs since muscle activity will be required for limb support in a standing position thus making them tire faster. Such vertebrates will rest sprawled on their ventral surface (amphibians and reptiles), crouch (rodents and rabbits), lie down (carnivores and many artiodactyls) or sit down (primates).

LOCOMOTION IN VERTEBRATES

The functional relationship between the skeletal and muscular systems together with the vertebrate group and the surrounding environmental conditions are important in understanding locomotion. The nature of vertebrate joints as well as ligaments present also plays a role in determining the type of movements possible in various body parts of the vertebrate body. The three major categories of locomotion performed by vertebrates include swimming, terrestrial locomotion and flying.

Swimming

Primary swimmers such as fish solely rely on swimming for locomotion and secondary swimmers such as some aquatic mammals have had to readapt partly or completely to an aquatic type of life. In both categories of swimming, vertebrates have had to adapt by developing streamlined bodies that offer less resistance to water, generating enough power to propel themselves in a dense medium and maintaining the proper orientation to steer the body. In terms of energy costs, swimming is cheapest, then flying and lastly walking or running.

Primary swimmers move by performing lateral undulatory muscular contractions in the body trunk and tail regions that are brought about by contraction of the segmented myomeres. Serial contractions of myomeres produce a smooth wave along the fish body due to the overlapping nature of myomeres. A greater proportion of fish body weight is muscle and it can be as high as 80% in some species when compared to 25-55% present in terrestrial mammals. The strong articulation of the vertebral column prevents the fish body from shortening during such contractions. The propulsive forces generated are transmitted along the trunk from the head to tail regions. The broad and flat caudal fin together with greater undulatory movements generated by the tail region of the body move more water than any other part of the body. Primary swimmers also generally have a fusiform body together with an integument that is closely attached to the underlying musculature by connective tissue. Such an integument adds to the compactness of a fish body.

Secondary swimmers have evolved appendages they use for performing paddle-like movements during oscillatory swimming. Although the appendicular skeleton has undergone some modifications towards reduction in size, the appendicular musculature is fairly well-developed. **Semi-aquatic** swimmers such as shrews, otters beavers, capybaras and the hippopotomus mainly use their limbs for swimming. Their bodies are usually long. **Solely aquatic** vertebrates such as cetaceans and sirenians have modified their forelimbs into flippers and their hindlimbs are vestigial whereas the hindlimbs of others such as seals have been modified into flippers. Cetaceans also have reduced the numbers of cervical vertebrae and lack a neck. The atlas and axis are fused and their tails have been modified into horizontal flukes that, together with the dorsal fins, lack a skeletal component and are supported by fibrous connective tissue. Secondary swimmers can temporarily streamline themselves while swimming in order to reduce the drag effects of water.

Terrestrial Locomotion

Terrestrial vertebrates have had to adapt to an environment that does not offer buoyancy to their body weight since air is not a dense medium. The vertebral column has undergone several changes such as ossification of its bodies (centra) in most species and presence of intervertebral disks between the bodies of adjacent vertebrae in order to support the weight of body trunk and transmit the same to the girdles and their limbs. Terrestrial vertebrates have evolved different types of movement to be able to cope with life in various terrestrial environments they occupy and include cursorial, saltatorial, scansorial, arboreal, fossorial and bipedal types of locomotion.

Cursorial (running) type of locomotion is demonstrated in fast tetrapods that run long distances such as carnivores and herbivores. The body of cursorial animals (cursors) is relatively long and their vertebral column plays an important role in lengthening the stride by stretching out (dorsoventral flexion) during forward propulsion. Carnivores possess flexible backs and herbivores, with limited trunk flexion, utilize their limbs entirely for locomotion. The stride length and rate determine the running speed. An increase in limb length adds to stride length whereas reduction in limb weight especially distally enhances stride rate. Other anatomical limb adaptations that enhance speed in cursorial locomotion include a short and muscular humerus, a dominant radius, reduction in number of digits, increase in tendinous and ligamentous elements, long metacarpals and metatarsals and adoption of **digitigrade** (only digits on ground) and **unguligrade** (hoof on ground) stances. These anatomical modifications ensure stability of cursorial tetrapod limbs over manipulative and rotational movements as well as keeping most of the limb weight proximally while reducing it distally. Reduction in amount of fleshy muscle distally favors reliance on passive mechanics.

Saltatorial (ricochetal) tetrapods (hoppers) such as kangaroos, primates such as ringtail lemurs; springhares and gerbils are bipedals whereas rabbits and hares are quadripedal. These saltators use their hindlimbs to jump or hop. As a result of convergent evolution, all saltators have many common features (Emerson, 1985). The body weight of saltatorial animals is shifted to the hind legs that tend to be long and powerful due to their large musculature. The center of gravity is aligned with the sacrum and the caudal vertebrae are fused to form long inflexible **urostyles**. There are several other adaptations to saltatorial locomotion. Saltators show a reduction in the number of hindlimb digits, stiffening of the spine as result of fusion of some cervical vertebrae, strong lumbar vertebrae and strong fusion between the sacrum and pelvic girdle to minimize whiplash, presence of a long tail and elastic ligaments in the hindlimb. The use of elastic storage mechanisms by elastic ligaments saves on energy expenditure. Saltatory animals also have long ligaments that run from cervical to lumbar vertebrae and act as shock absorbers.

Scansorial (climbing) animals are climbers with modified phalanges they use for clinging to vertical surfaces. Some scansorials are arboreal and spend most of their time in trees. Scansorials also have strengthened appendages as well as pectoral musculature. Large climbers tend to be slower and more cautious than small ones as the tendency to fall is higher and more severe in the heavier scansorials. Various scansorials have adapted to the habit of climbing by possessing features for this type of locomotion such as claws for digging into surfaces as is the case in squirrels, friction pads on hands and feet of primates, prehensile (capable of grasping) tails of opposum and suction cups in sucker-footed bats. The group of scansorials also shows other general adaptive features such as elongated forelimbs, expanded ribs that tend to overlap, a strong vertebral column, a long thorax and short lumbar region that limits movement between the pelvic girdle and ribs. Animals that display scansorial locomotion have feet that are more or less plantigrade.

Arboreal locomotion (brachiation or arm swinging) is found in gibbons and monkeys of the New World. They can rotate their forearms while retaining their hands in a fixed position due to the presence of a ball and socket joint in their wrists. They also possess opposable digits and some have prehensile tails.

Fossorials generally inhabit a subterranean environment and are adapted to digging. They live in burrows and their phalangeal claws and teeth are modified for digging. Fossorial tetrapods such as rodents, insectivores and marsupials have strong pectoral muscles and very flexible vertebral columns. Fossorials also have cylindrical bodies, flattened skulls and reduced eyes and ears.

Bipedal locomotion is seen in humans. Adjustments to this type of locomotion include an S-shaped spine, legs under the head and trunk, lengthened bones of the pelvic limbs, plantingrade feet, arched tarsals and parallel metatarsals and digits in relation to the ground. The toes are not opposable and knee joint are directed in a forward manner. The human being adapts a plantigrade stance during walking and a digitigrade stance while running.

Flying

Members of three animal groups have evolved true flight locomotion and include insects, birds and mammals. These groups are capable of generating a forward thrust that is independent of gravitational descent or air currents by flapping of their wings. Vertebrates display four generalized types ('levels') of flight locomotion that include parachuting, gliding, true flight and soaring.

Parachuting entails use of the body and limbs to increase the overall area to reduce a direct fall from the air and also escape from predators. Parachuting is moving down at an angle of greater than 45 degrees between the horizontal axis and the direction of descent. During parachuting, an animal is supposed to fall slowly as the surface area is supposed to be great in relation to weight. The greater the drag forces, the slower the descent. Some lizards and tree frogs (Fig. 5.20) are capable of spreading their limbs and flattening their bodies during parachuting. Vertebrates capable of flight occasionally parachute when it suits their needs.

Fig. 5.20 An arboreal frog *Rhacophorus* that is capable of parachuting. *Rhacophorus* has expanded toe membranes that enable it fall slowly as it leaps off trees during which it is capable of making turns in the air as it falls.

Gliding makes use of broad membranes that are attached to limbs thus increasing the body surface area. The gliding angle is more horizontal than parachuting and enables the gliding animal to travel for a longer distance. Streamlining is important during gliding as it reduces the drag effect. Gliders are to be found among several lizards, squirrels, marsupials and colugos (Fig. 5.21). Gliding enables the animal to move from one tree to another without descending and climbing again that could be expensive in energy costs.

True flight entails the flapping of wings to actively sustain movement in air. Unless a strong wind interferes, true flyers are capable of moving horizontally or ascending at a steady speed. True flight has evolved three times in vertebrates: first in the extinct pterosaurs, then birds and lastly in bats. The evolution of flight in these groups occurred independently. Modifications for flight have included reduction in body

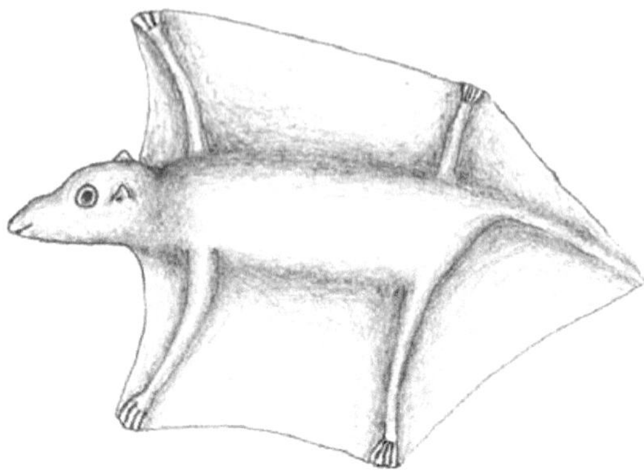

Fig. 5.21 A colugo ('flying lemur'). Colugos maneuvre around obstacles and can glide for a distance of upto 110 meters.

weight and increase in power production through development of pectoral musculature that can sustain active flight. Other modifications include fusion of bones, possession of hollow bones that are light, excretion of uric acid that reduces water loss from the body, efficient respiration and circulation, homeothermy and a high metabolic rate.

Soaring animals normally appear to be gliding as they donot flap their wings but keep them extended. Soaring makes use of energy of the surrounding air such as rising warm air to sustain an animal at a certain altitude. Expenditure of energy in a soaring animal occurs during take off, landing or making adjustments during soaring. Some large birds such as hawks and vultures are capable of soaring as well as the large extinct pterosaurs. Although not much energy is expended during soaring, some muscular activity is necessary to keep the wings extended to avoid the downward pull by gravity and the effect of the force of the rising air currents (Meyers, 1993).

ELECTRIC ORGANS

Unlike many tissues in the body that generate weak electrical currents, electric organs are modified muscle that generate, amplify, store and discharge electricity. Around 250 species of fish have evolved electric organs. In many of these species, weak electric signals are generated and aid the fish in navigating in murky water and also in species and sex recognition. During navigation, distortion of electric fields generated in fish by objects is detected as it differs from that of the water environment. The electric properties of objects, their form and their distance can be analysed and distinguished (Pusch et al., 2008). Some species of fish such as the electric eel *Electrophorus* (Fig. 5.22), the electric catfish of the Nile

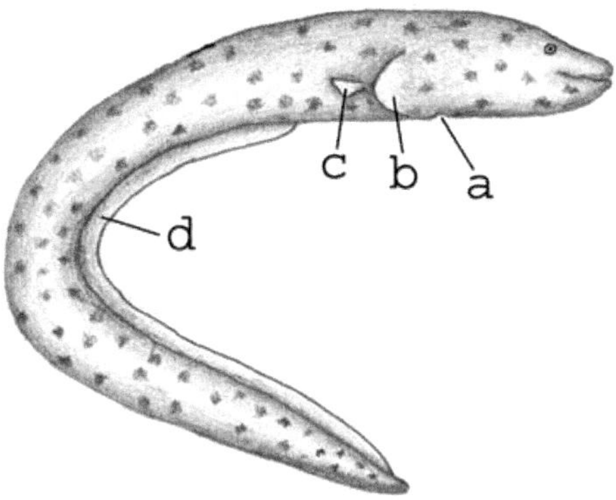

Fig. 5.22 A drawing of an electric eel, *Electrophorus electricus*. The skin is brown in color with tan spots. (a) anal opening, (b) operculum, (c) pectoral fin and (d) elongated anal fin. *Electrophorus* is found mainly in the river basins of the Amazon and Orinoco Rivers of South America and is not an eel but belongs to the order Gymnotiformes. Electric eels lack dorsal and pelvic fins and have elongated anal fins. *Electrophorus* can grow up to 2.5 m in length and weigh as much as 25 kg and its closest relatives are the catfish. The vital organs of electric eels are located in the anterior fifth of the body while the rest of the fish length is the electricity producing tail.

Melapterurus, a ray *Torpedo* and a skate *Raia* generate high voltage electricity that can be as high as 500 volts in the electric eel. Such high voltage discharges might be protective devices against predators as well as weapons to stun prey.

Electric organs consist of a number of electric plates (disks) known as **electroplaques**. Electroplaques are piled in either vertical or horizontal columns and form discharging units. Electroplaques normally discharge at the same time after induction by nerve fibers. The net effect of all the discharging electroplaques constitutes the electric output of the fish. Electric organs vary in their location and appearance in different fishes. In the electric eel, several long columns of electroplaques are found on each side of the body whereas in *Torpedo* an electric organ lies in each pectoral fin close to the gills. As the distribution of electric organs in fish is not systematic and the organs differ in their construction, electric organs could have evolved independently and are a result of convergent evolution.

Shark, skates, rays, lungfishes, some chondrosteans as well as some amphibians and the duckbill platypus possess electroreceptors in the skin. **Electroreceptors** detect weak electric signals that result from the muscular activity of other organisms in water. Objects in water also distort electric fields that are created by this group of vertebrates and such fields are detected by electroreceptors.

REFERENCES

Abbot, B.C. (1951). The heat production associated with the maintenance of a prolonged contraction and the extra heat produced during large shortening. J. Physiol. (London), 112:438-445.

Cannon, J.G. (1995). Cytokines in aging and muscle homeostasis. J. Gerontol. A Biol. Sci. Med. Sci., 50 Spec. No: 120-123.

Chen, C.C. (1955). The development of the shoulder region of the opossum, *Didelphis virginiana,* with special reference to the musculature. J. Morphol., 97:415-472.

Corbel, S.Y., Lee, A., Yi, L., Duenas, J., Brazelton, T.R., Blau, H.M. and Rossi, F.M. (2003). Contribution of hematopoietic stem cells to skeletal muscle. Nat. Med., 9(12):1528-1532.

Emerson, S.E. (1985). Jumping and leaping. In: M. Hildebrand, D.M. Bramble, K.F. Liem and D.B. Wake (eds), Functional Vertebrate Morphology. Harvard University Press, Cambridge.

Eriksson, E. and Myrhage, R. (1972). Microvascular dimensions and blood flow in skeletal muscle. Acta Physiol. Scand., 86:211.

Fung, Y.C. (1977). Rheology of blood vessels. pp. 299-324. In: G. Kaley and B.M. Altura (eds), Microcirculation. University Park Press, Baltimore-London-Tokyo.

Gregory, J.E., Morgan, D.L. and Proske, U. (2003). Tendon organs as monitors of muscle damage from eccentric contractions. Exp. Brain Res., 151(3):346-355.

Heike, T. and Nakahata, T. (2004). Stem cell plasticity in the hematopoietic system. Int. J. Hematol., 79(1):7-14.

Herrel, A., Meyers, J.J., Aerts, P. and Nishikawa, K.C. (2000). The mechanics of prey prehension in chameleons. J. Exp. Biol., 203:3255-3263.

Herrel, A., Meyers, J.J., Aerts, P. and Nishikawa, K.C. (2001). Functional implications of the superconducting muscle in the chameleon tongue retractors. J. Exp. Biol., 204:3621-3627.

Krogh, A. (1919). The number and distribution of capillaries in muscles with calculations of oxygen pressure heads necessary to for supplying the tissues. J. Physiol. (London), 52:409-415.

Krogh, A. (1922). The Anatomy and Physiology of Capillaries. Yale University Press, New Haven. Reprinted in 1959, Hafner, New York.

Mathieu-Costello, O. (1987). Capillary tortuosity and Degree of Contraction or Extension of Skeletal Muscles. Microv. Res., 33:98-117.

Mathieu, O., Cruz-Orive, L.M., Hoppeler, H. and Weibel, E.R. (1983). Estimating length density and quantifying anisotropy in skeletal muscle capillaries. J. Microscopy, 131:131-146.

Mathieu-Costello, O., Potter, R.F., Ellis, C.G. and Groom, A.C. (1988). Capillary configuration and fiber shortening in muscles of the Rat Hind Limb: Correlation between corrosion casts and stereological measurements. Microv. Res., 36:40-55.

Meyers, R.A. (1993). Gliding flight in the American kestrel *Falco sparverius:* An electromyographic study. J. Morphol., 215:213-224.

Pusch, R., von der Emde, G., Hollmann, M., Bacelo, J., Nöbel, S., Grant, K. and Engelmann, J. (2008). Active sensing in a mormyrid fish: electric images and peripheral modifications of the signal carrier give evidence of dual foveation. J. Exp. Biol., 211(Pt 6):921-934.

6

Integument

The integument (L. *integumentum,* covering) comprises the skin and its derivatives. Skin and its derivatives cover the entire vertebrate body and form 12-15% of the body weight in the human being. The integument forms the external boundary between the body and the environment. Since the integument is in contact with the external and internal environments it has various structural and functional adaptations that enable it to perform several vital functions at the interphase between these two different environments. The skin offers mechanical and physical protection to the body as it is made of several layers of cells that are continuously being replaced with regeneration of new similar cells. As the vertebrate body is exposed to various pathogens in the external environment, an intact skin offers a barrier to infection. The skin has evolved a highly immune defense system that comprises epidermal Langerhans cells, T-lymphocytes and keratinocytes that are the most distinct cells (Bjerke, 2002).

The skin plays an important role in thermoregulation in birds and mammals and to a lesser extent a similar role in reptiles. Many sensory organs that are vital to vertebrate responses are located in the skin. In many vertebrate species, the skin is an organ of osmoregulation and excretion. Skin derivatives such as nails, horns and poisonous glands are weapons of offense in some vertebrates. Some skin secretions are of nutritive value to young vertebrates as well as species recognition marks or may be offensive to predators and thus play a protective role. Unique coloration of the skin has played major roles in camouflage and behavioral responses during reproduction in some species. In most terrestrial vertebrates, the skin protects the body against dessication. Some species have used the skin for flight and gliding.

GENERAL STRUCTURE OF THE SKIN

The basic structure of the skin is the same in all vertebrates. Variations in structure occur in vertebrate species and even different parts of an individual. The skin has two major layers: the **epidermis** (Gr. *epi*, upon; *derma*, skin) that forms the outermost layer and a **dermis** that lies below (Fig. 6.1). A transitional organism between invertebrates and vertebrates, *Amphioxus*, has an epidermis comprising a single layer of columnar epithelium that is covered with a thin film of cuticle. The vertebrate epidermis is made up of a stratified squamous epithelium and originates from the ectoderm. Olivera-Martinez et al. (2004) have reviewed processes involved in skin formation. The accessory and keratinized structures of skin such as hair, feathers, horns and beaks develop from the epidermis. The exocrine glands that secrete their products via ducts to the skin surface are also of epidermal origin. The innermost layer of the epidermis, the **stratum basale** is made of actively dividing cells that are generally larger than other cells of the epidermis and rest on a basal lamina. This layer gives rise to the epithelial layers of the epidermis whose cells, also known as **keratinocytes**, are pushed superficially as they mature with time in a process that gives rise to several layers of cells in different stages of maturation. The cells become more flat as they get nearer to the body surface. The epithelial cells produce numerous peptides such as cathelicidin that have a direct antimicrobial activity and also signal the recruitment of circulating immune cells such as neutrophils (Baff et al., 2005).

There is deposition of the protein keratin (Gr. *kerat*, horn) in the superficial layers of the epidermis. **Keratin** is a fibrous and insoluble

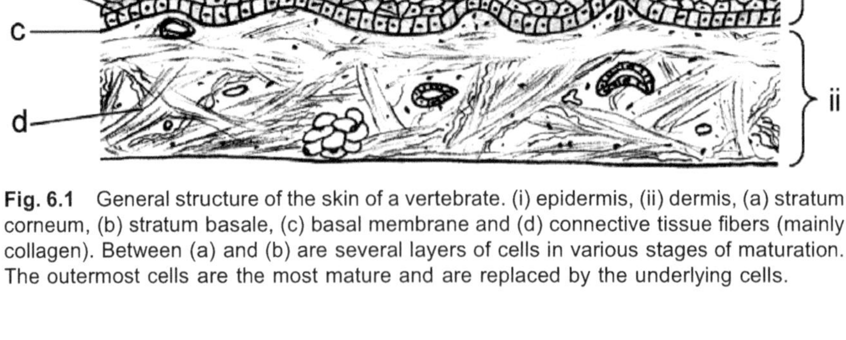

Fig. 6.1 General structure of the skin of a vertebrate. (i) epidermis, (ii) dermis, (a) stratum corneum, (b) stratum basale, (c) basal membrane and (d) connective tissue fibers (mainly collagen). Between (a) and (b) are several layers of cells in various stages of maturation. The outermost cells are the most mature and are replaced by the underlying cells.

protein. Most skin and its derivatives contain α-keratin that is relatively rich in the sulfur-containing amino acid cysteine as well as most of the common amino acids and contains many disulfide cross bridges. The proportion of cysteine in hard and keratinized structures such as claws, hooves and horns can be as high as 22% whereas it varies from 10 to 14% in more flexible keratins of skin and hair. The scales, claws and beaks of reptiles and birds contain β-keratins that mainly contain small side chain amino acids mainly glycine, alanine and serine and lack cysteine. While α-keratins stretch on exposure to heat, β-keratins do not.

Keratinization leads to cell death. Keratized (cornified) skin provides mechanical protection to skin and also reduces water loss considerably. The **stratum corneum** is the outermost layer of the epidermis and can be made of several layers especially in areas of the body that are constantly under pressure. The outermost cells of the epidermis are continuously shed off the body of most vertebrates. The epidermis also has several non-epithelial cells known as **chromatocytes** or **chromatophores**. Capillaries are not present in the epidermis of vertebrates except in some salamanders.

The **dermis** lies below the epidermis and the two layers are separated by a basal lamina. At this junction, dermal papillae protrude into the epidermis bringing this layer closer to the body surface. Among all animals, the dermis is only found in vertebrates. Most of the dermis develops from dermatomes that are of mesodermal origin. The dermis is generally much thicker than the epidermis. In many vertebrates, the dermis acts as a template for production of dermal bone in skin through intramembranous ossification. Such a process was well pronounced in extinct ostracoderms and occurs during formation of some of the vertebrate bones such as the flat bones of mammals. The dermis is fibrous connective tissue that has a lot of collagen and elastic fibers. Fibroblasts as well as other cells such as fat cells, macrophages and pigment cells are also present in the dermis. Between these cells and fibers are blood vessels, sensory receptors and large molecules such as proteoglycans that bind water. In the deeper layer of the dermis or **stratum compactum**, collegen fibers are more tightly packed than is the case with the more superficial **stratum laxum** where the fibers are loosely packed. The dermis is the basis of leather in the leather industry.

The **hypodermis** or **subcutis** lies below the dermis and above the superficial fascia. It consists of a loose network of irregular bundles of collagen and elastic fibers. The composition of the hypodermis or its presence varies in vertebrates considerably and determines the degree of movement of the skin over underlying tissue. In birds and mammals, the hypodermis has a considerable amount of adipose (fat) cells and a rich blood supply.

SKIN PIGMENTATION

There are various types of pigments in the skin of vertebrates (except albinos) that are normally located in cells known as **chromatophores**. Originally of neural crest origin, chromatophore cells are usually located in the upper part of the dermis of fishes, amphibians and reptiles but they migrate to the epidermis in birds and mammals. There are various types of chromatophores that bear names according to pigment produced. **Melanophores** of ectothermic vertebrates or **melanocytes** of birds and mammals produce the dark brown pigment **melanin**. Melanin is passed on to other cells of the epidermis and feathers and hair of birds and mammals (respectively) through the many long processes of the pigment producing cells while the pigment is retained by melanophores in other lower vertebrates. Melanin granules are capable of movement in melanophores under hormonal control. The movement of the granules can alter the color of the vertebrate.

Other pigment producing cells include **xanthophores, erythrophores** and **iridophores** that contain yellow, red and silvery pigments respectively. Sometimes several of these pigments may be present in an area and as a result impart various colors depending on the combination of pigments present. Several factors can cause changes in colors that are caused by pigments of individual vertebrates. Solar radiation is responsible for the suntan seen in humans of lighter complexion. The rate at which the pigments are synthesized in chromatophores varies depending on different factors such as nervous or hormonal stimulation. Sympathetic stimulation causes aggregation of melanosomes in some fish whereas norepinephrine (noradrenalin) from the adrenal gland causes pigment aggregation in amphibians and reptiles. Melanophore stimulating hormone causes pigment dispersal in amphibians and reptiles.

Skin pigmentation serves various purposes in the vertebrate body. In several vertebrates including humans, pigmentation protects body tissues from the effect of ultraviolet light. Pigmentation is also useful in species recognition, camouflaging for purposes of avoiding predators or offense. As some reptiles depend a lot on solar radiation for thermoregulation, pigmentation plays an important role in the process. Such reptiles increase the radiation received by coming out in open sunlight and dispersing melanin in melanophores.

Skin of Fishes

In addition to the other general roles played by skin, the skin of fish is an important respiratory organ and also houses structures of living light in some species. The skin also contains mucous glands that secrete a slimy liquid that is important in shedding microorganisms from the skin surface and in reducing friction during swimming. The skin of the electric catfish (*Melapterurus*) contains electric organs whose discharge is capable of stunning other organisms. Some fish possess venomous glands in their skin.

Structure

The skin of fish in various fish groups has a similar general structure apart from the structure of scales. This has resulted as fish are faced with similar challenges in an aquatic environment. The **epidermis** of fish skin consists of several layers of moist living cells. The epidermis is relatively thinner when compared to that of other vertebrate groups. There are numerous unicellular glands in the epidermis. These glands together with other epithelial cells produce mucus. Mucus forms a coat that covers the surface of fish and plays an important role in reducing the drag effect and surface microbial load when it sloughs off the body. The microbial load could be harmful to fish if allowed to accumulate. Since fish odors are present in mucus, the secretion could be a source of communication among members of a group of fish. In some fish species, mucus coagulates and precipitates suspended material in water.

The **dermis** contains blood vessels, nerves, sense organs and connective tissue. Connective tissue is rich in collagen fibers that are arranged in layers that are perpendicular and also bind skin tightly to underlying muscles. The dermis plays a major role in formation of scales. Tubular or flask-shaped multicellular glands are found in the dermis of some fish such as hagfish that possess large slime glands that produce a lot of slime. Other fish also possess poison glands that open to the surface of the skin and produce toxins. Light producing photophores are found in the dermis of some fish.

Fish Scales

Most fish species possess scales. Lampreys and members of catfishes lack scales. Other fishes possess scales in a few areas of their bodies such as paddlefishes of central North America and the freshwater swordbill of China. The chemical composition of the scales varies depending on quantities of bone, dentine and enamel present. The scales are laid down by bone forming cells known as osteoblasts. Some scales still contain these cells and retain the structure of cellular bone but most scales of teleosts lack such cells and the normal bone structure and are known as **acellular scales**. Dentine and enamel are normally deposited on scales to various degrees in different fish groups. The dynamics of scale development and evolution have been presented and reviewed by Alibardi (2004) and Sire and Akimenko (2004).

The various scale patterns seen in fish are related to the body segments. Scales normally overlap in a similar manner to shingles on a roof. The embedded part of the scale is found anteriorly whereas the free end points caudally (Fig. 6.2). There are various types of scales that differ in shape, chemical composition and distribution in different fish groups (Fig. 6.3). The dermal bone of fishes includes the scales that represent the remnants of the early ancestral armor. **Cosmoid scales** (Gr. *kosmios*, well ordered) were found in extinct crossopterygians and were composed of an inner layer of compact bone, a middle spongy layer and an outer layer that had dentine-like material known as cosmine. Cosmine consists of dentine tubules that are grouped into radiating tufts. Keratin covered cosmine. The body of the coelacanth is covered with modified and thinner cosmoid scales. The **placoid scales** or **dermal denticles** (Gr. *derma*, skin; L. *denticulus*, small tooth) are found in sharks and their relatives (elasmobranchs) and have a structure that is similar to that of teeth. Placoid scales have an ectodermal cap with a cusp that is made up of an enamel-like substance known as vitrodentine. The cap covers a disk-like body of dentine in the dermis that bears a pulp cavity. Placoid scales do not increase in size with fish growth as happens with other scales. Growth in size of elasmobranchs is accompanied by an increase in the number of placoid scales.

Non-placoid scales include ganoid, ctenoid and cycloid scales. **Ganoid scales** (Gr. *ganos*, sheen) are found in bichirs, garpikes and reedfishes. Ganoid scales are shiny, hard and diamond shaped. The scales are made of a series of lamellar bone that is covered by layers of

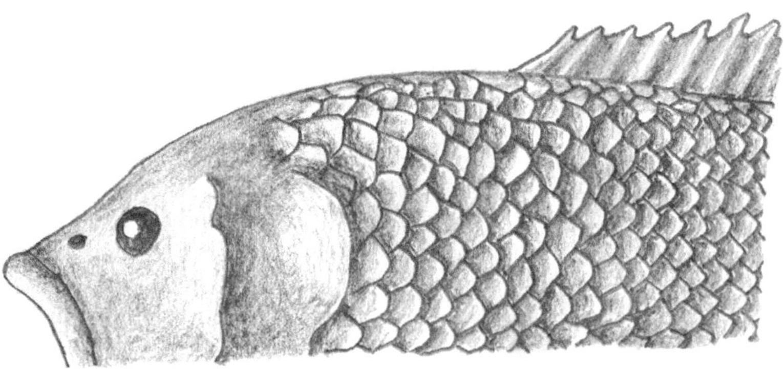

Fig. 6.2 A diagram showing the general arrangement of scales in fish. Such an arrangement of scales reduces resistance to smooth swimming movements by fish.

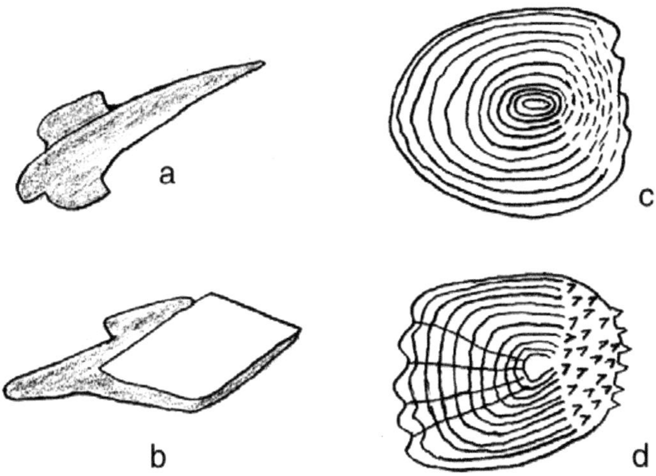

Fig. 6.3 Scales of bony fish. (a) placoid, (b) ganoid, (c) cycloid and (d) ctenoid.

enamel or ganoine. The **ctenoid** and **cycloid scales** (Gr. *ktenoeides*, like a cock's comb; *kyklos*, circle) of actinopterygians are very thin and overlap. The scales are made of acellular bone with fibrous material underneath. The scales are quite flexible and have growth rings. Ctenoid scales have a series of spiny projections at their outer posterier end. Cycloid scales tend to be circular in outline with a smooth outer edge. Cycloid and ctenoid scales add concentric layers as they grow.

Scale Derivatives and Uses

Throughout the evolutionary history of vertebrates, scales have evolved into various structures. Placoid scales have been modified into the jaw teeth of sharks and other higher vertebrates. Bony scales have evolved into dermal bones and lepidotrichia.

Fish scales have been used for classifying and ageing fish. Specific scales are found in some groups of fish and not others. The age of fish has been determined from annuli on scales. Scales are also useful fossil deposits for purposes of classification and in the study of the feeding habits of piscivorous fish since they can remain undigested in the gastro-intestinal tract.

Other Derivatives of Fish Skin

Chromatophores

Fish display quite a variety of colors. Some of these colors are permanent whereas others are temporary and depend on season of the year such as during the spawning period. Some fish are uniformly colored whereas others are a mosaic of various colors. Most colored fish are found in tropical coral reefs such as angelfishes, wrasses, butterfly fishes and parrotfishes. Coloration in fishes is due to pigments that are produced by chromatophores, iridiocytes that reflect colors from the environment as well as physical configuration (schemachromes) (Fox, 1953). Pigments present in fish include melanins, carotenoids, porphyrins, flavines, indigoids, chromolipoids and flavines. These pigments are produced by melanophores, xanthophores, leucophores, iridophores and cyanophores. The distribution of these pigments in the body varies. Melanins, carotenoids, flavines and purines are found in the skin mainly; porphyrins occur in blood and muscle whereas flavines have widespread distribution in many body systems.

Coloration serves various purposes in fish including identification, communication and camouflage. Coloration plays a role in breeding of fish. In guppies (*Poecilia reticulata*), the female fishes prefer brightly colored males (Pilastro et al., 2004) whereas carotenoid colored males are preferred by females of the live bearing relative of the guppy *Poecilia parae* for mating (Bourne et al., 2003). The diversity of colors seen in more than 1,000 species of cichlids (more than 10% of the world's freshwater fish species) of Lakes Malawi and Victoria in eastern Africa have occurred over millions of years as a result of replicated radiations into the same color types resulting in fish that are phenotypically very different yet closely related (Allender et al., 2003). Communication by means of visual signals has been seen in fish. Rapid darkening has been observed in juvenile Atlantic salmon that were losing territorial encounters whereas the victors retained their original coloration (O'Connor et al., 1999). The aggression level between the fighting fish decreased as soon as the losing fish became darker.

Some fish are capable of gradual or rapid color changes such as trout and characins. Physiological color changes occur rapidly and are due to motile responses by chromatophores whereas morphological changes are attributed to the morphology and density of chromatophores and are responsible for long-term changes. Mechanisms responsible for color changes include complex endocrine and nervous interactions as well as some paracrine factors such as endothelins (Fujii, 2000). Chromatophores have also been known to differentiate and die by apoptosis under the influence of factors that regulate motile responses (Sugimoto, 2002)

Barbels and Flaps

There are various structures that protrude from the skin of fishes including barbels and flaps (Fig. 6.4). Barbels are thin and long extensions

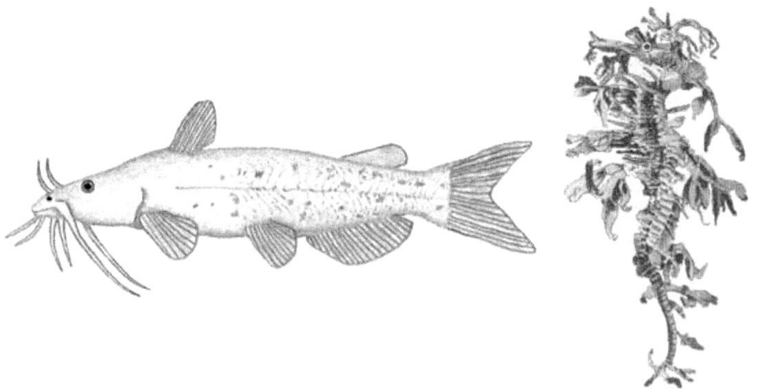

Fig. 6.4 The channel catfish (*Ictalurus punctatus*) (left) showing barbels protruding from the head and the sea dragon (*Phycodurus eques*) (right) with long leaf-like flaps all over the body. Catfish are a diverse group of slightly over 3,000 species of fish that are named so because of the prominent barbels that resemble the whiskers of a cat. The sea dragon, also known as the leafy sea dragon, is related to the seahorse and inhabits shallow and temperate waters around southern and western Australia. The flaps act as a camouflage in fish that possess these skin protrusions.

that possess sensory organs and act as accessory feeding structures. Barbles mainly extend from the head and are found in species such as catfishes, sturgeons and goatfishes. Flaps are thin and wide extensions that resemble leaves in appearance and are found in some fish such as the sea dragon and sargassum fish. Flaps are camouflaging structures in these fish.

Light Organs

The evolution of bioluminescence (production of light by living structures) has occurred independently many times and the genes responsible are not related. Bioluminescence is found in various organisms including bacteria, algae, coelenterates, beetles and some marine fishes. Organs responsible for production of light are found in the skin of such fish. The light organs or **photophores** are well developed in benthic and mid-water species. Light organs are important in species recognition and luring of prey. The sexual dimorphism that has been associated with some photophores has implicated bioluminescence in attraction of fish to their mates (Herring, 2000). Deep-sea fish are able to recognize their own bioluminescence that has a certain spectral waveband which is invisible to other organisms (Douglas et al., 1998).

Luminescence in fish arises from the presence of luminous cells on fish or luminous bacteria (Harvey, 1957). Photophores produce light from either cells or luminous material that is secreted by the skin. Such skin secretions are found in the searsiid. Fish that are capable of generating light from their own cells include the lanternfishes, toadfishes, deepsea scaly dragonfishes and some sharks and electric rays. Various members of cods, sea basses, deepsea anglerfishes, cardinalfishes and grenadiers produce light of bacterial origin. Most of the light produced by bioluminescent fish is blue or green.

Generally, production of light involves reaction of oxygen with various compounds known as luciferins and luciferases. These exothermic reactions produce photons of visible light whose color is determined by various factors such as the structure of luciferins and the amino acid sequence of luciferase (Wilson and Hastings, 1998).

Skin of Amphibians

The amphibian skin has a thin stratum corneum that is one or two cells thick and shows a low level of keratinization. The aquatic larval stages of amphibians have skin that is similar in structure to that of fish and does not show keratinization. As the stratum corneum is thin, the skin of amphibians can be used as a respiratory organ as well as providing some protection against desiccation. Some salamanders that lack lungs depend on the skin entirely for respiration. The earliest and now extinct amphibians possessed scales like their fish ancestors but most extanct amphibians lack scales apart from some caecilians and anurans. The topmost layers of skin in amphibians are sloughed off in large sheets periodically that are normally eaten by these vertebrates.

Frogs and salamanders have mucous and poison glands in their dermis (Fig. 6.5). These glands are multicellular and alveolar in nature. Most of the glands are mucous glands and their secretions keep the skin surface moist as well as reducing water loss. The dermis is well vascularized. The parotid gland of toads secretes irritating liquid that keeps off potential predators. Highly colored frogs possess poison glands that produce very toxic poison. Such frogs are found mainly in the

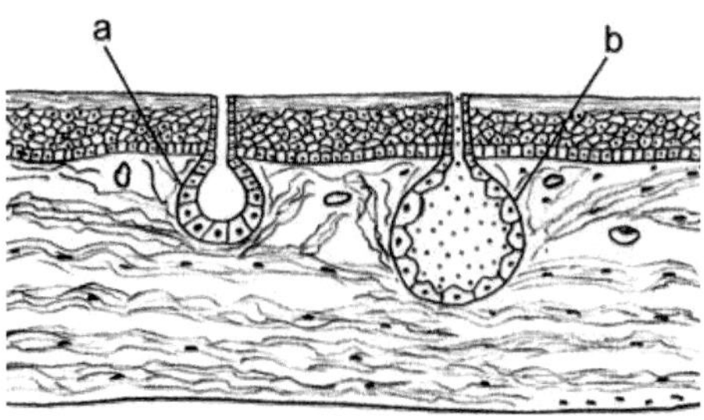

Fig. 6.5 A vertical section through an amphibian skin. (a) mucous gland and (b) granular gland that secretes toxic substances.

tropics. **Osteoderms** (dermal bones supporting the epidermis) are present in some anurans and caecilians. They comprise a few bony nodules that are located in grooves of the skin. Coloration in amphibians is due to presence of chromatophores. The main chromatophores present include melanophores, xanthophores and iridophores.

Skin of Reptiles

The reptilian skin is extensively keratinized and has a thick stratum corneum. This is an adaptation to terrestrial life as a high degree of keratinization and presence of phospholipids greatly reduces water loss and also offers protection against abrasion. The dead superficial cells have evolved into **epidermal horny scales** with a hinge between them (Fig. 6.6) in snakes and lizards and **horny plates** in turtles. During **molting** (ecdysis), the outer layer of cells is shed off as a unit and is replaced by the deeper layer of cells. The intraepidermal shedding layer results from temporary deposition of β-keratin in this epidermal layer (Alibardi, 2003). In most turtles, the shell is covered by horny scales known as scutes and wears off at the surface mainly. Newly formed horn grows around the borders of old plates resulting in a series of rings at the margins of these plates. Horny scales are lacking in some turtles including the leatherback sea turtle and their shells are covered by leathery skin.

Reptiles with limbs have claws on their most distal digits. These claws are strong and are derivatives of the epidermis. They are reinforced by the presence of calcium salts. The claws can be quite long in some lizard species.

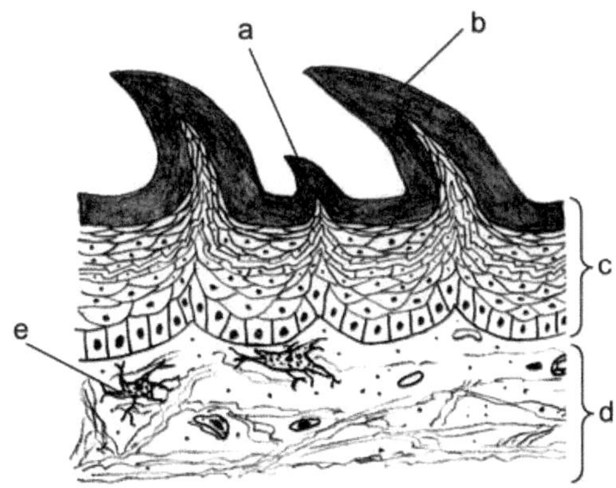

Fig. 6.6 A section through reptilian skin. (a) hinge, (b) horny scale, (c) epidermis, (d) dermis and (e) chromatophore.

In some reptilian species, scales have been modified for several functions. The scales on the ventral body surface of snakes are modified and can become erect due to the action of the **costocutaneous muscle** in the dermis that attaches to these structures. The scales provide friction for locomotion and prevent a backward slip of the body during undulatory motion. The tips of digits in some reptiles such as geckos have modified their scales into suction pads. The presence of such pads enables geckos scale vertical surfaces or crawl upside down on horizontal objects.

There are few glands in the reptilian skin. Mucous glands are lacking. Some scent glands are present and produce secretions that are important during courtship. Scent glands are located in specific areas of the body depending on species such as femoral region in lizards and cloaca and lower jaw in crocodiles and turtles. The skin also has a few sense organs. **Apical pits** bear small hairlike structures that serve tactile functions and are located posterior to epidermal scales. Osteoderms are present in some reptilian groups such as crocodiles and tuataras and are arranged into long and hanging rib-like structures known as **gastralia** (Gr. *gaster*, stomach) that are located in the ventral and lateral wall of the abdomen between the sternum and pubis (Fig. 6.7). Gastralia are often referred to as ventral, abdominal, belly or gastric ribs.

Many of the more than 100 chameleon types have a body color that suits their habitats and camouflages them. Chameleons do change their body color, sometimes in as short a time as 20 seconds. The color changes

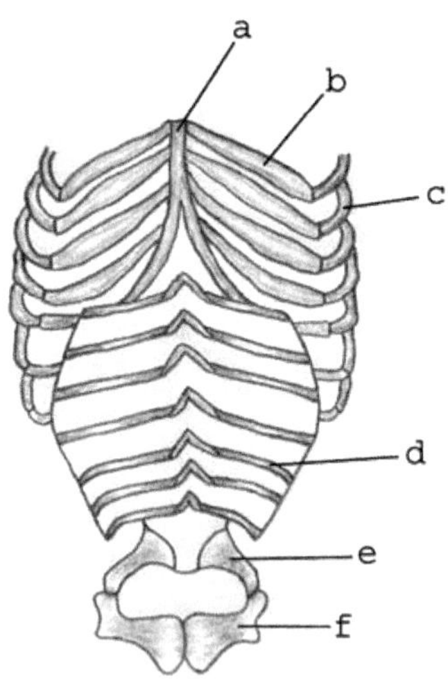

Fig. 6.7 Part of the ventral side of a crocodile. (a) sternum, (b) costal cartilage, (c) vertebral rib, (d) gastralia, (e) pubis and (f) ischium. Gastralia were also present in saurischian dinosaurs especially the theropods where they are thought to have played a role in respiration.

are brought about by light, temperature and mood and enable the chameleon to communicate with other members of the species. Chromatophores are responsible for such color changes. The outer skin of the chameleon is transparent and the top layers of chromatophores contain the red or yellow pigment whereas the inner layers bear the blue or white pigments that reflect light. White can shine through red and yellow colors enhancing them in the process while blue reflects through yellow giving it the green color. Each chromatophore is surrounded with a muscle that is capable of contracting on receiving the appropriate signals from the brain. Contraction of the chromatophore muscle pushes the pigment to the outer part of the cell that also flattens out to become wide and thus shield other relaxed chromatophores. The overall skin color will depend on the type of chromatophores that have contracted and those that are in the relaxed state. Skin color can also darken or lighten by altering the amount of melanin in the long tentacle-like processes of melanophores. The chameleon is thus able to display various colors and patterns by contracting and relaxing its chromatophores.

Skin of Birds

The skin of birds produces and bears feathers. The feathers protect the relatively thin skin from abrasion. Skin in feathered parts of the body shows little keratinization. Skin is thicker in non-feathered parts of the body such as feet and claws, combs, wattles and the beak. Horny scales are found on the feet of most birds. The skin of birds is easily movable on the developed hypodermis and this is important in movement of feathers.

The epidermis in feathered body parts consists of a few layers generally. The thin stratum corneum is constantly being desquamated. Melanocytes are found in the epidermis to which they have migrated from the dermis and are responsible for various colors seen in birds such as yellow or red coloration of feet or beak that is caused by lipochromes. The bird skin displays several colors such as ultraviolet, blue, green or yellow that result from coherent scattering by several collagen fibers located in the dermis (Prum and Torres, 2003).

The superficial part of the dermis has collagen and elastic fibers that form ridges that are responsible for epidermal patterns which are found in areas of skin with scanty or no feathers. **Papillary bodies** are well developed in areas without feathers such as the balls of the feet. The deeper part of the dermis consists of bundles of collagen and elastic fibers that are interwoven together with feather follicles and the non-striated muscles **arrectores plumarum** (mm.pennarum) that attach to feather follicles. The elastic tendons of these muscles extend into the follicles.

Arrectores plumarum is innervated by the sympathetic nervous system. The dermis also has striated muscle in areas that are covered by feathers. These muscles contract voluntarily and bring about folding of skin and erection of feathers. The voluntary muscles are also important in movement of primary wing feathers and tail steering feathers.

The hypodermis has adipose tissue whose color depends on the species. It is normally yellowish in the duck and goose. Adipose cushions are located in digital pads and are well developed regardless of the nutritional status of the bird. The hypodermis and the dermis have many blood vessels. Areas of skin with great blood supply include feather follicles, ear lobes, wattles, the comb and the snood (frontal process) of the turkey.

Brood patches are located on the ventral part of the breast and show varied levels of prominence in different birds. These areas have scanty feathers and a high blood supply at the time of brooding. The bird's body heat is transferred to incubating eggs in this region.

The skin of birds has few glands. Sebaceous glands are sparsely distributed and **ceruminous** or **wax glands** are located in the skin of the outer ear. The **preen** or **uropygial gland** is located above the end of the tail (pygostyle) and consists of two bilaterally symmetrical halves. This branched alveolar gland is covered with a connective capsule. The sebaceous secretions of each gland are similar to those of mammals and pass out through a single pointed papilla. The bird spreads the fatty and waxy secretions of the gland over its feathers during preening using its beak. The film of fat over the feathers acts as a water repellent. The preen oil differs in its composition between birds of different ages, sex and diet (Sandilands et al., 2004). Preen glands are well developed in aquatic birds but are lacking in a few species such as the ostrich. The skin of birds lacks sweat glands.

The skin of a bird's foot is adapted to performing the specific functions of a species. In many aquatic birds, the second, third and fourth digits are joined for most of their length by a web that is made of a double layer of skin. The digits of birds bear claws at their tips. These claws keep growing at a rate that is similar to that of regular wear. Claws are quite long, strong and pointed in predaceous birds such as hawks, eagles and owls. The bony pointed spurs that are found on the tarsometatarsal bones of some male and old female birds are covered with horny sheaths. The soles of metatarsophalangeal and interphalangeal joints are lined with cushions of adipose tissue that is located in the hypodermis and forms pads.

All birds possess feathers that play an important role in body temperature regulation as well as flight in most birds. Feathers entrap air and reduce its flow over the skin. The air forms an insulating layer that reduces water loss by evaporating and also heat loss. Birds are able to maintain a high and stable body temperature as a result. Several non-avian dinosaurs possessed filamentary appendages in their skin that may have been similar to the bristles of the wild turkey beard (Sawyer and Knapp, 2003; Sawyer et al., 2003). An Early Cretaceous dinosaur that possessed a filamentous integumentary covering was discovered in China and could provide evidence that such dinosaurs had protofeathers (Xu et al., 2004). Feathery appendages have also been found in therapods of Late Jurassic and have been associated with the evolution of higher metabolic rates, improved locomotory abilities and visual communication (Kundrat, 2004).

Apart from the down feathers, each feather (Fig. 6.8) has a **shaft** or **quill** that is divided into the **calamus** or **barrel** that lacks a vane and the **rachis** that possesses a vane. The calamus is cylindrical and hollow and lies in the feather follicle surrounded by the dermis. The follicle consists of epidermal cells and the inner sheath that is in contact with the calamus. The proximal end of the calamus has an opening known as **inferior umbilicus** through which the vascular dermal papilla is able to enter the proximal part of the shaft. The dermal papilla plays a role in the growth of a feather but recedes when growth is complete leaving a calamus that

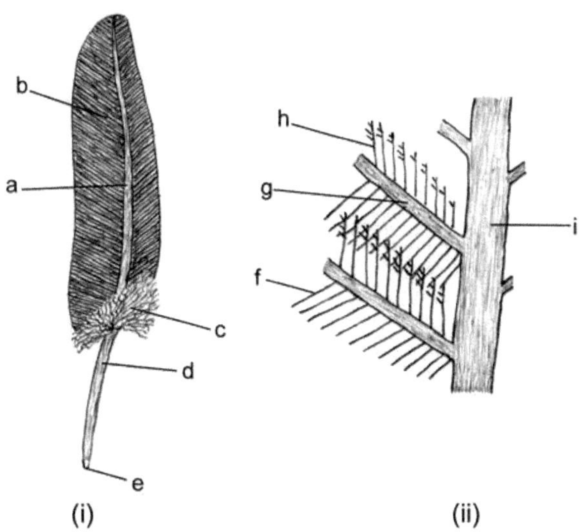

Fig. 6.8 General structure of a feather (i) and (ii) relationship of various structures of the vane. (a) rachis, (b) vane, (c) afterfeather, (d) calamus, (e) inferior umbilicus, (f) proximal barbule, (g) barb, (h) distal barbule with hooklets and (i) rachis.

is composed of keratinized epidermal cells and air. On the lower side of the calamus is a groove that runs up to the beginning of the vane known as **superior umbilicus**. The **afterfeather** comprising a small second vane arises from the superior umbilicus in some bird feathers. The **vane** comprises **barbs** that attach to the rachis obliquely, and the smaller proximal and distal **barbules** that attach to barbs. The distally directed barbules bear hooklets into which fit the proximal barbules that lack hooklets. Such an arrangement enables the proximal and distal barbules maintain a firm but flexible union. The outer layers of the calamus and rachis are fibrillar in nature due to the presence of keratinized epithelial cells that are arranged into long threads. The inner cells of the rachis possess polygonal cells containing air.

The various types of feathers of birds include contour, down and filoplumes. **Contour feathers** cover the body surface of birds. A type of stronger and larger contour feather that plays a role in flying is known as the **flight feather**. The body's outline is determined by contour feathers as they vary in size and shape and cover the trunk evenly. The barbs of **down feathers** (Fig. 6.9i) arise from the distal end of the quill and are fluffy and soft since they lack hooklets. **Filoplumes** (bristles) (Fig. 6.9ii) are short and stiff and have an almost vaneless rachis. Their distribution is limited in the body to such areas as around the mouth, nostrils and eyes.

Feathers show various colours depending on pigmentation and their physical structure. Pigmentation is largely due to melanin. Melanoblasts normally migrate into immature feathers where they will produce

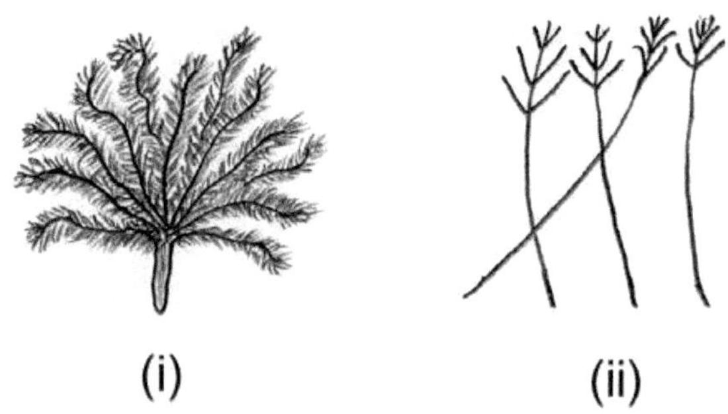

Fig. 6.9 Other feather types. (i) down feather and (ii) filoplumes. Down feathers are covered by the tougher contour feathers and are good thermal insulators since they trap small pockets of air that act as thermal barriers. Down feathers are stuffed in products such as bedding material, sleeping bags, pillows and jackets and are the only feathers that cover very young birds. Filoplumes are arranged around other feathers and are thought to be pressure and vibration receptors that can sense the arrangement of other feathers so that they are adjusted to their proper position.

melanin granules. The granules are deposited in the non-keratinized epidermal cells of the feather. The colors normally seen are red, yellow brown or black and depend on the intensity of pigment present. Pigmentation of feathers can also result from ingestion of pigments such as lipochromes that are contained in food. Lipochromes impart yellow, orange or red colors to feathers. The physical structure of a feather together with its light scattering properties renders white, blue and iridescent colors to hair.

Feathers are considered to have evolved from the horny scales of reptiles and the two structures are considered to be homologous. The development of these two structures is similar. The dermis influences the epidermis and there are specific areas where morphoregulatory molecules may be exchanged between the two layers. These areas shift their positions with development and results in production of different appendages depending on the genetic composition of the epidermis of each vertebrate species (Alibardi, 2004).

Mature feathers that have ceased to grow are dead structures that are subject to mechanical forces of wear. Replacement of worn out feathers is a continuous process. Partial or complete loss of feathers is known as a molt and is controlled by hormones of the thyroid and pituitary glands. The cells of the dermal papillae are in a quiescent state after feathers have matured and can be activated to form more feathers when the need arises. The old feathers are normally discarded during such a replacement process.

Skin of Mammals

The mammalian skin (Fig. 6.10) comprises an epidermis that rests on a basement membrane and a dermis that lies over a hypodermis. The skin possesses structures such as sensory organs, hair, sebaceous and sweat glands and blood vessels.

The **epidermis** consists of several layers of squamous epithelium that varies in thickness and degree of keratinization depending on location and pressure encountered by each part of the body. There are various transitional stages between the stratum basale and stratum corneum that comprise of many layers of keratinocytes in various stages of maturity. The hair-bearing parts of skin tend to have a thin epidermis and show low keratinization. The highest level of keratinization is found in digital organs such as claws, hooves, nails and horns.

Branched or **dendritic cells** are also found in the mammalian epidermis. Under normal conditions, dendritic cells are located between the cells of stratum basale. These cells can be stimulated to produce

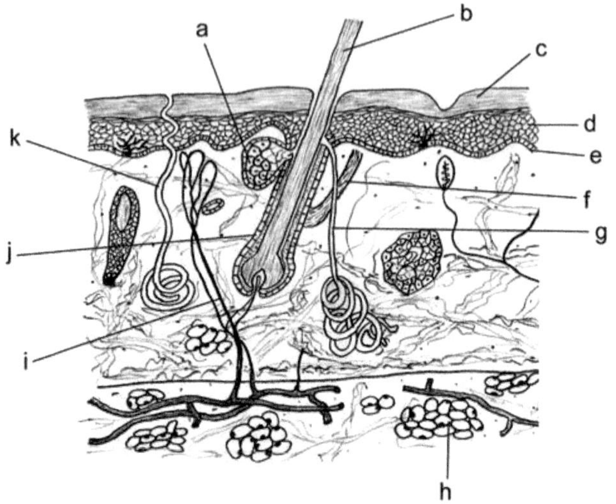

Fig. 6.10 Mammalian skin. (a) sebaceous gland, (b) hair shaft, (c) stratum corneum, (d) epidermal cells in different stages of maturation, (e) stratum basale, (f) arrector pili muscle, (g) duct of apocrine sweat gland, (h) adipose tissue in the hypodermis, (i) blood vessel, (j) hair follicle and (k) duct of eccrine sweat gland.

melanin and are known as melanophores or melanocytes. Other cells located in the epidermis are **non-specific dendritic** and **Langerhans cells**.

The **dermis** varies in thickness depending on the species and the area of body. The dermis produces dermal bones in some mammals such as the dermal plates of armadillos. The mammalian dermis has two main layers; a more superficial papillary layer and a deeper reticular layer. The **papillary layer** forms the dermal papillae that project into the epidermis and is made of a loose connective tissue with collagen and elastic fibers. On the palms of the human being, the thin epidermis conforms closely to the contours of the dermal papillae and thus gives rise to the individually specific ridges on the surface of skin in these areas that are important in gripping surfaces and in fingerprinting. The papillae are supplied by hairpin-like capillary loops that supply nutrients to the epidermis and sensory nerve endplates. The dermal papillae increase the surface area that is available for nutrition of the epidermis. The deeper **reticular layer** has a dense network of fibers. The thick and dense cords of mainly collagen fibers run in various directions and are responsible for the firmness of leather. Between these cords are elastic fibers that render the property of elasticity to skin. Several sensory receptors to pain, pressure, touch and temperature are located in the dermis of most areas of skin.

Derivatives of the Epidermis

Various epidermal structures have evolved in mammals and include hair, horns, antlers, baleen, claws, nails, hooves and glands.

Hair

Hair is a distinct mammalian feature. Hair is derived from epithelial cells of the epidermis and is a flexible horny fiber. Hair normally lies obliquely in skin. A thick coat of hair can trap air thus forming a good insulating layer. A good cover of hair also affords mechanical protection to a mammal. Structurally, hair consists of a **shaft** that lies outside the skin and a **root** that lies obliquely in a hair follicle. The hair root is enlarged at its proximal end into a hair **bulb** that contains a conical hair papilla. The bulb is nourished by the blood vessels of the papilla that also contains nerve endings.

The **hair follicle** comprises mainly of epidermal cells that grow into the dermis and surround the shaft. The inner part of the epidermal follicle is made of keratinized cells. The epidermal part of the follicle is surrounded with a fibrous connective tissue sheath that comprises outer and inner follicular layers. The inner layer has fibers that run in a circular manner whereas the outer layer consists of longitudinal fibers. It has been shown that the hair follicle contains a reservoir of stem cells of suprabasal or follicular bulge region that can regenerate the epidermis and hair follicles during wound healing (Blanpain et al., 2004; Christiano, 2004). These cells are of neural crest origin (Fernandes et al. 2004). The same cells express nestin that could play a role in generation of blood vessels (Amoh et al., 2004).

Hair growth occurs above the apex of the papilla (growth point) that possesses epithelial cells of the stratum basale. Hair forming cells multiply and are pushed to the base of the hair shaft and upwards in the follicle while undergoing keratinization. A hair shaft (Fig. 6.11) consists of a medulla, cortex and cuticle. The **medulla** is a central cord of polygonal, cuboidal or longitudinally flattened dead cells that sometimes contain air between and even within the cells. The middle **cortex** comprises layers of keratinized and spindle-shaped cells that contain varying amounts of melanin. This layer is responsible for most of the strength of hair. The **cuticle** is thin and contains scaly cells that overlap like roof shingles and covers a hair shaft. The free borders of these cells face the tip of the hair. The pattern of these cuticular cells, whose cell borders form a system of delicate lines, varies in different mammals and can be used, in addition to other features, in identifying the mammalian species the hair belongs to. The shape and thickness of the cortex and

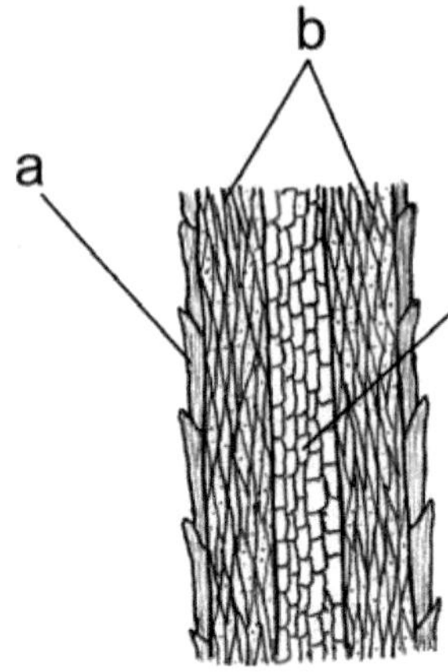

Fig. 6.11 Hair shaft. (a) cuticle, (b) cortex and (c) medulla. Special qualities of hair including elasticity and curl are due to the composition of the cortex. The keratin fibers in the cortex are low in sulfur and are compressed into bundles of larger fibers.

medulla varies in different mammalian species and the type of hair present in a species. The medulla is the most variable part of hair. For forensic purposes, the relative thickness of the cortex and medulla, the structure of medullary cells and the lines that form the free borders of cuticular cells have been used before the advent of molecular methods in identifying the mammalian species to which the hair belongs.

In each hair shaft, adjacent hair fibrils are linked to each other by adjacent sulfur bonds that are formed between amino acids with sulfhydryl groups such as cysteine. The number and arrangement of these disulfide linkages determines the hair shape. This arrangement can temporarily be altered and thus change the hair shape by some chemicals and heat. The normal shape of hair is restored with new growth.

Hair varies in color in major mammalian groups, species and even within individuals. Hair color is chiefly determined by the nature of pigment within the cells of the cortex. Interaction between air in the cortex and quantity of pigment present also determines the color of hair. The color of many mammals is determined by the color of their skin hair which does not have to be the same as that of skin. The blue color seen in some mammals such as mandrill, vervet monkey and some species of opossum has been attributed to the presence of color producing dermal collagen arrays that are anatomically and structurally identical to structures such as the tapetum of some mammals and the cornea of some fishes (Prum and Torres, 2004).

Seasonal changes sometimes affect the color of animals. Many birds and mammals show different colors of hair and feathers depending on the time of the year. Changes in day length and temperature trigger a hormonal response that stimulates production of different biochromes.

Associated with hair is a smooth muscle, **arrector pili**. This muscle originates from the superficial part of the dermis and inserts to the lower distal part of a hair follicle. Contraction of the muscle makes hair 'stand on end'. The muscle contracts on stimulation by the sympathetic nervous system as a result of various factors such as low temperatures leading also to a 'goose skin' which results from depression of the epidermis by the muscle. Erect hair traps a layer of air that insulates the animal. The muscle can also be stimulated when an animal senses danger thus making an animal appear larger than normal and frightening off potential enemies.

There are several types of hair. A dense coat of hair is known as **fur** (**pelage**). **Outer hairs** form the coat of most mammals and determine the color of an animal. They vary in length and may be wavy and soft. **Wool** hair is usually wavy and is located between and below outer hairs. **Bristles** are stiff outer hairs with split tips and include the body hair of pigs, nostril hair, eyelashes, tragi and the moustache hair of the horse. **Tactile** (**sinus**) hairs or **vibrissae** are long hairs that are found in the oro-nasal region and are very sensitive to touch. The root of a tactile hair is within a blood sinus whose covering has many nerve endings. Pressure of tactile hair is amplified by hydraulic effect of the sinus and detected by its sensory nerves.

The life span of hair varies. Hair growth normally ceases after a while when root cells become quiescent. During such a time, the base of the old hair will be detached from the root. When growth begins again, the old hair is pushed out of the follicle. Before birth, the fetus in many mammalian species is covered with fine and soft hair that is relatively more pigmented known as **lanugo** (L. *lana*, wool) that is mostly lost before birth. Such hair contributed to the high value of fur from the fetus of the Asian karakul sheep before the practice of harvesting fetuses from these sheep was discouraged.

Skin Glands

Skin glands are numerous and are found in most mammals. The secretions of these glands perform various roles that include regulation of body temperature, scent organs, excretion of metabolic wastes, production of milk and greasing of the skin and hair.

Sebaceous glands (L. *sebum*, tallow) are branched and saccular in nature and are normally linked to hair follicles. These glands produce a fatty secretion (sebum) by disintegration of entire cells (holocrine secretion). The secretion is passed into hair follicles and skin surface where it lubricates and waterproofs hair and skin thus protecting these structures from effects of water and air. Secretions of sebaceous and apocrine glands also play a role in thermoregulation in cold-and heat-stressed hunters (Porter, 2001). During hot conditions, secretions from these glands emulsify eccrine sweat leading to formation of a layer of sweat that reduces loss of water from skin. In cold conditions, sebum changes into a water repelling substance thus protecting skin and hair. True or free sebaceous glands occur in certain regions of the body such as nipples and open directly to the surface of the skin. Such glands can develop into scent organs that produce specific secretions that are important during the mating season, in recognition of members of the same species or group or serve as a warning to the presence of predators in an area. The eyelids possess long sebaceous glands known as **tarsal glands** that open on the edge of the eyelids. Their secretions protect the surface of the eyeball.

Sweat glands are tubular glands whose ducts may be coiled. The terminal parts of sweat glands are surrounded with myoepithelial cells and contraction of these cells leads to discharge of sweat. Sweat glands are widely distributed in human beings and chimpanzees but are lacking in some mammals such as cetaceans and sirenians. Higher mammals possess eccrine and apocrine sweat glands.

Eccrine sweat glands do not branch and open directly on the skin surface. Their ducts lead from the terminal parts that lie in the dermis and pass through the epidermis in a corkscrew manner. The glands are mainly found in areas of the body with little hair and produce a watery secretion (sweat) that is slightly acidic and sometimes contains urea and salts. Evaporation of sweat leads to cooling of the body surface. Secretion of sweat also plays a role in water and salt balance as well as inhibiting the growth of bacteria and fungi. The hippopotamus produces colorless and viscous sweat that gradually turns red and then brown as the non-benzoid aromatic compounds present in the perspiration polymerize (Saikawa et al., 2004). These compounds are acidic and have antibiotic and sunscreen activity.

Apocrine sweat glands are older than eccrine sweat glands phylogenetically. Their terminal ends branch and are located in the dermis. They discharge their thicker odoriferous secretion that is characteristic of individual mammals into hair follicles. The distribution of these glands is limited to the inguinal and axillary areas in humans. Aporcine glands can be modified into scent glands that are located in various parts of the body. In higher primates including humans, apocrine glands of **Moll** are found at eyelid margins and their secretions that have a defense action to the eye (Stoeckelhuber et al., 2004). **Ceruminous glands** are large sebaceous and apocrine glands that are located in the skin of the external auditory meatus. The apocrine glands of this region are simple coiled and tubular in nature and either open directly on to the surface of skin covering the external auditory meatus or into hair follicles together with sebaceous glands. These produce earwax or cerumen that is a mixture of their secretions and epidermal scales. Earwax and hair at the entrance to the meatus known as **tragi** minimize entrance of foreign bodies into the ear.

Mammary Glands

Mammary glands are modified sweat glands that produce milk for nourishing mammalian young. They are fully developed in the female mammals after puberty and represent an important secondary sexual character. Mammary glands also develop in males but do not become functional unless there are hormonal disturbances. The number of mammary glands present and their location varies in mammals. For example, the glands are located in the thoracic region in humans, monkeys and the elephant whereas they are found in the inguinal region of ruminants and equine species.

Superficially, the mammary gland (Fig. 6.12) consists of two parts: the hemispherical glandular part and the teat that is papilla-like in shape. Monetremes lack teats. The size of the gland depends largely on the

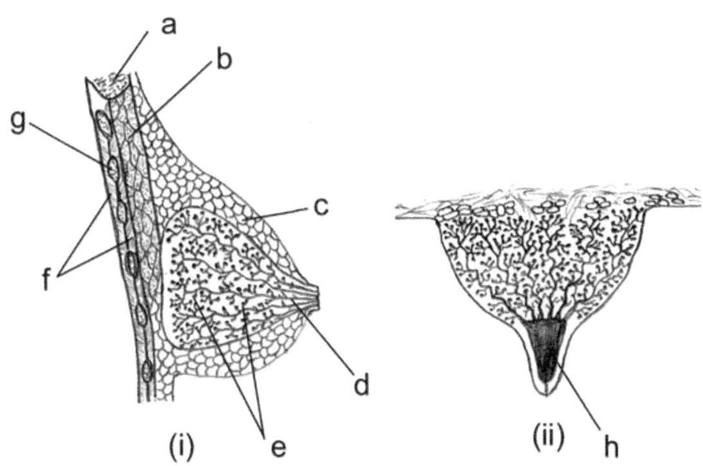

Fig. 6.12 Mammary glands of (i) the human being and (ii) small ruminant. (a) clavicle, (b) pectoralis major muscle, (c) adipose tissue, (d) lactiferous duct, (e) alveoli, (f) external and internal intercostals muscles, (g) rib and (h) teat cistern.

amount of fat around the glandular tissue than the glandular tissue itself. Mammary gland development depends on estrogens and progesterone. Estrogens stimulate the development of the duct system whereas progesterone controls development of the secreting cells.

The milk secreting cells (alveoli) are arranged in grapelike clusters that are surrounded by myoepithelial cells. The clusters lead to small ducts. The many ducts present will unite and eventually lead to lactiferous ducts that lead to teat openings at the apex of the teat. The number of teat openings present varies in mammals. There is a single opening per teat in many ruminants whereas the human being has 15 to 20 openings per nipple.

Lactation could have evolved before mammals from apocrine-like glands that were associated with hair follicles (Oftedal, 2002) as is evident in monotremes. Stem cells responsible for mammary gland development are present in the terminal end buds. The proliferation of these stem cells is under the control of hormones and other local mechanisms (Dulbecco et al., 1982). Lactation is important not just for nutritive purposes but also for imparting immunity to the young since mammals show a physiological delay in production of immune factors (Goldman et al., 1998). The long period of contact between mothers and the young that is brought about by suckling especially in primates enables the young to learn more and might be responsible for the high level of intelligence in this group (Peaker, 2002).

Hairless Skin Organs

Such organs lack hair and glands and are characterized by a high degree of keratinization of the epidermis. Hairless skin organs (Fig. 6.13) include digital pads, digital organs, horns, antlers and baleen of whales.

Digital pads are located at limb extremities of many mammals such as carnivores. The epidermis of the pads is greatly thickened and comprises of a soft and elastic keratinous layer. The hypodermis is well developed with firm adipose cells and a meshwork of elastic fibers. The pads act as shock absorbers due to the presence of structural fat. Some mammalian species such as carnivores have apocrine glands in the wall of the pads and produce secretions that act as markers. The **ergot** and **chestnut** of the equine species and the **dewclaw** and **carpal pads** of the dog family are non-functional rudiments of digital pads and do not touch the ground when these animals are in a standing position.

Digital organs include claws, hooves and nails. Nails are found on the most distal phalanges of primate fingers and toes. Each nail has a visible body and a root that is covered by a cuticle made of a fold of skin.

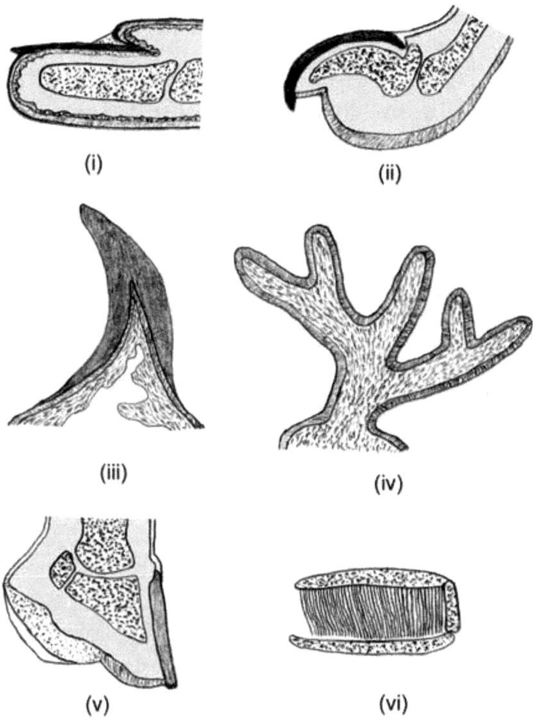

Fig. 6.13 A paramedian section through some of the hairless skin organs. (i) nail, (ii) claw, (iii) horn, (iv) antler, (v) hoof and (vi) baleen. (a) distal phalanx, (b) sole, (c) horny plate, (d) cuticle, (e) digital pad, (f) frontal sinus, (g) cornual process, (h) velvet, (i) distal sesamoid bone and (j) baleen fibers.

The proximal part of the nail has a crescent shaped white colored lunula that is the nail producing part. The stratum basale lies beneath the lunula. Below the nail is the nail bed that has a rich blood supply. The nail is thicker at its free border than at its root.

Hooves are found in ungulates and are a protective cover of horn for the terminal phalanges. They also serve as shock absorbers on concussion with the ground and can be used as weapons in some species. The ligaments of the toes also play a role in shock absorption due to their springy nature that allows for separation of toes when weight is placed on limbs. The thick epidermis and dermis are firmly united by the presence of lamellae and secondary lamellae of the dermis. Lamellae and secondary lamellae are highly developed in the equine family.

Claws are found in carnivores and cover the pointed terminal phalanges. Claws are the oldest digital organs of mammals. The epidermis of the cone shaped claw is quite hard.

Horns are found in many herbivores and comprise a hollow bony (cornual) process of the frontal bone that is covered by a highly keratinized skin. Horns vary in shape depending on species and environmental conditions such as nutrition. Some species have horns with superficial grooves as a result of uneven growth in response to environmental conditions. In many horned herbivores, the horns taper to the tips and vary in length.

Antlers consist of a solid bone that is covered with a sensitive non-glandular skin known as the '**velvet**' during the growing period. The branched antlers of male and female reindeer and caribou and other male cervidae differ in many ways. The number of branches on an antler varies with age and is useful in determining a deer's age.

The antlers are shed off yearly, normally when the mating season is over. The antlers take about two to three months to grow to full size. Antlers are the only mammalian appendage that is capable of re-growth after being shed off. Sex steroids regulate the cell-cycle progression and cellular differentiation in antlers (Price and Allen, 2004). The rich blood supply of the velvet nourishes the growing antler. When mature, a ring forms at the base of each antler thus cutting off blood supply to the velvet that will wither and eventually fall off. Antlers play a role during the mating season as they attract the opposite sex. Male deer use antlers to fight off potential rivals and secure more female mates. Antlers can be used as defensive weapons and in marking territories.

Baleen (L. *balaena* or Gr. *phallaina*, whale), also known as **whalebone**, is a sieve-like structure found in large and toothless baleen whales (Mysticeti) and is used to trap organisms including krill and plankton from water. Each plate of baleen is found on each side of the upper jaw and is a series of stiff, flexible and keratinized hair-like fibers that also contain small quantities of calcium phosphate. Baleen is used to filter krill from water in the mouth. The separate fibers are cemented together at their bases. As the lower ends of baleen are continuously wearing out, baleen grows throughout life.

Cutaneous Scent Glands

Such glands are located in the mammalian skin and are specific depending on the mammalian species or sex of the individual animal. Scent glands produce secretions with a specific scent whose principle role is recognition of members of the same species. Quite often, the scent glands are an invagination of the epidermis. The glands are coiled tubular, apocrine or modified sebaceous in nature. The glands are located in various parts of the skin with little hair such as the head, ano-genital, proximal part of the tail and distal parts of the limbs. The high level of muscles, myoepithelial cells and blood capillaries in these glands can exert pressure on the secretory cells at various times such as during excitement leading to secretion of their products onto the skin surface.

The odoriferous products of scent glands are pheromones that are disseminated by direct contact with the skin or by the longer hairs that are found in such areas in some mammalian species. The scent glands that are located in the ano-rectal region discharge their secretions into the rectum and anal opening where they will mix with feces and will be discarded to the environment.

Generally, scent glands with a rich supply of sebaceous glands such as anal glands produce secretions that last for a longer time in the environment. Apocrine glands produce secretions that are more volatile than those of sebaceous glands and last for a shorter time in the environment such as the secretions of the glands of the distal parts of limbs. Such highly volatile secretions are important in communication between the mother and its young, between sexual partners or migratory species.

Scent glands produce various odoriferous substances and are differentiated into marking, rutting, recognition and stink glands. **Marking glands** are found in many mammals and their secretions are important in communication within the same species whereas **recognition glands** play a role in identification of members of the same species. **Rutting glands** are active at breeding seasons and are important in attraction of sexual partners as is seen in the deer family. **Stink glands** such as those of the skunk produce a powerful evil smelling secretion that is used as a defense weapon.

Blood Supply to Skin

Arteries that supply blood to the mammalian skin are mainly branches of vessels that supply the superficial muscles of the body. These blood vessels supply the hypodermis while others traverse this layer and follow the course of connective tissue fibers and may take a meandering course in areas with loose skin to the dermis. Sweat and scent glands receive their blood supply from branches of these vessels that form a wide mesh. Hair follicles and sebaceous glands also receive their blood supply from branches of the cutaneous network. The dermal papillae receive their blood supply from end arteries. Capillaries form an extensive meshwork and loops in the dermis. Veins normally follow the course of arteries. The dermal blood vessels show a high degree of arteriovenous anastomoses that bypass capillary beds and shunt blood flow straight from arterioles to venules on contraction of pre-capillary sphincters. The epidermis lacks a blood supply and is nourished through diffusion and active transport by the dermis.

Nerve Supply to Skin

There are numerous somatic sensory receptors in the skin. These receptors make skin an important sense organ that overlies the entire body surface save for natural openings. All major layers of skin contain sensory and sympathetic nerves. Generally, areas of skin with less hair have a higher level of such nerves than the more hairy regions. Sympathetic nerves that are found around blood vessels in the dermis and hypodermis are not found in the epidermis. Most of the sensory nerve fibers are located in the dermis where they originate as sensory touch receptors while they form free nerve endings in the deeper layers of the epidermis. The various receptors present including touch, pain, pressure and temperature receptors are activated by their respective stimuli and enable the body respond to changes occurring in both external and internal environments.

Vitamin D Production by Skin

The mammalian epidermis contains a pro-vitamin known as 7-dehydrocholesterol that is converted into cholecalciferol under the influence of ultraviolet light that is naturally present in sunlight. Cholecalciferol is transported by blood to the liver and kidney where it is converted into vitamin D under the influence of parathyroid hormone thus the name 'sunshine' vitamin. Vitamin D influences various reactions in the body including increased absorption of calcium from the intestine and its deposition in bone. Vitamin D thus plays a role that qualifies it to be classified as hormone. The rate of vitamin D synthesis depends on the level of skin pigmentation and the amount of sunshine available. A lot of pigmentation reduces the amount of ultraviolet light that penetrates the skin.

Skin in Homeostasis of Body Temperature

Birds and mammals are homeothermic and therefore have a stable body temperature that shows very little variation with changes in environmental temperature. Such homeostasis of body temperature is vital to the survival of these homeotherms since biochemical pathways are mainly dependent on enzymes that show a narrow range of temperature at which they function. The body normally maintains a balance between the amount of heat that is generated and lost. Such a balance minimizes the overall body temperature changes of birds and mammals.

The large surface area of the skin and the structures present enable this system to play an important role in homeostasis of body temperature. The rate at which heat is lost from the body surface is regulated by the amount of blood that flows through the dermal blood vessels. Such blood flow is regulated by the temperature-regulating center in the hypothalamus that influences the control of the sympathetic nervous system on dermal blood vessels. A rise in body temperature resulting from activities such as exercise and fever are accompanied by a reduction in sympathetic tone and relaxation of pre-capillary sphincter muscles. Such action is accompanied by an increase in blood flow through the extensive dermal capillary network that allows rapid loss of heat from the body by radiation, conduction and convection.

Radiation involves exchange of heat between two body surfaces without contact. Heat normally radiates from warmer to cooler bodies. In an environment that is cooler than that of the body, radiation is responsible for greater heat loss than evaporation and conduction combined. Heat loss by radiation ceases when the environmental temperature exceeds that of the body in which case heat is gained by the body through radiation from the warmer structures. During **conduction**, heat is transferred to other substances by contact with the body. Conduction is responsible for relatively low heat loss from the body. **Convection** is responsible for loss of heat from the body by movement of air or liquid. The sweat glands of mammals increase their output of sweat that takes up more heat from the body as it evaporates. **Evaporation** is quite important in mammals when the environmental temperature is high as it is the only method used in heat loss by the skin.

When heat production is less than the body's requirements, sympathetic tone is increased leading to contraction of pre-capillary sphincters. Blood will flow directly from the arterioles to venules through arteriovenous anastomoses thus bypassing capillaries in the process. There is also reduction in sweating. These processes lead to conservation of body heat. The level to which hair or feathers are elevated also determines the amount of air that is trapped or lost to conserve or lose body heat. Skin also plays a role in body **blood pressure** regulation by adjusting its circulation in response to a decrease or an increase overall pressure. The same mechanism that is applied during thermoregulation is employed in responding to blood pressure changes. Should the body temperature continue to fall, **thermeogenesis** (heat generation) through increased metabolism and shivering will be mobilized.

The skin also plays a role in homeostasis by continuously renewing adult tissues such as the epidermis. This layer is sustained by epidermal stem cells that are a small population of undifferentiated basal keratinocytes that multiply to give rise to daughter amplifying cells that form most of the proliferating basal cells of the epidermis (Bickenbach and Grinnel, 2004).

Hypodermis

The hypodermis (subcutis) lies below the dermis and is above superficial fascia and thus binds skin to underlying tissue. It comprises 'loose' connective tissue with irregular bundles of sheets of collagen and elastic fibers and also fat cells. The hypodermis shows varying levels of development in different parts of the body. It is extensively mobile due to the presence of several sheets that are maintained by elastic fibers (Kawamata et al., 2003) and may function to transmit mechanical signals between the abundant fibroblasts, immune, vascular and neural cells present (Iatridis et al., 2003). The hypodermis is most developed in birds and mammals and plays a major role in insulation of the body. In marine mammals, especially those of the temperate world and other polar mammals (Fig. 6.14), the hypodermis can be quite thick and is known as **blubber**. Blubber can account for about 40% or more of body weight in

Fig. 6.14 A polar bear *Ursus maritimus*. The polar bear is one of the largest carnivores on Earth and can grow up to 3.0 m in length while weighing as much as 770 kg. The thick layer of blubber together the thick woolly fur that is close to the skin help the bear keep warm. The hollow tube-like guard hairs stick out and protect the bear from getting wet as well as directing the sun's energy directly to the bear's skin. A marine mammal with feet that are partly webbed for swimming and toes with sharp claws, the polar bear is a formidable predator that hunts for fish and seals.

some large whales. Studies conducted on bottlenose dolphins show that blubber is not a homogeneous tissue through its depth and varies in its morphology and lipid content with the life history of the dolphin (Struntz et al., 2004). The blubber is an important heat insulator and metabolic energy store in the temperate world mammals and also provides buoyancy and helps streamline the body of whales. When whales migrate to their breeding grounds and feed less frequently, blubber serves as a source of fuel.

REFERENCES

Alibardi, L. (2003). Adaptation to the land: The skin of reptiles in comparison to that of amphibians and endotherm amniotes. J. Exp. Zoolog. B. Mol. Dev. Evol., 298(1):12-41.

Alibardi, L. (2004). Dermo-epidermal interactions in reptilian scales: speculations on the evolution of scales, feathers, and hairs. J. Exp. Zoolog. Part B. Mol. Dev. Evol., 302(4): 365-383.

Allender, C.J., Seehausen, O., Knight, M.E., Turner, G.F. and Maclean, N. (2003). Divergent selection during speciation of Lake Malawi cichlid fishes inferred from parallel radiations in nuptial coloration. Proc. Natl. Acad. Sci. U.S.A., 100(24):14074-14079.

Amoh, Y., Li, L., Yang, M., Moossa, A.R., Katsuoka, K., Penman, S. and Hoffman, R.M. (2004). Nascent blood vessels in the skin arise from nestin-expressing hair-follicle cells. Proc. Natl. Acad. Sci. U.S.A., 101(36):13291-13295.

Bickenbach, J.R. and Grinnell, K.L. (2004). Epidermal stem cells: interactions in developmental environments. Differentiation, 72(8):371-380.

Bjerke, J.R. (2002). The skin as an immunological organ. Tidsskr. Nor. Laegeforen., 122(8):793-796.

Blanpain, C., Lowry, W.E., Geoghengan, A., Polak, L. and Fuchs, E. (2004). Self-renewal, multipotency, and the existence of two cell populations within an epithelial stem cell niche. Cell., 118(5):635-648.

Bourne, G.R., Breden, F. and Allen, T.C. (2003). Females prefer carotenoid colored males as mates in the pentamorhic livebearing fish, *Poecilia parae*. Naturwissenschaften, 90(9):402-405.

Braff, M.H., Zaiou, M., Fierer, J., Nizet, V. and Gallo, R.L. (2005). Keratinocyte production of Cathelicidin provides direct activity against bacterial skin pathogens. Infect. Immun., 73(10):6771-6781.

Christiano, A.M. (2004). Epithelial stem cells: stepping out of their niche. Cell., 118(5):530-532.

Douglas, R.H., Partridge, J.C. and Marshall, N.J. (1998). The eyes of deep-sea fish. I: Lens pigmentation, tapeta and visual pigments. Retin. Eye Res., 17(4):597-636.

Dulbecco, R., Henahan, M. and Armstrong, B. (1982). Cell types and morphogenesis in the mammary gland. Proc. Natl. Acad. Sci. USA., 79(23):7346-7350.

Fernandes, K.J., McKenzie, I.A., Mill, P., Smith, K.M., Akhavan, M., Barnabe-Heider, F., Beirnaskie, J., Junek, A., Kobayashi, N.R., Toma, J.G., Kaplan, D.R., Labosky, P.A., Rafuse, V., Hui C.C. and Miller, F.D. (2004). A dermal niche for multipotent adult skin-derived precursor cells. Cell Biol., 6(11):1082-1093.

Fox, D.L. (1953). Animal Biochromes and Structural Colors. Cambridge University Press, Cambridge.

Fujii, R. (2000). The regulation of motile activity in fish chromatophores. Pigment Cell Res., 13(5):300-319.

Goldman, A.S., Chheda, S. and Garofalo, R. (1998). Evolution of immunologic functions of the mammary gland and the postnatal development of immunity. Pediatr. Res., 43(2): 155-162.

Iatridis, J.C., Wu, J., Yandow, J.A. and Langevin, H.M. (2003). Subcutaneous tissue mechanical behavior is linear and viscoelastic under uniaxial tension. Connect. Tissue Res., 44(5):208-217.

Harvey, E.N. (1957). The luminous organs of fishes. In: E.M. Brown (Ed). The Physiology of Fishes. Academic Press, New York. 2:345-366.

Herring, P.J. (2000). Species abundance, sexual encounter and bioluminescent signalling in the deep sea. Philos. Trans R. Soc. Lond. B Biol. Sci., 355(1401):1273-1276.

Kawamata, S., Ozawa, J., Hashimoto, M., Kurose, T. and Shinohara, H. (2003). Structure of the rat subutaneous connective tissue in relation to its sliding mechanism. Arch. Histol. Cytol., 66(3):273-279.

Kundrat, M. (2004). When did therapods become feathered? Evidence for pre-*Archaeopteryx* feathery appendages. J. Exp. Zoolog. Part B Mol. Dev. Evol., 302(4):355-364.

O'Connor, K.I., Metcalfe, N.B. and Taylor, A.C. (1999). Does darkening signal submission in territorial contests between juvenile Atlantic salmon, *Salmo Salar?* Anim. Behav., 58(6):1269-1276.

Oftedal, O.T. (2002). The mammary gland and its origin during synapsid evolution. Mammary Gland Biol. Neoplasia, 7(3):225-252.

Olivera-Martinez, I., Viallet, J.P., Michon, F., Pearton, D.J. and Dhouailly, D. (2004). The different steps of skin formation in vertebrates. Int. J. Dev. Biol., 48(2-3):107-115.

Peaker, M. (2002). The mammary gland in mammalian evolution: a brief commentary on some of the concepts. J. Mammary Gland Biol. Neoplasia, 7(3):347-353.

Pilastro, A., Simonato, M., Bisazza, A. and Evans, J.P. (2004). Cryptic female preference for colorful males in guppies. Evolution Int. J. Org. Evolution, 58(3):665-669.

Porter, A.M. (2001). Why do we have apocrine and sebaceous glands. J.R. Soc. Med., 94(5): 236-237.

Price, J. and Allen, S. (2004). Exploring the mechanisms regulating regeneration of deer antlers. Philos. Trans. R. Soc. Lond. B Biol. Sci., 359(1445):809-822.

Prum, R.O. and Torres, R. (2003). Structural coloration of avian skin: convergent evolution of coherently scattering dermal collagen arrays. J. Exp. Biol., 206(Pt.14):2409-2429.

Prum, R.O. and Torres, R.H. (2004). Structural coloration of mammalian skin: convergent evolution of coherently scattering dermal collagen arrays. J. Exp. Biol., 207(Pt. 12): 2157-2172.

Saikawa, Y., Hashimoto, K., Nakata, M., Yoshihara, M., Nagai, K., Ida, M. and Komiya, T. (2004). Pigment chemistry: the red sweat of the hippopotamus. Nature, 429(6990):363.

Sandilands, V., Powell, K., Keeling, L. and Savory, C.J. (2004). Preen gland function in layer fowls: factors affecting preen oil fatty acid composition. Br. Poult. Sci. 41(1):109-115.

Sawyer, R.H. and Knapp, L.W. (2003). Avian skin development and the evolutionary origin of feathers. J. Exp. Zoolog. B Mol. Dev. Evol., 298(1):57-72.

Sawyer, R.H., Washington, L.D., Salvatore, B.A., Glenn, T.C. and Knapp, L.W. (2003). Origin of archosaurian integumentary appendages: the bristles of the wild turkey beard express feather-type beta keratins. J. Exp. Zoolog. B Mol. Dev. Evol., 279(1):27-34.

Sire, J.Y. and Akimenko, M.A. (2004). Scale development in fish: a review, with description of sonic hedgehog (shh) expression in the zebrafish (*Danio rerio).* Int. J. Dev. Biol., 48(2-3):233-247.

Stoeckelhuber, M., Stoeckelhuber, B.M. and Welsch, U. (2004). Apocrine glands in the eyelid of primates contribute to the ocular host defense. Cells Tissue Organs, 176(4): 187-194.

Struntz, D.J., McLellan, W.A., Dillaman, R.M., Blum, J.E., Kucklick, J.R. and Pabst, D.A. (2004). Blubber development in bottlenose dolphins (*Tursiops truncatus*). J. Morphol., 259(1):7-20.

Sugimoto, M. (2002). Morphological color changes in fish: regulation of pigment cell density and morphology. Microsc. Res. Tech., 58(6):496-503.

Wilson, T. and Hastings, J.W. (1998). Bioluminescence. Annu. Rev. Cell Dev. Biol., 14:197-230.

Xu, X., Norell, M.A., Kuang, X., Wang, X., Zhao, Q. and Jia, C. (2004). Basal tyrannosauroids from China and evidence for protofeathers in tyrannosauroids. Nature, 431(7009):680-684.

7

Nutrition and Digestion

Nutrition is vital to the survival of vertebrates as it entails provision of nutrients that are necessary for the complex metabolic processes performed and also provides the building blocks for body compounds and as a result vertebrates are **heterotrophs**. Ingested material is normally in a form that cannot cross the cell membrane of the digestive system and has to undergo some form of digestion before absorption occurs. The energy content of foodstuffs and the rate at which vertebrates feed varies considerably. Some small birds and mammals have to eat almost continuously to maintain their high metabolic rates whereas some snakes such as large rattlesnakes feed a few times in a year. Vertebrates have evolved diverse feeding habits with certain groups feeding on a variety of foodstuffs whereas others show extreme specialization such as the koala that feeds only on a few species of the plant *Eucalyptus*.

NUTRITION

Nutrition deals with the food that is eaten and the nutrients it contains. Foodstuffs provide the energy that is necessary for metabolic processes and molecules for anabolic reactions. In order to perform the complex body activities, vertebrates have had to diversify their nutritional requirements when compared to other simpler organisms. The major food groups of vertebrates are carbohydrates, proteins and fats that form three of the major organic substances in the vertebrate body. These food groups are digested into simpler components that are metabolized through oxidation to produce energy mainly as adenosine triphosphate (ATP) and heat or can be used in the synthesis of large and complex molecules that form structural or functional units of the body. Excess energy in the form of nutrients is normally stored as lipid regardless of the original composition of food consumed.

Birds and mammals that are exposed to natural conditions in temperate parts of the world generally require a higher intake of the energy supplying foods than those in warmer parts of the world as they have to maintain their body temperature by producing extra body heat. Endotherms have higher body temperatures and metabolic rates than ectotherms and as a result need a higher intake of food to maintain the higher body temperatures. Tropical ectotherms tend to have higher metabolic activities than those of their temperate counterparts as a result of higher environmental temperatures. A 10°C rise or fall in body temperature results in doubling or halving of the metabolic rate.

The other major group of organic substances in the body is the nucleic acid group that comprises of deoxyribonucleic acid (DNA) that stores hereditary information and ribonucleic acid (RNA) that plays a major role in protein synthesis, gene regulation and other cellular activities. In addition to the major food groups, certain minerals and vitamins are essential for the survival of vertebrates.

CARBOHYDRATES

Carbohydrates occur in many forms such as polysaccharides starch and glycogen that is sometimes referred to as animal starch. Carbohydrates are broken down in the digestive system into simple sugars such as glucose that can be absorbed into the body. **Cellulose** is a carbohydrate that forms a major part of plant tissues and cannot be digested by vertebrates as they lack the enzyme **cellulase**. Many herbivorous vertebrates have colonies of microorganisms in parts of their digestive system that can synthesize cellulase. Cellulose is an important 'dietary fiber' or roughage as it thickens the ingesta and facilitates its movement within the digestive system and also plays a role in mixing of digestive tract contents.

In many vertebrates, the main carbohydrate that is used in production of chemical energy is glucose. Excess energy in the form of heat is important in the maintenance of body temperature especially in endotherms. When the level of glucose is inadequate to meet the energy requirements of cells, more can be synthesized from protein or glycerol of fats in the liver during **gluconeogenesis**. Fats and proteins can also be catabolized to produce energy. Oxygen is necessary for the complete breakdown of glucose to yield the maximal amount of ATP molecules. Anaerobic respiration (glycolysis) is the only process that generates energy for body cells in the absence of oxygen and yields only a few molecules of ATP (approximately 5%) as the rest of the energy is held in the lactic acid bonds. Excess glucose in the body is used in the synthesis of glycogen or **glycogenesis**. Glucose is then stored in the form of glycogen mainly in the liver and muscle. Glucose can also be used in the synthesis of fat. The pentose (five-carbon) carbohydrates that comprise deoxyribose and ribose serve a structural role since they are part of the nucleic acids DNA and RNA respectively.

LIPIDS

The main lipids in foodstuffs are triglycerides that consist of a glycerol unit to which are attached three fatty acids. The physical and chemical properties of fatty acids depend on the level of their saturation (extent to which the available bonds in the hydrocarbon chain are filled) and this is dependent on the type of fatty acids present. Saturated fatty acids such as stearic acid are straight and inflexible molecules. Lipids with such fatty acids tend to be solid at room temperature. Unsaturated fatty acids including linoleic and linolenic acid have carbon that is not completely saturated with hydrogen. The fatty acid chains of these molecules show more flexibility and movement and lipids containing such fatty acids are soft and oily at room temperature. Other important lipids of foodstuffs are phospholipids and cholesterol that are **structural components** of cell membranes. Cholesterol is also a part of steroid hormones such as corticosteroids, testosterone and estrogen. The amount of triglycerides present in each type of food and the proportion of saturated and unsaturated fatty acids present varies considerably. Animal tissue has a higher lipid content than plant tissue.

Catabolism of a unit weight of lipids (**ketogenesis**) yields more energy than an equivalent amount of carbohydrates. For example, catabolism of 1 g of fat yields 9.0 kcal whereas catabolism of a similar amount of carbohydrates yields 4.1 kcal. A calorie is the amount of energy required to raise the temperature of one gram of water by 1°C. There are 1,000 calories in 1 kcal. Lipids are the main source of energy for muscle tissue. Lipid anabolism in the vertebrate body leads to the synthesis of triglycerides, cholesterol and prostaglandins. These lipids are synthesized from fatty acids and glycerol or glucose and amino acids. Most of the body's energy reserves are stored in adipose tissue as triglycerides. Although most fatty acids can be synthesized in the body, **essential fatty acids** such as linoleic and alpha linolenic acid must be obtained from the diet. These polyunsaturated fatty acids are important in the maintenance of the cell membranes and synthesis of prostaglandins. Prostaglandins are composed of a 20-carbon unsaturated fatty acid and act as 'local hormones' in the body by regulating various processes such as activities of other hormones, inflammation, blood clotting and secretion of digestive juices.

PROTEINS

Animal tissues such as meat are rich in proteins and contain the essential amino acids that are part of the 20 or so protein amino acids. Essential amino acids form about half of these amino acids and must be obtained from the diet of most vertebrates since they are not manufactured by vertebrates. Nonessential amino acids are normally synthesized in the vertebrate body. Some microorganisms are able to synthesize essential amino acids in the digestive system of certain animals such as ruminants and vertebrates with well-developed caeca. The primary metabolic role of proteins is anabolism and proteins are tissue building blocks as opposed to carbohydrates and fats that are mainly catabolized to generate energy. Although the protein requirements of vertebrates vary depending on age, species, sex and season, younger and fast growing vertebrates require a higher protein intake than mature ones.

Structural proteins such as the fibrous and stable collagen are found in many structures of the vertebrate body such as tendons, ligaments and bone where they strengthen these structures and render some degree of resilience. Other **functional proteins** of the body including antibodies that play a major role in the body's defense mechanism are globular in structure and are soluble and active. Although a lot of protein in the body is synthesized in the liver and glands, each cell of the body synthesizes its own structural proteins and enzymes. Protein catabolism starts off in the liver during deamination (splitting off of an amino group from an amino acid). The amino group is converted to ammonia, urea or uric acid that will be excreted from the body. The remaining keto acid can be oxidized for generation of energy or can be converted to glucose or fat.

VITAMINS

Information on the dietary requirements for vitamins in many animals is lacking. Most of the studies on effects of vitamins in vertebrates have been carried out in human beings and some mammals and birds. Vitamins are organic molecules that are needed in small quantities in the diet of most vertebrates. A lot of vitamins are not synthesized by the vertebrate body itself and have to be obtained from food eaten or vertebrates have to eat molecules that can be converted to vitamins. The fat-soluble vitamins include A, D, E and K and can be stored in the liver for later use whereas those soluble in water such as B and C have to be continuously supplied by the diet in animals that lack the ability to synthesize them. Many vitamins such as the B series are used in the manufacture of coenzymes that act as cofactors which are necessary for enzymes to be catalytically active. Vitamin C (ascorbic acid) can be ionized to ascorbate that acts as an antioxidant (reducing agent) when it is oxidized to dehydroascorbic acid. Vitamin C is also necessary for the activity of prolyl hydroxylase that plays a major role in the synthesis of 4-hydroxyproline, an amino acid that is needed in the synthesis of collagen (the most abundant protein in the animal body). Hydroxyproline makes the triple helix of collagen stable by forming hydrogen bonds between the polypeptide strands. Vitamin C is not synthesized by primates and is normally acquired by the group from their diets.

The fat-soluble vitamins play various roles in the vertebrate body. Vitamin A (retinol) is the precursor of retinal that is sensitive to light in visual pigments such as rhodopsin. The same vitamin is necessary for growth as it activates certain genes that mediate growth and development. Vitamin D regulates the metabolism of calcium and phosphorus. Vitamin E (α-tocopherol) serves as an antioxidant in the body by attracting and neutralizing molecules with unpaired electrons such as free radicals that oxidize and damage DNA molecules and electron dense molecules of cell membranes such as phospholipids. Vitamin K is necessary for normal clotting of blood by modifying glutamic acid through carboxylation thus enhancing its Ca^{2+} chelating properties.

MINERALS

Minerals are naturally occurring inorganic elements. Several minerals are vital to the proper functioning of enzymes and other organic molecules. Macroelements present in the vertebrate body include oxygen, hydrogen, carbon and nitrogen that make up about 96% of vertebrate body weight as well as calcium, phosphorus, sodium, potassium, magnesium, chlorine and sulfur that make up about 4% of body weight. Minerals such as sodium and calcium are necessary for proper conduction of nervous impulses and muscle contraction respectively whereas calcium and phosphorus are important building material for the skeletal system. Minerals are normally acquired in various ways including normal consumption of food and geophagia (ingestion of earth to obtain minerals). Mineral requirements of vertebrates vary depending on the species, age and season. Proper quantities of minerals are required by vertebrates otherwise too much consumption of certain minerals can result in toxicity and a deficiency of others could impair body functions. Trace elements such as iron, fluorine, copper, zinc, selenium and iodine occur in amounts that are less than 0.1% of the body weight and are also important for the proper functioning of the body. It is the complex interrelationship of all the elements found in the body and not just some elements that is responsible for the proper functioning and survival of vertebrates.

DIGESTION

Since vertebrates feed on a variety of foods that occur in different forms, they have evolved various food capture strategies. The digestive system has adjusted to the variety of foods eaten and shows a number of similarities and differences among the various species as a result. The most remarkable variation in the digestive system of vertebrates is seen in the stomach (Smith et al., 2000). Considered to belong to the external environment because of the continuous nature of the system with the outside world, the primary role of the digestive system is nutrition of the body and includes various processes such as prehension, mastication in many vertebrate groups, digestion, absorption and egestion of undigested and excess material. The digestive system can also act as a temporary storage system for food that is undergoing digestion and the nutrients that result from the process in some vertebrates.

Prehension

As vertebrates inhabit various environments and have varied feeding habits, their methods of acquiring food is quite diverse. The jawless ammocoetes larvae of lampreys are **filter** or **suspension feeders** that employ several methods such as expansion and contraction of buccopharyngeal cavity to move water through the mouth and pharynx. The larvae also produce mucus that traps suspended food particles and together with ciliary action, the food particles are moved into the esophagus. Many fish employ the filter feeding method by straining plankton and other organisms in water in the oral cavity. Such a method selects food by size and not type. Other fast swimming fish are **ram feeders** and swim with their mouths wide open, out-swimming their prey in the process. As an adaptation to straining of organisms, fish have evolved gill rakers (Fig. 7.1) that comprise a series of elongated structures that could be quite close to each other and long in some groups. The two largest fish (whale shark and basking shark) have evolved long rakers that play a role similar to that of baleen found in some whales for straining plankton from water in the mouth.

Parasitism is seen in some fish species such as lampreys and hagfish that attach their circular mouths and horny teeth to suck body fluids. An extreme case of obligatory parasitism is seen in the males of some deep-sea anglerfishes (*Ceratias*) that attach permanently to fleshy papillae on the body surface of females and remain quite small in relation to their

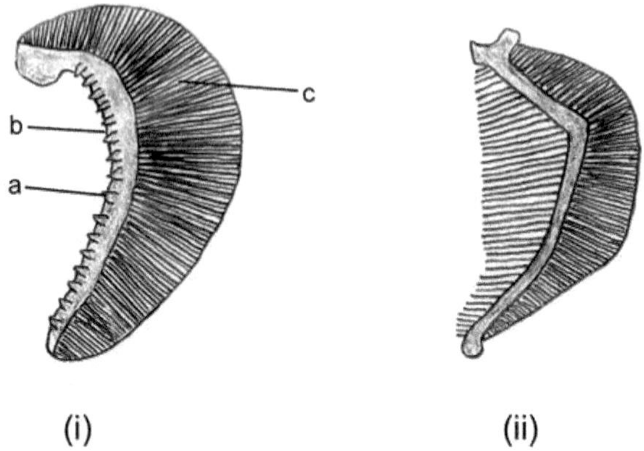

Fig. 7.1 The gills of (i) a tilapia (*Oreochromis niloticus*) and (ii) blackfin cisco (*Coregonus nigripinnis*). (a) gill raker, (b) gill arch and (c) gill filaments. The blackfin cisco, a strainer, has many elongated gill rakers that are close-set.

female hosts. The evolution of teeth and jaws has enabled many vertebrates to feed on larger prey as they can bite off smaller pieces. Predaceous vertebrates normally chase after their prey that they normally subdue.

Feeding on land is quite different from feeding in water as air is a less dense medium than water and offers less buoyancy to food. This has resulted in prehension being one of the major changes that have occurred in the transition of vertebrates from water to land. The jaw and tongue movements in terrestrial vertebrates are more enhanced as they replace the role played by the more dense water in prehension. In amphibians, reptiles and birds, there is a transverse joint across the skull roof that enables movement of the front part of the skull relative to the cranium in what is known as cranial kinesis. **Cranial kinesis** has been lost in mammals and is quite advanced in some snakes such as the pit vipers. Snakes are able to swallow prey that is much wider than their heads after constricting them or disabling the prey with poison. The oral cavity of snakes can be quite wide relative to the normal size of the head as the loss of the temporal roof with evolution has freed the squamosal bone thus enabling it to move rostrally and laterally. The joint between the quadrate and squamosal bone is quite movable and the rostral ends of the left and right mandibles are united by a ligament that enables the mandibles to move in either direction.

Some amphibians display tongue prehension using their **ballistic tongues** such as plethodontid salamanders whereas some frogs possess hydrostatic tongues. Prehension in amphibians can also involve the coordination of various body parts such as the tongue and jaws in some toads and frogs. The ballistic tongue projection mechanism is also unique among lizards and is highly developed in the chameleon. The chameleon can eject its tongue in less than one sixteenth of a second and 600% of its resting length that is about two times the chameleon's length. The tip of the tongue contains sticky saliva and is used to capture prey. Chameleons normally ambush their prey and as they feed infrequently normally capture relatively large prey.

Many reptiles use their jaws for prehension. Squmates use their jaws for prehension though their ancestors relied on tongue prehension (Schwenk and Throckmorton, 1989). The slow acting ancestors of squamates used to ambush their prey that they could detect using sight as they had poorly developed chemosensory systems. They could capture their prey using their tongues. These features of ancestral squamates are still present in the group of iguanas that includes iguanids, agamids and chameleons. The tongue of squamates has evolved into a chemical sampler that can transfer material into the mouth to be detected by the vomeronasal (Jacobson's) organ.

Snakes that feed on eggs of birds mainly including the African egg-eating snake that feeds on weaverbird eggs normally have to swallow eggs that are much wider than their mouth. The snakes move their heads over the egg that is sometimes held in position by the snake coils. Egg eating snakes lack teeth but have several folds of gum tissues arranged in a series in their mouths that act as suction cups. When the egg reaches the esophagus, the fish will bend its neck considerably, crushing the egg using the blunt modified internal spines of the vertebrae. The empty eggshell is then regurgitated to the outside. Snakes that occasionally eat eggs normally carry them in their digestive system where the shell is digested.

In birds, the long neck, the beak and feet are used in prehension of food. Parrots (order Psittaciformes) use the beak, tongue and feet for prehension. Large parrots use their feet to grasp food to the beaks. The beak together with the tongue is used to break the hull of the seed. Mammals use their forelimbs, lips, tongues and jaws to grasp their food. Predators have to chase or ambush their prey whereas scavengers have to feed on carrion mainly.

Mastication

Vertebrates either swallow their prey whole or tear it into smaller pieces before swallowing. Mastication (chewing) is seen in most mammals and some reptiles. The African lungfish chews its food in a manner that is rare among fish using the two teeth at the front of the upper jaw and several crushing teeth on the lower jaw. Mammals have specialized teeth (heterodont dentition) that perform several functions. Fish, amphibians and reptiles generally have homodont dentition whereas birds and turtles lack teeth.

Mastication leads to an increase in the surface area of the food and increases its exposure to digestive enzymes and also softens the food thereby changing it into a size that can be swallowed. Mastication also assists in the lubrication of food by mixing it with saliva. Chewing can be both a reflex and voluntary action.

Teeth

The teeth of vertebrates could have evolved from scales that were located on the lips of vertebrates. A good example of such a change is seen in sharks where the placoid scales on the lips gradually change into teeth on jaws and both structures are made of enamel, dentine and a pulp cavity. The teeth of other vertebrates are thought to have evolved from bony dermal scales. The teeth of bony fishes are normally found in places with ectoderm in the oral cavity and as a result are located on jaws, on the roof of the oral cavity, gill arches (pharyngeal teeth) or the tongue depending on species. The function of such fish teeth is to hold the prey. In amphibians, teeth are located on jawbones mainly although some may be found on the palate. The reptilian and mammalian teeth are located on the pre-maxilla (incisive), maxilla and mandibles.

Vertebrates show variation in the way their teeth are attached to jaws and surrounding structures (Fig. 7.2). The bases of **acrodont teeth** (Gr. *acro*, end; *odous* or *odont*, tooth) are fused to the surface of jaws and such teeth are found in most teleosts and sharks. Since acrodont teeth are not firmly anchored they are often lost and replaced. The teeth of many sharks are arranged in rows. The anterior teeth are lost and are replaced by the posterior row of teeth. Since such teeth are continuously replaced, acrodont teeth are **polyphylodont**. **Pleurodont teeth** (Gr. *pleur*, side) are

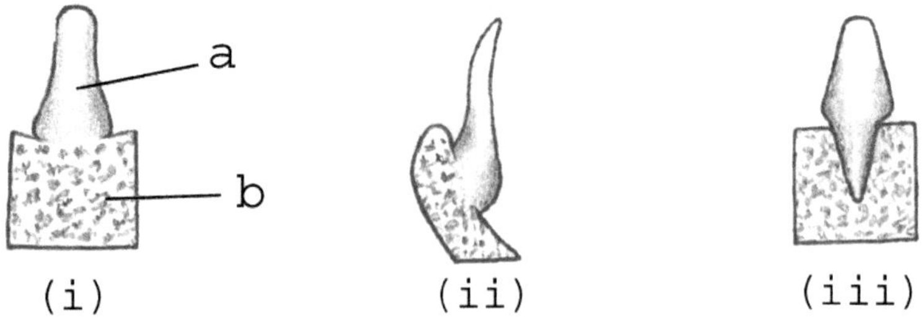

Fig. 7.2 Attachment of teeth to jaws. (i) acrodont, (ii) pleurodont and (iii) thecodont teeth. (a) tooth and (b) jaw.

fused on one side to the inner surface of the jaw. These teeth are also polyphylodont and are found in frogs, lizards and *Necturus*. Toads lack teeth. Reptilian and mammalian teeth are **thecodont** (Gr. *theke*, cup or case) and their roots lie in sockets (alveolar processes) found on bones of the jaws. These teeth are anchored in the sockets by the periodontal membrane. The teeth of reptiles are single rooted and some are more complex and are used for piercing, crushing and cutting. Some snakes have specialized teeth that are needle shaped and function as poison fangs.

The teeth of fishes, amphibians and reptiles have a relatively constant shape in any individual and are referred to as **isodont** or **homodont** (Gr. *iso* or *homo*, equal). Generally, mammalian teeth are composed of enamel, dentine and a pulp cavity (Fig. 7.3). **Enamel** is of epidermal origin and covers the tooth surface. It is the hardest substance in the body. **Dentine** lies below enamel and is of dermal origin. Dentine is similar to bone structurally though harder. The **pulp cavity** too is of dermal origin and is surrounded by dentine. Many mammalian teeth such as premolars and molars have several roots. Mammalian dentition is generally **heterodont** (Gr. *heteros*, different) and the teeth are modified in shape and size to perform specialized functions such as cutting, piercing, grinding and crushing and shearing by incisors, canines, premolars and molars and carnassials respectively. Mammalian teeth are adapted to their food habits depending on whether they are herbivores, carnivores, omnivores or insectivores. The two tusks of the male and female African elephant as well as the male Asian elephant are upper incisors and are the only incisors present in the elephant.

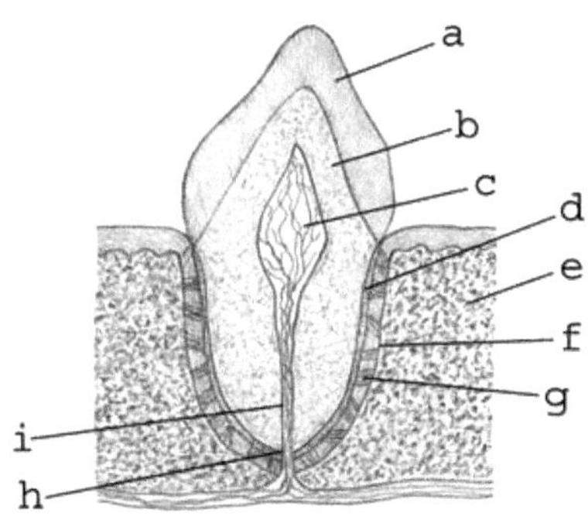

Fig. 7.3 Diagram of a longitudinal section through a mammalian tooth. (a) enamel, (b) dentine, (c) pulp cavity with nerves and blood vessels, (d) cementum, (e) alveolar bone, (f) periodontal membrane, (g) periodontal ligament, (h) apical foramen and (i) root canal.

Tongue

In jawed fish and lower amphibians, the tongue is crescent-shaped and is normally moved by contraction of underlying muscles. Most amphibian tongues have hypobranchial musculature. The reptilian and mammalian tongue has more hypobranchial musculature than that of amphibians whereas the tongue of birds lacks intrinsic muscle. The tongue in turtles, crocodilians, some birds and whales is not mobile. In snakes and lizards that feed on insects, amphibians and some birds, the tongue can be quite long and moves in and out of the mouth. The mammalian tongue can be extended out of the oral cavity. The tongue of vertebrates can be used in prehension, taste, swallowing, thermoregulation, grooming, movement of food in the oral cavity and speaking in the human being.

Saliva

Salivary glands (Fig. 7.4) produce saliva. Fish, apart from the parasitic lampreys, lack salivary glands. Salivary glands, when compared to other body glands, have evolved quite rapidly as they could be a major structure animals use in colonizing new territories or adapting to changing environments. Saliva produced by various vertebrates varies considerably and is adapted to their dietary needs. Vampires that suck blood possess anticoagulants in their saliva; the saliva of the Komodo dragon (lizard) of Indonesia has more than 15 poisonous chemicals that subdue its prey and the giraffe produces thick mucus-rich saliva that protects its oral cavity from the thorny shrubs it feeds on. The secretory units of salivary glands are a group to cells known as **acini** (L. *acinus*, berry) that originate from the oral ectoderm. The secreted fluid is passed into ducts and contains water, electrolytes, mucus and enzymes. The small ducts lead into larger ducts that will eventually end up in the main ducts that empty their contents into the oral cavity. As the secretion flows along the ducts, its composition is altered. A lot of the sodium present is actively reabsorbed and potassium and bicarbonate are secreted.

Mammals normally have three large salivary glands that include **parotid**, **mandibular** (**submaxillary**) and **sublingual**. Parotid, sublingual and mandibular glands produce serous, mucous and both secretions respectively. The **minor salivary glands** are numerous small clusters of seromucous secretory units that are located in the submucosa of the oral cavity. These clusters bear names according to their location as they are found to the inside of lips (labial) and cheeks (buccal), tongue (lingual) and palate (palatine) among other structures. The volume and type of saliva secreted is under the control of the autonomic nervous system. The

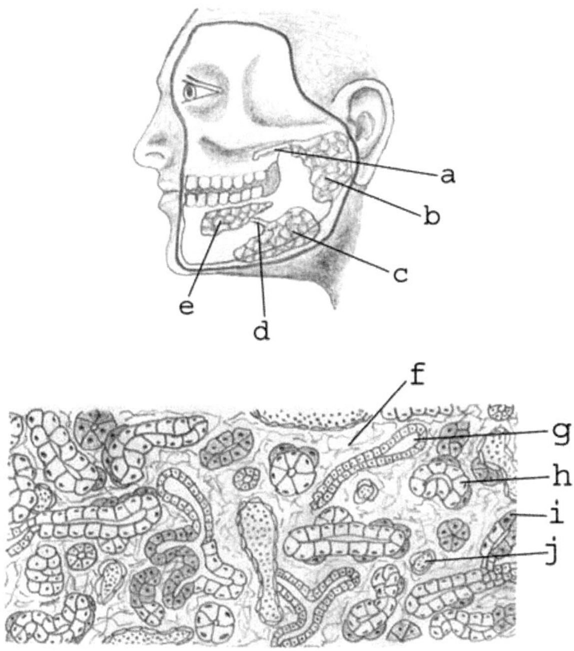

Fig. 7.4 Major salivary glands (above) and fine structure of a salivary gland (below). (a) parotid duct and (b) gland, (c) submandibular gland and (d) duct, (e) sublingual gland, (f) connective tissue, (g) excretory duct, (h) mucous cell, (i) serous cell and (j) blood vessel.

striated ducts of the main salivary glands play a role in electrolyte homeostasis as well as secreting organic products (Tandler et al., 2001).

The roles played by saliva vary depending on the species but generally include lubrication and binding of food into a slippery bolus by mucus, turning food into a liquid, starch digestion due to the presence of alpha-amylase that lacks generally in carnivores or cattle and the buffering effect due to the presence of bicarbonates. Buffering is important in ruminants due to the presence of large quantities of volatile fatty acids in the fore stomachs. Vertebrates with poorly developed sweat glands such as dogs depend to a great extent on the effect of evaporative cooling of saliva during panting. Saliva is also important in oral hygiene as it lessens the microbial load by flushing away food remnants. The enzyme lysozyme that is also present in tears is found in saliva and causes bacterial lysis.

In some vertebrates such as many snakes, digestion begins before ingestion. Most snakes inject their **venom** from their venom glands during **envenomation** into their prey. Venom glands are modified salivary glands. Venom is approximately 90% protein. About 20 types of toxic enzymes have been identified in venom. No single snake has all these toxins but a certain number. Some of these enzymes are digestive while others paralyze the victim and include cholinesterase, amino acid oxides, proteinases and adenosine triphosphatase. Snakes with long venom fangs also inflict mechanical injury to their prey. The venom of snakes is secondarily used as a defense system. Some snakes such as rattlesnakes strike their prey then set them free to die. The snakes then track the prey by using the chemical trail of their venom which they detect using their vomeronasal organ.

The oral cavity of mammals is separated from the nasal cavity by a secondary palate comprising a rostral hard and caudal soft palates. The posterior mid-border of the soft palate has a cone-shaped process known as the **uvula** in the human being. The uvula is made of muscular tissue, serous and seromucous glands and large excretory ducts and is capable of producing large quantities of saliva in a short time and could possibly be an accessory organ of speech (Finkelstein et al., 1992).

Swallowing

During swallowing (deglutition), food passes from the oral cavity to the esophagus. In aquatic anamniotes, food is transported with water currents from the mouth to the pharynx. The currents are started with repeated movements of jaws and the hyoid. In many birds, food is moved from the mouth to the esophagus by movement of the head backwards and forwards. Parrots use their muscular tongues to move food from the mouth to the esophagus. Mammals use their tongues to push food backwards into the pharynx and this is a voluntary action. The pharynx is a common passage way for both digestive and respiratory systems (Fig. 7.5). As the bolus is passed into the esophagus, alternative routes of escape are closed. The epiglottis moves backwards to cover the laryngeal opening and the larynx is squeezed. The opening of the esophagus is made larger by the forward and downward pulling of the larynx. The tongue is finally pressed backwards and the peristaltic contraction of the pharynx pushes the food into the esophagus. A wave of peristaltic contractions of the esophagus anterior or superior to the bolus in a posterior or inferior direction will move food into the stomach.

The roof of the oral cavity of vertebrates is the palate. Fishes, amphibians, reptiles and birds have a primary palate since the oral and nasal cavities are represented by a single cavity. There are two internal nasal openings at the rostral part of the primary palate in these vertebrate groups known as **choanae** (Gr. *choane,* funnel). Choanae form the posterior opening of the nasal cavity into the pharynx in mammals as the

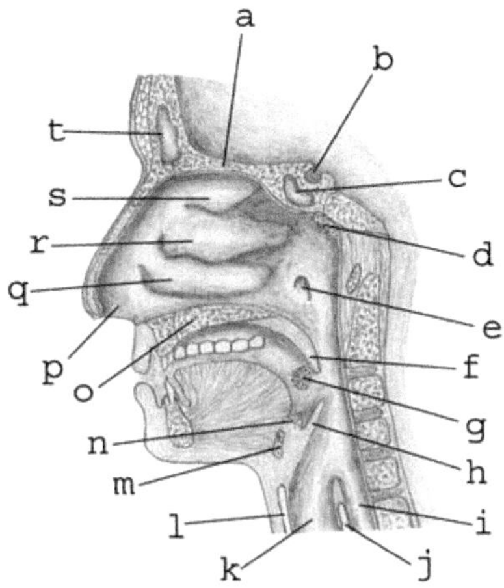

Fig. 7.5 A section showing the pharynx of the human being. (a) cribriform plate of ethmoid bone, (b) sella turcica, (c) sphenoid sinus, (d) pharyngeal tonsil, (e) opening of internal auditory (eustachian) tube, (f) uvula, (g) palatine tonsil, (h) epiglottis, (i) esophagus, (j) cricoid cartilage, (k) larynx, (l) thyroid cartilage, (m) hyoid bone, (n) lingual tonsil, (o) hard palate, (p) vestibule, (q) inferior, (r) middle and (s) superior nasal turbinates and (t) frontal sinus.

mammalian oral cavity is separated from the nasal cavity by a secondary palate that resulted from the union of palatine processes that project into the primitive oral cavity in the mid-line. The rostral part (hard palate) of the secondary palate is bony and the posterior part forms the soft palate. The hard palate of mammals has transverse ridge-like structures known as palatine ridges that may be cornified in many mammals. The ridges represent a rough surface that can hold food during chewing. Many fishes have palatal teeth.

Tonsils represent an accumulation of lymphocytes (lymph nodules) in the mucosa of the posterior oral and nasal as well as pharyngeal cavities. Tonsils are normally surrounded by a capsule and play a role in the defense system of the body.

Alimentary Canal

The alimentary canal (Fig. 7.6) is a tubular structure that extends from the beginning of the esophagus to the anal opening. The system is enlarged in some parts such as the stomach and receives secretions from the liver and pancreas. The wall of the canal has three layers that comprise an inner mucous membrane (mucosa) that normally has many glands, a middle muscular coat and an external layer of connective tissue (Fig. 7.7).

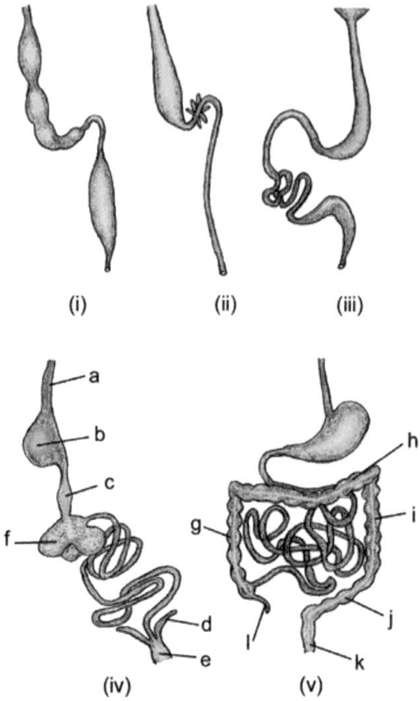

Fig. 7.6 The alimentary canals of various vertebrates. (i) shark, (ii) teleost, (iii) frog (iv) bird and (v) human being. (a) esophagus, (b) crop, (c) proventriculus, (d) caecum, (e) cloaca (f) gizzard or venriculus, (g) ascending, (h) transverse, (i) descending and (j) sigmoid colon segments, (k) rectum and (l) vermiform appendix.

The mucous membrane is the most complex of the three layers. It has an inner epithelium that is derived from the endoderm, a lamina propria and an internal thin muscular layer. All glands of the digestive system develop from the epithelial lining. The lamina propria is a layer of loose connective tissue and supports the epithelium. Most of the glands of the digestive system are located in this layer as well as blood and lymphatic vessels and lymphoid material in the form of lymph nodules in many parts of the canal. The thin internal muscular layer is the outermost layer of the mucous membrane and contains a few layers of smooth muscle fibers.

The submucosa lies beneath the mucosa. The layer is composed of collagen and elastic fibers, many blood and lymphatic vessels, nerves and nervous (Meissner's) plexuses of the autonomic nervous system. In parts of the system that lack the muscular layer of the mucosa, lamina propria and the submucosa blend together without a clear boundary.

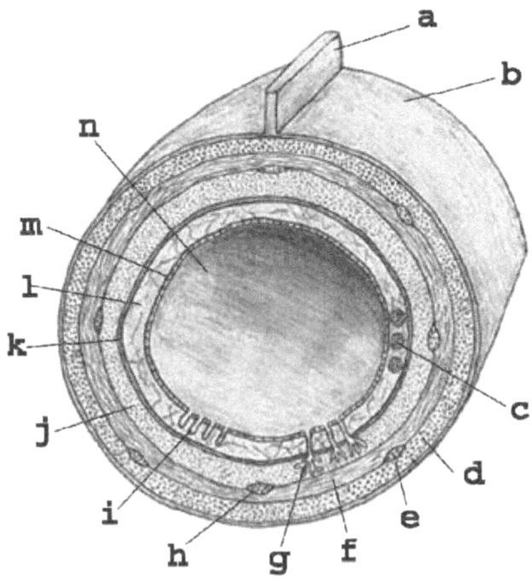

Fig. 7.7 Diagram of a cross section through the alimentary canal showing the general design of the system. (a) mesentery, (b) serosa, (c) lymph nodule, (d) longitudinal smooth muscle layer of muscularis externa, (e) myenteric or Auerbach's plexus, (f) circular smooth muscle layer of muscularis externa, (g) submucosal gland, (h) submucosal or Meissner's plexus, (i) mucosal gland, (j) submucosa, (k) muscularis interna or mucosa, (l) lamina propria, (m) epithelium and (n) lumen.

To the outside of the submucosa is an external layer of smooth muscle that comprises inner circular and outer longitudinal layers. Between the two layers of smooth muscle are myenteric (Auerbach's) plexuses of the autonomic nervous system that coordinate the activities of the muscles. The contraction of the external layer of smooth muscle is responsible for the peristaltic movement of ingesta in the alimentary canal. In cetain parts of the canal, the external smooth muscle layer is quite thick and forms sphicters such as the pyloric sphincter between the stomach and the duodenum.

The serosa (serous membrane) is a thin membrane that forms the outermost lining of the digestive tract. It is made of the areolar connective tissue and an epithelial lining (mesothelium) in most of the alimentary canal.

Esophagus

The esophagus (Gr. *oisophagos*, gullet) is a distensible and muscular tube that runs from the pharynx to the stomach or intestines in species that lack a stomach such as agnathans. The esophagus is short in fish and amphibians but is much longer in the other vertebrates. It is often quite distensible in fish and some amphibians and reptiles. The esophagus has all the layers of the tubular part of the alimentary canal. It is lined with a stratified squamous epithlium that can be keratinized in many birds and herbivorous mammals. The lamina propria has collagen and elastic fibers. The musculature in the wall of the esophagus is mainly skeletal especially in the anterior part of the tube. Mucous and serous glands are normally present in the submucosa. When relaxed, the esophagus usually has longitudinal folds as a result of the loose nature of the submucosa.

The distal part of the esophagus may have a diverticulum known as a **crop** in some bird species and some other vertebrates. The crop can store ingested food for a short time. Food can undergo fermentation and softening in the crop. The crop of the pigeon and dove of both sexes contains mucous glands that produce a milky secretion ('crop milk') that comprises mucus and desquamated epithelial cells that have undergone fatty change and are fed to the young.

Stomach

The stomach is a muscular chamber that stores food for sometime. It also produces hydrochloric acid that kills microorganisms and pepsin, an enzyme that breaks down proteins into shorter chain amino acids. The stomach is hardly evident as a circular structure in cyclostomes as it resembles the esophagus. It is less developed in lungfishes, chimaeras and minnows. The stomach shows increasing levels of specialization in other fish, amphibians and reptiles.

The fish stomach varies in shape. Piscivorous fishes such as barracudas and bowfin have sac-shaped stomachs similar to those of mammalian carnivores whereas other stomachs are modified into grinding organs known as **gizzards** with thick muscular and connective tissue walls such as stomachs found in the gizzard shad, sturgeons and mullets.

A muscular valve is normally found at the junction between the stomach and intestines. Pyloric caeca that are blind sacs are also found at the same level in some fish. These sacs have digestive, absorptive or both functions. Ducts from the liver and pancreas connect to the alimentary canal at the duodenum.

The glandular stomach of other vertebrates is generally J-shaped with certain variations depending on the species. The glandular part of the single stomach of rodents and mammalian herbivores that feed on coarse food is located between the anterior non-glandular part of the stomach into which the esophagus opens and the duodenum at the posterior end of the glandular segment. In slender and long vertebrates such as snakes and some lizards, the stomach tends to be a straight structure. The point at which the esophagus enters the stomach may be thickened as a result of the presence of a lot of smooth muscle in the outer muscular layer forming a **cardiac sphincter**. **Pyloric sphincter** (Gr. *pyloros*, gatekeeper) is found at the posterior end of the stomach where it joins the duodenum. The sphincter can contract to retain food in the stomach. The J-shaped stomach has greater and lesser curvatures, a body, pylorus and a fundus near the esophageal entrance in mammals.

The non-glandular part of the stomach is found in vertebrates that feed on coarse food and is lined by a stratified squamous epithelium whereas the glandular part has a simple columnar epithelium with numerous gastric glands. The skeletal muscle of the wall of esophagus gradually changes to smooth muscle of the stomach. The epithelium of the glandular stomach contains numerous goblet cells and branched tubular glands that produce mucus continuously. Mucous coats the stomach lining and together with bicarbonate protect the lining from selfdigestion. Deeper gastric gland cells (chief and parietal cells) lack a mucous barrier yet they are protected from autodigestion because they have an intracellular pH that is uniquely resistant to extreme degrees of gastric luminal acidification since their luminal barriers are impermeable to material such as ammonia and carbon dioxide (Waisbren et al., 1994). The gastric glands of the body and fundus are branched tubular and possess **parietal cells** that produce hydrochloric acid and **chief** or **peptic cells** that secrete pepsinogen in mammals (Fig. 7.8). Parietal cells are found between the chief cells and the basal lamina. Other cells present in gastric glands include **mucous neck cells** that produce a lot of mucus and **endocrine cells** that secrete their products such as gastrin that regulates gastric activity directly into the bloodstream.

In most vertebrates, hydrochloric acid and pepsinogen are secreted from the same type of cell. Pepsinogen is transformed into pepsin by hydrochloric acid in the lumen of the stomach. Young mammals also produce rennin in their stomachs. This enzyme curdles milk thereby retaining the milk protein for a longer time in the stomach so that it can be digested by pepsin. A number of amphibians, reptiles, birds and some mammals have the ability to digest chitin, the structural polysaccharide that forms the cuticular exoskeleton of insects, spiders, lobsters and shrimps. Such vertebrates produce the enzymes chitinase in their stomachs.

The esophagus of birds leads to a glandular stomach, the **proventriculus** (Fig. 7.9) that is generally spindle-shaped. The inner lining of the proventriculus bears longitudinal folds and its epithelium is simple columnar. The mucosa also bears compound tubular glands that open in large ducts on papillae that are visible with the naked eye at the

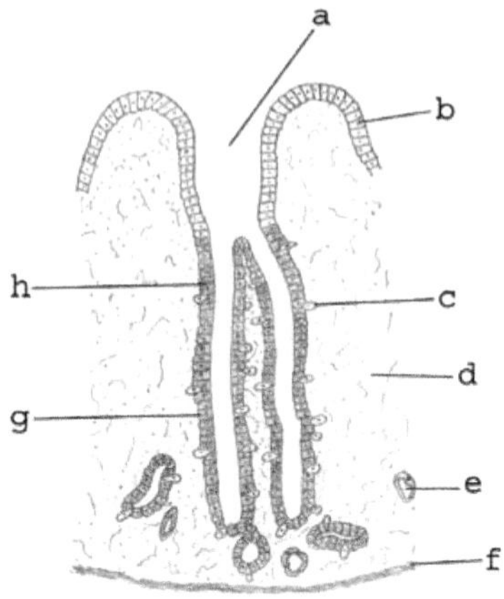

Fig. 7.8 A drawing of gastric glands that are mainly found in the body and fundus of the glandular stomach. (a) gastric pit, (b) surface epithelial cell that produces mucus, (c) parietal or oxyntic cell, (d) lamina propria, (e) capillary, (f) muscularis mucosa, (g) chief cell and (h) mucous neck cell. The secretory product of gastric glands is known as **gastric juice** and contains among other products water, mucus, pepsinogen and hydrochloric acid. Pepsinogen is an inactive proenzyme that is converted into the enzyme pepsin by hydrochloric acid. Pepsin digests proteins.

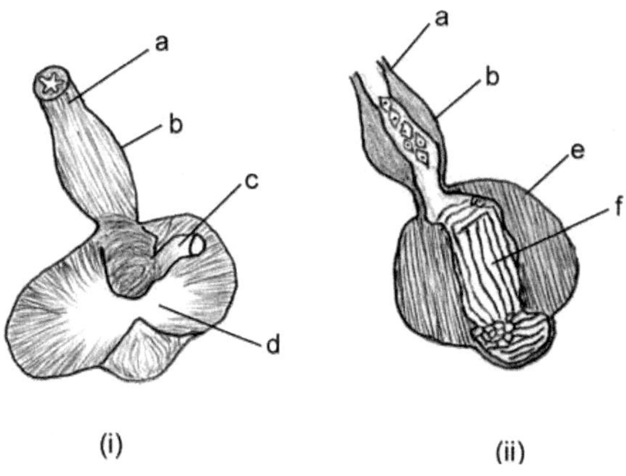

Fig. 7.9 (i) surface view and (ii) inside of a gizzard. (a) esophagus, (b) proventriculus, (c) duodenum, (d) gizzard, (e) thick muscular wall of a gizzard and (f) mucosa of body of a gizzard.

surface of the mucosa. The glands produce gastric secretions that mix with food while it is ground up in the gizzard. Analysis of this juice has shown it to be of similar basic composition to that of mammals. The proventriculus leads to the muscular stomach or **gizzard** also known as **ventriculus** (L. small belly). There is a clearly defined constriction between these two structures. The gizzard is designed according to the feeding habits of the bird and resembles a biconvex lens superficially. The wall of the gizzard has a lot of smooth muscle.

The mucous membrane of the gizzard has a simple columnar epithelium that forms a series of ridge-like folds. The lamina propria contains tubular glands that continuously secrete a product that solidifies on the luminal surface of the gizzard forming a hard mucosal covering known as the **keratinoid layer**. Together with stones that have been swallowed, this layer of keratinoid serves as a powerful grinding surface as a result of the powerful grinding muscles. The ability to grind objects is so great that even lead shots used in hunting can be broken into pieces. This has led to restrictions in the use of lead shots for hunting as the bullets can be ingested accidentally by birds such as ducks and geese and this can result in lead poisoning. The ground food moves to the pylorus that leads into the duodenum and is located close to the junction between the proventriculus and the gizzard.

Although many mammals possess simple stomachs, herbivorous mammals show various modifications in their stomachs that tend to make this organ larger and able to cope with the large quantities of forage that is eaten and is low in protein and energy when compared to animal sources. Digestion of plant material requires a longer time when compared to animal food sources and so herbivores generally have long digestive tracts. For example, the digestive tract of sheep is 27 times as long as the animal's body length. Herbivores with simple stomachs including the horse, rabbit and gorilla have simple stomachs and relatively large caeca. The anterior part of the stomach is normally non-glandular and is lined with a stratified squamous epithelium in these mammals.

The ruminant (L. *ruminare*, to chew cud) stomach such as that of cattle, sheep, goats, camels and antelopes is much larger and more complex than that of other herbivorous stomachs and comprises four chambers (Fig. 7.10). The stomach occupies most of the abdominal cavity, especially the left half, leaving only part of the right half of the cavity for the intestines. The first three chambers that comprise the **rumen**, **retriculum** and **omasum** are non-glandular and constitute the fore-stomach or proventriculus whereas the fourth chamber, the **abomasum**, is the glandular stomach. The non-glandular chambers are covered with a stratified squamous epithelium whereas the glandular one is covered with a simple columnar epithelium. The rest of the walls of the four chambers are similar to those of other parts of the digestive tube. Some skeletal muscle fibers that are a continuation of the longitudinal muscle

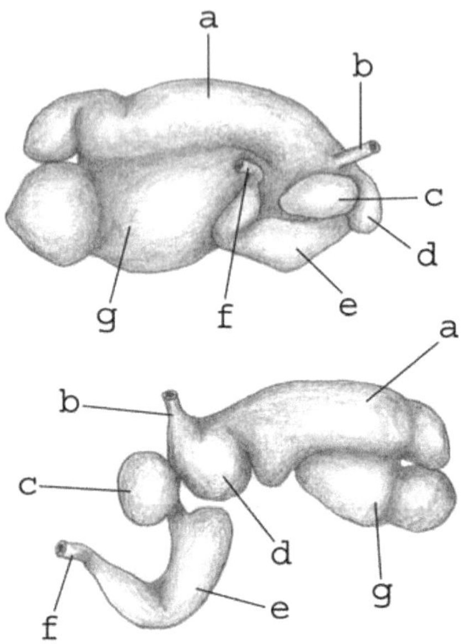

Fig. 7.10 The right side of the ruminant stomach (top) and the stomach compartments spread out (bottom). (a) dorsal sac of rumen, (b) esophagus, (c) omasum, (d) reticulum, (e) abomasum, (f) duodenum and (g) ventral sac of rumen.

layer of the esophagus continue into the wall of the reticulum and rumen. The largest of these chambers is the rumen and has a capacity of about 102-148 liters in adult cattle.

The internal surfaces of the ruminant stomach chambers differ. The grooves that divide the rumen externally form ridges (pillars) internally that contain smooth muscle. The surface of the lumen contains large tongue-like or conical papillae and is greenish yellow or brownish in color due to plant dyes and accumulation of tannic acid in the epithelial layers of the rumen. The papillae greatly increase the size of the luminal mucosa and their density is low on the dorsal surface of the rumen. Bacterial and protozoan fermentation of cellulose present in plant material leads to production of volatile fatty acids such as acetic and butyric and also gases including methane and carbon dioxide in the rumen. The glucose released is either utilized by the microbes or is absorbed in the rumen. Ruminants are sometimes referred to as **foregut fermenters** as a result of such activities in the anterior part of the digestive system. There is also regurgitation of the more coarse ingesta for rechewing. The bacteria synthesize their own proteins using urea that results from breakdown of proteins. The bacterial proteins contain the essential amino acids and will eventually be digested in the glandular stomach for the benefit of the ruminant.

The contents in the digestive system move to the reticulum from the rumen. The reticulum is circular or ovoid in shape and has a lot of smooth muscle in its wall. The internal surface of this chamber bears honeycomb like structures (reticular cells) that project into the lumen and bear tiny papillae. The reticular cells are further subdivided by secondary crests. Small and coarse pieces of food are assembled into cuds to be regurgitated to the oral cavity for rumination in the reticulum. Fine ingesta moves to the spherical omasum from the reticulum. The internal structure of the omasum bears leaf-like folds of varying sizes that are parallel to each other. These folds bear small papillae. The three non-glandular chambers absorb the fatty acids and water from the ingesta.

The glandular abomasum lies between the omasum and duodenum and is J-shaped. The abomasum like other glandular stomachs contains gastric glands and its surface bears prominent permanent folds that run in an oblique manner to the longitudinal axis of the chamber. The microorganisms that manufacture protein in the rumen are partly digested in the abomasum by the enzyme **lysozyme**. The efficient stomach lysozyme system of advanced ruminants such as cattle, sheep and deer has 10 lysozyme genes when compared to a single one in the pig and this has been associated with the success of this group of ruminants. Changes in rates of stomach lysozyme evolution have been attributed to selective pressure and not mutation rate (Yu and Irwin, 1996). Some leaf-eating monkeys also have lysozyme that has resulted from convergent evolution.

There is a gastric groove in the ruminant stomach that starts at the cardia and runs through the reticulum, omasum to the abomasums at points these chambers are closest to each other. This groove is of significance in suckling young ruminants. The presence of liquid in the pharynx and anterior part of the esophagus brings together the lips of the groove thereby forming a tube that runs from the cardia to the abomasum. Milk and other liquids are able to bypass the other chambers of the stomach in the process. The gastric groove reflex becomes weaker as the ruminant grows older.

Intestines

Intestines start from the stomach up to the cloaca, rectum or anal opening and take various courses in the body cavity ranging from being almost straight to being quite coiled. Generally, predaceous species of vertebrates tend to have shorter intestines when compared to herbivores where these segments can be quite long and enlarged. Intestines have also undergone other adaptations to increase their surface area that is major in absorption of nutrients including folding of the intestinal lining and presence of villi and microvilli on their inner surfaces (Fig. 7.11).

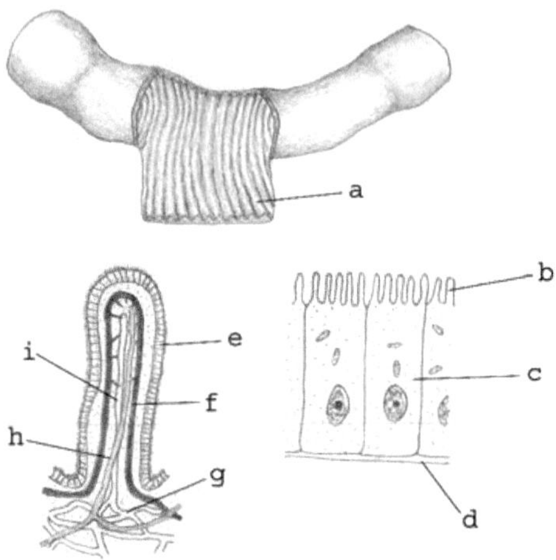

Fig. 7.11 Some of the structures that increase the surface area of intestines. (a) part of the lining of the small intestine (top), a villus (bottom left) and epithelial cells of the small intestine (bottom right). (a) circular mucosal fold (plica), (b) microvillus, (c) columnar epithelial cell, (d) basal lamina, (e) simple columnar epithelium with microvilli, (f) venule, (g) lymphatic plexus, (h) arteriole and (i) lacteal. The lacteal (L. *lac*, milk) is a blind ending lymphatic capillary and main route for the uptake of fat molecules that are absorbed in the small intestine. The arteriole forms an arteriovenous loop and a capillary network below the epithelium.

The mucous membrane of the fish intestines is similar to that of the stomach. There are numerous glands that secrete mucus and digestive enzymes. The small and large intestines play an absorptive role and their surface areas are enlarged by the presence of various structures including spiral valves (Fig. 7.12), longitudinal and transverse folds and villi depending on the species. The intestine of fish shows great variability in length depending upon the diet. In some predaceous species, the intestine can just be as long as the body cavity whereas in some herbivorous fish the intestine forms several coils and is several times the length of the entire fish. The intestines of cartilaginous fish and primitive bony fishes such as lungfishes and sturgeons are relatively short.

The small intestines of amphibians are coiled and lead into short and fairly straight large intestines that terminate at the cloaca. Reptiles and birds generally have coiled small intestines and relatively short large intestines that lead to a cloaca. The small intestines of the bird comprises the duodenum, jejunum and ileum (in succession). Small birds and bats have significantly shorter small intestines and enhanced intestinal

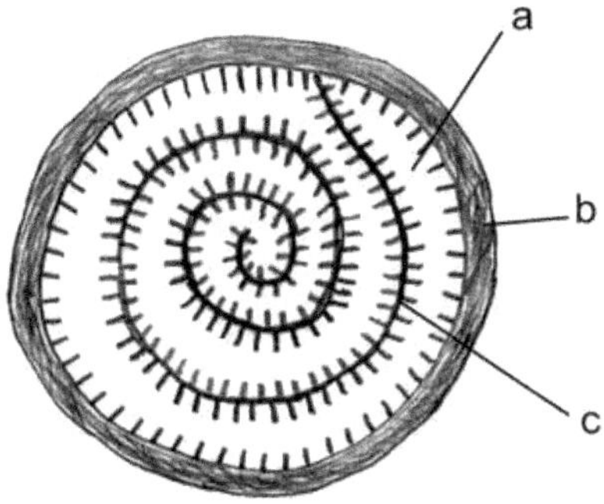

Fig. 7.12 A spiral valve. (a) lumen, (b) wall of intestine and (c) valve.

paracellular absorption than similarly sized nonflying vertebrates leading to a reduction in intestinal volume and mass of digesta carried thus reducing energetic costs of flight (Caviedes-Vidal et al., 2007). The large intestine has two caeca and a short colon that opens into a cloaca. Intestinal length including that of its segments varies in length in birds depending on the species, breed, age and diet. At the point where the caecum opens into the transition between the ilium and colon lies a circular fold of mucous membrane.

The cloaca of birds (Fig. 7.13) is divided into three compartments. The colon leads to the first cloacal compartment, the **coprodeum** (Gr. *kopros*, dung; *hodaion*, way). This compartment holds egesta temporarily before passing it to the **urodeum** (Gr. *ouron*, urine). Two ureters, the seminal ducts of the male and oviducts of the female bird open into this compartment. The last compartment, the **proctodeum** (Gr. *proktos*, anus), houses the male copulatory organ in aquatic birds. The **bursa of Fabricius**, an important structure in the bird's immune defense system as

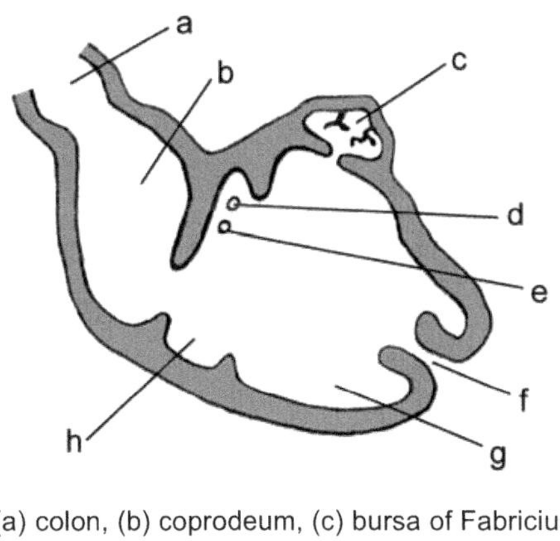

Fig. 7.13 A cloaca. (a) colon, (b) coprodeum, (c) bursa of Fabricius, (d) orifice of ductus deferens or oviduct, (e) ureter, (f) anus, (g) proctodeum and (h) urodeum.

it contains lymphoid follicles, is located above the proctodeum with which it communicates. The proctodeum leads to the anal opening that is surrounded by an anal sphincter.

In mammals, the intestines extend from the pylorus of the stomach to the anus. The small intestines are divided into the duodenum, jejunum and ileum. The ileum joins the large intestine at its terminal end. The large intestine comprises the caecum, colon and rectum that runs into the anal canal. The colon of carnivores and the human being show a relatively simple topography and can be divided into ascending, transverse, and descending segments of the colon. In the human being, the descending colon leads into a sigmoid colon that joins the rectum. In herbivores, the colon shows marked modification and diversity and is much longer and larger than that of carnivores in relative terms. For example, in ruminants, the ascending colon resembles a disk and in the horse it forms a larger and very long horse shoe-shaped structure that doubles on itself (Fig. 7.14). The blind ending caecum is smaller in carnivores and can be quite large in herbivores where it might be sacculated.

The **vermiform appendix** (L. *vermis*, worm; *forma*, shape) is present in the human being and many primates and is a blind ending structure that arises from the caecum. The appendix is not considered to be a vestigial organ by some scientists. There is evidence that it first evolved in certain Old World monkeys and developed progressively in other primates, including anthropoid apes (Scott, 1980). The appendix is believed to play a role in the body's mucosal immune system via the B-lymphocyte responses (Zahid, 2004). Rabbits and hares also have an appendix.

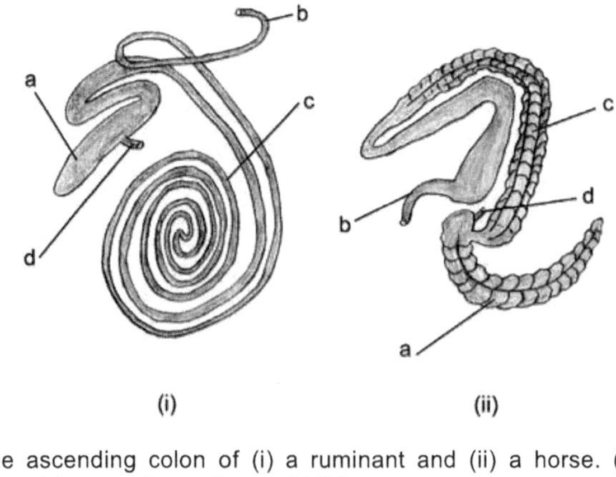

Fig. 7.14 The ascending colon of (i) a ruminant and (ii) a horse. (a) caecum, (b) transverse colon, (c) ascending colon and (d) ileum.

The anal canal forms the terminal part of the digestive system. The mucous membrane of the canal is covered with a stratified squamous epithelium. The canal ends at the anal opening that is guarded by internal and external sphincters. The internal sphincter is made of thickened smooth muscle that is a continuation of the circular smooth muscle layer of the rectum. The control of this sphincter is involuntary. The voluntary external sphincter muscle has skeletal muscle that arises from the caudal vertebrae and surrounds the anal opening. Some of these muscle fibers continue lateral to the anal opening to join the urogenital muscles.

The intestines are a major site for digestion and absorption of nutrients. The digesta is mixed with digestive juices from the pancreas, liver and intestinal mucosa. Mixing enables the products of digestion to make contact with the intestinal wall where absorption takes place.

Chemical digestion involves hydrolysis of food to produce the simple building blocks that can cross the wall of the alimentary canal into the internal environment and starts in the mouth of some vertebrates where starch is broken down by salivary amylase to two saccharide compounds (disaccharides). Some digestion of proteins also occurs in the stomach where proteases, mainly pepsin, break down the polypeptide chains of proteins.

Carbohydrate digestion entails the hydrolysis of polysaccharides such as starch and glycogen that contain many saccharide groups to disaccharides such as sucrose, lactose and maltose that contain two saccharide molecules. Disaccharides are further hydrolyzed to monosaccharides icluding glucose, galactose and fructose that are single saccharide molecules and can cross the intestinal wall. The hydrolylsis of polysaccharides is catalyzed by amylases that are found in saliva and pancreatic juice that is secreted into the duodenum. Disaccharide units are broken down by enzymes that are specific to these sugars such as sucrase, maltase and lactase which are located in the cell membrane of epithelial cells that line the intestinal lumen. The monosaccharide sugars that result from such digestion are thus at the site of absorption.

During protein hydrolysis, the long polypeptide chains are catalytically hydrolyzed by pepsin in the stomach, trypsin and chymotrypsin from pancreatic juice in the intestines and peptidases of the intestinal epithelium. Because each of the proteases catalyzes the breakdown of specific peptide bonds and there are various kinds of peptide bonds in a polypeptide chain, several proteases are needed to completely break down the chain. The resulting amino acids are absorbed in the intestines.

Fats are insoluble in water and are emulsified (spread into small droplets) by the phospholipid lecithin and bile salts that are present in bile which is secreted into the duodenum from the liver. Emulsification increases the surface area of lipids thus facilitating their chemical digestion. Fats are mainly digested by lipases that are present in pancreatic juice. The main lipids, triglycerides, are hydrolyzed into fatty acids, glycerol and monoglycerides that are absorbed in the intestines. The pancreas also produces nucleases that hydrolyze nucleic acids.

Transport of digested material across the intestinal lining involves diffusion, facilitated diffusion and co-transport. Fatty acid and fat-soluble vitamins diffuse through the lipid plasma membrane based on their concentration gradient across the diffusion barrier. Water, including water-soluble material such as glycerol diffuse through the protein lined pores in the membrane. Facilitated diffusion occurs where some minerals bind to a carrier molecule that is capable of moving across the membrane. During co-transport, several products bind to the same carrier molecule and are transported across the plasma membrane. Amino acids, monosaccharides and sodium ions can bind to a particular carrier molecule at the same time.

Apart from occurring in the fore-stomachs of ruminants, fermentation also occurs in the large (ascending) colon and caeca of the horse, rhinoceros, elephant, rodents, rabbits and hares. These species are sometimes referred to as **hindgut fermenters**. The bacteria present here break down food and the resulting products are absorbed in this part of the digestive system. A lot of the protein that is synthesized by bacteria is lost in feces as protein digestion occurs mainly in the stomach. Some species such as rodents and lagomorphs practise **coprophagy** (Gr. *kopros*, dung; *phagein*, to eat) whereby they ingest their own feces and re-cycle it through the digestive system. Such a behavior ensures that the proteins synthesized in the lower parts of the digestive system are properly digested in the stomach and are utilized by the body.

Undigested material together with bacteria that multiply in the digestive system, water, pigments and mucus are eliminated as feces.

THE LIVER

The liver is the largest gland in the body. The adult human liver weighs between 1.0 to 2.5 kilograms and is divided into lobes in some vertebrates. Lobes are further subdivided into lobules by connective tissue. A classic **liver lobule** is made of hepatic (Gr. *hepar*, liver) cells or hepatocytes that are arranged around a **central vein** (Fig. 7.15i) that is a branch of the hepatic vein which carries blood away from the liver. The liver cells are arranged in branching plates or laminae that run from the side of the lobule to its central part and surround sinusoids. **Sinusoids** drain into the central vein of the lobule. Sinusoids are like capillaries but lack a complete endothelial lining. Blood circulating through the sinusoids comes in direct contact with liver cells and the reticuloendothelial phagocytes that lie along the lining of sinusoids. In cross-section, lobules appear roughly hexagonal. **Portal areas** are triangular shaped areas that comprise of connective tissue, branches of the hepatic portal vein, hepatic artery, lymph vessels, bile ducts and nerves and are located peripheral to each lobule. There are portal areas at three to six angles of each hepatic lobule. Branches of the hepatic portal vein carry blood from the stomach, intestine and spleen and contain nutrients that have resulted from digestion of food whereas branches of the hepatic artery contain oxygenated blood.

A **portal lobule** forms the functional unit of a liver and is an area around a small bile duct (ductule) that lies in the portal area (Fig. 7.15ii). The lobule is triangular in shape and consists of the parenchyma of three adjacent lobules whose bile flows into the bile ductule of the portal area. Hepatocytes normally absorb the bile pigment bilirubin that results from the breakdown of hemoglobin of red blood cells. Bilirubin is conjugated and secreted as part of bile. Bile also contains salts, cholesterol and protein and is normally secreted into **bile canaliculi** that are located between hepatic cells. Bile will eventually flow to the bile ductules, hepatic ducts and the common bile duct or the gall bladder where it is stored and concentrated. A gall bladder is mainly found in vertebrates that do not feed continuously and is connected to the common bile duct by a cystic duct (Fig. 7.16). The common bile duct leads to the duodenum.

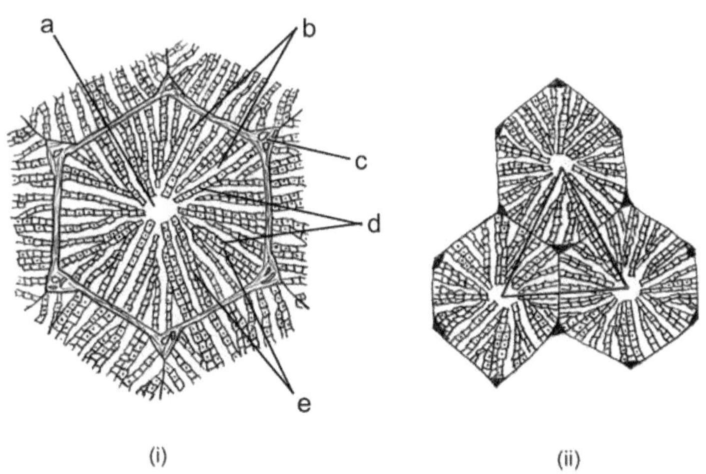

Fig. 7.15 A liver or hepatic lobule (i) and portal lobule (ii). (a) central vein, (b) sinusoids, (c) portal area, (d) liver cells (hepatocytes) and (e) bile canaliculi. The triangle in the center of three liver lobules in (ii) designates a portal lobule.

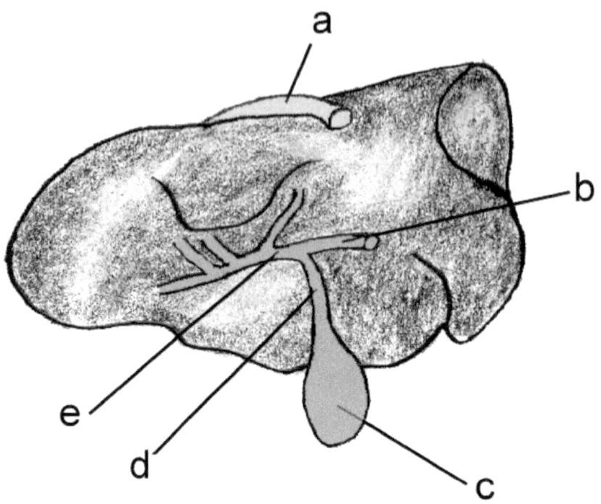

Fig. 7.16 The liver of an ox. (a) caudal vena cava, (b) common bile duct, (c) gall bladder and (d) cystic and (e) hepatic duct.

The liver serves various functions in the vertebrate body including detoxification, secretion of bile, storage of vitamins such as A, D, E and K as well as iron and the manufacture of many plasma proteins. The liver is also an important site for the metabolism of carbohydrates, proteins and fats. The liver acts as a site for hemopoiesis (production of blood cells) during embryonic development in all vertebrates and also adult fishes and amphibians. Carbohydrates are stored in the liver in the form of glycogen. Fats and small amounts of proteins can also be stored in the liver.

The liver can regenerate itself under normal conditions. In humans, living donor liver transplantation has been carried out since 1989 and involves removal of part of the liver from a living person to replace the entire liver of the recipient. About 20% of an adult's liver is required as a liver allograft for a small child. Adult to adult liver transplantion has also been carried out recently.

THE PANCREAS

Although all vertebrates possess a pancreas, the structure is not grossly distinct in lampreys, lungfishes and teleosts as it is embedded in the wall of the intestines and the liver or is scattered in the mesentery. Histologically, the pancreas is similar to the parotid salivary gland and is a compound acinar gland that has both exocrine and endocrine secretory units. Most of the pancreas is exocrine and this portion produces several enzymes including amylase, lipase and trypsin. These enzymes are secreted into the duodenum by the main pancreatic duct that in some vertebrates also unites with the common bile duct before joining the duodenum.

Within the exocrine portion of the pancreas are embedded many tiny clusters of pancreatic islets of Langerhans that form about 2% of the pancreas in the human being. The islets are mainly made up of alpha and beta cells. Alpha cells secrete the hormone glucagon that increases glucose levels in the blood by activating liver enzymes that breakdown glycogen to glucose. Beta cells produce insulin that lowers blood sugar levels by increasing the permeability of the plasma membrane of certain cells to glucose where it is used to synthesize glycogen mainly in muscle and liver or fats in adipose tissue.

Not all vertebrates possess both alpha and beta cells. The endocrine tissue also differs in the way it is arranged in the pancreas. The beta cells evolved before the alpha cells and are found in some earlier vertebrates such as lampreys where they are located in the intestinal wall.

METABOLISM

Metabolism includes all the interactive chemical processes that occur in the body to sustain life. It involves the use of the nutrients that have been absorbed after digestion by the body cells. The nutrients are normally assimilated by the body cells in a process that involves entrance of nutrients into cells that is followed by several chemical changes. Metabolism varies in the different cells of the body depending on the nature of activities and products that are synthesized or broken down. The two major metabolic processes are catabolism and anabolism and occur at the same time.

During catabolism, food molecules are broken down into simpler molecules in a process that releases energy. Some of the nutrient energy is stored in the energy rich bonds of ATP and is referred to as chemical energy. The rest of the energy is heat and is important in maintenance of body temperature or is lost to the environment. ATP supplies energy to the many metabolic processes occurring in the vertebrate body. Anabolism involves synthesis of products from nutrient molecules and is an energy consuming process.

Carbohydrates can be metabolized by catabolism and anabolism. Glucose is the main source of energy in many vertebrate cells. Within the cell, glucose reacts first with ATP to form glucose-6-phosphate during glucose phosphorylation before undergoing further metabolic processess. During **glycolysis** (first process of carbohydrate catabolism) in the cytoplasm, one glucose molecule is broken into two pyruvic acid molecules (each of which contains three carbon atoms) during anaerobic respiration that does not require oxygen. The breaking of the chemical bonds in glucose during this process of glycolysis releases about 5% (two ATP molecules) of the energy stored in a glucose molecule. Glycolysis is followed by the **citric acid** or **tricarboxylic acid (TCA) cycle**. Pyruvic acid is converted to acetyle-CoA (coenzyme A) that enters the TCA cycle that occurs in mitochondria. During the electron transfer process in the mitochondria, small amounts of energy are released that will eventually lead to oxidative phosphorylation whereby a phosphate group joins adenosine diphosphate (ADP) to form ATP. Oxidative phosphorylation depends on the presence of oxygen since it is the final receptor of electrons and hydrogen ions and leads to the complete breakdown of glucose molecules leading to production of about 36 ATP molecules per molecule of glucose catabolized.

During glycolysis, the maximal amount of energy is not obtained from the glucose molecules as it is still held in the pyruvic acid bonds. In the absence of oxygen, pyruvic acid is converted to lactic acid and not acetyl CoA. Lactic acid can either be converted back to pyruvic acid or diffuses into the bloodstream and transported to the liver where it will be converted to glucose. Both processes require energy since anaerobic respiration incurs an oxygen debt and oxygen is required later for ATP production that is necessary for the conversion of lactic acid to pyruvic acid or glucose.

In glucose anabolism, the excess glucose molecules are used in the synthesis of glycogen (glycogenesis) that is stored mainly in the liver and muscle. When levels of glucose are low, it can also be synthesized from proteins and less commonly glycerol from fats during gluconeogenesis in the liver. The liver thus plays a major role in blood glucose homeostasis.

Lipid catabolism occurs mainly when triglycerides are hydrolyzed to yield fatty acids and glycerol. Before glycerol can enter glycolysis, it is converted to glyceraldehyde-3-phosphate. Fatty acids are normally broken into acetyl-CoA that can then join the TCA cycle. The energy generated during lipid catabolism is much more than that released during breakdown of carbohydrates. Lipogenesis (lipid anabolism) is a process that leads to the synthesis of various lipids in the body, chiefly triglycerides, phospholipids, cholesterol and prostaglandins. Structural lipids such as phospholipids and cholesterol and triglycerides are synthesized from glycerol and fatty acids. They can also be synthesized from excess glucose and proteins. Although most fatty acids are synthesized in the body, essential fatty acids must be obtained from food that is eaten. Such fatty acids are important in the synthesis of vital lipids such as prostaglandins.

Amino acids form major building blocks of the body and as a result protein anabolism is their chief metabolic pathway in the body. Cells of the body are capable of synthesizing their structural proteins that are important in many processes including growth, reproduction and regeneration or repair of damaged tissues from amino acids. The liver and glands, in addition, synthesize proteins for use by other body cells such as plasma proteins and hormones respectively. Breakdown of proteins occurs under several conditions including low levels of carbohydrates and lipids. Before proteins can be utilized in energy production, they undergo **deamination** in a process that removes the amino group from amino acids in the liver. The amino group forms ammonia that can be converted to urea or uric acid in the liver according to the species. The three forms of nitrogenous wastes are eliminated from the body. The remaining keto acid can enter the TCA cycle or can be converted to glucose during gluconeogenesis or fat (lipogenesis).

REFERENCES

Caviedes-Vidal, E., McWhorter, T.J., Lavin, S.R., Chediack, J.G., Tracy, C.R. and Karasov, W.H. (2007). The digestive adaptation of flying vertebrates: high intestinal paracellular absorption compensates for smaller guts. Proc. Natl. Acad. Sci. USA, 104(48):19132-19137.

Finkelstein, Y., Meshorer, A., Talmi, Y.P., Zohar, Y., Brenner, J. and Gal, R. (1992). The riddle of the uvula. Otolaryngol. Head Neck Surg., 107(3):444-450.

Schwenk, K. and Throckmorton, G.S. (1989). Functional and evolutionary morphology of lingual feeding in squamate reptiles: phylogenetics and kinematics. J. Zool. Lond., 219:153-175.

Scott, G.B. (1980). The primate caecum and appendix vermiformis: a comparative study. J. Anat., 131(3):549-563.

Smith, D.M., Grasty, R.C., Theodosiou, N.A., Tabin, C.J. and Nascone-Yoder, N.M. (2000). Evolutionary relationships between the amphibian, avian and mammalian stomachs. Evol. and Dev., 2(6):348-359.

Tandler, B., Gresik, E.W., Nagato, T. and Phillips, C.J. (2001). Secretion by striated ducts of mammalian major salivary glands: review from an ultrastructural, functional and evolutionary perspective. Anat. Rec., 264(2):121-145.

Waisbren, S.J., Geibel, J.P., Modlin, I.M. and Boron, W.F. (1994). Unusual permeability properties of gastric gland cells. Nature, 368:332-335.

Yu, M. and Irwin, D.M. (1996). Evolution of stomach lysozyme: the pig lysozyme gene. Mol. Phylogenet. Evol., 5(2):298-308.

Zahid, A. (2004). The vermiform appendix: not a useless organ. J. Coll. Physicians Surg. Pak., 14(4):256-258.

8

Respiratory System

The primary function of the respiratory system is to deliver oxygen from the environment to the mitochondria in body cells where various cellular processes occur leading to the production of energy, carbon dioxide and water. The respiratory system has to adapt to meet the metabolic requirements of individual vertebrates. In many organisms, including the human being, mechanical respiration and circulation are regulated to correspond to the demand of the moment (Krogh, 1941).

Exchange of oxygen and carbon dioxide between the outer environment and the circulatory system of gills, lungs or other respiratory structures for transport to body tissues is known as **external respiration**. At the tissue level, there is exchange of these gases between the circulatory system (capillaries) and tissue cells. In both cases exchange of oxygen and carbon dioxide occurs by diffusion that is a physical process and involves movement of molecules down a partial pressure gradient. The difference in concentration of a substrate between two compartments of similar composition is the driving force for diffusion as is stated in Fick's first law of diffusion. The substrate should have the same capacitance (solubility) coefficient in both media in the two compartments as the rate of diffusion depends on differences in partial pressure between the two media. When the two media are not identical, Fick's law does not apply. **Internal respiration** is a process that uses up oxygen for oxidation of mainly glucose to produce energy in the form of adenosine triphosphate (ATP), carbon dioxide and water.

The different structures used by vertebrates for gaseous exchange with the environment have varying permeation coefficients that are products of diffusion and solubility coefficients. Air and water as respiratory media also differ mainly in their oxygen content, density and viscosity. The quantity of oxygen in water is much less than in air because of the low solubility of oxygen in water. Surface water contains about 5 ml of oxygen per liter of water when compared to 210 ml of oxygen per liter of air. The quantity of oxygen in water also depends on temperature, barometric pressure and concentration of dissolved solids such as chlorides. The evolution of a respiratory system with large surface areas and short diffusion distances facilitates diffusion of gases across respiratory barriers. The presence of a circulatory system and development of specific respiratory carriers for oxygen such as hemoglobin in vertebrates further enhances diffusion of gases across diffusion barriers as the convection and binding of oxygen increase the diffusion gradient.

Aquatic homeotherms have to spend more energy to cover an equivalent distance during swimming in comparison to walking on land by their terrestrial counterparts since water, as a more-dense medium than air, offers greater resistance to movement. Swimming close to the water surface requires much more energy when compared to swimming in deep water as the drag effect is greatest at the water surface or close to it (Blake, 1983). The water surface has waves and other edge effects that are much reduced in deeper waters. Aquatic birds and mammals have devised strategies to minimize energy expenditure while swimming. Penguins and dolphins leap out of the water surface rapidly and in the process reduce the overall drag effect and increase locomotory efficiency (Blake, 1983). Sea otters normally dive fairly deep into the water where they spend most of the time, remaining in water below the surface zone with a greater drag effect (Williams, 1989).

RESPIRATORY SYSTEM OF FISHES

Evolution of the fish respiratory system has enabled this major group of vertebrates to live in most aquatic habitats on Earth with varying oxygen levels. Colder water contains more oxygen than warm water and seawater contains less oxygen than freshwater. Similarly, stagnant water when compared to moving and stirred water contains less oxygen. Other factors such as presence of organic matter and pollutants promote microbial growth thus reducing further the quantity of oxygen present in a body of water. Despite the various constraints fish have faced in their history on Earth, the gills of most bony fishes are capable of taking up between 80 and 95% of oxygen in water that ventilates these respiratory structures. Apart from gills, fish have evolved other structures that are used in gaseous exchange including lungs, skin, swim bladder and other accessory air-breathing organs as a result of the diverse environments in which they live.

Fish Gills

Gills are the primary gas exchange organs of fish and are found in all fish species. Most fish depend on **internal gills** entirely for gaseous exchange but some fish have to supplement the role of gills in respiration by depending on accessory gas exchange structures. Gills also play the important roles of excretion of nitrogenous wastes and osmoregulation. **External gills** are found in the larvae of many fish species. Such gills have many filaments with large surface areas and are highly vascularized. The filaments attach laterally on the head surface between the gill slits. The external gills are quite distinct in the larvae of the spiny dogfish (*Squalus acanthians*) and the torpedo (*Torpedo marmorata*). Adult fish normally have internal gills that are located in the pharyngeal region.

Gills of Agnathans

The large branchial pouches (Fig. 8.1i) of lampreys and hagfish possess **gill filaments** (primary gill lamellae) that are also known as **pouched gills**. The pouches are supported by the irregularly arranged cartilaginous branchial basket (latticework) that represents the visceral skeleton. The short filaments have **secondary lamellae** that amplify the gas exchange surface area. Lampreys can feed and respire at the same time as their pharynx is divided longitudinally into a dorsal esophagus and a ventral respiratory tube that ends blindly (Fig. 8.1ii).

During inspiration, water enters the oral cavity then flows through the ventral respiratory tube to the branchial pouches through the pore-shaped internal gill slits by tidal flow. Each internal gill slit is surrounded with a sphincter. Expiration occurs when the branchial compressor muscles of the branchial basket contract. Water will flow to the outside through the external gill slits due to contraction of the sphincters of the internal slits that close these openings.

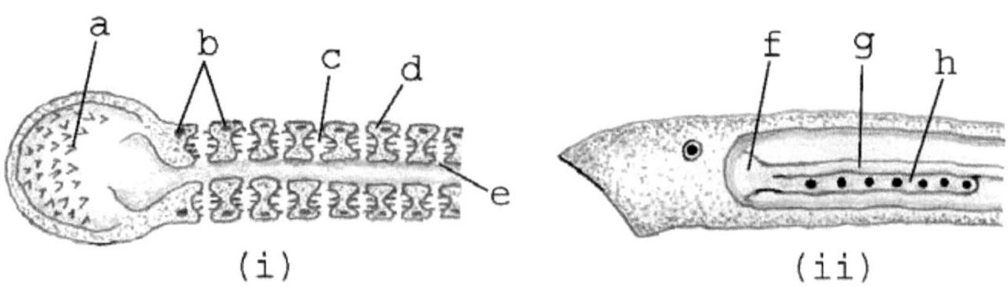

Fig. 8.1 (i) Frontal and (ii) sagittal sections through the pharynx of a lamprey. (a) horny tooth, (b) part of the branchial basket, (c) a pouch with gills, (d) interbranchial septum, (e) internal gill slit, (f) oral cavity, (g) esophagus and (h) respiratory tube with external gill slits.

Hagfish display two modes of respiration depending on whether they are feeding or not. When hagfish are not feeding, water enters the gill pouches from the nasopharyngeal cavity and then flows to the outside. When feeding, water flows in and out of the pouches in an alternate manner through the esophageocutaneous duct that opens to the outside posterior to the last pouch.

Gills of Elasmobranchs

Elasmobranchs have five and rarely six or seven vertical gill slits (branchial pouches) that are narrow and are located in front of the pectoral fins on the lateral side of the body in sharks and ventrally in rays and skates. Between the slits are **interbranchial septae** that have cartilaginous support and extend to the body surface (Fig. 8.2). **Septal gills** are found on either side of an interbranchial septum and gills on one side of the septum constitute a hemibranch. The two **hemibranchs** of each septum form a **holobranch**. Gills are made of filaments that are arranged in a perpendicular manner on septae and bear secondary lamellae. The latter are arranged transversely on filaments. The gills do not extend to the external part of the septum.

Deoxygenated blood flows to the gills from the ventral aorta through the **affarent branchial artery**. The latter artery gives off two branches at the level of each gill filament known as **affarent filament arteries**. These two vessels run on each side of the interbranchial septum (between the

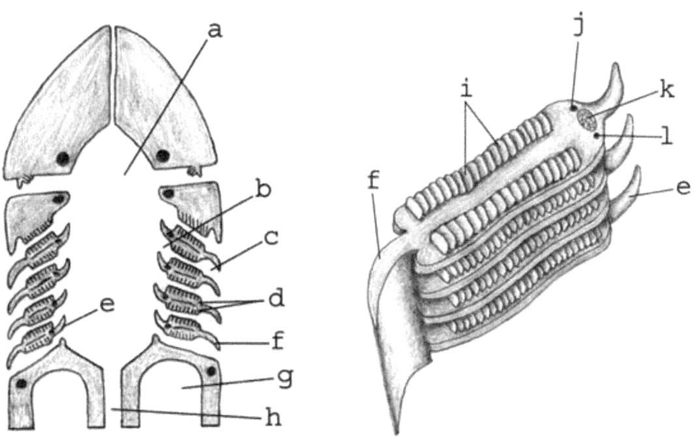

Fig. 8.2 Frontal section of the pharynx of an elasmobranch (left) and part of the holobranch of a shark (right). (a) pharynx, (b) internal and (c) external gill slits, (d) holobranch, (e) gill raker, (f) interbranchial septum, (g) coelom, (h) esophagus, (i) secondary lamellae, (j) pretrematic artery, (k) cartilaginous branchial arch skeleton and (l) posttrematic artery.

septum and secondary lamellae) and give off smaller vessels or **blood channels** to secondary lamellae. Blood flows from the secondary lamellae to the **efferent filament arteries** that join **pre-** and **posttrematic arteries** that are located on the anterior and posterior sides of the internal part of the interbranchial septum. Pre- and posttrematic arteries eventually unite into the **efferent branchial artery** that carries blood to the dorsal aorta.

During **inspiration** in cartilaginous fish, the mouth and the valves that close the spiracle open (Fig. 8.3). Contraction of the **coracomandibularis** and the **rectus cervicis** (**sternohyoideus**) muscles increases the volume of the pharyngeal cavity while lowering its pressure in relation to the surrounding water. Since the external water pressure is higher than the pharyngeal pressure, the external ends of the interbranchial septae will close the gill slits. Water will then flow into the pharyngeal cavity through the oral opening and the spiracles and into the internal vertically long **gill slits**. The cartilaginous gill rakers located medially on the interbranchial septae prevent food particles from entering the **branchial chambers** between the vertical interbranchial septae and their gills. Whereas water flows between neighboring secondary lamellae of one side on a filament towards the interbranchial septae from branchial chambers, blood in the blood channels of secondary lamellae flows away from the septum towards the branchial chambers. This is a **countercurrent system** of gas exchange (Fig. 8.4) since blood flows in a direction which is opposite that of water. The countercurrent system of gas exchange ensures maximal diffusion of oxygen from the water into the gills. Water then flows from the branchial chambers to the parabranchial chambers.

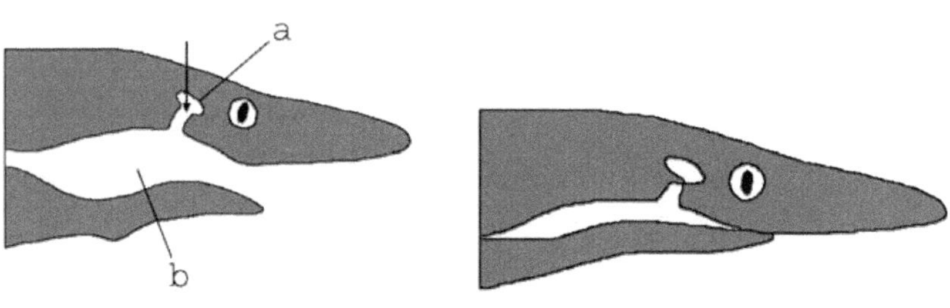

Fig. 8.3 Breathing movements in a cartilaginous fish. During inspiration (left), the mouth and spiracle (a) are open, so water rushes into the pharyngeal cavity (b) that has negative pressure in relation to the surrounding water. Expiration (right) is accompanied by closure of the mouth and spiracle. Arrows indicate the direction of water flow. The spiracle is the reduced first gill pouch that is present in most cartilaginous and early bony fishes.

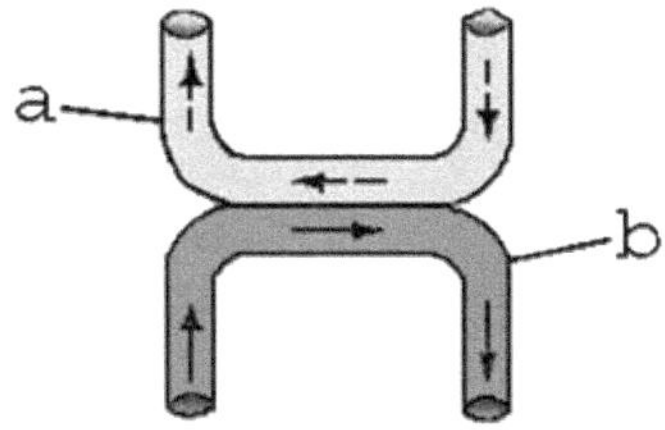

Fig. 8.4 The countercurrent system of gas exchange. Directions of water flow over gills (a) and blood flow within the secondary lamellae (b).

During expiration, the mouth and spiracles close. The **adductor mandibulae** and **preorbitalis** muscles contract at the same time decreasing the volume of the pharyngeal cavity and increasing its pressure relative to the surrounding external water. The external gills slits open and water flows from the parabranchial chamber to the outside. This system of ventilating the gills requires energy.

Some fast swimming sharks and teleosts attain high swimming speeds while their mouths are open during **ram ventilation**. Water is able to flow over the gills in the process without employing the principles of respiration that require energy to push water over gills when the fish are swimming at low speeds. Sharks that practice ram ventilation have much smaller spiracles or they are lacking altogether. The bottom-swimming rays and skates have large spiracles as most of the water that ventilates the gills enters the pharyngeal cavity through these openings.

Gills of Bony Fishes

The gills of bony fishes are arranged in a manner that generally resembles that of cartilaginous fishes even though the two structures evolved independently. The gills of these two major groups have slits that separate each gill from another for most of the gill length. Bony fishes normally have four slits between the gills that are covered with a bony operculum. There is a common opercular cavity and a gill slit that opens to the outside on each side of the head (Fig. 8.5). The spiracle is lacking in bony fishes. The two hemibranchs that form a holobranch show varying degrees of independence from each other depending on the amount of tissue (gill septum) present between the two rows of filaments. Teleosts have reduced gill septae (Fig. 8.6).

Inspiration is brought about by the contraction of the sternohyoideus muscle that leads to an increase in the volume of the orapharynx. The opercular cavity is closed at the same time. The lowering of pressure in the oral cavity results in an influx of the surrounding water into the cavity. Further contraction of the dilator muscles lowers pressure further

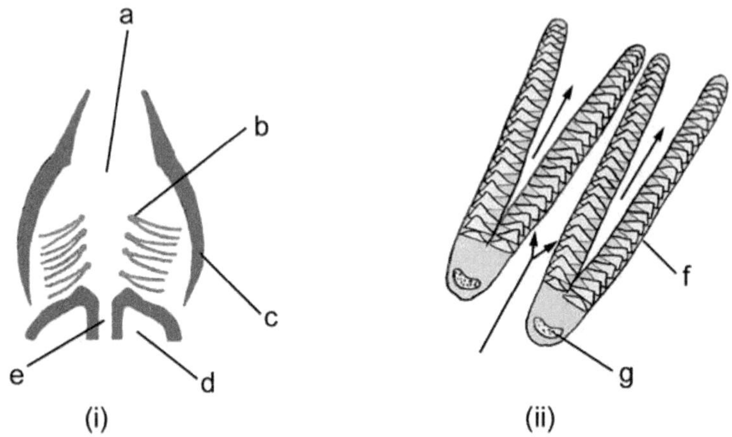

Fig. 8.5 (i) Frontal section of the pharynx and (ii) part of gills of a bony fish. (a) oral cavity, (b) gill, (c) bony operculum, (d) coelom, (e) esophagus, (f) gill filament and (g) bony branchial arch skeleton. Arrows indicate the direction of water flow from the oral cavity to the outside. At the posterior end of the operculum is a gill slit that leads into the opercular cavity that lies between the gills and the bony operculum.

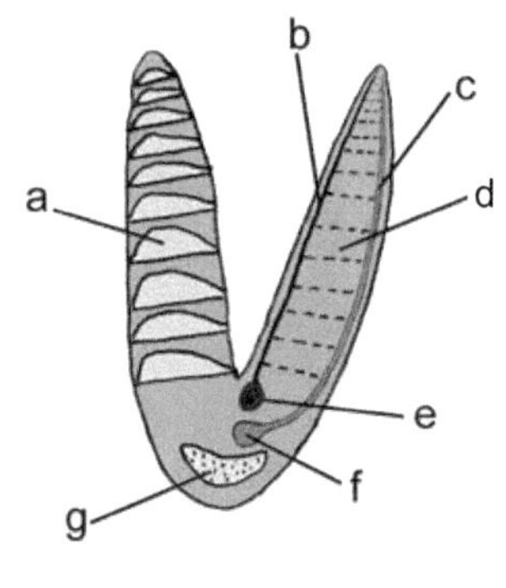

Fig. 8.6 Two gill filaments. To the right is a horizontal section through a gill filament. (a) secondary lamella, (b) affarent and (c) efferent filament arteries, (d) filament, (e) affarent and (f) efferent branchial arteries and (g) bony branchial arch skeleton. Dotted lines represent the blood channels of the secondary lamellae that run from affarent to efferent arteries. The gill septum is located between the two rows of filaments on a gill and is quite reduced in this figure.

in the opercular cavity by increasing its volume. Water will flow from the oropharynx to ventilate the gills then move to the opercular cavity. Contraction of the adductors and geniohyoideus muscles with the mouth closed reduces the volume of the oropharynx. The resulting increase in pressure leads to the flow of more water into the opercular cavity. The opercular cavity will thus open leading to flow of water to the outside. The countercurrent system of gas exchange also applies in the gills of bony fishes. The surface area and oxygen diffusion distances of the gill secondary lamellae that are sites of gas exchange vary depending on the metabolic requirements of fish. Active pelagic tuna have a secondary lamellar surface area of 3151 mm^2 per gram of body weight (Muir and Hughes, 1969) as compared to the much less active coelacanth (*Latimeria chalumnae*) of 18 mm^2 per gram of body weight (Hughes, 1980). Active fish also have short gill diffusion distances that in the yellow-fin tunny and mackerel are 0.17-1.13 m and 0.60-3.63 μ respectively (Hughes, 1970). The same distances are 5.0-6.0 m in the coelecanth (Hughes, 1972).

Secondary Lamellae

These are the exchange sites of gills and develop on both sides of each filament and are arranged in a perpendicular manner to the long axis of filaments. The secondary lamellae of the superior surface of a lower filament fit between those of the inferior surface of the higher filament. In teleosts, new secondary lamellae are formed at the tips of filaments so that the oldest lamellae are those at the base of the filament. The number and shape of secondary lamellae vary with each species. The number of secondary lamellae per fish increases with body size and activity. The greatest proportion of the surface of a lamella is found towards the edge at which water enters the gill from the oropharynx and is the leading edge of a secondary lamella. This distribution maximizes the lamellar area at the inlet side and produces the best gas exchange situation in counter and co-current flow systems (Hughes, 1984). The secondary lamellae are probably supported in their functional position by their blood channels, hydrostatic pressure, stiffness of red blood cells, the pillar cell system of the basement membrane and collagen fibers (Hughes and Weibel, 1972; Hughes, 1984).

The epithelium of secondary lamellae (Fig. 8.7) is the first layer of the water/blood pathway and is made of two layers of squamous cells that overlap and interdigitate with neighboring cells. The layers may be separated in some areas by lymphoid (intercellular) spaces that are frequently occupied by macrophages that may have a function that is similar to that of mammalian alveolar macrophages but due to the water current along the respiratory surface in fish, they are confined to the lymphatic tissue gaps. The epithelium rests on a basal membrane. Below this membrane are pillar cells and their flanges that surround blood channels. The tips of secondary lamellae contain the marginal channels that have endothelial cells with flattened nuclei. Pillar cell flanges contain actomyosin filaments and probably prevent the walls of the blood

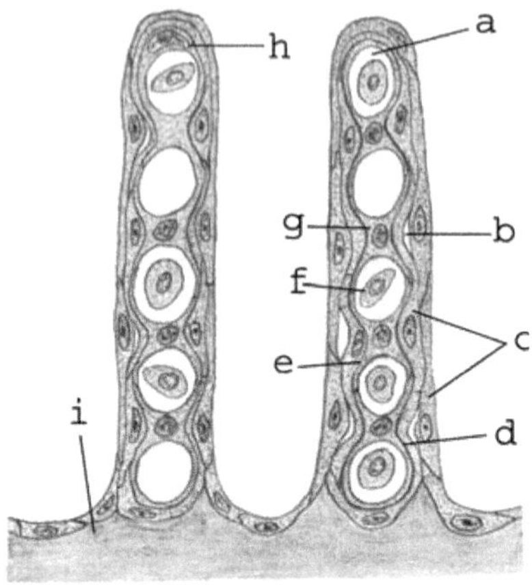

Fig. 8.7 Diagram of a cross section through secondary lamellae. (a) marginal channel, (b) intercellular (lymphoid) space, (c) two-layer epithelium, (d) basement membrane, (e) pillar cell flange, (f) red blood cell in a blood channel, (g) pillar cell (h) endothelial cell of marginal channel and (i) gill filament or primary lamella.

channels from extending too much when the secondary lamellae fill with blood or collapsing when blood bypasses them. Blood flow through the blood channels of secondary lamellae can be shunted to the marginal channels and the anastomosing vessels between the affarent and efferent filament arteries when the need arises.

Chloride cells (ionocytes) are located in inter lamellar regions and non-lamellar areas of gill filaments and are in close association with neighboring epithelial cells. Their apex has microvilli and the cells excrete chloride, potassium and sodium ions. Mucous cells are found in the gill arch and filament epithelium. They discharge mucus in response to violent changes in the water. The mucus covers the surface of the gills.

Since water and salts are lost or gained through the gills depending on the concentration of solutes in the surrounding water and body fluids, fish are faced with the challenge of ventilating their gills to a level that is only necessary to meet their oxygen requirements. The process of preferential perfusion of gills hence partial use in diffusion of oxygen has been noted in some fish (Randall et al., 1972; Booth, 1978; Nilsson, 1986). The surface area of secondary lamellae ventilated can also be adjusted by abductor muscles of gills as well as vascular shunts diverting some blood directly from affarent to efferent branchial arteries without flowing through the secondary lamellae. When the adductor muscles contract, water flows between the rows of filaments of adjacent gills thus avoiding the secondary lamellae. The gill area that is exposed to ionic and water gain or loss is thus reduced. The opposite takes place when these muscles relax and the abductor muscles contract.

Accessory Respiratory Organs

Accessory respiratory organs have enabled bony fishes belonging to more than 50 genera breathe oxygen directly from the air. These structures could have evolved as a result of the inability of gills to meet the oxygen requirements of fish in water with low oxygen tension such as warm pools, swamps with a lot of decaying matter or stagnant water. Several changes have occurred with transition from aquatic to aerial respiration including morphological and physiological changes in gas exchange, regulation of ions, excretion of nitrogenous wastes and acid-base balance. An obligate air-breathing fish such as *Arapaima gigas* of the river Amazon that undergoes transition from water to air-breathing during development and the accompanying changes in gills offers a good model system to understanding this transition (Brauner et al., 2004).

Lungs

The three surviving dipnoan genera (*Neoceratodus* of Australia, *Protopterus* of Africa and *Lepidosiren* of South America) that live in tropical rivers have evolved a pair of long and hollow sacs in the dorsal part of the body cavity. Cranially, the two sacs merge to form a common chamber that is connected to the glottic sphincter of the pharynx. The inner wall of the sacs bears septae (trabeculae) that form a spongy network that increases the respiratory surface area and demarcate incompletely closed alveoli. The trabeculae contain smooth muscles and blood vessels. These lungs are thought to have evolved in the Devonian when bodies of water could become stagnant and hypoxic and would periodically dry up and so fish needed an accessory respiratory organ to supplement the gills in gaseous exchange. *Protopterus* and *Lepidosiren* survive dry seasons by forming a cocoon that is surrounded by mucus and has a small tube that leads to the surface of a dried up riverbed or pond for breathing. The metabolic rate of these fish is also lowered during which time they excrete urea instead of ammonia. Another fish with lungs is the primitive ray-finned bichir (*Polypterus*) of Africa whose right lung is much longer than the left one.

When using lungs for breathing, a lungfish comes to the surface of water and inhales air by opening its mouth and expanding the buccal cavity. The fall in pressure in the cavity is followed by an influx of air into the buccal cavity. The slit-like glottis is then opened and air is exhaled from the lungs to the buccal cavity. The elastic recoil together with contraction of smooth muscles of lungs causes expiration. There is mixing of inspired and expired air in the buccal cavity and some of this air escapes to the outside. The mouth is then closed and the buccal cavity compressed to push air into the lungs. The rate at which lungfishes ventilate their lungs is low due to their low metabolic rates. This leads to accumulation of the highly soluble carbon dioxide in their bodies that is eliminated by the gills.

Although *Neoceratodus* is a facultative air-breather that uses its lungs to supplement the gills when the water becomes hypoxic, *Protopterus* and *Lepidosiren* are obligatory air-breathers due to the thickness of their gill epithelia and filament organization (Laurent, 1996). These fish rely on their lungs for about 90% of their oxygen supply (Johansen, 1970). The internal lung septae are more developed in *Protopterus* as it has lost its first two gills.

Lungfishes have evolved pulmonary circulation while still retaining branchial circulation. Such a system differs from that of fish that entirely rely on their gills for respiration and represents a transition from which the tetrapod mode of respiration evolved. The presence of a ductus arteriosus, pulmonary artery vasomotor segments and gill shunts enables the lungfishes perfuse the lungs and gills preferentially. The cardiorespiratory morphology and physiology during aerial and aquatic respiration particularly features that distinguish *Neoceratodus* from *Protopterus* and *Lepidosiren* as well as lungfishes from aquatic vertebrates have been reviewed by Burggren and Johansen (1986).

Swim (Gas) Bladder

The swim (gas or air) bladder and lungs of fishes are structures that have resulted from convergent evolution. The swim bladder could have evolved in fish that lived in oxygen rich waters and where the need for a hydrostatic organ rather than lungs was favored. The swim bladder is important in overcoming problems associated with water pressure. An increase in pressure from the surrounding water that is associated with deep swimming compresses the gas in the swim bladder and leads to sinking of fish. In shallow or surface water, the low pressure leads to expansion of gas in the swim bladder resulting in floating of fish. When a fish is neutrally buoyant (neither sinking nor floating), it expends less energy to maintain itself at the required depth in water.

Most of the swim bladder has a poor blood supply and is not permeable to gases as it is lined with sheets of guanine crystals. The wall of the organ also contains elastic and loose fibers and smooth muscle. The swim bladder occupies about 5 to 6% of the fish volume in freshwater fish and 7% to 10% in seawater fish. The buoyancy densities of fresh and seawater are 1.0 and 1.026 respectively.

In fish species that retain the ancestral **physostomous bladder** (Gr. *physa*, bladder; *stoma*, mouth), a pneumatic duct connects the caudal end of the esophagus to the swim bladder. Physostomous fish include carp, trout, salmon and esocids. A physostomous swim bladder has undergone structural and physiological modification to be used as an air-breathing organ that supplements the gills in gaseous exchange. Fish with a physostomous bladder are found near the water surface and normally gulp air that they pass to the bladder by buccal contraction. A **physoclistous bladder** (Gr. *kleiein*, to close) is not connected to the esophagus as a pneumatic duct is lacking. Physoclistous fish include walleye, perch, bass and many panfishes. A physoclistous bladder has a **gas gland** and a **rete mirabile** (L. *rete*, net; *mirabile*, wonderful) that secrete gas from blood and the oval that absorbs gas back to the blood stream (Fig. 8.8). The rete mirabile has parallel countercurrent affarent and efferent capillaries. When blood enters the gas gland, there will be release of oxygen by hemoglobin as a result of the presence of carbonic acid and lactic acid in blood that are secreted by the gland. Oxygen will accumulate in the blood and will diffuse into the swim bladder that can also act as a storage organ of the gas. Removal of oxygen from the bladder is by the **oval**. Relaxation of the sphincter that connects the oval to the bladder results in entry of oxygen into the oval and diffusion into blood.

The rate at which adjustments are made to the swim bladder varies in different fish. Physoclists with highly developed gas glands and retia mirabilia make great adjustments to the organ in a short time. Fish with poorly developed glandular tissue in the gas gland or those that lack a rete mirabile such as the salmonids are unable to replace air lost from the swim bladder when denied access to surface air. Physostomous fish can

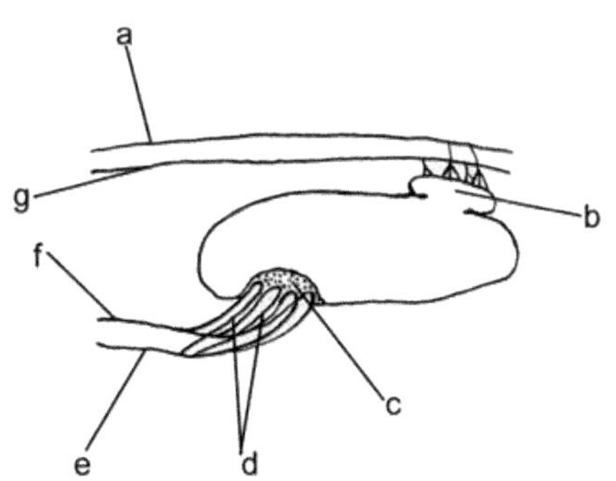

Fig. 8.8 The swim bladder. (a) dorsal aorta, (b) the oval (c) gas gland, (d) rete mirabile, (e) venous and (f) arterial blood and (g) posterior cardinal vein.

make buoyancy adjustments faster than physoclistous fish as they can expel air from their swim bladders that are directly connected to the esophagus with a duct.

Other Accessory Respiratory Organs

These structures (Fig. 8.9) have developed as extensions of pharyngeal, branchial or opercular chambers. The pharyngeal chambers have undergone diverse modifications in various species of fish. The air-breathing organs of various species of the snakehead of Asia, *Channa,* have developed on the dorsal part of the pharynx as a pair of extensions known as suprapharyngeal chambers. These chambers participate in both air and at times water breathing. The surface of the chambers bears numerous vascular papillae that are also found on the buccopharynx, palate and even the tongue. The dome-shaped papillae bear spiral or wave-like capillaries below the epithelium. The snakehead has been known to leave poor quality water and move on land for three to four days out of water in search of suitable water. Air sacs that are extensions of the pharynx and are located on the lateral sides of the head are found in the cuchia (*Monopterus cuchia*). These air sacs are covered with the opercula.

The air sac catfish of Asia, *Heteropneustes fossilis,* has evolved four pairs of gill fans that have resulted from fusion of gill filaments and two air sacs that are extensions of the suprabranchial chambers into the body trunk. The air sacs are embedded in the myotomes (Munshi and Choudhary, 1994). The gill fans bear lamellae that have microvilli. In the African catfish, *Clarias mossambicus,* the accessory air-breathing organs are a labyrinthine structure and a suprabranchial chamber membrane. This obligate air breather inhabits rivers and swamps in Africa that are liable to drying up. Another obligate air-breather, the climbing perch

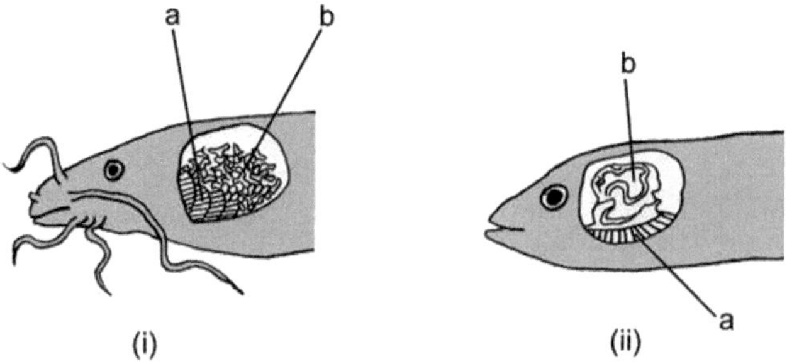

Fig. 8.9 Some of the accessory air-breathing organs of (i) the catfish *Clarius* and (ii) the climbing perch *Anabas*. (a) gills and (b) air-breathing organ.

(*Anabas testudineus*) of Asia, has labyrinthine organs consisting of many plates covered with the suprabranchial chamber membrane.

The opercular chamber has evolved into air-breathing organs in some estuarine fishes of the family Gobiidae. In the mudskipper, *Periophthalmus,* opercular bones are usually elastic with a thin epithelium that is richly vascularized. These fish usually expand their opercular chambers with air for diffusion to occur into the opercular vessels. *Electrophorus* of the Amazon River has a highly vascularized mouth.

Part of the gastro-intestinal tract has been modified in some fish into a respiratory organ. Air is normally passed into the system through the mouth and out the mouth or anus. The segment of the system used in gaseous exchange is modified into a thin walled structure. The armored catfishes such as Loricariidae use their stomachs and loaches (Cobitidae) use the middle and posterior parts of their intestines for gaseous exchange.

The skin of some fish such as the eel *Anguilla anguilla* functions as an accessory respiratory organ to complement the gills and other air-breathing organs. Such fish normally lack scales on their skin that is highly vascularized. The skin also plays an important role in elimination of carbon dioxide from the body.

RESPIRATORY ORGANS OF AMPHIBIANS

Amphibians use gills, lungs and the skin for gaseous exchange. Amphibian gills are external and are mainly fimbriae that protrude from gill slits. The gills are present in the aquatic larvae and adult neonetic groups such as the mudpuppy (*Necturus)* and axolotl that retain gills throughout life. Throat pulsations cause water turbulence near the gills. The lowering of the floor of the oral cavity increases the volume of the cavity while decreasing its pressure. Water enters the cavity through the oral opening and or the nostrils. Raising the floor of the cavity forces the water to flow to the pharynx and gills to the opercular chamber that is covered by an extensive opercular fold before passing through the spiracle to the outside. Some external gills contain muscle that contracts to bring about some motion. Tadpole gills lack these muscles. Many amphibian larvae possess valves that prevent water from flowing out of the oral cavity through nostrils. The larval gills are lost during metamorphosis with development of lungs that will be used in respiration. The larvae also use the skin and buccopharyngeal mucosa in respiration.

Adult amphibians possess a respiratory tract that leads from the outside to lungs (Fig. 8.10). The lungs can be rudimentary in groups that

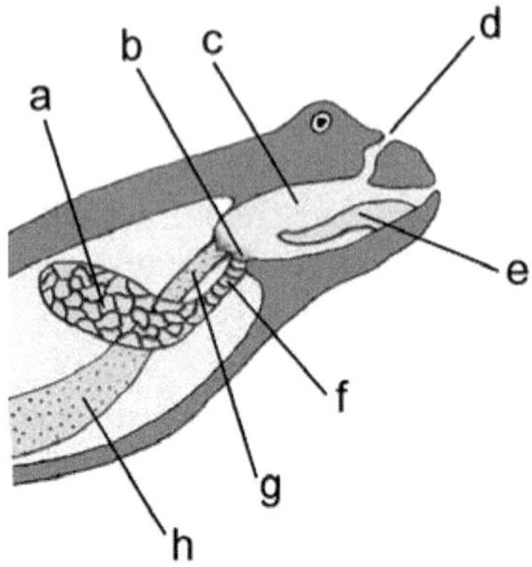

Fig. 8.10 The respiratory system of a frog. (a) lung, (b) glottis, (c) pharynx, (d) external nasal opening, (e) tongue, (f) laryngotracheal chamber, (g) esophagus and (h) stomach.

rely mainly on gills for gaseous exchange in adult life such as axolotl. Amphibians have nostrils that bear valves externally. These nostrils open into the short nasal cavities that lead to the choanae (internal opening) at the rostral part of the primary palate. The pharynx is a common opening for both respiratory and digestive systems. The **glottis** is the opening into the **laryngotracheal chamber** that is ventral to the esophagus and is surrounded by lateral laryngeal cartilages. The chamber has vocal cords and is lined internally by cilia as well as mucous and serous secreting cells. In many amphibians, the larynx serves the role of a check valve as it prevents collapse of lungs during diving since amphibians lack a ribcage. The chamber leads to the sac-shaped or elongated lungs depending on the species. The laryngotracheal chamber is lined with mucous and serous cells and also bears cilia. The lungs are subdivided by trabeculae (septae) into compartments that increase the surface area of the gas exchange surface (Fig. 8.11). The central part of the lungs lacks such divisions and is hollow.

During inspiration, the mouth is closed and the floor of the pharynx is lowered and this action is followed by a fall of pressure in the buccopharyngeal cavity. Air from outside is drawn into the cavity through the nostrils. The glottis then opens and air is expired from the lungs by the elastic recoil of the lungs and contraction of the lower body trunk muscles. There will be some mixing of the inspired and expired air in the pharynx but the latter is exhaled over the former. The nostrils are then closed and the floor of the pharyngeal cavity that acts as a **buccal pump** is raised. Air flows into the lungs after which the glottis is closed. Due to the low levels of metabolism, the rate of ventilating lungs in amphibians is low.

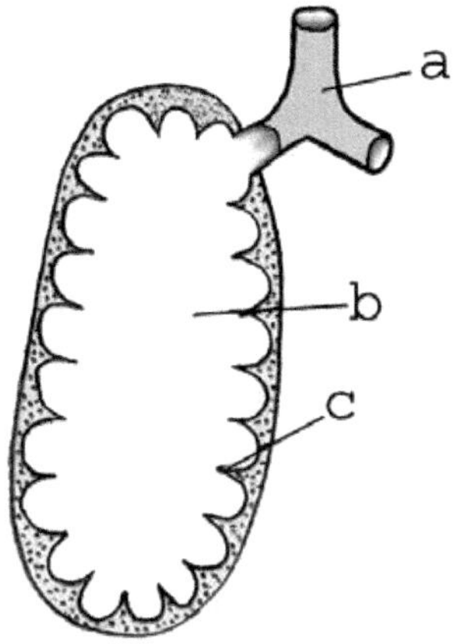

Fig. 8.11 Drawing of a section through an amphibian lung. (a) laryngotracheal chamber, (b) hollow central part of lung and (c) trabecula. The trabeculae of some amphibians including frogs occasionally bear secondary trabeculae that form a series of pockets that further increase the gas exchange areas.

Cutaneous respiration occurs in many amphibians with thin, moist and vascular skin. Such a skin is also important in diffusion of carbon dioxide from an amphibian since the rate at which lungs are ventilated is low. Most toads that are terrestrial have a drier skin and do not use it for cutaneous respiration. Due to the problem of dehydration facing many amphibians, the group prefers a moist environment.

Male frogs possess vocal sac-like diverticula on the lateral wall of the pharynx that can be filled with air. The expulsion of air from these sacs causes the vocal cords to resonate. Such sounds are for communication during breeding seasons and attract female frogs to males.

RESPIRATORY ORGANS OF REPTILES

The reptilian embryos as well as embryos of birds and monotremes exchange gases through the well vascularized allantois that unites with the chorion and the eggshell. The allantois arises within the body from the hindgut and is partly intra-embryonic. The membrane also stores excreted material, absorbs albumen and calcium from the shell for nutritional requirements of the embryo. The shell also provides mechanical protection to the embryo, food and water within the egg.

The reptilian neck is relatively longer than that of amphibians and this has been accompanied by elongation of the laryngotracheal chamber that is divided into an anterior larynx and a posterior trachea. The wall of the reptilian larynx is supported with more laryngeal cartilages and the larynx lacks vocal cords. The crocodilian larynx has a structure that is analogous to the mammalian epiglottis that is located in the nasopharynx and protects the lower part of the respiratory system from water while the reptile drowns its prey. The larynx of the snake is located in the oral cavity. It can be extended beyond the lower teeth and be used for breathing while the snake swallows its prey. The tracheal wall has ring shaped cartilages. The trachea divides into two central bronchi that lead to each lung and are located in the central part of these organs (Fig. 8.12). The central bronchi further divide into various secondary bronchi that bear alveolar sacs. The reptilian lung has thus more compartments and is larger than that of amphibians. Reptiles show variation in the structure of their lungs. The long reptiles including snakes and many lizards normally lose one lung. In the same reptilian groups, the posterior part of the lungs shows less compartmentalization and vascularization in relation to the anterior part and could store air for ventilating the lungs.

Inspiration is brought about by contraction of trunk muscles that bring ribs closer to each other while expanding the volume of the pleuroperitoneal cavity. The fall in pressure within the cavity is followed by expansion of the lungs and fall in pressure within the airways. Air moves into the lungs from the outside. The glottis is then closed until the reptile takes its next breath that could be after a long time. The reverse process occurs during expiration that is enhanced by contraction of smooth muscle fibers in the wall of the lung. Such a way of ventilating the lungs is known as an **aspiration** or **suction pump** and transfers air to the lungs in one movement as opposed to the amphibian buccal pump that does the same action in two movements.

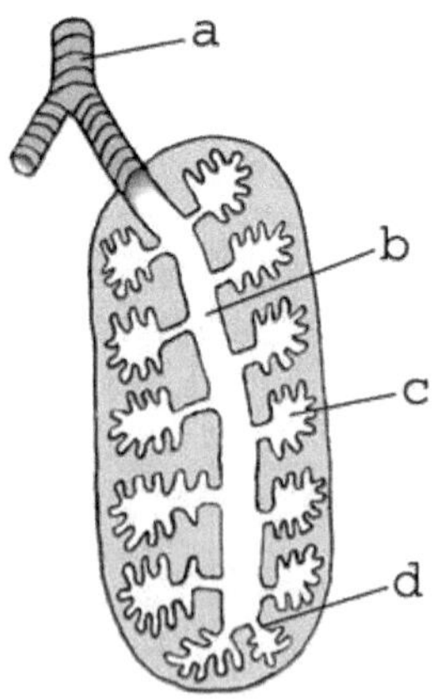

Fig. 8.12 The lung of a reptile showing the internal structure. (a) trachea, (b) central bronchus, (c) alveolar sac and (d) secondary bronchus. The lung of a reptile has more compartments and is relatively larger than that of an amphibian.

In crocodiles, the action of the **diaphragmatic muscle** that runs from the caudal aspect of the liver to the cranial part of the pelvic girdle (Fig. 8.13) plays a major role in respiration. Contraction of this muscle pulls the liver caudally thus increasing the volume of the two separate pleural cavities and creating space for the lungs to expand resulting in inspiration. The diaphragmatic muscles also pull the movable crocodilian pubis backwards on contraction. Relaxation of the muscle and contraction of abdominal muscles pushes the liver forwards and increases pressure in the pleural cavities leading to expiration. Crocodilians also rotate their pubis to increase tidal volume (Claessens, 2004). Movements of the pelvic girdle also contribute to inspiratory inflow in birds although the mechanism is quite different from that of crocodiles and might indicate that the pelvic musculoskeletal system may have played a role in the breathing of the extinct archosaurs (Carrier and Farmer, 2002).

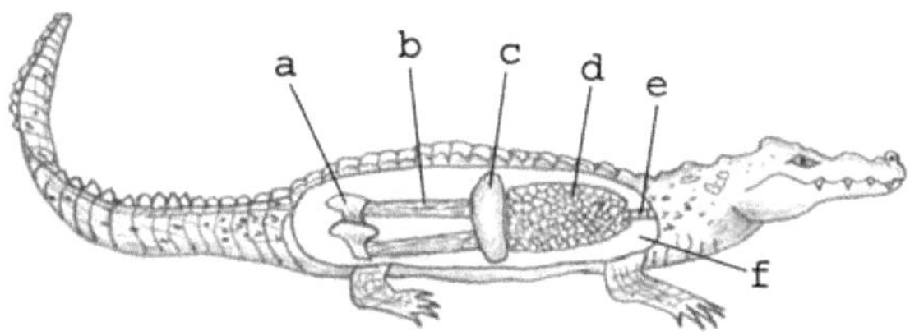

Fig. 8.13 The body cavity of a crocodile showing some of the structures that play a role in respiratory movements. (a) pelvic girdle, (b) diaphragmatic muscle, (c) liver, (d) lung, (e) trachea and (f) thoracic part of body cavity. The attachment of the diaphragmatic muscle to the liver has also been referred to as the 'hepatic piston'.

A **proto-diaphragm** or **post-hepatic membrane** is present in some lizards such as the tegu lizard of South America and separates the lungs and the liver from the other organs of the body cavity. The proto-diaphragm is a sort of a primitive diaphragm and does not divide the body cavity into two parts completely. Although the proto-diaphragm does not contract, it acts as a mechanical barrier by preventing the movement of viscera in a cranial manner during physical exertion thus allowing for greater inflation of lungs (Klein et al., 2003).

The skin of reptiles, due to the high degree of keratinization, is not used in respiration in most species. Some aquatic turtles use the skin, buccopharyngeal cavity and cloaca for gaseous exchange. The cloaca of many aquatic turtles has a pair of evaginations known as **accessory bladders** or **cloacal burasae** (Fig. 8.14). Whereas these accessory bladders can be extruded from the cloaca for gaseous exchange in several

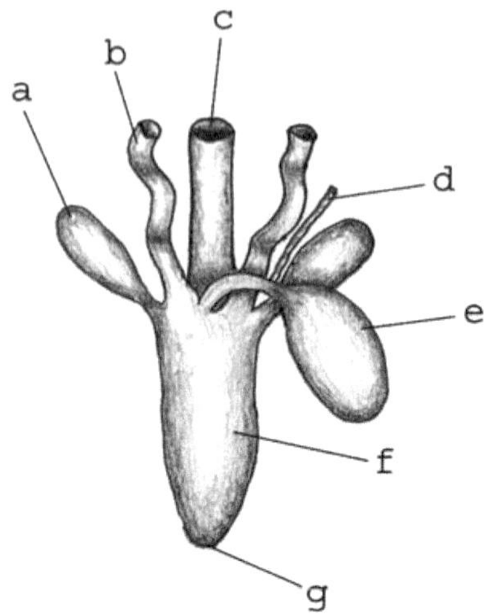

Fig. 8.14 A ventral view of some organs found in the posterior part of the body cavity of a turtle. (a) accessory bladder, (b) oviduct, (c) intestine, (d) ureter, (e) urinary bladder, (f) cloaca and (g) anal opening.

pleurodires, the bladders are used primarily in buoyancy control and secondarily in nesting and ion transport in cryptodires (Peterson and Greenshields, 2001).

Locomotion and Breathing

Amphibians and reptiles normally flex their body trunks when they move. The lung to the side that is flexed is compressed whereas the one on the other side of the body is expanded. Since these vertebrates are unable to run and breathe simultaneously, they cannot run for a long time without stopping to breathe as lactic acid builds up to high levels in their bodies. The largest lizard, the Komodo dragon or lizard of Indonesia that measures up to 3 meters, can hardly sustain a continuous run for more than 10 meters.

RESPIRATORY ORGANS OF BIRDS

The respiratory organs of birds have adapted to the demands of flight and to a certain extent sound production. There is also a reduction of the bones of the splanchnocranium and evolution of a large incisive bone. There is the **cranial larynx** that plays a role in respiration and a **syrinx** or **caudal larynx** that is concerned with sound production. The lungs of birds are fairly rigid in comparison to those of other vertebrate groups and are used in exchange of gases. They have thin-walled **air sacs** that act as bellows and ventilate lungs during inspiration and expiration since airflow in lungs is unidirectional. The air sacs are not used for gaseous exchange.

The nasal cavity of birds opens to the outside through two nasal openings. The left and right nasal cavities are separated from each other by a cartilaginous nasal septum. The roof of the oral cavity has a wide cleft that is narrowed by mucous membranes which form the palatine or choanal cleft that communicates with the nasal cavity. There are two nasal turbinates (conchae) in each nostril. Air flows from the nasal cavity through the choanal cleft into the cranial larynx that is supported by cartilages. As the epiglottis is lacking in birds, the width of the laryngeal opening is controlled with two pairs of muscles that dilate and constrict the opening. The larynx continues as the trachea in a posterior manner. In about 60 species of birds with long necks including the whooping crane (*Grus americana*), whooper swan (*Cygnus cygnus*), black swan (*Cygnus atratus*), helmeted curassow (*Crax pauxi*) and white spoonbill (*Platalea leucorodia*), the trachea is elongated and forms **tracheal loops** or **coils** within the enlarged and hollow keel of the sternum or thorax (Fig. 8.15). Since tracheal length is generally correlated to the body size of a bird, it has been suggested that elongation of the trachea serves to exaggerate the apparent size of a vocalizing bird (Fitch, 1999).

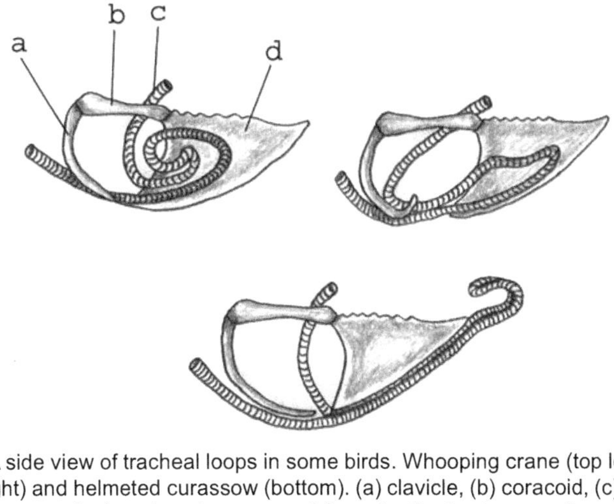

Fig. 8.15 A side view of tracheal loops in some birds. Whooping crane (top left), whooper swan (top right) and helmeted curassow (bottom). (a) clavicle, (b) coracoid, (c) trachea and (d) keel of sternum. In the whooping crane of Canada, the trachea can be as long as five feet and is straight along the neck but coils greatly forming two loops inside of the sternum before entering the thoracic cavity. Such a long trachea partly contributes to the loud resonating sound or 'whooping' call of the bird that can be heard as far as five miles away.

Cartilaginous rings are found in the wall of the trachea and tend to ossify early in many bird species. The trachea divides into two main branches (bronchi). Part of the lower trachea and beginning of the two main bronchi is modified into the syrinx. The **syrinx** (Fig. 8.16) lies between the cartilaginous rings and is surrounded with one or several **tympanic membranes** that are activated by muscle and air to produce various sounds like the vocal chords of mammals. The sounds are important in communication and reproductive behavior. The musculature is highly differentiated in singing birds. The main bronchi run through the center of the lungs.

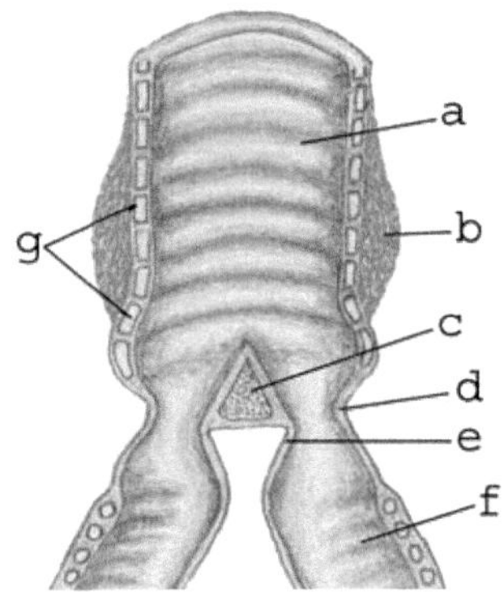

Fig. 8.16 Drawing of a frontal section of the syrinx. (a) trachea, (b) syringeal muscles, (c) cartilage known as pessulus that connects the dorsal and ventral extremeties of the first pair of bronchial cartilages, (d) external tympaniform membrane, (e) internal tympaniform membrane, (f) main bronchus and (g) tympanum or drum. The cartilaginous rings are represented by the lighter parts in the wall of the airways.

The lungs of birds are relatively small and are not lobed. They lie close to the wall of the thorax dorsolaterally and bear rib impressions. There are no pleural cavities. The main bronchus or **mesobronchus** continues caudally in each lung to the **abdominal air sac** (Fig. 8.17) at the caudal end of the lung. The main bronchus gives off secondary bronchi that are known as ventro-, dorso-, and laterobronchi. **Ventrobronchi** branch ventrally. The first ventrobronchus is connected to the cervical air sac and also supplies cranially the ventral and medial parts of the lung together with the second ventrobronchus. Apart from supplying the caudal half of the lung ventrally, the third ventrobronchus also give off branches to the clavicular and cranial thoracic air sacs. The small fourth ventrobronchus supplies the caudal part of the lung ventrally. The main bronchus also gives off seven to ten **dorsobranchi** from its caudal part. These bronchi spread out in the dorsal and lateral parts of the lungs.

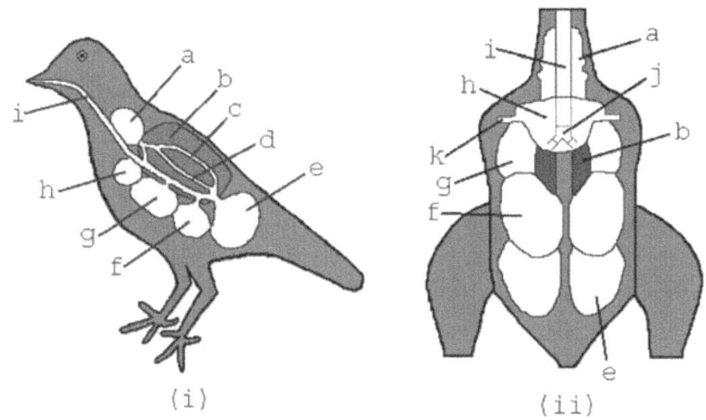

Fig. 8.17 (i) Longitudinal and (ii) frontal views of air sacs of the avian respiratory system. (a) cervical, (e) abdominal, (f) posterior and (g) anterior thoracic and (h) clavicular or interclavicular air sacs; (b) lungs, (c) dorsal and (d) ventral bronchi, (i) trachea (j) syrinx and (k) diverticulum to pneumatic part of humerus.

Color image of this figure appears in the color plate section at the end of the book.

Laterobironchi branch off the main bronchus and run ventrally and along the lateral part of the lung. The first laterobronchus supplies the caudal thoracic air sac.

Parabronchi or **tertiary bronchi** are analogous to the mammalian alveolar ducts and branch from the secondary bronchi in a series. Neighboring parabronchi branch and unite to form a meshwork of tubes that occupy more than half of the lung volume. The internal surface of parabronchi bears openings or air vesicles known as **atria** that project radially and lead to air capillaries (Fig. 8.18). The vesicles are surrounded by ridge-like septae that join to form a spongy-like appearance internally. The ridges are covered with a cuboidal epithelium and contain smooth muscle and elastic fibers. Air capillaries from neighboring parabronchi branch and interconnect with each other giving a very large gas exchange surface area. The air capillaries are surrounded with a dense network of

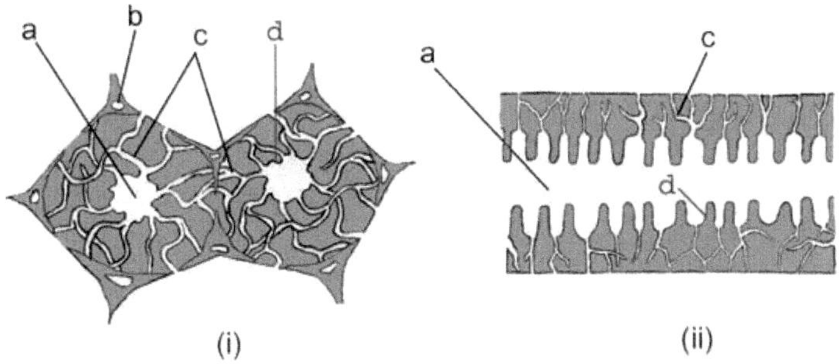

Fig. 8.18 (i) Transverse and (ii) longitudinal sections of parabronchi. (a) lumina of parabronchi, (b) blood vessel, (c) air capillaries and (d) septae.

blood capillaries. The air capillaries are lined with a single layer of low epithelial cells.

The walls of the nine air sacs (four paired and one unpaired) are very thin and contain fibrous and connective tissues and muscle fibers. Their internal surface is lined with a simple squamous epithelium. Air sacs have a poor blood supply and are not used in gaseous exchange. The main function of air sacs is to ventilate the lungs, together with the respiratory muscles, in a continuous manner during inspiration and expiration (mainly by caudal thoracic and abdominal air sacs). Air sacs partly contribute to a reduction of the specific gravity of birds and regulation of body temperature. Air sacs have extensions to the pneumatic bones of the trunk and limbs.

Breathing Cycle of Birds

The alternate expansion and compression of air sacs that is brought about by sternal movements ventilates the lungs. Such movements are possible due to the presence of joints between the vertebral (dorsal) and sternal (ventral) ribs. Respiratory muscles such as the intercostals are involved in the movement of the ribcage. There is also coordination between flight and respiration in birds. Contraction of flight muscles brings about movement of the sternum, ribs and the clavicle. Suprapubic muscles of the tail also participate in respiration by acting at the notariosynsacral junction by elevating the pelvis during inspiration and the infrapubic muscles depress the pelvis and the uropygium leading to compression of the thoracoabdominal cavity during expiration (Baumel et al., 1990). A complete cycle of respiration in birds is brought about with two inspiratory and two expiratory movements. The glottis is therefore open for most of the time since the lungs are ventilated continuously.

The initial inspiration draws in air through the main, the ventral and lateral bronchi into the abdominal and posterior thoracic air sacs. The first expiration moves the air to the dorsobronchi and the parabronchi. Air will then move from the dorsobronchi and parabronchi to the cervical, clavicular and anterior thoracic sacs during the second inspiration. The second expiration will then push air to the outside. Air will be flowing through the parabronchi during each movement of the respiratory cycle in one direction. Blood flow in capillaries is transverse to air flow in the air capillaries. Such a **cross current** system of gas exchange (Fig. 8.19) ensures that air leaving the parabronchi to be exhaled contains less oxygen than the blood leaving the system. The type of respiration that occurs in birds ensures that the volume of air in the avian lung is almost constant at all times. The avian respiratory system is more efficient than the mammalian system at extracting more oxygen from inhaled air.

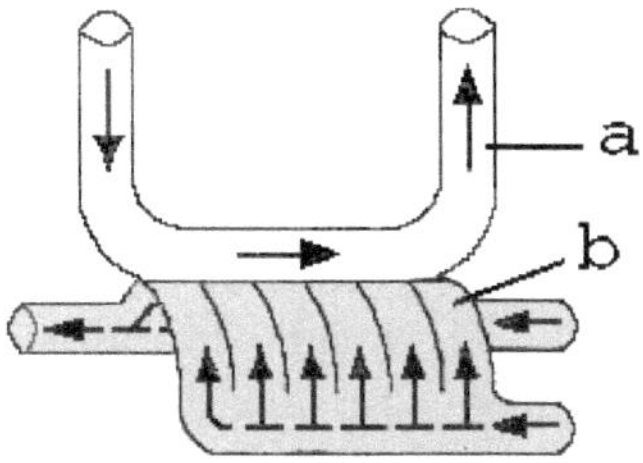

Fig. 8.19 The avian cross current system of gas exchange. (a) air and (b) blood capillaries. Arrows indicate the directions of flow of air and blood.

RESPIRATORY ORGANS OF MAMMALS

Respiratory movements in mammals are brought about by contraction of respiratory muscles, chiefly the diaphragm that is unique to this group of vertebrates. The mammalian respiratory system has evolved independently and differs from that of birds but these two groups of homeotherms have efficient systems that meet their high metabolic demands. Mammals have **alveoli** in their lungs with very thin walls where gaseous exchange takes place. The separation of the oral and nasal cavities by a secondary palate enables mammals to perform functions such as chewing and breathing at the same time. Most of the respiratory passageway is lined with a pseudostratified columnar epithelium that bears cilia and mucus producing goblet cells. The wavelike movements of cilia either move particles towards the nostrils from where they are expelled from the body or towards the pharynx, from the lower part of the system, where they are then swallowed. Areas of the system that are exposed to wear including nostrils and parts of the larynx are lined with a stratified squamous epithelium and the terminal part of the system has a very low epithelium.

The nasal openings lead to two relatively large nasal cavities. Each cavity is connected to the air filled **paranasal sinuses** of some of the skull bones. The respiratory mucosa of the nasal cavity has an extensive vascular network. The amount of blood flowing through these vessels can be regulated. Inspired air is thus warmed and the secretions of seromucous glands can be vaporized. The mixture of inspired air with water is important in smelling. The surface area of each nasal cavity is increased by the three scrolls of bone known as **turbinates** or **conchae**. The caudal or superior part of the cavity bears the **olfactory** or **smell region**. The vomeronasal organ is found on the floor of the nasal cavity on each of the nasal septum in many mammals.

Inspired air passes to the **nasopharynx** from the nasal cavity through the paired **choanae** that form the posterior openings of the nasal cavity.

The nasopharynx is separated from the lower oropharynx by the **soft palate**. The common pharynx is a passage for both air and food. Air enters the larynx through the glottis from the pharynx. The glottis is covered by the **epiglottis** (one of the laryngeal cartilages) during swallowing of food. The glottis is normally open since mammalian lungs are ventilated continuously. The larynx contains vocal cords and consists of several different cartilages that are mainly lined internally with a mucous membrane. The cartilages are connected to each other, the hyoid bone and the trachea by various ligaments and muscles. The elastic **vocal ligaments** or **cords** are responsible for sound production by expired air. The air passages in the head as well as the oral cavity determine the nature of the sound. The human being uses the tongue and other oral structures to produce coordinated sounds that form speech.

The trachea runs from the larynx to the root of the lung where it divides into the right and left thick and short **primary** or **principal bronchi** (Fig. 8.20). The wall of the trachea is supported by a series of cartilaginous rings that prevent the collapse of the tube. The primary bronchi enter the lungs to divide into **lobar** bronchi that ventilate each of the lobes of the lung. The lobar bronchi further subdivide into several **segmental** bronchi. There will be further subdivisions of the segmental bronchi that will eventually end in respiratory bronchioles that bear some **alveoli** (L. *alveolus*, a small hollow or cavity) in their walls. Respiratory bronchioles lead to **alveolar ducts** that end in alveoli (Fig. 8.21) that are the last structures of the bronchal (respiratory) tree. Gaseous exchange takes place in the walls of alveoli that are very thin and are lined by mainly flattened simple squamous cells or **pneumocyte type I cells** that are surrounded by a basal layer. Between the tightly packed alveoli are capillaries.

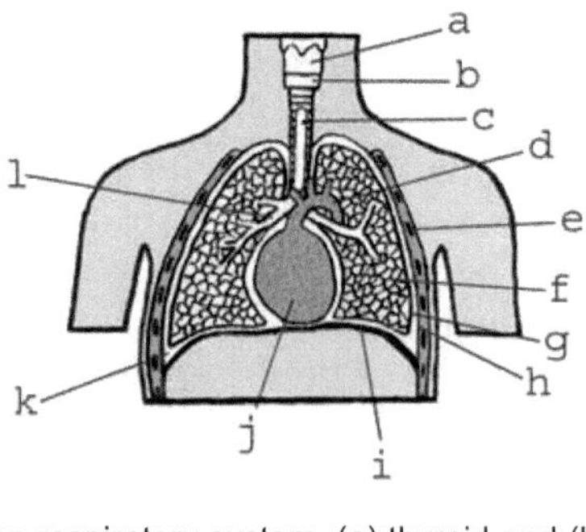

Fig. 8.20 The mammalian respiratory system. (a) thyroid and (b) cricoid cartilages of the larynx; (c) trachea, (d) aorta, (e) cross-section of a rib, (f) lung, (g) visceral pleura, (h) parietal pleura, (i) diaphragm, (j) heart surrounded by the pericardial cavity, (k) pleural space and (l) one of the two main (principal) bronchi.

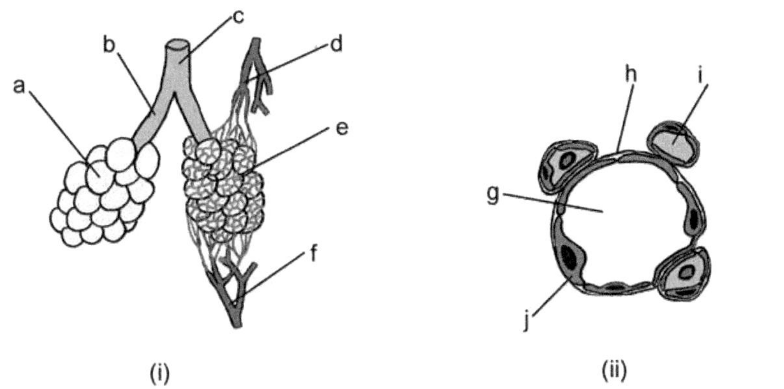

Fig. 8.21 (i) Alveoli appearing in clusters and (ii) a section through an alveolus. (a) alveolus from the surface, (b) respiratory and (c) terminal bronchioles, (d) venous and (f) arterial blood systems, (e) capillary network around alveoli, (g) lumen of alveolus, (h) basal lamina, (i) capillary and (j) endothelial cell.

BREATHING CYCLE OF MAMMALS

Inspiration is brought about by contraction of the diaphragm, external intercostal and other muscles of inspiration. The dome shaped diaphragm flattens on contraction, increasing the length of the thoracic cavity. Contraction of external intercostal muscles pulls the ribs anteriorly or upwards (in bipeds) and outwards. The sternum is also pushed outwards. The size of the thorax increases followed by a fall in intrapleural (intra-thoracic) and intra-alveolar pressure. Air rushes into the lungs due to the resulting pressure gradient between the atmosphere and the interior of the lungs. Expiration follows inspiration during which inspiratory muscles relax. The elastic recoil of lungs together with contraction of abdominal and internal intercostal muscles lead to a build up in intra-alveolar pressure. The positive pressure gradient created from the alveoli to the atmosphere results in flow of air from the lungs to the outside.

Lung surfactants are important in the proper functioning of mammalian lungs as a result of the large alveolar surface area. Surfactants are detergent-like phospholipids that prevent alveoli from collapsing as air moves in and out of the lungs since there is a high surface tension at the air/water interface. The surfactants reduce this surface tension and coat the inner surface of alveoli. Lung surfactants are secreted by **pnemocyte type II** cell that forms part of the wall of an alveolus. It has been suggested that the surfactant system that has been demonstrated in the swim bladder of teleosts and lungs of air-breathing fish could have originated in epithelial cells lining the pharynx (Daniels et al., 2004). Surfactant-like proteins have also been found in intestines of actinopterygian fish and this shows that the proto-surfactant was already present before evolution of lungs (Bourbon and Chailley-Heu, 2001). The composition of surfactant is highly conserved despite the diversity in lung structure seen in various vertebrates. Sullivan et al. (2003) have reviewed mechanisms controlling pulmonary surfactant maturation in amniotes. Lung breathing could not take place in the absence of surfactant. **Alveolar macrophages** or **phagocytes** are also found in alveoli and engulf inhaled particles or microbes and are sometimes referred to as dust cells.

Gas Exchange in Mammalian Lungs

Alveoli are surrounded with a dense network of capillaries. Since alveoli are continuously ventilated, there is always fresh air moving into the lungs and this ensures that the partial pressure of gases in alveoli does not change much at any moment. The total functional alveolar surface area and the oxygen pressure gradient across the diffusion barrier between alveoli and capillaries among other factors, determine the amount of oxygen that diffuses into blood per unit time. The blood-gas barrier is extremely thin while remaining strong due to the presence of collagen type IV in the basement membrane (West, 2003). The gas exchange system in the mammalian blood is the **uniform pool** or **concurrent system** (Fig. 8.22). In such an exchange system, blood in capillaries flows in various directions around the blind ending alveoli. The diffusion distances are quite short and measure up to a few microns. Despite the rapid rate of diffusion, the gases in alveoli and the capillaries do not have enough time to be in equilibrium due to the short capillary transit time. The amount of oxygen in the blood leaving the lungs will be lower than that present in alveoli and the level of carbon dioxide in the alveoli will be less than that in capillaries leading from the lungs.

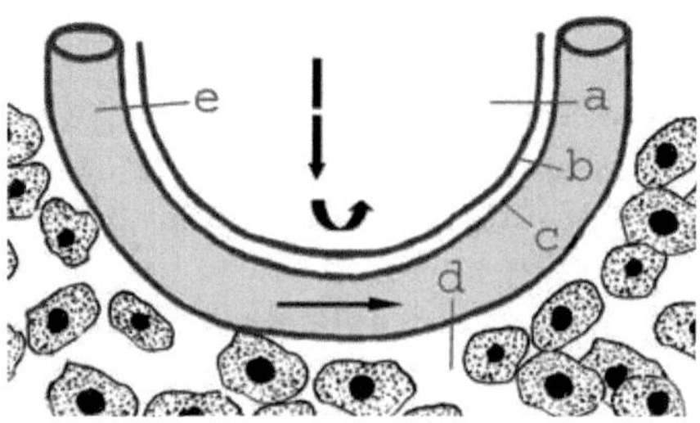

Fig. 8.22 The uniform pool or concurrent system of gas exchange in mammals. (a) alveolar space, (b) and (c) are outer limits of an alveolus and capillary respectively, (d) interstitial space and (e) capillary. The thicker arrows show the course taken by inhaled air while the thinner one indicates the direction of blood flow.

Transport of Gases by Blood

Transport of oxygen and carbon dioxide by blood occurs mainly in either solute form or in combination with other chemicals. Oxygen and carbon dioxide normally dissolve in the plasma. Most of the gases will unite with other molecules (chemically) including hemoglobin, water or plasma proteins. Since transport of gases is a gradual and continuous process, the diffusion gradient across gas exchange barriers determines the rate and direction of diffusion of gases in lungs and tissues down a partial pressure gradient.

Transport by Hemoglobin

The red pigment hemoglobin is found in red blood cells of vertebrates and is a protein with four polypeptide chains that are linked to an iron containing heme group. Two of the polypeptide chains are alpha and the other two are beta chains. Oxygen combines with iron in each heme group whereas carbon dioxide can combine with amino acids of alpha and beta polypeptide chains. Hemoglobin is thus capable of absorbing both oxygen and carbon dioxide in blood. Oxygen diffuses into red blood cells of alveolar capillaries and combines with hemoglobin to form **oxyhemoglobin**. Saturation of hemoglobin by oxygen is fast and quite high. Increasing the blood partial pressure of oxygen accelerates the rate of combination with hemoglobin. A little oxygen is transported in dissolved form in the plasma.

About 10% of carbon dioxide is transported in the dissolved form in the plasma while most of it (about 65% or more) diffuses into red blood cells and reacts with water in these cells to form **carbonic acid** (H_2CO_3) that dissociates to form carbonate and hydrogen ions. The reaction of CO_2 with water is reversible and is catalyzed by the enzyme **carbonic anhydrase** that is present in the red blood cells. Formation of carbonic acid lowers the level of CO_2 in red blood cells thus creating a gradient with CO_2 in the plasma that will increase diffusion of CO_2 further into red blood cells. Carbon dioxide is then transported in the blood as bicarbonate ions. The continued conversion of CO_2 to bicarbonate also creates a diffusion gradient between the plasma and tissues and leads to more CO_2 diffusing to the plasma from tissues thus enhancing the CO_2 carrying capacity of blood. Bicarbonate ions diffuse out of red blood cells into the plasma as a result of an increase of these ions in the cells. This movement is counteracted by the chloride shift that involves movement of chloride ions into red blood cells to prevent an imbalance in charge. In lungs or gills where the levels of CO_2 are lower, the hydrogen and bicarbonates ions combine to form carbonic acid that dissociates into CO_2

and water. CO_2 then diffuses out of the blood capillaries and into the alveoli. About a fifth of CO_2 in the blood combines with NH_2 (amine) groups of amino acids that form the polypeptide chains of hemoglobin and other plasma proteins to form **carbamino compounds**. This is a reversible reaction and is governed by the partial pressure of CO_2 in the blood. Most of the CO_2 combining with amino acids is transported in red blood cells.

At the tissue level, blood circulating in capillaries from arterioles has about 100 mmHg of oxygen. The oxygen tension in intercellular and intracellular fluids is lower than this figure and as a result oxygen will then diffuse to tissues. Increase in activity lowers intracellular oxygen tension resulting in a high oxygen gradient that leads to greater rates of diffusion of the gas to tissues. Diffusion of CO_2 from tissues to capillaries is governed by the same principle. The ability of fetal hemoglobin or hemoglobin F (HbF) to bind oxygen is higher than that of adult hemoglobin. Fetal hemoglobin is composed of two alpha and two gamma polypeptide chains. The protein myoglobin of muscles has an even higher oxygen binding ability.

REFERENCES

Baumel, J.J., Wilson, J.A. and Bergren, D.R. (1990). The ventilatory movements of the avian pelvis and tail: Function of the muscles of the tail region of the pigeon (*Columba livia*). J. Exp. Biol., 151:263-277.

Blake, R.W. (1983). Energetics of leaping in dolphins and other aquatic animals. J. Mar. Biol. Assoc. U.K., 63:61-70.

Booth, J.H. (1978). The distribution of blood flow in gills of fish: application of a new technique to rainbow trout (*Salmo gairdneri*). J. Exp. Biol. 83:31-39.

Bourbon, J.R. and Chailley-Heu, B. (2001). Surfactant proteins in the digestive tract, mesentery and other organs: evolutionary significance. Comp. Biochem. Physiol. A Mol. Integr. Physiol., 129(1):151-161.

Brauner, C.J., Matey, V., Wilson, J.M., Bernier, N.J. and Val, A.L. (2004). Transition in organ function during the evolution of air-breathing: insights from *Arapaima gigas*, an obligate air-breathing teleost from the Amazon. J. Exp. Biol. 207 (Pt.9):1433-1438.

Burggren, W.W. and Johansen, K. (1986). Circulation and respiration in lungfishes. J. Morphol. Suppl., 1:217-236.

Carrier, D.R. and Farmer, C.G. (2000). The evolution of pelvic aspiration in archosaurs. Paleobiol., 26(2):271-293.

Claessens, L.P. (2004). Archosaurian respiration and the pelvic girdle aspiration breathing of crocodyliforms. Proc. R. Soc. Lond. B Biol. Sci., 271(1547):1461-1465.

Daniels, C.B., Orgeig, S., Sullivan, L.C., Ling, N., Bennett, M.B., Schurch, S., Val, A.L. and Brauner, C.J. (2004). The origin and evolution of the surfactant system in fish: insights into the evolution of lungs and swim bladders. Physiol. Biochem. Zool., 77(5):732-749.

Fitch, W.T. (1999). Acoustic exaggeration of size in birds via tracheal elongation: Comparative and theoretical analyses. J. Zool., 248:31-48.

Johansen, K. (1970). Air-breathing in fishes. pp. 361-411. In: W.S. Hoar and D.J. Randall (eds.). Fish Physiology, Academic Press, New York.

Klein, W., Andrade, D.V., Abe, A.S. and Perry, S.F. (2003). Role of the post-hepatic septum on breathing during locomotion in *Tupinambis merianae* (Reptilia: Teiidae). J. exp. Biol., 206:2135-2143.

Krogh, A. (1941). The Comparative Physiology of Respiratory Mechanisms. University of Pennsylvia Press, Philadelphia.

Hughes, G.M. (1970). Morphological Measurements on the Gills of Fishes in Relation to their Respiratory Function. Folia Morphologia, Prague, 18:78-95.

Hughes, G.M. (1972). Aspects of the Respiration of *Latimeria chalumnae.* Proceedings of the International Congress of Physiological Sciences, Sydney, No. 136.

Hughes, G.M. (1980). Ultrastructure and morphometry of the gills of *Latimeria chalumnae* and a comparison with the gills of associated fishes. Proc. Roy. Soc. London B208:309-328.

Hughes, G.M. (1984). General anatomy of the gills. pp. 1-72. In: W.S. Hoar and D.J. Randall (eds.). Fish Physiology, Vol. XA: Academic Press, New York.

Hughes, G.M. and Weibel, E. R. (1972). Similarity of supporting tissue in fish gills and the mammalian reticuloendothelium. J. Ultrastruct. Res., 39:106-114.

Laurent, P. (1996). Vascular organization of lungfish, a landmark in ontogeny and phylogeny of air-breathers. pp. 47-56. In: J.S. Datta Munshi and H.M. Dutta (eds.). Fish Morphology, Horizons of New Research. Science Publishers, Inc. Lebanon, U.S.A.

Muir, B.S. and Hughes, G.M. (1969). Gill dimensions for the three species of tunny. J. Exp. Biol., 51:271-285.

Munshi, J.S.D. and Choudhary, S. (1994). Ecology of *Heteropneustes fossilis* (Bloch): An Air-breathing Catfish of Southeast Asia. Monograph Series. Freshwater Biological Association of India, Bhagalpur.

Nilsson, S. (1986). Control of gill blood flow. pp. 87-101. In: S. Nilsson and S. Holmgren (eds.). Fish Physiology: Recent Advances. Croom Helm, London.

Peterson, C.C. and Greenshields, D. (2001). Negative test for cloacal drinking in a semi-aquatic turtle (*Trachemys scripta*), with comments on the functions of cloacal bursae. J. Exp. Zool., 290:247-254.

Randall, D.J., Baumgarten, D. and Malyusz, M. (1972). The relationship between gas transfer across the gills of fishes. Comp. Biochem. and Physiol. 41A:629-637.

Sullivan, L.C., Orgeig, S. and Daniels, C.B. (2003). The role of extrinsic and intrinsic factors in the evolution of the control of pulmonary surfactant maturation during development in the amniotes. Physiol. Biochem Zool., 76(3):281-295.

West, J.B. (2003). Thoughts on the pulmonary blood-gas barrier. Am. J. Physiol. Lung Cell Mol. Physiol., 285(3):L501-513.

Williams, T.M. (1989). Swimming by sea otters: adaptations for low energetic cost locomotion. J. Comp. Physiol. A 164:815-824.

9

Reproduction and Early Developmental Biology

Survival of any species of a living organism lies in its ability to regenerate itself as each species has a limited period of life on Earth. Age at maturity is typically proportional to adult life span and mammals that live longer for their body size, such as bats and primates, also tend to have a longer developmental time for their body size (de Magalhães et al., 2007). The reproductive system of vertebrates is designed to play this major role of procreation. Most vertebrates regenerate themselves through sexual reproduction apart from a few that multiply asexually by parthenogenesis. Methods involved in sexual reproduction by vertebrates are quite diverse and have contributed greatly to the success of certain species. Male or female reproductive organs normally occur in separate individual vertebrates except in cases of hermaphroditism.

The start of life is brought about by the union of male and female reproductive cells (gametes) that contain half the number of chromosomes of normal somatic cells (Gr. *somatikos*, bodily). Reducing to half the number of chromosomes in each gamete is brought about by meiosis which differs from mitosis that involves multiplication of somatic body cells while retaining the normal diploid number of chromosomes. Meiosis occurs in male and female gonads (Gr. *gone*, seed) and leads to production of spermatozoa (Gr. *sperma*, seed) and ova (L. *ovum*, egg) respectively.

PARTHENOGENESIS

The process by which an individual female organism is able to reproduce itself without fertilization of its eggs is known as parthenogenesis

(Gr. *parthenos*, virgin; *genesis*, birth). Parthenogenesis is more common in lower animals than vertebrates. Among vertebrates, this form of reproduction is seen in some fishes, amphibians, reptiles and a few birds. Parthenogenesis is rarely seen in mammals. Analysis of the geographical distribution of cases of animal parthenogenesis has revealed that most of the affected species exist in isolation from closely related species and most live in natural disclimax communities (climax communities that have been disturbed by various influences) (Cuellar, 1977). Parthenogenetic groups are females that give rise to offspring that are genotypically identical or nearly identical to the parent. Parthenogenesis has some advantages over sexual reproduction. Since a single individual is required to start a new colony, the rate of population increase is twice under ideal conditions. Such a higher rate of population increase results due to the presence of twice as many females in a parthenogenetic population as compared to half the number of the same sex in a sexually reproducing group as the other half comprises males. Parthenogenesis can also be disadvantageous. Should a genotype be defenseless against a particular pathogen, predator or change in the physical environment, the entire group of a parthenogenetic lineage could be eliminated. Some members of a sexual reproduction group could survive under similar conditions as a result of new genotypes that have resulted due to genetic recombination.

Artificial parthenogenesis has been induced in the major vertebrate groups and usually results in incomplete and abnormal development. Research is being conducted to establish parthenogenetic stem line using oocytes (Chen and Li, 2004). Nonhuman primate eggs have been parthenogenetically activated to the blastocyst stage. The formed cells have been used as stem cells to provide a potential source for autologous cell therapy in the female thus eliminating the need for creating a competent embryo (Vrana et al., 2003).

HERMAPHRODITE VERTEBRATES

Hermaphrodites (derived from Hermaphroditus who was the son of Hermes and Aphrodite and who fused with a nymph in Greek mythology) contain male and female sexual organs at some point during their lives. Hermaphroditism generally occurs in invertebrates. In vertebrates, hermaphrodism occurs in a fair number of fish species and to a lesser extent in other vertebrates and is a normal part of the life cycle in many such species. The condition also occurs in many individual vertebrates including humans where it is not part of their normal life cycle. The term **intersexual** is used to describe such cases. In intersexuals, one set of sex organs is normally incompletely developed. A **sequential**

hermaphrodite is born as one sex then later on changes into another sex.

Cases of **protandry** (Gr. *protos*, first; *andr*, man) involve a change in sex from a male to female later in life. Protandry is seen in sea basses of the family Serranidae that are highly fished for food. The fishing pressure on the fish is altering the time change in sex takes place. It has been observed in a marine protandrous hermaphrodite teleost *Sparus aurata* that sex specific cell proliferation patterns in the hypothalamus are associated with changes in sex of the fish and this might indicate the role of a central cellular mechanism in sex reversal (Zikopoulos et al., 2000, 2001). In the black porgy (*Acanthopagrus schlegeli*) that is a marine protandrous hermaphrodite, an estrogen and gonadal aromatase that converts testosterone into estrogen are closely associated with the occurrence of sex reversal (Chang et al., 1997).

In **protogyny** (Gr. *gyn*, woman), an organism starts as a female then later on changes into a male. Wrasses of the family Labridae are protogynous. Various studies have shown that estrogen plays a major role in sex change of the protogynous honeycomb grouper and treatments with aromatase inhibitor inhibit endogenous production of estrogen leading to oocyte degeneration and inhibition of estrogen synthesis resulting in change of sex from female to male (Bhandari et al., 2004a and b).

Simultaneous or **synchronous hermaphrodites** have both male and female sexual organs at the same time as adults. Such individuals practise self-fertilization. The hamlets of the family Haemulidae normally take turns in exchanging their male or female roles during mating. These fish are evolving into other species quite rapidly.

MEIOSIS

Meiosis is a type of cell division that reduces the number of chromosomes in each daughter cell to half the number present in each parent cell and other somatic cells of the vertebrate body that normally have a diploid number of chromosomes (Fig. 9.1). The process of meiosis is controlled by several genes that are activated at the right time. Each individual vertebrate species has a certain number of chromosomes that vary and limit breeding between species. Meiosis occurs only in the sex cells of the body and involves the nucleus of the parent cell undergoing two divisions that result in formation of four cells with each containing half (haploid state) as many chromosomes as the parent cell. The resulting daughter cells then differentiate into the male and female gametes or sperm cells and oocytes respectively. Each gamete also has one chromosome from each homologous pair that is present in each parent

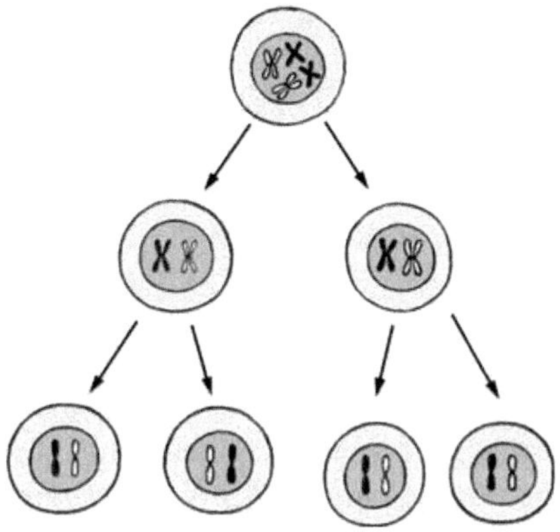

Fig. 9.1 The process of meiosis that leads to production of gametes with a haploid number of chromosomes from a parent cell that contains a diploid number of chromosomes.

cell. The union of male and female gametes restores the normal diploid number of chromosomes of an individual animal cell.

The first cell division of meiosis (meiotic division I) leads to replication of each DNA molecule to form a pair of **chromatids** that constitute a chromosome and are attached to each other at the **centromere**. Chromosomes do not divide during meiotic division I as is the case during mitosis. However, homologous chromosomes will separate into each of the two cells resulting from this division. Each of these cells contains half (haploid) the number of chromosomes present in each parent cell. Each chromosome still consists of two chromatids that are united at the centromere. Meiotic division II is similar to mitosis and involves division of each of the two chromatids resulting in formation of two more cells with each containing the haploid number of chromosomes.

During **spermatogenesis** (formation of spermatozoa) in sperm producing **seminiferous tubules** (L. *semen*, seed) of male gonads (testis), the undifferentiated germinal stem (primordial) sex cells or **spermatogonia** are transformed into mature sperm cells at the onset of sexual maturity. Spermatogonia are normally connected with cytoplasmic bridges to ensure synchronous development of each clone. Spermatogonia undergo several mitotic divisions to replace themselves and also form cells that will differentiate into primary spermatocytes. Formation of primary spermatocytes from spermatogonia is knowm as **spermatocytogenesis**. Each spermatogonium undergoes several mitotic divisions to give rise to several **primary spermatocytes**. One primary spermatocyte undergoes meiotic division I to form two **secondary spermatocytes** that each contains a haploid number of chromosomes.

Each secondary spermatocyte then undergoes meiosis II that involves separation of two chromatids from each chromosome present to result in two **spermatids**. Each primary spermatocyte thus gives rise to four **spermatids**, each of which contains a haploid number of chromosomes. Spermatids will then differentiate into mature **spermatozoa** that have heads and tails for movement in a process known as **spermeogenesis**. In mammals generally, only 2-3 out of a possible 10 spermatozoa are produced from each differentiated type A1 spermatogonium (Franca et al., 2005). The molecular mechanisms that control spermeogenesis have been summarized and involve regulation of gene expression during changes in chromatin structure when protamine replaces histone (Tanaka and Baba, 2005). Protamines are low molecular weight proteins that are normally found in fish sperm and are rich in arginine whereas histone is a protein that is rich in basic amino acids and is found in nuclei of cells in association with nucleic acids of eukaryotes. The whole process of spermatogenesis takes about 60 days in male humans. Spermatogenesis depends on the action of follicle stimulating hormone (FSH) and androgen and ablation of either hormone has deleterious effects on Sertoli cell activity and the progression of germ cells through spermatogenesis (Abel et al., 2008).

Oogenesis is the process of formation of mature **ova** (female germ cells) in the female gonad (ovary). During embryological development, **oogonia** (immature sex cells) undergo several mitotic divisions to form **primary oocytes**. Most primary oocytes develop to the first of four stages of meiosis I at this time and remain in this stage till sexual maturity. Before birth in many vertebrates, all oogonia that will be needed for reproduction are formed. At onset of sexual maturity, a primary oocyte undergoes further division (depending on species) to give rise to a **secondary oocyte** and the first polar body (small cell that contains half the number of chromosomes). The secondary oocyte undergoes further development to the second phase of meiosis II and remains in this state up to the time of fertilization during which the secondary oocyte completes meiosis (in the released oocyte) on penetration by a sperm cell to form a **zygote** and a second polar body. Research has shown that cyclic AMP under the control of phosphodiesterases plays a role in regulation of meiosis (Conti et al., 1998). Should fertilization fail to occur, the secondary oocyte degenerates.

Sex Determination

The way the sex of an individual vertebrate is determined depends on various factors. In birds and mammals, the distribution of sex chromosomes determines the sex of an individual at the time of fertilization. In mammals, males determine the sex of the offspring as they are **heterogametic** since the spermatozoa have X and Y chromosomes. These chromosomes normally segregate in males during meiosis to form X and Y spermatozoa. Female mammals are **homogametic** and produce ova with only X chromosomes. The sex chromosome combination for normal female mammals is XX whereas males possess the XY combination. In birds, the females are heterogametic with the sex chromosome combination being W and Z whereas males are homogametic with two Z sex chromosomes. Hormones can also alter the sex of an individual even though this is determined at conception.

Most other vertebrates lack sex hormones. In many snakes and lizards, either the males or females are heterogametic. The sex chromosomes of the few species of fish that have been studied are not much differentiated from the **autosomes** and are not easily distinguished from them. When sex hormones are lacking, environmental factors and genes of autosomal chromosomes determine the sex of an individual. The temperature at which eggs are incubated is known to determine sex in some reptiles that lack sex chromosomes.

THE MALE REPRODUCTIVE ORGANS

The male reproductive organs of vertebrates comprise the sex glands or gonads, the duct system and accessory (secondary) glands whose secretions are vital to the proper functioning of spermatozoa. The male gonad is the testis and is either located within or outside the body cavity.

The fish testes are located in the dorsal part of the body cavity and are long structures that are suspended by lengthwise mesenteries known as **mesorchia**. Lampreys and teleosts have a single testis whereas the structure is paired in other fish. The testes are composed of **follicles** or **ampullae** in which spermatozoa are formed (Fig. 9.2). The size and color of the testis varies depending on the stage of sexual maturity and ripeness. On average the testes weigh about 12% of the body weight. This proportion increases considerably just before spawning. The testes have a smooth surface and are creamy white in color most of the time. In most fish species, the inactive spermatozoa from the testis pass through convoluted sperm ducts during which secretions form the ducts are added to the spermatozoa to form **milt** that is exuded through the urogenital pore at the time of spawning. The secretions from the ducts are important in making the sperm cells active and mobile.

Salmons and trout lack sperm ducts and their spermatozoa rupture into the body cavity to exit through a spermatic opening posterior to the anus. Viability of sperm cells depends on the nature of the substrate into

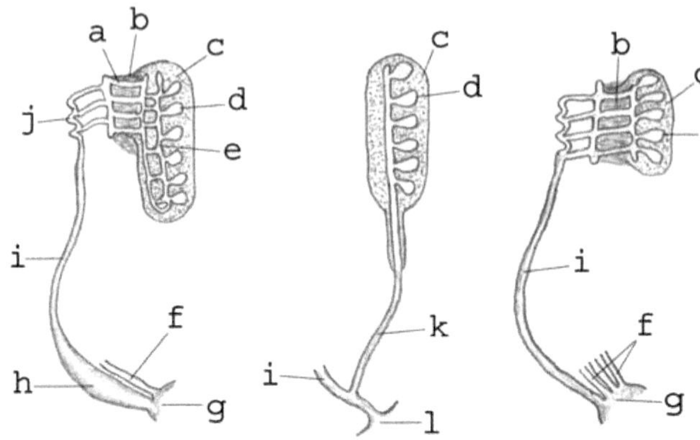

Fig. 9.2 Male reproductive organs of some anamniotes. Shark (left), teleost (middle) and salamander (right). (a) efferent ductule, (b) mesorchia, (c) testes, (d) seminiferous ampullae or follicles, (e) testis canal, (f) accessory urinary ducts, (g) cloacae, (h) seminal vesicle, (i) archinephric ducts, (j) duct of epididymis, (k) sperm duct and (l) body surface. The archinephric duct of a male shark widens posterioly into a **seminal vesicle** that contributes to the seminal fluid and also stores sperm.

which they are deposited. Conditions in the various substrates including temperature and salt concentration vary considerably. Although most fishes deposit their spermatozoa into water, sharks have modified parts of their pelvic fins into **claspers** while other fishes such as the mosquito fishes and guppies have **gonopodia** as parts of their anal fins (Fig. 9.3). Claspers and gonopodia are intromittent organs that are used for internal fertilization. External fertilization entails production of enormous amounts of spermatozoa to ensure successful fertilization.

Apart from the testis of frogs, those of other amphibians consist of follicles in which sperm cells are formed. The testes of frogs contain long coiled **seminiferous tubules** (L. *semen*, seed) (Fig. 9.4) that play the role of the follicles in the production of spermatozoa. The testicular ducts of amphibians converge to form a **testis canal** that is located at the center or the side of the testis. Several **efferent** or **connecting tubules** lead from the testis canal to the lateral kidney canal when present or to the kidney tubules. The latter join the **archinephric duct** (see Chapter 13) that ends up at the cloaca. In some immature amphibians, the anterior part of the gonad resembles an ovary and the posterior part the testis. Such an undifferentiated state of the gonad can undergo sex reversal when the right conditions prevail. The testis of urodeles is composed of lobes that show levels of increasing maturity caudally. Such a group of vertebrates have an annual sexual cycle that is dependent on germ cell progression. Factors regulating germ cell progression make the amphibian testis a good model for studying germ cell changes (Pierantoni et al., 2002).

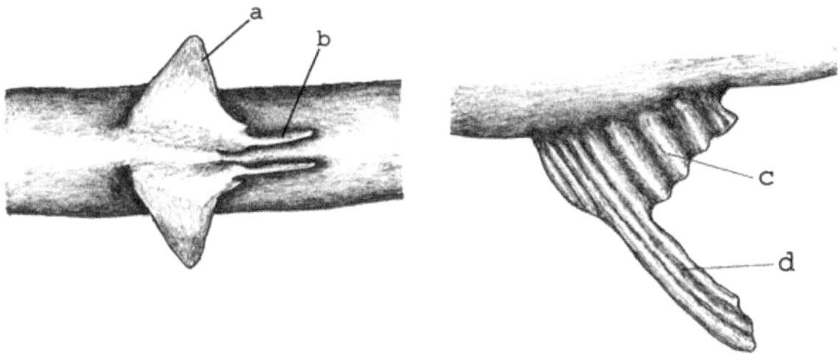

Fig. 9.3 The ventral side of part of a shark showing claspers (left) and an anal fin with an intromittent organ (right). (a) pelvic fin, (b) clasper, (c) anal fin and (d) gonopodium. Claspers are modified parts of the medial side on pelvic fins of male sharks and rays. At the time of mating, the erectile claspers bend in an anterior direction. The male inserts one clasper at a time into the female and muscular contraction pushes seminal fluid through a groove into the female oviduct. A gonopodium is a trough- or tube-like structure that is formed by the 3rd, 4th and 5th rays of the anal fin and transfers sperm into female mosquito fish, guppies and other live bearing fishes such as the tooth carp. The gonopodium can be as long as 50% of the total body length in some fish species.

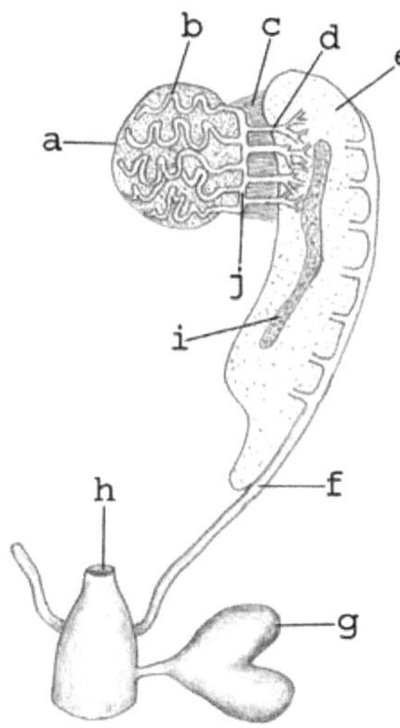

Fig. 9.4 Reproductive parts of a male frog. (a) testis, (b) seminiferous tubule, (c) mesorchium, (d) efferent ductule, (e) kidney, (f) archinephric duct, (g) urinary bladder, (h) large intestine, (i) adrenal gland and (j) testis canal.

The general pattern of reproductive ducts of amniotes is similar (Fig. 9.5). The testis of amniotes are made of long and highly coiled seminiferous tubules most of which end blindly and form the bulk of the testis (Fig. 9.6). Sperm cells are manufactured in these tubules that

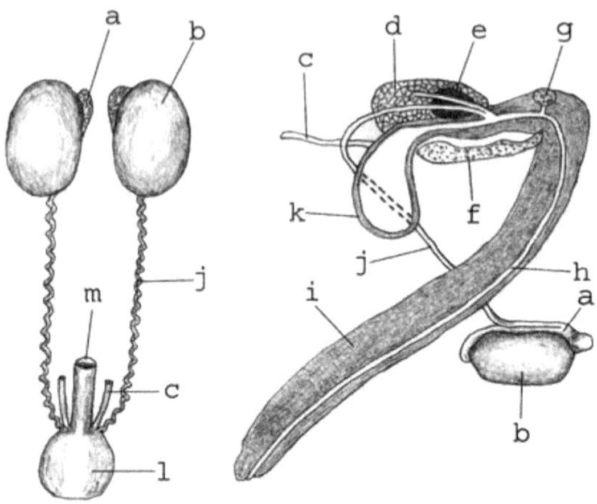

Fig. 9.5 Male reproductive organs of a bird (left) and horse (right). (a) epididymis, (b) testis, (c) ureter, (d) seminal vesicle, (e) prostate gland, (f) pelvic symphysis, (g) bulbourethral gland, (h) erethra, (i) penis, (j) vas or ductus deferens, (k) urinary bladder, (l) cloaca and (m) large intestine.

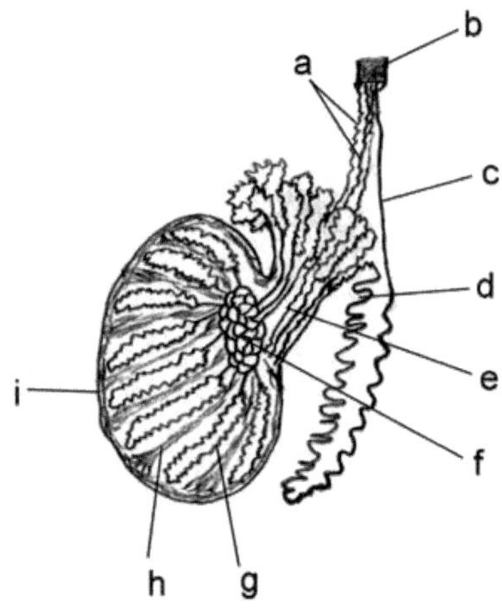

Fig. 9.6 Structure of the testis. (a) blood vessels, (b) spermatic cord, (c) vas deferens, (d) epidididymis, (e) efferent ductules, (f) rete testis, (g) seminiferous tubule, (h) trabecula and (i) tunica albuginea.

contain the spermatogonia (sperm cell precursor cell). The testes develop in the body cavity of all vertebrates and in most, with the exception of many mammals, the testis will remain at or close to these sites throughout the entire life of a vertebrate. In most mammals, the testes descend into the **scrotum** (L. *scrotum*, pouch). The testis of elephants, monotremes, cetaceans, sirenians and edentates remain in the abdomen throughout life. In some mammalian groups such as many carnivores, some ungulates, bats, lagomorphs and rodents, the testis descend to the scrotum during the mating season after which they are moved to the abdomen.

Each testis is surrounded with a fibrous capsule known as **tunica albuginea**. In many amniotes, this capsule sends out septae into the interior part of the testis dividing it into lobules. In mammalian species with testes that descend into the scrotum, each testis together with the sperm duct, blood vessels and nerves are surrounded by a sheath that lies to the inside of the scrotum known as **tunica vaginalis** that is an out-pocketing of the parietal peritoneum and passes into the scrotum through the inguinal canal. A small striated muscle that is a detachment of the internal oblique abdominal muscle known as **cremaster muscle** attaches to the tunica vaginalis and can alter the distance of the testis from the body trunk by contracting and relaxing.

The seminiferous tubules normally unite to form a plexus known as **rete testis**. Several sperm ducts known as **efferent ductules** (L. *ex*, out; *ferens*, to carry) lead from the rete testis to the outside of the testis through the tunica albuginea. The efferent ductules lead to the **epididymis** (Gr. *epi*, upon; *didymos*, testis) that is quite coiled and lies on the surface of the testis. Spermatozoa undergo further maturation and are also stored in the epididymis. The distal part of the epididymis leads into the **vas deferens** or **deferent duct** that also stores spermatozoa and connects the epididymis to the cloaca or the mammalian urethra. In many amniotes, the secondary sex glands that include the **prostate** (Gr. *prostates*, one who stands before), **vesicular** (L. *vesicus*, small bladder) and **bulbourethral glands** (L. *bulbus*, a bulbous root; *ourethra*, urethra) empty their secretions in the distal part of the vas deferens. Spermatozoa and secretions from these glands constitute **semen**. The secondary glands of the male reproductive system are well developed in mammals.

Stages in the development of sperm cells can be traced from spermatogonia that lie close to the outer limit of seminiferous tubules to the center of the tubules and constitute the **seminiferous epithelium** (Fig. 9.7) that is surrounded by a lamina or tunica propria that contains contractile myeloid cells. The action of the myeloid cells brings about peristaltic waves that propel spermatozoa and testicular fluid out of the tubules. The lumen of a seminiferous tubule is filled by fluid. A certain group of germ cells develops in synchrony that is coordinated as a result of interaction between cells of the same group (Eddy, 2002). Between the developing spermatozoa are irregularly shaped **Sertoli cells** that are thought to be important in the interaction as well as playing a nutritive role to these maturing germ cells. Sertoli cells, through their tight

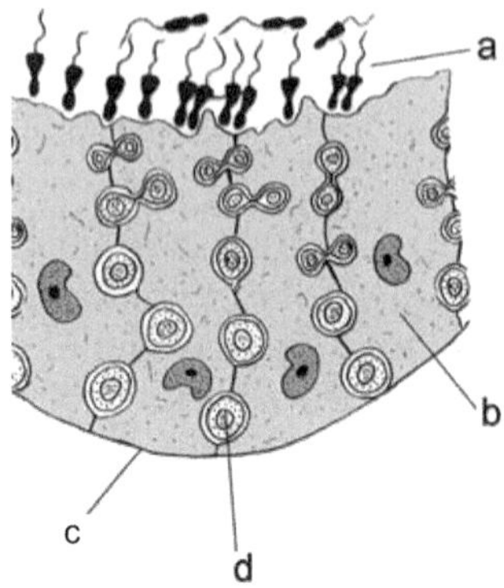

Fig. 9.7 The seminiferous epithelium showing stages of spermatogenesis. (a) spermatozoa in the fluid-filled lumen, (b) Sertoli cell, (c) basement membrane and (d) spermatogonium. Between the spermatogonia and spermatozoa are various stages of germ cells that show increasing maturity towards the lumen.

junctions, are also part of the blood testis barrier that protects the haploid spermatozoa that are antigenically foreign from an autoimmune reaction by antibodies in circulation. The number of Sertoli cells present in testes is established just before the onset of sexual maturity and this determines the potential for sperm production (Franca et al., 2005). Maturation of sperm cells is supported by testosterone and follicle stimulating hormone (FSH). Holdcraft and Braun (2004) have reviewed recent advances in hormonal regulation of spermatogenesis. Seminiferous tubules are normally located in lobules and are surrounded by loose connective tissue that contains **interstitial cells of Leydig**. These cells normally lie in small groups or individually and secrete androgens, chiefly testosterone. The function of the interstitial cells is influenced by various trophic factors including luteinizing hormone (LH), FSH and growth hormone.

Many male amniotes possess intromittent organs for transferring semen into the female reproductive system. The organ can be single as is the case in most amniotes or can be bifid as found in kangaroos. The organ is known as a penis in mammals and is made of erectile tissue that changes its turgidity depending on amount of blood present in most species.

FEMALE REPRODUCTIVE ORGANS

In female vertebrates, the gonad is the ovary that produces ova and the duct system shows various levels of organization in different species. Vertebrates that bear live young are viviparous or ovoviviparous and those that lay eggs are oviparous. In live bearing vertebrates, the duct system has evolved various structures that support embryological

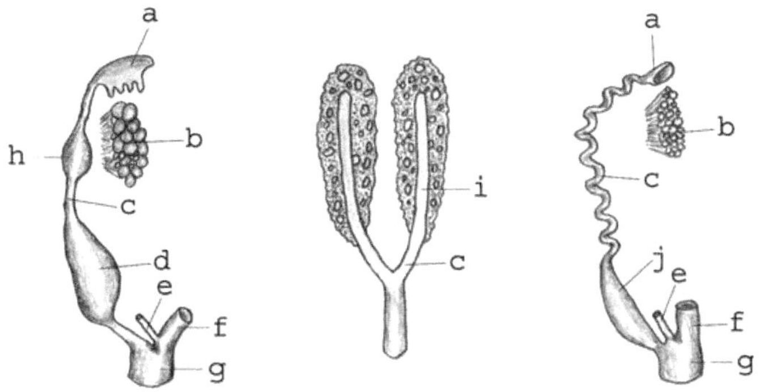

Fig. 9.8 Female reproductive parts of a shark (left), teleost (middle) and salamander (right). (a) infundibula, (b) ovaries, (c) oviducts, (d) uterus, (e) archinephric ducts, (f) large intestines, (g) cloacae, (h) nidamental gland, (i) cavity of ovary and (j) ovisac. The caudal end of the oviduct of some amphibians is expanded into an ovisac that stores eggs temporarily before they are laid.

development within the body when compared to development in oviparous species where it occurs outside the body within the egg. Female reproductive organs of some anamniotes are shown in Fig. 9.8.

Fish

Fish ovaries are long and paired but are often fused to varying degrees. The ovaries are suspended from the dorsal sides of the body cavity by a pair of mesenteries known as **mesovaria**. The ovaries lie below the swim bladder when one is present. The size of the ovaries varies depending on the stage of sexual maturity of the female fish. When fully ripe, the ovaries can make up to 50% of the body weight thereby distending and occupying most of the body cavity. The color of ovaries is whitish in young fish, greenish in immature ones and yellowish in ripe adults. The ovaries of adult fish are grossly granular with eggs of varying sizes depending on their stage of development. The size of eggs laid varies and can be as large as 5.0 cm or more in diameter as is the case in some sharks. During oogenesis (egg development) in fishes, the surrounding epithelial cells in an ovary provide the developing egg cell with granular yolk that is composed of protein and oil droplets. Normally large quantities of eggs are laid to ensure successful fertilization and development.

Cyclostome eggs are passed into the body cavity from the ovary and will pass through the abdominal pore that leads to the urogenital sinus at the posterior part of the fish to the outside. Cartilaginous fish have an **oviduct** (L. *ductus*, to lead) with a funnel-shaped entrance at the anterior part of the egg. The ovarian capsule is not continuous with the oviduct in this **gymnovarian condition** (Gr. *gymnos*, naked). The oviduct of egg-laying chondrichthyans is modified anteriorly into a **shell** or **nidamental gland** (L. *nidamentum*, nesting material). In live bearing species, the posterior part of the oviduct is enlarged into a **uterus** that retains the embryo for further development. Such cartilaginous fish produce eggs with thin shells that are later reabsorbed. The degree to which the embryo and the uterus are associated varies depending on the level of development that occurs before birth. The embryo's vascular yolk sac is in contact with the uterus forming a **yolk sac placenta**. The oviduct leads to the cloaca through a genital opening. The path taken by the egg in the oviduct of lungfishes, sturgeons and bowfin (*Amia*) is similar to that of cartilaginous fish.

Most teleosts have **cystovarian ovaries** (Gr. *cystis*, bladder) that contain a large central cavity that results from the folding of the ovary against itself or the body wall. Eggs arc ovulated into this cavity which is continuous with a tube-like extension of the ovary that plays the role of the oviduct. The two oviducts unite posteriorly to open to the outside through genital papillae. Salmons and trout lack oviducts. Eggs rupture into their body cavities and pass out of the body through pores at the posterior part of the body. These pores open at the time of reproduction and may represent greatly shortened oviducts.

Teleosts generally lay a lot of eggs, sometimes millions at a time, but the number laid varies depending on the group and even within individual fish. Factors including age and size of fish, temperature of water and food availability also play a major role in determining the number of eggs spawned. Species that do not accord care to their eggs generally produce a lot of eggs when compared to those that care for their eggs and young. Species that provide parental care for their eggs make elaborate nests and guard the developing eggs.

The reproductive cycle of most fish is yearly and once it has started will continue on that basis till death. Some fish have very long reproductive cycles. The sea lamprey (*Petromyzon marinus*) and freshwater eels (*Anguilla anguilla*) have cycles of 5 to 6 years and 10 to 14 years respectively. Where external fertilization occurs, proximity of two individuals of different sexes for spawning is the most frequent method used. Some form of courtship is seen in some fish. In many species, eggs and spermatozoa are emitted at the same time.

Amphibians

The paired amphibian ovaries lie in the anterior part of the body cavity and normally occupy a large proportion of the cavity during the breeding season. The long and convoluted oviducts start from the anterior part of the ovaries as expanded funnel-shaped structures known as **infundibula** (L. little funnels). The infundibulum opens into the body cavity and is not continuous with the ovary. The oviducts extend posteriorly in the body cavity to open into a cloaca. After ovulation, eggs are swept towards the opening of the infundibulum by cilia on the peritoneum. Egg movement in the oviduct is enhanced by ciliary action and muscular contraction of the oviduct. The layers of the egg are laid down as the eggs move posteriorly in the oviduct.

Most amphibians lay their eggs in water but some lay eggs on land in a moist environment. The eggs are sprayed with semen from males as they are laid. The jelly-like layers that surround the eggs imbibe water to swell and become sticky. The eggs will thus stick to each other and surrounding objects. Male salamanders will store their semen in mucus casings known as **spermatophores** (Gr. *phoros,* bearing) that are deposited for female animals to pick using their cloaca. A few amphibians including some caecilians are viviparous.

Reptiles

Reptiles have two ovaries that are separate from the long and tortuous oviducts (Fig. 9.9). During ovulation, the eggs first pass to the body cavity before entering the ostium (L. entrance or mouth) of the oviduct. The reptilian egg has more yolk and albumen than that of an amphibian. The presence of more nutrients in such eggs ensures that the reptilian embryo is able to undergo more development within the egg than is the case with amphibians. The additional laying down of a calcium-impregnated shell that covers the egg enables embryonic reptiles to bypass the aquatic stage to lay their eggs on land. The shell and shell membrane are added to the egg in the lower part of the oviduct. Oxygen can diffuse across through the pores in the shell as the eggs undergo incubation. Some snakes and lizards bear live young.

Birds

During the early developmental stages in birds, both left and right gonads are present. Only the left ovary is able to reach full development. The mature left avian ovary has numerous follicles in different stages of development and resembles a bunch of grapes. The large follicles hang from their stalks into the body cavity. At ovulation, the follicles are released into the funnel-shaped infundibulum of the oviduct. In wild birds and many domestic birds, maturation of follicles and ovulation stop at the time of brooding and molting. In birds that have been bred to eliminate the brooding gene including domestic chicken laying breeds, egg production is interrupted only by the molt.

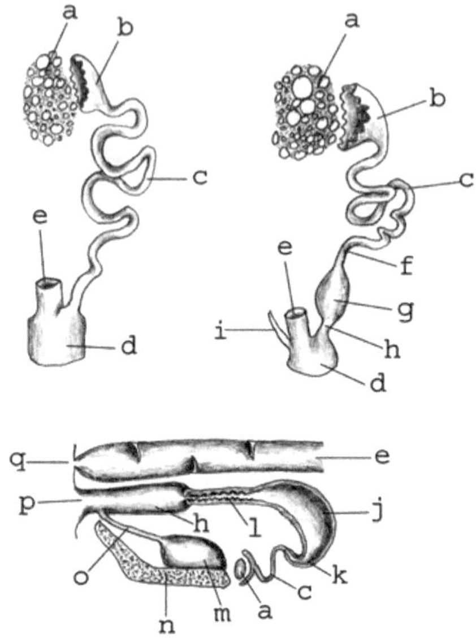

Fig. 9.9 Female reproductive organs of a reptile (top left), bird (top right) and a quadriped mammal (bottom). (a) ovaries, (b) infundibula, (c) oviducts, (d) cloacae, (e) large intestines, (f) isthmus, (g) shell gland, (h) vagina, (i) vestigial right oviduct, (j) uterine body, (k) uterine horn, (l) cervix, (m) urinary bladder, (n) pelvic symphysis, (o) urethra, (p) vestibule and (q) anal opening.

The rudimentary right ovary consists of medullary tissue that was suppressed from undergoing further development by female hormones produced in the cortex of the left ovary. Sometimes in old hens, the left ovary stops producing the female hormones leaving the remnants of the right ovary to undergo further development into a testis. The same condition can occur when the left ovary has undergone destruction by disease. The resulting **sex reversal** leads to development of male characteristics such as behavior and plumage. Some of these males are capable of fertilizing females.

The avian egg (Fig. 9.10) consists of the fertilized or unfertilized egg cell together with yolk that was produced in the ovary and is surrounded with a delicate yolk membrane.

The yolk is surrounded with three layers of **egg albumen** or **egg white**. Attached to the yolk membrane are cord-like and twisted structures that lie in the albumen known as **chalazae** that suspend yolk in the albumen so that it can move within the egg. There are two shell membranes below the shell. An **air chamber** is found between the two membranes at the blunt end of the egg. The outer shell has more calcium than that found in the reptilian egg and its surface is covered with cuticle.

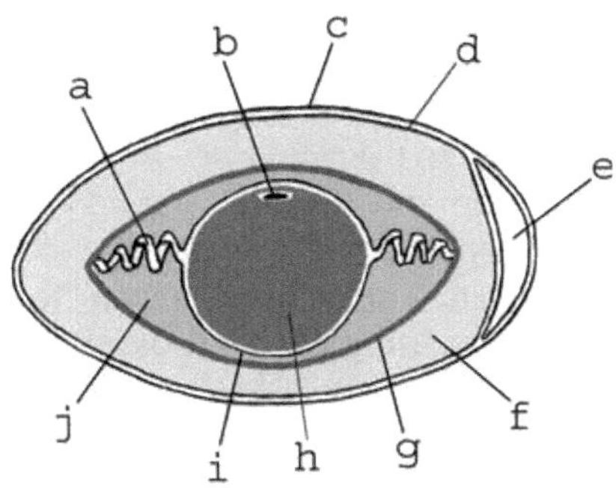

Fig. 9.10 Structure of an avian egg. (a) chalaza, (b) germinal disc or egg cell, (c) calcium shell (outer part) and outer shell membrane to its inner part, (d) inner shell membrane, (e) air chamber, (f) outer and (j) inner less viscous layers of egg albumen, (g) viscous layer of albumen, (h) yolk and (i) vitelline membrane (L. *vitellus*, yolk) that surrounds the plasma membrane of the ovum. The plasma membrane surrounds the yolk and the germinal disk. Both plasma and vitelline membranes give rise to the yolk sac membrane during embryonic development.

During ovulation, the muscular infundibulum grasps the follicle that is then pushed into the oviduct. As it moves down the oviduct, the follicle is covered with a delicate double-layered membrane. The albumen, chalazae and the shell membranes are laid down as the eggs moves towards the uterus. The chalazae result due to the revolving nature of the egg in the oviduct and addition of fibrous material. Shell formation occurs in the uterus. Abnormally large eggs are normally double yolked and could have originated from the same follicle in which case they are surrounded with a single yolk sac membrane or from two follicles that were ovulated close to each other and have independent membranes.

Monotremes

The egg-laying monotremes have the mammalian type of reproductive system. The two ovaries are partly surrounded with infundibula that continue as much narrower uterine tubes. Each uterine tube leads to a uterus that is much thicker. The two uteri open independently into a **urogenital sinus** that also receives two ureters one from each of the kidneys. The urogenital sinus leads into the cloaca. The fertilized egg stays in the uterus for some time where there is additional deposition of nutrients and laying down of the membranes and shell that covers the egg. There is low deposition of salts in this parchment type of egg with a thin permeable shell. Embryological development starts when the egg is still in the uterus. Female monotremes incubate their eggs upto hatching time.

Marsupials

Marsupials are viviparous but their young are born relatively early in an undeveloped state. The marsupial ovaries are not continuous with the infundibulum that is wide and leads to a much narrower uterine tube (Fig. 9.11). The uteri are paired and open independently into the **vaginal sinus**. The paired vagina also opens into the same sinus. The reproductive system of marsupials is thus termed **didelphous**. The vaginal sinus leads into a **pseudo-vaginal canal** in advanced marsupials. In primitive marsupials including the opossum, the **two vaginae** are located lateral to the ureters and the pseudo-vaginal canal is lacking. These lateral vaginae conduct the migration of spermatozoa and the medial pseudo-vaginal canal functions as the birth canal in more advanced marsupials such as kangaroos.

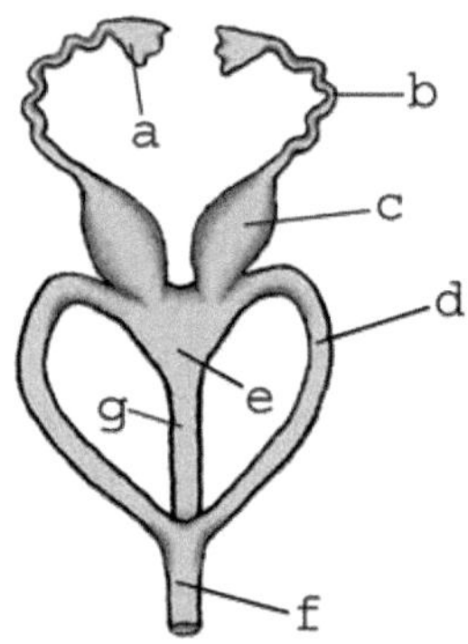

Fig. 9.11 The reproductive system of a female marsupial. (a) infundibulum, (b) uterine tube, (c) uterus, (d) vagina, (e) vaginal sinus, (f) urogenital sinus and (g) pseudo-vaginal canal.

After fertilization, the marsupial zygotes are coated with mucus and a shell membrane as they move down the uterine tubes towards the uterus. During the short gestation period, the embryo develops in the shell membrane for most of the time. Exchange of nutrients and gases occurs through the shell membranes before it is reabsorbed by the end of the second third of gestation. By this time the yolk sac is large enough and is richly supplied with blood to make contact with the chorion to form a **yolk sac placenta**. In most marsupials, the yolk sac placenta is superficial and is loosely attached to the uterus. Some marsupials such as koala develop a **chorioallantoic placenta**.

After the brief gestation period that can last up to a month or slightly longer depending on species, the young, known as **joey**, are born and will have to crawl as well as being assisted by abdominal contractions of the mother to be attached to a mammary gland nipple that is mainly located in a pouch or **marsupium**. Since the newborn marsupials lack the ability to suckle, milk is passed into the mouth of the young by contraction of maternal mammary glands. Most young marsupials are dependent on maternal care for a long time. Kangaroos and koalas carry the newly hatched young in their pouches for about 10 and 8 months respectively. As the newly born marsupial has undeveloped lungs, it has to rely on its skin for gaseous exchange.

Placental Mammals

The eutherian mammalian female reproductive system has various segments found in marsupials although there is quite some variation in the way the two systems are arranged. The vaginal section of placental mammals has united into a single tube that runs from the vestibule to the outer part of the cervix. Four uterine types are present among female placental mammals based on the relationship of uterine horns (branches) into duplex, bipartite, bicornuate and simplex uteri (Fig. 9.12). In a **duplex uterus** that characterizes lagomorphs, most rodents and elephants, the uterine horns are completely separated and have separate cervices that open into the vagina independently. The uteri of other placental mammals show varying degrees of uterine horn fusion thus

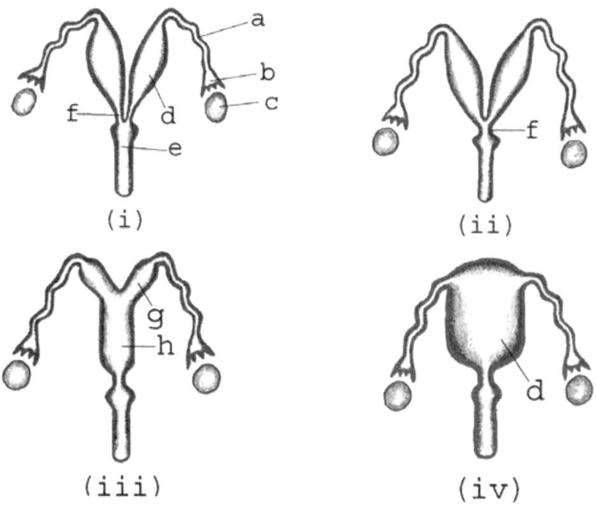

Fig. 9.12 Various types of mammalian placentas. (i) duplex, (ii) bipartite, (iii) bicornuate and (iv) simplex. (a) uterine tube, (b) infundibulum, (c) ovary, (d) uterus, (e) vagina, (f) cervix, (g) uterine horn and (h) uterine body. The funnel-shaped infundibula lead to tortuous uterine tubes that open into the expanded uterine horns or bodies that are connected to vaginae via cervices.

giving rise to a uterine body and uterine horns. Most carnivores have a **bipartite uterus** in which the horns are separated for the greater part of their length but join the vagina via a single cervix. Many ungulates and whales have evolved a **bicornuate uterus** that has a uterine body that is longer than the uterine horns. In armadillos and higher primates, the uterine horns of the **simplex uterus** are lacking as they have united entirely to form a single chamber. Uterine tubes open into this chamber. Development of the placental mammalian fetus takes place in uterine horns or body.

MAMMALIAN BREEDING CYCLES

The mammalian type of breeding, level of organization and parental care of young seen in most species is relatively advanced when compared to that of other vertebrates. Reproductively mature female mammals show a sexual cycle of events known as the **estrous cycle** that is controlled by hormones produced by the pituitary gland and ovaries. Many mammals show **estrus** (Gr. *oistros,* frenzy) or 'heat' phenomenon that coincides with ovulation. During estrus, the female mammal is receptive to the male for breeding purposes. Before estrus is **pro-estrus** that is a period during which ovarian follicles undergo maturation. Follicle stimulating hormone (FSH) that is produced by the anterior pituitary gland controls the process of follicle development. As they mature, follicular cells start producing estrogen that stimulates the proliferation of the endometral layer of the uterus. Later on there is a sudden surge in the level of estrogen that is followed some hours later by an increase in the level of luteinizing hormone (LH) from the anterior pituitary. LH triggers off ovulation shortly afterwards. The ruptured ovarian follicle is transformed into a golden body, the **corpus luteum** (L. *corpus,* body; *luteus,* yellow) that secretes progesterone. Progesterone slows endometral movements in readiness for implantation by the developing egg (blastocyst) after fertilization.

When the ovum is not fertilized, **met-estrus** follows and is characterized by the reproductive tract preparing itself to begin the cycle again. Sometimes met-estrus is followed by involution of the reproductive tract that is characterized by a period of quiescence known as **an-estrus**. Mammals that breed less frequently have long periods of an-estrus whereas those that breed frequently have shorter such periods. An-estrus is normally followed by pro-estrus.

Higher primates have a **menstrual cycle** that differs from the estrous cycle since they experience **menstruation** once implantation fails to take place due to lack of fertilization of the ovum by a spermatozoon. The corpus luteum also stops producing progesterone in the absence of implantation. A drop in production of progesterone causes constriction of muscles in the walls of the tightly coiled arterioles, an action that is followed by a reduction in blood supply (ischaemia) to the endometrium, death of tissue, sloughing off of uterine lining and bleeding.

Newly born mammals behave differently depending on species. **Precocial young** such as ungulates are fairly well developed and are able to start walking soon after birth. **Altriacial young** are usually born blind and in a helpless state as is seen in carnivores and primates. Precorcial young are generally born after a relatively long gestation period. The mammary glands of marsupials and placental mammals bear nipples whereas those of monotremes lack such structures. In monotremes, the young lap up milk that is secreted by mammary glands.

MAMMALIAN POPULATION STRUCTURE

There are generally three major groups in any population of a certain species that include pre-reproductive, reproductive and post-reproductive stages. The structure and dynamics of a population depend on how these three major groups interact, the relative length of time individuals spend in each group, rate of reproduction and recruitment into the group and rate of death or emigration from the group. At one extreme end are mammals that become sexually mature when they are a few weeks old and breed several times a year during which they produce several offspring per breeding season and at the other end are those that start breeding when they are several years old and have long gestation periods that give rise to a single individual per breeding season as happens with the African elephant. Interaction between various mammalian species such as predators and their prey has an effect on the general population groups of certain species. Population growth occurs within a certain species when individuals are able to survive to reproductive age.

Since mammals have to feed their young on milk for a considerable length of time, contact between mother and offspring is constantly maintained. This contact allows for a period of training that enables the young to learn non-genetic information that has enabled mammals develop sophisticated skills that are unknown in other vertebrates and have contributed to the success of mammals on earth. Some mammals are solitary in behavior and only join others during the breeding season. In social mammals, there is hierarchy in dominance setup. Such an order could be established after physical confrontation during which the winning member dominates the group or it could result from instinctive behavior. Social mammals tend to show relatively varied sexual dimorphism. In most cases the dominant males are the largest or most aggressive and endowed with power to conquer others. The dominant males will have several females each or will subdue other males when females are ready for breeding.

EARLY DEVELOPMENTAL BIOLOGY

The ability of a living organism to regenerate itself ensures its continued survival in the world. Throughout vertebrate evolution, many species have emerged while others have become extinct, partly due to their varying abilities to regenerate themselves. The giant panda of China, a bear, has low reproductive ability and this factor, coupled with destruction of the bamboo shoots it feeds on, has led to a decline in numbers of this animal. The giant panda has been declared an endangered species as a result of its low population. Species with a short generation interval such as many rodents have to be kept in check otherwise they will wreck havoc all over. Nature has ways of keeping such vertebrates under control. Human interference has also had its effect in elimination and preservation of certain vertebrates.

Life of a new individual normally starts at conception where the diploid number of chromosomes is restored after the two haploid sets each from a male and female gamete (Gr. *gamet*, spouse) have united. Gametes are among the most highly specialized and species-specific cells of vertebrates. The male gametes are the tail-bearing **spermatozoa** (Fig. 9.13) and vary in their head structure and size in various vertebrates. Each sperm cell is much smaller than an ovum and contains little cytoplasm. The sperm cell has a head, middle piece and a long tail that moves from side to side in a lash-like manner similar to the tail of a tadpole in a pond of water. The genetic material from the male vertebrate is contained in the head and is capped by the **acrosome** (Gr. *akron*, tip; *soma*, body) that is derived from Golgi apparatus and contains enzymes

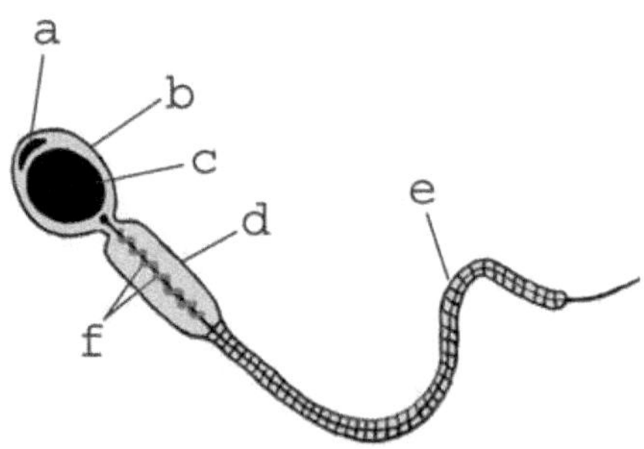

Fig. 9.13 A sperm cell. (a) acrosome, (b) head, (c) nucleus, (d) mid-piece, (e) tail piece and (f) mitochondria.

including hyaluronidase and acrosin for lysing the outer covering of the ovum. The normally cylindrical middle piece contains mitochondria that are arranged in a helical manner from end to end around a central axis. The mitochondrial enzymes are necessary for generation of energy needed for motion by the flagellum-like tail during swimming. The tail bears a principal piece and a short end piece.

The female ovum has at least enough energy reserves to be able to undergo the first few stages of embryonic development that involve a series of cell divisions before relying on other sources for nutrition. The amount of yolk present in the eggs varies considerably and greatly influences the level of development the embryo has to undergo while relying on its own source of nutrients. **Macrolecithal eggs** (Gr. *makros*, large; *lekithos*, yolk) of most fish, reptiles, birds and monotremes have relatively high quantities of yolk and their embryos undergo considerable development in eggs before hatching into young that resemble adults. A great reduction of yolk in eggs of eutherian mammals is related to the dependence of the embryo on the uterus for nourishment during gestation. **Mesolecithal eggs** (Gr. *mesos*, middle) of lampreys and amphibians have an intermediate amount of yolk and their young hatch into larvae that do not resemble adults. The distribution of yolk in an egg is not even. In **telolecithal eggs** (Gr. *telos*, end), the side of the egg with most yolk is known as the **vegetal pole** whereas the **animal pole** has less yolk. Yolk is evenly distributed in the mammalian **homolecithal eggs** (Gr. *homos*, same). The survival of the ovum depends on whether or not it undergoes further development after ovulation. Further development is brought about by fertilization or other chemical or physical stimuli that promote parthenogenesis in some species of vertebrates. An ovum that fails to undergo development soon disintegrates.

Fertilization

Fertilization is a process that involves the fusion of male and female gametes and is the starting point of a new cycle of life. Spermatozoa have to undergo capacitation before they are capable of fertilizing an ovum. **Capacitation** is a series of reactions that involve enzymatic alteration of the surface proteins and an increase in motility of sperm cells. The sperm cell has to penetrate through various layers of the ovum including the **vitelline** and **plasma membranes**. The vitelline or yolk membrane surrounds the plasma membrane and is secreted by follicular cells. In mammals, the glycoproteinous vitelline membrane is known as **zona pellucida** (Gr. *pellucidus*, transparent, clear). Zona pellucida, also known as **oolemma** (Fig. 9.14), is a transparent, elastic, thick and solid layer that serves as a protective coating or 'shell' that keeps the cells of the early

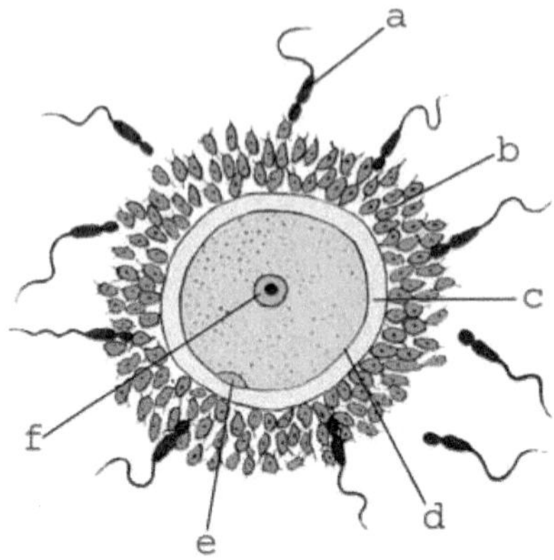

Fig. 9.14 An ovum. (a) spermatozoon, (b) corona radiata, (c) zona pellucida, (d) plasma membrane, (e) first polar body and (f) nucleus. An ovum measures about 100 to 140 µm in diameter and is a secondary oocyte since the second meiotic division has not been completed. The stimulus of fertilization is necessary for the completion of the second meiotic division that results in production of a zygote and a second polar body.

embryo together until the blastocyst hatches through it in preparation for implantation. Zona pellucida is surrounded by two or three layers of follicular cells referred to as **corona radiata**.

During the process of fertilization, a sperm cell normally binds to a receptor on the zona pellucida believed to be the glycoprotein ZP3. The **acrosome reaction** is initiated in the process and involves production of enzymes from the acrosome that break down the barriers of the ovum. The head and neck of the sperm cell fuse with the ovum as the tail drops off. It has been shown in the human being that the paternal messenger RNA is also passed to the egg at the time of fertilization and these transcripts could be important during early development (Ostermeier et al., 2004). Fusion of the plasma membranes of the head and neck of a spermatozoon with an ovum is followed immediately by a series of reactions within the ovum that lead to inhibition of entry by another sperm cell into the ovum. Vesicles that lie below the plasma membrane of the ovum produce enzymes that inactivate sperm receptors on the zona pellucida that then transforms into a **fertilization membrane** which cannot be penetrated by other sperm cells. Penetration of more than one sperm cell into an egg is known as **polyspermy** and is normally lethal.

The ovum is a secondary oocyte as further meiosis had stopped in the second meiotic division. Penetration into an ovum by a sperm cell leads to continuation of meiosis. The chromosomes of the ovum are then divided between the egg and the second polar body. The haploid number of chromosomes from the egg and spermatozoon will unite to restore the diploid number of chromosomes in the fertilized ovum or **zygote** (Gr. *zygon*, yoke or union). The zygote is the first cell of a new individual. After fertilization, there is redistribution of material within the ovum. Fertilization starts off the **prenatal period** (L. *natus*, born) that will last till hatching or birth.

In vitro fertilization (L. *in vitro*, within a glass) has been practised in some vertebrates including the human being. The world's first 'test tube' baby, Louise Brown, was born in Oldham, England on July 25, 1978. In vitro fertilization occurs outside the body in a temperature controlled environment. The developing zygote undergoes several mitotic divisions before it is transferred to the mother's uterus. The method has about 50% success rate and has been used by most women with fallopian tubes that hinder either sperm from fertilizing the ovum or movement of the dividing zygote to the uterus.

Cleavage

The formation of a zygote is followed immediately by cleavage that involves a series of rapid cell divisions. The cell divisions result in formation of many cells that decrease in size with each division known as **blastomeres** (Gr. *blastos*, bud; *meros*, part). The cells in this multicellular embryo are much smaller than the zygote since cellular divisions are not accompanied with growth at this time. The pattern of cleavage that occurs depends on the quantity of yolk present in the dividing zygote (Fig. 9.15). **Meroblastic cleavage** (Gr. *meros*, part) is seen in reptilian and bird eggs with large quantities of yolk that prevents complete division of the zygote. Cleavage is partial under such circumstances. In placental mammals and other vertebrates with low quantities of yolk such as amphibians, cleavage is total or **holoblastic** (Gr. *holos*, whole).

In macrolecithal eggs, the zygote comprises yolk and the embryonic disk that has a nucleus and other cytoplasmic organelles. Cleavage starts at the embryonic disk and during the process, several cell divisions occur with incomplete **cytokinesis** (cell division) but complete **karyokinesis** (nuclear division) resulting in formation of a **syncytium** (multinucleated cell) since the cell membranes do not fully separate the cells from each other. With further cleavage in birds, complete cytokinesis occurs in the cells of the more centrally located blastomeres. The rapidly multiplying blastomeres then separate from the underlying yolk leading to the creation of a **subgerminal cavity**. The embryo will develop from the **central blastomeres** that lie above the subgerminal cavity. The **marginal blastomeres** lie more peripheral to the central blastomeres and will divide and migrate slowly over the surface of the remaining yolk.

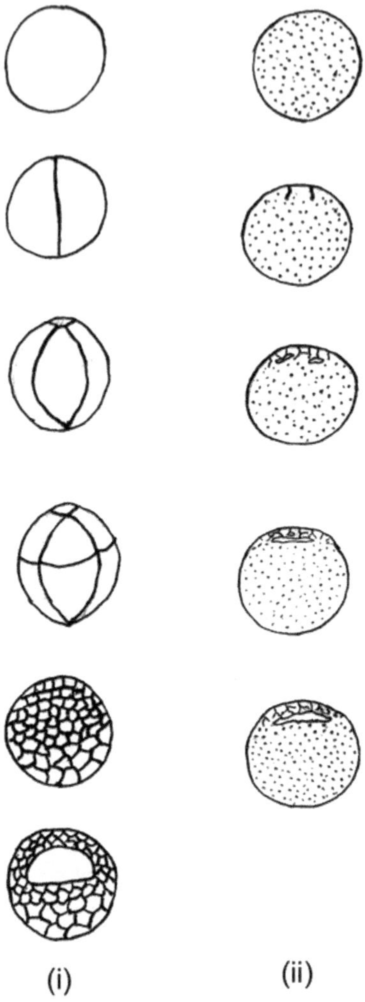

Fig. 9.15 Cleavage in (i) mesolecithal and (ii) macrolecithal eggs. The large quantities of yolk in macrolecithal eggs prevent complete cleavage of the zygote resulting in formation of partly divided cells at the embryonic disk. The central blastomeres above the subgerminal cavity of macrolecithal eggs later on undergo complete cleavage while marginal blastomeres continue to undergo incomplete cleavage and spread around the surface of the yolk.

Marginal blastomeres are trophoblastic and also form some of the extra-embryonic membranes.

The area of the central blastomeres is clearer than that of the surrounding blastomeres and is referred to as **area pellucida**. Towards the end of the cleavage stage, the central blastomeres divide into two groups of cells. The larger cells that also contain more yolk separate from the smaller cells and move to form the roof of the subgerminal cavity in a process known as **segregation**. With further development, **delamination** (L. *de,* from; *lamina,* small plate) occurs and involves some of the larger cells that are more deeply located breaking off the smaller cells and moving into the subgerminal cavity (Fig. 9.16). The larger cells then come together to form a layer of cells known as the **hypoblast** leading to the creation of a blastocoel above while the subgerminal cavity lies below this new layer. The roof of the blastocoel is formed by the smaller and more numerous cells of the **epiblast**. Gastrulation follows after the epiblast and hypoblast have been formed.

The first cleavage in chordates with low quanties of yolk is in the vertical plane and runs from the animal to the vegetal poles thus dividing

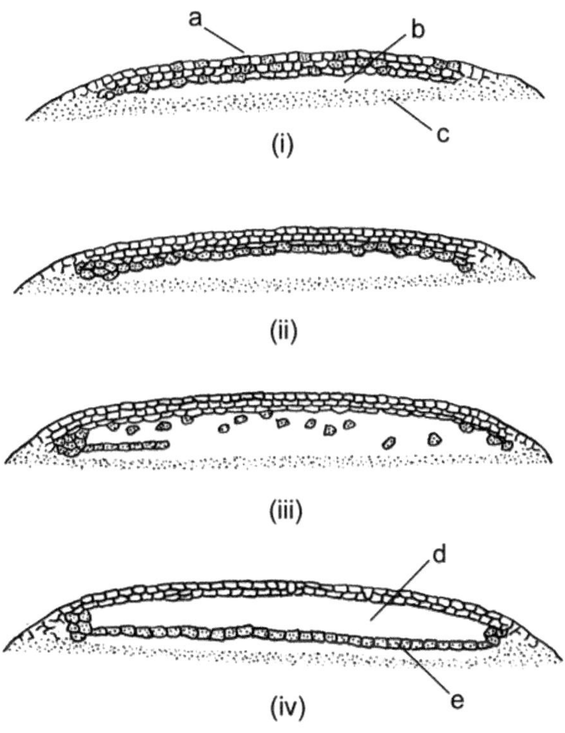

Fig. 9.16 Formation of a blastocoel in a macrolecithal egg. The cells with dots contain more yolk and are larger than other blastomeres. These cells segregate from other cells to form the roof of the subgerminal cavity then delaminate to form the hypoblast. (i) early blastula stage, (ii) segregation, (iii) delamination and (iv) formation of the hypoblast. (a) epiblast, (b) subgerminal cavity, (c) yolk, (d) blastocoel and (e) hypoblast.

the zygote into two cells. The second cleavage also occurs in a similar manner and results in formation of four cells. The third cleavage occurs in a horizontal plane and results in formation of eight cells. This horizontal cleavage occurs halfway between animal and vegetal poles in zygotes with low quantities of yolk but lies closer to the animal pole in very yolky zygotes. The yolky zygote has blastomeres of unequal size as a result with the cells closer to the vegetal pole tending to be larger. Cleavage normally alternates along vertical and horizontal planes. In many mammals, cleavage is mainly asynchronous resulting in formation of embryos with an odd number of cells after the four-cell stage. Any of the developing embryonic cells up to the four-cell stage in many mammals is capable of developing into an embryo and is said to be **totipotent**. Identical twins develop when totipotent cells separate and develop separately. Cleavage results in formation of a cluster of cells known as a **morula**.

The blastomeres of a morula undergo **compaction** in a process that involves loss of their spherical appearance and becoming tightly apposed to each other. Blastomeres start producing secretions that accumulate to form a fluid-filled cavity known as a **blastocoel** (Gr. *koilos*, hollow). Such an embryo is known as a **blastocyst** or **blastula**. The blastocyst undergoes a marked increase in size after breakdown of the zona pellucida. The cells that surround a blastocoel are known as the **blastoderm**. A group of cells occupying a small part of the blastocyst become slightly larger than the rest of the cells surrounding the blastocoel and are known as the **embryonic disk**. The embryo develops from the embryonic disk. The cells at the peripheral part of the blastocyst constitute the **trophoblast cells** (Gr. *trophe*, nourishment) whose role is to absorb nutrients early in development. Trophoblast cells participate in the formation of extra-embryonic membranes of amniotes and the placenta of eutherian mammals.

In mammals, delamination occurs during the blastocyst stage when cells from the inner layer of the embryonic disk detach and expand laterally below the trophoblast. The thin sheets of cells continue to multiply and spread out in a ventral manner until they eventually join ventrally to form a tube of hypoblast that is surrounded by the trophoblast to the outside (Fig. 9.17). The cavity of the hypoblast is the archenteron and most of it remains outside the embryo forming the yolk sac.

Gastrulation

The process of formation of the gastrula (Gr. *gastrula*, little stomach) from the blastocyst is known as gastrulation. During this process, the primitive

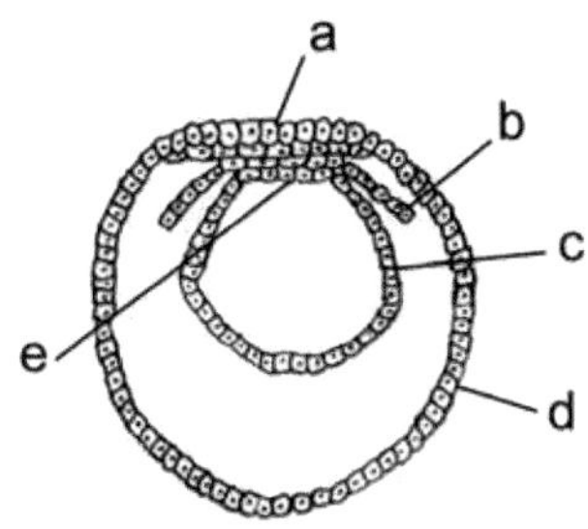

Fig. 9.17 Cross section through a mammalian embryo after delamination. (a) embryonic disk, (b) mesoderm, (c) hypoblast, (d) trophoblast and (e) endoderm.

gut will be formed. The rate of cell division is much lower during gastrulation when compared to the cleavage stage and there is little change in the overall volume of the embryo. Since the amount and distribution of yolk varies, the process of gastrulation differs between various vertebrate species though the general principle is the same. During gastrulation, the cells with potential for developing into different parts of the body are arranged in their appropriate parts in the embryo thus establishing the basic body plan. The three germ layers (L. *germen*, bud) are the ectoderm, mesoderm and endoderm and are laid down during the process of gastrulation. Most mammalian embryos establish contact with the uterine wall during gastrulation. Formation of the three layers results from different rates of cell division, movement of cells to new positions and changes in the shape and size of cells. The endoderm is the innermost of the three layers and is formed when some cells from the surface of the embryo move to the interior and replicate to form an invagination known as **hypoblast** or **mesendoderm**.

Lower Chordates

The large blastocoel of microlecithal gastrulation of lower chordates such as amphioxus provides a lot of space for internal movement of the surface cells (Fig. 9.18). The primitive gut that is formed as a result of this migration of cells has one opening at the posterior part of the gastrula known as **gastropore**. The axial part of the blastopore has a signaling center known as the **organizer** that patterns the germ layers and also regulates cell movement during gastrulation. The inward migrating cells form the **endoderm** that lines the primitive gut or **archenteron** (Gr. *arche*, origin or beginning; *enteron*, intestine or gut) internally and will eventually give rise to the inner lining of the tubular digestive system and the duct system of associated glands.

The outermost layer of cells of the gastrula is the **ectoderm** and will give rise to the epidermis of the integumentary system as well as the

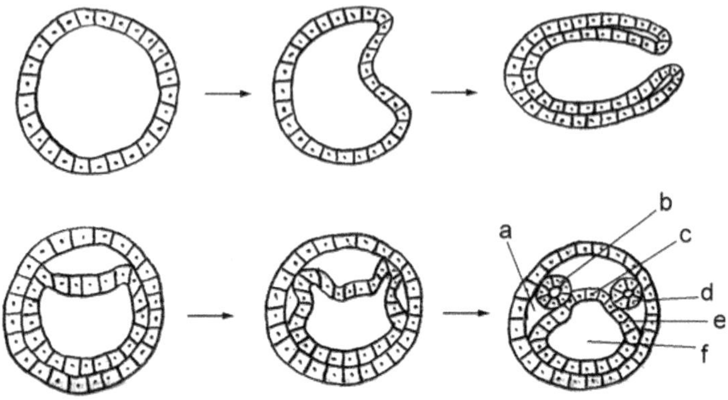

Fig. 9.18 Gastrulation in a microlecithal egg with very little yolk as is the case in amphioxus from the sagittal section (top row) and in cross section (bottom row). (a) coelom, (b) mesoderm, (c) future notochord, (d) ectoderm, (e) endoderm and (f) archenteron.

nervous system. The **mesoderm** lies between the ectoderm and endoderm and results from movement of future mesodermal cells from the dorsal part of the developing primitive gut. Out-pockets (evaginations) are formed on each side of the dorsal part of the hypoblast to form mesodermal (enterocoelic) pouches that will eventually cut off from the endoderm to form the mesoderm. Cavities within the mesoderm are known as **coelomic cavities**. The mesoderm will develop into the skeletal, muscular and cardiovascular systems, the dermis and many other body parts. Cells that lie centrally on the dorsal part of the hypoblast (above the primitive gut) are referred to as **chordamesoderm** and will give rise to the notochord. Cells will also bud off the hypoblast and migrate into the blastocoel to form **mesenchymal cells** (Gr. *mesos,* middle; *enchein,* to pour in).

Amphibians

The multiplication of amphibian eggs into independent larvae that do not resemble adults is rapid since the eggs develop externally and have little yolk. The cells at the vegetal pole of the amphibian mesolecithal embryo are larger than those at the animal pole. Some of the cells surrounding the blastocyst proliferate to the inside of this cavity to form a cleft-like invagination (Fig. 9.19). The cleft grows further inwards as more cells on the dorsal and lateral aspects of the embryo continue to proliferate and grow towards the blastopore and inwards to create more archenteron during this process of **involution** (L. *involutus,* rolled up). Other cells above the blastopore proliferate and multiply anteriorly as a sheet beneath the ectoderm to give rise to the mesoderm. The enlarging

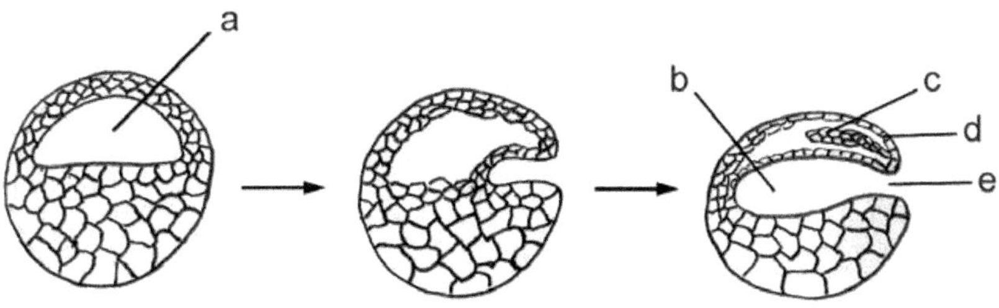

Fig. 9.19 Gastrulation in a mesolecithal egg such as that of amphibians. (a) blastocoel, (b) archenteron, (c) mesoderm, (d) ectoderm and (e) blastopore.

primitive gut replaces the blastocoel. Ectodermal cells also multiply from the animal pole to grow over the surface of the more yolky cells of the vegetal pole in what is known as **epiboly** (Gr. *epibole*, act of throwing on). Cell movements during epiboly bring about expansion and thinning of the new germ layers as the embryo elongates. The chordamesoderm cells that will develop into the notochord separate from the mid-dorsal part of the archenteron roof while mesodermal cells that are located more laterally multiply and move anteriorly as a sheet between the ectoderm and endoderm. The endodermal ends remaining after formation of the notochord unite to complete the roof of the primitive gut.

Fishes, Reptiles and Birds

As the eggs of most fishes, reptiles and birds are macrolecithal, the process of gastrulation in these vertebrates differs from that of mesolecithal embryos. The large quantities of yolk present in macrolecithal eggs restrict blastula formation. The blastula is disk shaped in bird embryos with a blastocoel and upper (epiblast) and lower (hypoblast) layers of cells around this cavity. The cells of the hypoblast proliferate to the margins of the embryonic disk and continue to grow over the yolk and under the proliferating marginal blastomeres. The more cranially located cells of the epiblast undergo expansion resulting in movement of cells caudally towards the central part of the embryonic disk. More cells of the epiblast move towards the midline of the embryonic disk resulting in the formation of an elongated and thick ridge-like structure known as the **primitive streak** (Fig. 9.20). The long axis of the primitive streak runs in a longitudinal manner to that of the body axis. The primitive streak increases in length as a longitudinal furrow known as the **primitive groove** develops in its central part.

An increase in the number of cells growing towards the primitive streak is followed by an inward migration of cells into the streak and spreading out of the cells peripherally from the primitive streak below

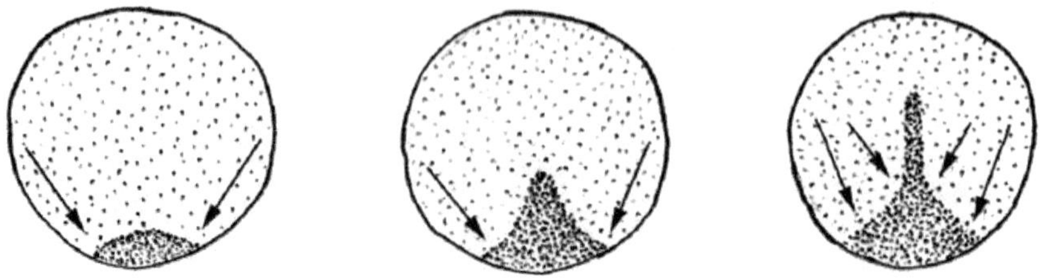

Fig. 9.20 Formation of the primitive streak from the dorsal surface view during the late blastula and early gastrula stages in an avian embryo. Arrows show the direction in which cells of the epiblast move towards the midline where the the primitive streak forms (darker region).

the epiblast and above the underlying endoderm in a process known as **involution** (Fig. 9.21). The migrating cells also displace cells of the hypoblast to form the endoderm. The primitive groove marks the area at which cells are moving away from the primitive streak. The layer of migrating cells between the epiblast and the developing endoderm will give rise to the mesoderm. Mesoderm is formed from the whole length of the primitive streak. Mesodermal cells continue to expand in a lateral and anterior manner until the lateral margins of these cells meet in the front of the embryo at the midline. The last mesodermal cells to develop from the epiblast remain close to the notochord and give rise to paraxial mesoderm while the first such cells to form are extraembryonic and play a role in the formation of extra-embryonic membranes. The primitive streak is considered a homologous structure to the blasopore since cell movements that lead to formation of the endoderm and mesoderm spread from these points. The process of involution proceeds in an anterior-posterior manner to the long axis of the embryo.

The primitive streak regresses (shortens) later when the cells of the primitive groove migrate away and are not replaced by the arrival of additional cells from the epiblast. Regression of the primitive streak proceeds in a similar direction to that of involution and leaves an invaginated mesoderm. The notochord is formed by the first group of cells of the epiblast to involute at the cranial part of the primitive streak and grows in length by intrinsic growth as involution of epiblast cells decreases. The position previously occupied by the primitive streak is occupied by the elongating notochord. A thin area next to the notochord on either side lacks mesodermal cells. The cells of the epiblast that do not participate in involution but cover the embryo become the ectoderm. Gastrulation comes to an end after regression of the primitive streak.

Gastrulation of the mammalian embryo that has very little yolk is very similar to that of reptiles and birds. The main differences between

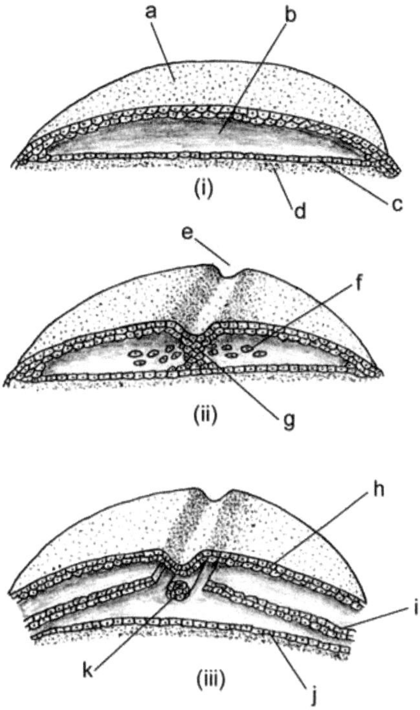

Fig. 9.21 Involution in a macrolecithal embryo. (i) beginning of gastrulation, (ii) involution as cells of the epiblast move inwards at the primitive streak before spreading outwards below this layer of cells and (iii) end of gastrulation when the three germ layers have been formed. (a) epiblast, (b) blastocoel, (c) hypoblast, (d) yolk, (e) primitive groove, (f) migrating cell, (g) primitive streak, (h) ectoderm, (i) mesoderm, (j) endoderm and (k) notochord.

these groups of vertebrates are related to the need to develop extra-embryonic membranes very early in mammalian embryos. Many mammalian embryos undergo implantation (make contact with the uterine wall) at the beginning of gastrulation. The mammalian epiblast occupies a small area of the blastocyst since most of the embryo is occupied by the extraembryonic trophoblast. The primitive streak forms in the epiblast in a similar manner to that of the bird. Cells of the epiblast that move inwards at the primitive streak eventually give rise to the endoderm, mesoderm and notochord.

Neurulation

Neurulation is part of organogenesis and is the process of neural tube formation from the ectoderm that begins towards the end of gastrulation. During neurulation the embryo elongates into a tubular form with a distinct front and back. Neurulation is induced by the notochord that secretes soluble growth factors. The different parts of a neural tube are formed by primary and secondary neurulation (Fig. 9.22). During **primary neurulation**, the ectoderm above the notochord thickens then flattens to form a longitudinal neural plate. The margins of the neural plate are then raised to form neural folds. A neural groove is then formed between the two folds. The margins of the folds continue to grow dorsally and will eventually meet in the middle to form a neural tube that is longitudinal to the long axis of the body. The folds will then separate from the ectoderm above in a craniocaudal manner.

Secondary neurulation starts with the formation of a solid cord of cells or **medullary cord** from the cells of the neuroectoderm and endoderm. The cord of cells condenses before separating to form cavities that will later on coalesce to form a neural tube. Secondary neurulation normally occurs in the posterior part of most vertebrates. At the end of neurulation neural tubes that result from both primary and secondary neurulation will unite. Neuroepithelial cells of a neural tube are capable

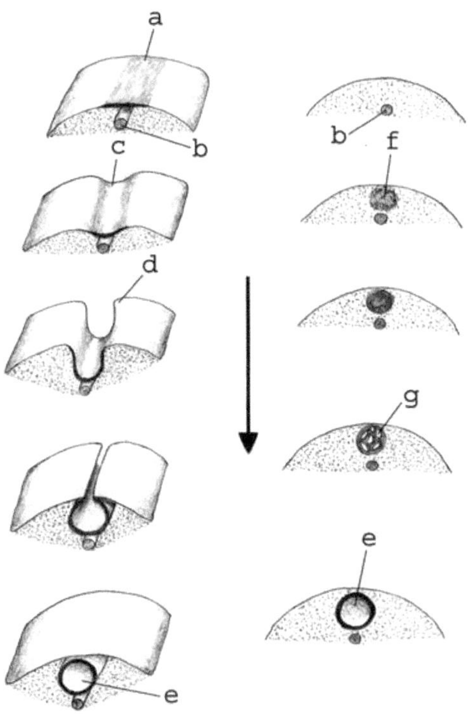

Fig. 9.22 Stages of primary (left) and secondary (right) neurulation. (a) neural plate, (b) notochord, (c) neural groove, (d) neural fold, (e) neural tube, (f) medullary cord and (g) cavity in medullary cord. Arrow indicates changes that occur during neurulation with time.

of developing into neurons and neuroglial cells. The neural tube will develop into most of the nervous system. The canal in the tube will develop into the cavity of the nervous system.

Neural crest cells are a group of ectodermal cells that break loose from each neural fold in vertebrates. They lie at the dorsolateral border of the neural tube. The cells will later migrate to other locations in the body and become part of the star-shaped mesenchymal cells. Migration of neural crest cells starts cranially and then extends caudally in a gradual manner and is determined by intrinsic properties of the cells and the nature of the surrounding environment such as signals that guide them along specific routes to their final destinations. Migration of neural crest cells normally stops when the cells encounter barriers. Neural crest cells will differentiate into many structures in the body including peripheral nerves, glial cells, epinephrine producing cells of the adrenal gland, pigmented cells of the epidermis, parts of the cranium and teeth and sensory structures.

Derivatives of the Mesoderm

Homeotic genes (**Hox**) influence the differentiation of anterioposterior structures of vertebrates. The genes regulate the expression of other genes that are responsible for the development of each body segment by turning them on and off in a sequence. All vertebrate embryos show marked segmentation including the segmented somites of the mesoderm that will develop into various body structures. Other genes including **Pax** and **hedgehog** influence the dorsoventral development.

Towards the end of gastrulation up to the end of neurulation, the mesoderm starts to differentiate into various structures such as somites and their derivatives (Fig. 9.23). On each side of the notochord and the neural tube, mesodermal cells proliferate to form thick columns known as **paraxial mesoderm** that together with the notochord extend cranially to where the pituitary gland is developing from the diencephalon. After several hours, the column undergoes further proliferation into distinct segments known as **somites** beginning from the cranial end of the embryo and proceeding caudally. Each somite consists of three parts that will later on contribute to the formation of the musculoskeletal system and the dermis. The dorsolateral part of a somite is the **dermatome** that will contribute to the formation of the dermis. The middle component is the **myotome** that eventually develops into axial musculature and the ventromedial section is the **sclerotome** that will form vertebrae and the axial skeleton.

Cranial to the spinal cord, paraxial mesoderm develops into four pairs of somites known as **occipital somites** (Fig. 9.24). Paraxial

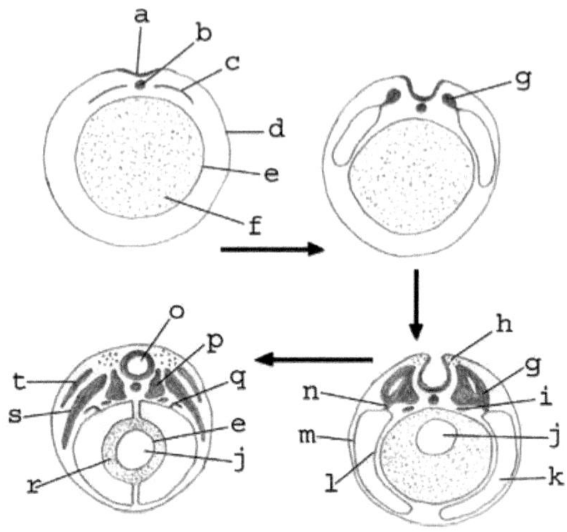

Fig. 9.23 Differentiation of the mesoderm. (a) neural groove, (b) notochord, (c) paraxial mesoderm, (d) ectoderm, (e) endoderm, (f) yolk, (g) somite, (h) neural crest cells, (i) dorsal aorta, (j) archenteron, (k) coelom, (l) lateral plate (splanchnic layer), (m) lateral plate (somatic layer), (n) intermediate mesoderm, (o) neural tube, (p) sclerotome, (q) nephric ridge that develops from the intermediate mesoderm, (r) muscles, connective tissue of gastrointestinal system wall and visceral peritoneum, (s) myotome and (t) dermatome. Somatic mesoderm will later give rise to the body wall and limb buds while splanchnic mesoderm will develop into mysentery, wall of the digestive tract and the heart.

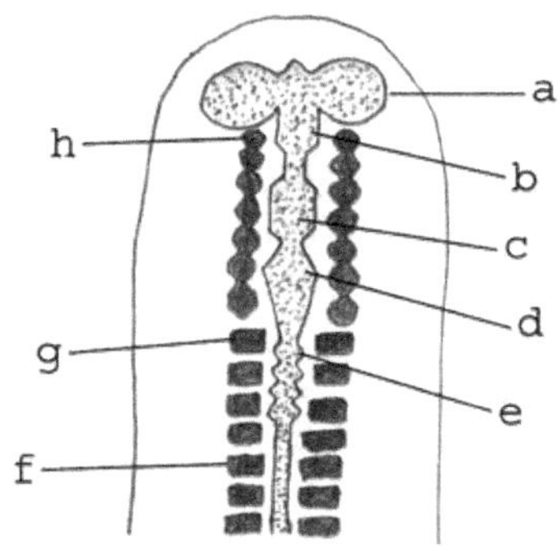

Fig. 9.24 Diagram of a frontal section of the anterior part of a developing embryo showing the relationship of somitomeres and somites with the developing brain. (a) telencephalon, (b) diencephalon, (c) mesencephalon, (d) metencephalon, (e) myelencephalon, (f) 1st cervical somite, (g) 1st occipital somite and (h) 1st somitomere.

mesoderm that is found anterior to these somites develops into seven pairs of rudimentary somites known as **somitomeres** that are linked to each other. The occipital somites and somitomeres develop into the skeleton of most of the braincase and skeletal muscle of the jaws.

The **intermediate mesoderm** or **nephrogenic plate** is a flat sheet that comprises of several cells and develops from the lateral proliferation of the paraxial mesoderm. It is segmented anteriorly but is continuous posteriorly. The mesoderm is continuous medially with the somites. The intermediate mesoderm ultimately develops into parts of the urinary system, a portion of the reproductive system and the adrenal cortex. The broad and unsegmented lateral mesoderm (plate) occupies the remaining area between the ectoderm and endoderm and grows from the intermediate mesoderm. The lateral mesoderm eventually splits into two thin layers that proliferate ventrally. The outer somatic (parietal) layer is adjacent to the ectoderm and will develop into the body wall. The inner (splanchic or visceral) layer remains next to the endoderm and will contribute to the formation of part of the body wall of organs. The space between the outer and inner mesodermal layers is known as the **coelom** and will later on contribute to the formation of the major body cavities. The ectoderm and somatic mesoderm are known as **somatopleura** whereas the endoderm and splanchnic mesoderm are called **splanchnopleura**. The lateral mesoderm also contributes to the formation of extra-embryonic membranes. At this stage, only the dorsal surface of the embryo has formed but not the embryonic body.

During early development in eggs of reptiles and birds, the developing germ layers are adjacent to the yolk and albumen and absorb nutrients directly to meet the metabolic requirements of the embryo. With further development, the nutritive requirements of the embryo increase and a vascular system is necessary for the transport of nutrients. The vascular system is thus one of the first to develop and function in the embryo.

Extra-embryonic Membranes of Reptiles and Birds

The evolution of the **cleidoic egg** of reptiles and birds has enabled these vertebrate groups to bypass the aquatic larval stage of amphibians. The egg can be deposited on land and under the right conditions will undergo development into viable young. Fetal or extra-embryonic membranes (Fig. 9.25) are derived from tissues that do not form part of the embryo but are of importance to the survival of the embryo before hatching or birth. The **amnion** and **chorion** are derived from somatopleura whereas the **allantois** and **yolk sac** are from splanchnopleura. In eutherian mammals, the chorion, allantois and yolk sac unite with the uterine mucosa to form the placenta that is used in exchange of material between the fetus and the mother.

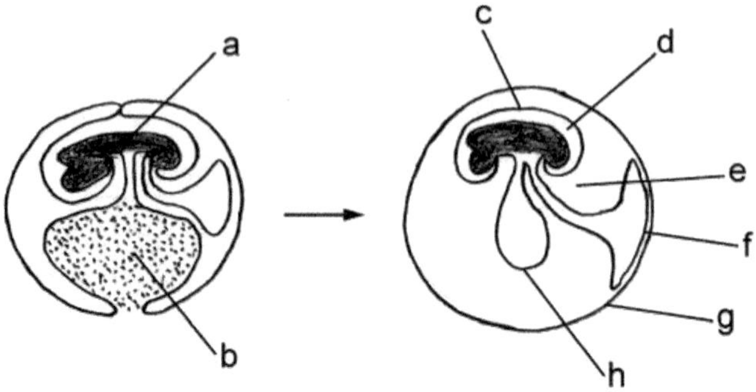

Fig. 9.25 Extra-embryonic (fetal) membranes. (a) embryo, (b) yolk, (c) amnion, (d) amniotic cavity, (e) extra-embryonic coelom, (f) allantois, (g) chorion and (h) yolk sac. The yolk is covered by the endoderm and splanchnic mesoderm that together form the yolk sac.

Yolk Sac

The yolk is a source of nutrients to the embryo through the membrane and its blood vessels. During gastrulation, extra-embryonic germ layers grow to surround the yolk that was previously surrounded by the cell membrane of the zygote and the vitelline membrane. The endoderm and the adjacent splanchnic mesoderm that cover the yolk form the yolk sac. Yolk is digested and absorbed by the endoderm then passed to the vitelline blood vessels (L. *vitellus*, yolk) in the splanchnic mesoderm. Blood in these vessels flows to vitelline veins along the yolk stalk that eventually supply the embryo. Late in incubation, the yolk sac and remaining undigested yolk are drawn into the abdomen in birds and will be absorbed entirely soon after hatching.

Amnion and Chorion

These membranes start forming when folds emerge from the somatopleura known as **chorioamniotic folds** (Fig. 9.26). The folds initially form cranial to the head and increase in length as they grow over the head of the embryo in a caudal manner. Similar folds develop laterally and caudally to the embryo and grow medially and anteriorly over the embryo respectively. The chorioamniotic folds grow until they meet and fuse in the middle over the back of the embryo. The final result is the presence of two layers that originated from the somatopleure over the embryo. **Extra-embryonic coelom** separates the two layers of somatopleure. The inner membrane of the somatopleura is the amnion and the outer one is the chorion. Between the amnion and the embryo is the amniotic cavity that is filled with amniotic fluid that serves to buoy and provide a protective cushion to the embryo. Amniotic fluid also

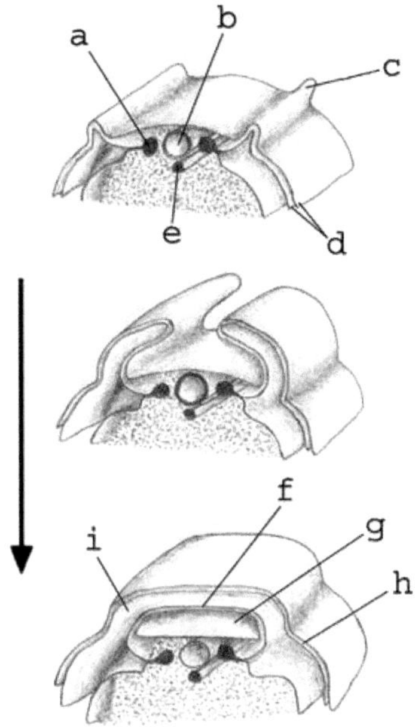

Fig. 9.26 Development of the chorion and amnion. (a) somite, (b) neural tube, (c) chorioamniotic folds, (d) somatopleure, (e) notochord, (f) amnion, (g) amniotic cavity, (h) chorion and (i) extra-embryonic coelom.

provides an environment in which the embryo can move its body and limbs with development. The reptilian and avian fetuses are thus surrounded with an environment that is aquatic in nature as that of fishes and amphibian larvae even though development is taking place in a cleidoic egg on land. Amniotic fluid is initially produced by the amniotic ectoderm but later on with development the fluid is also produced by fetal kidneys, respiratory tract and glands of the oral cavity. Smooth muscle fibers develop in the somatic mesoderm of the amnion with further development and their contraction causes movement of amniotic fluid. The chorion expands quickly to cover the inner part of the eggshell and together with the allantois participates in gas and water exchange between the embryo and the surrounding environment.

Allantois

The allantois develops from the hindgut as an outgrowth that grows out of the abdominal cavity and expands to fill the extra-embryonic coelom. The allantois makes contact and fuses with the yolk sac, amnion and chorion. When the allantois fuses with the chorion, its vessels that branch from the two caudal aortae vascularize the chorion. These vessels lie just under the shell and are used for gaseous exchange. The allantois also severs as a site for deposition of urinary excretory wastes that are mainly in form of the water insoluble uric acid.

Eutherian Mammals

The development of the mammalian embryo is similar in many ways to that of reptiles and birds. The low level of yolk present enables cleavage to take place throughout the mammalian zygote. Placental mammals undergo gestation whereby the developing zygote will implant to the uterus to undergo further development. Before implantation, the developing blastocyst obtains nutrients from fluids of the uterine cavity known as **histotrophe** that is secreted by uterine mucosal glands. During implantation, the trophoblast of the blastocyst attaches to the wall of the uterus. Small and large animals tend to have eccentric and central implantation respectively. **Interstitial implantation** is where the trophoblast invades and partially erodes the myometrium and is found in rodents, many carnivores and primates.

Placental Barrier

The placenta (L. flat cake) forms the contact between the chorioallantois and maternal uterine tissues. It has been stated before that the placenta shows more variation in structure than any other mammalian organ. The first placental structures can be traced back to the fossil remains of icthyosaurs that lived about 170 million years ago. Several structures that resemble the placenta are also found in squamate reptiles. Some of the diverse placentas found in mammals could have originated separately. It is probable that there could be some unmodified structures in placentas of any of the living mammals and more studies on the various types of placentas should improve the understanding of placental function as well as interrelationships among mammals (Enders and Carter, 2004).

The level at which the blastocyst attaches to the uterine lining varies greatly depending on the species. The superficial or trophoblast cells of the embryo develop villi that interdigitate with those of the uterus. The trophoblast cells normally digest maternal cells for the nourishment of the embryo. Such action leads to erosion of endometrial tissue (Gr. *endon*, within; *metra*, womb) at the point of attachment. In primates, the fetal trophoblast erodes even maternal blood vessels and its villi are in contact with blood from the maternal circulatory system (Fig. 9.27). In some cases, the maternal connective tissue cells are stimulated to grow around the embryo. Both the allantois and yolk sac grow and make contact with the chorion. The yolk sac and allantois contribute to the formation of the

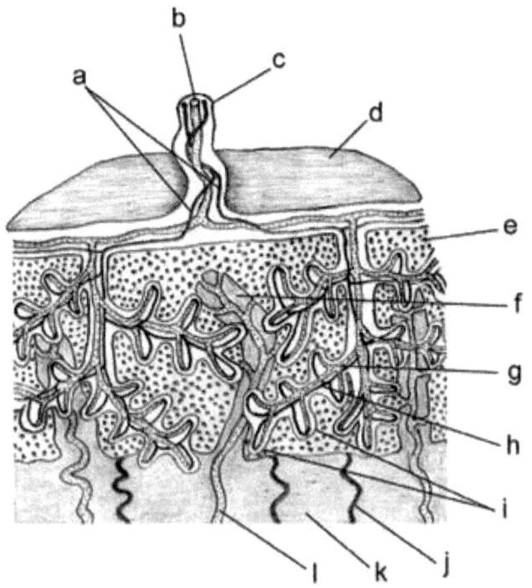

Fig. 9.27 Diagram showing part of a primate placenta. Despite the reduced number of layers between fetal and maternal circulatory systems resulting from erosion of the endometrium by the fetal trophoblast, blood from the two systems does not mix. (a) umbilical arteries, (b) umbilical vein, (c) umbilical cord, (d) luminal surface of the placenta, (e) maternal blood, (f) endometrial villus, (g) fetal venule, (h) fetal arteriole, (i) chorionic villi, (j) maternal arteriole, (k) endometrium and (l) maternal venule.

cord-like umbilical cord by which the embryo is connected to the placenta. The blood vessels of the two membranes will also grow to make contact with the chorion. The number of major blood vessels present in the umbilical cord varies. When two vessels are present—one is the umbilical artery and the other is the umbilical vein. There is no or very little mixing of maternal and embryonic blood in normal pregnancies. Arteries of the embryo transport wastes to the placenta while veins carry nutrients and oxygen to the embryo. At the placenta, blood flow in the vessels of the embryo and uterus is in opposite directions (countercurrent flow).

The placenta serves various functions including transport of nutrients to the fetus and gaseous exchange. Diffusion is responsible for the transport of most of the small molecules. The placenta also stores lipids, glycogen and proteins, synthesizes substances such as estrogen, progesterone and gonadotropins and anchors the embryo in the uterus. Transport of immunity from the mother to the fetus is through the placenta. The placenta, especially the fetal membranes, serves as a barrier that prevents mixing of maternal and fetal blood. Although mammalian fetal membranes are similar to those of reptiles and birds, the mammalian yolk sac is not well developed since the developing zygote lacks or has little yolk. The yolk sac is also a site of hemopoiesis in the early stages of the embryo.

Male hormones are not synthesized in the placenta as this could produce masculine features in the embryo. In some mammalian species such as some herbivores, development of embryos of different sexes at the same time in the uterus results in the blood systems of the two chorions making contact with each other leading to mixing of blood from the two sexes. The male hormones circulate to the female embryo causing development of a masculanized and sterile female known as a **freemartin**. Such females retain female sexual characteristics but have male secondary characteristics.

Types of Placenta

Placentas are classified according to the location of chorionic villi, the number of layers forming the placental barrier and the quantity of placental tissue that is shed off after parturition. There are various types of placentas according to the location of their villi (Fig. 9.28). In diffuse placentas, villi are evenly spread over the chorion. **Diffuse placentas** are found in horses, pigs and whales. **Cotyledonary placentas** are found in ruminants and here the uterus develops several cotyledons (Gr. *kotuledon*, cup-shaped cavity) to which placental **caruncles** that represent groups of villi attach. Caruncles and cotyledons constitute **placentomes**. A **zonary placenta** forms a zone or belt of villi around the middle of the oval shaped chorion of carnivores. Incomplete zonary placentas are found in some carnivores such as walruses and mink. Reddish lines are seen at the edges of a zonary placenta as the mother's blood circulates into some sinuses of the placenta where it is phagocytized by cells of the

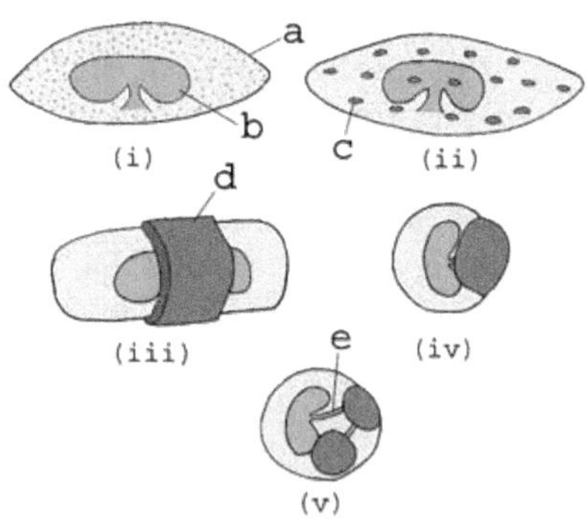

Fig. 9.28 Various types of placentas based on distribution of chorionic villi and their attachment sites to the maternal uterus. Diffuse (i), cotyledonary (ii), zonary (iii), discoid (iv) and double discoid (v) placentas. (a) chorion, (b) embryo, (c) caruncle, (d) placenta and (e) developing umbilical cord. In the double discoid placenta of monkeys, blood vessels connect the first and second disks.

embryo. Rabbits, bats and most primates including the human have a **discoid placenta**. The villi in a discoid placenta are arranged in a circular plate occupying one spot of the chorion. Monkeys have a **double discoid** placenta where there are two circular plates that are connected by blood vessels. The embryo is connected to one of the plates.

At its maximum, the placenta has six layers separating maternal and embryonic blood. The six layers between the maternal and fetal circulatory systems include fetal and maternal capillary endothelial linings and the surrounding connective tissues plus the chorion and endometrium (Gr. *metra*, womb). According to the number of layers between the two circulatory systems of the uterus and the embryo, there are four main types of placentas (Fig. 9.29). In the **epitheliochorial placenta**, the six layers are present and such a placenta is found in horses, pigs and whales. **Syndesmochorial placentas** of ruminants have five layers as the maternal epithelial layer has been eroded. The four-layered **endotheliochorial placenta** comprises three layers of the embryo and one endothelial maternal layer. An endotheliochorial placenta is found in carnivores. Further erosion of the maternal capillary endothelium gives the **haemochorial placenta** found in primates. Maternal blood pours

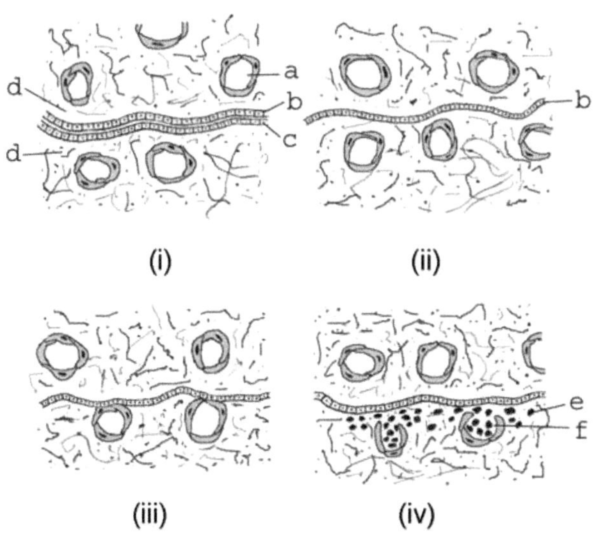

Fig. 9.29 Placental types based on the number of layers between fetal and maternal circulatory systems in various mammalian species. (i) epitheliochorial, (ii) syndesmochorial, (iii) endotheliochorial and (iv) haemochorial placentas. The epitheliochorial placenta has the maximum number of six layers between the two circulatory systems while the other placentas show various levels of maternal tissue erosion. (a) fetal capillary, (b) chorion, (c) endometrium, (d) connective tissue, (e) eroded maternal capillary and (f) blood cell.

from vessels into spaces that are in contact with the chorion. Fetal cells sometimes engulf maternal blood.

After parturition, the placenta is normally discarded shortly afterwards. A **deciduate placenta** (L. *decidere*, to fall off or down) is discarded entirely after parturition since the fetal chorion has invaded the maternal uterus to varying degrees. When only fetal tissues are shed off while maternal tissues that comprise the placenta are retained to be part of the uterus, the placenta is known as **non-deciduate** or **adeciduate**. Epitheliochorial placentas are non-deciduate. Marsupials have a **contra deciduate placenta** where all fetal membranes rupture to free the young to be delivered. The fetal membranes that remain in the kangaroo are reabsorbed in the uterus.

Nursing of Young Mammals

Mammalian young are born in an altricial state (immature and unable to obtain food on their own). They are nursed with the mother on milk that is produced by the mammary glands and are also taken care of for some time. The milk is able to meet the nutritional requirements of the newborn until it starts feeding on other foods. Milk that is produced for the first few days after birth by the mother is known as **colostrum**. Colostrum has immune and growth factors that are vital for the survival of the newborn. It also has a laxative effect on the newborn. Immunoglobulin A (IgA) that is present in colostrum is important in areas which are prone to infection such as the mucous membranes of the pharynx, lungs and intestines. The intestines of the newborn are highly permeable and colostrum provides a barrier against this.

Lactation is the most energetically demanding part of reproduction and the mother invests a lot, nutritionally, in this process. In kangaroos where the young are born quite immature after a few weeks in the uterus, lactation is a major part of the reproductive process as the young spend several months in the pouch attached to a nipple with their lips. Lactation can last for a few days as is the case with some seals or may continue for several years as happens in some bears. During lactation, there is an interrelationship between the mother and the offspring during which the two are close to each other and the young is able to learn some practices from the mother. The mammary gland undergoes substantial structural development during the gestation period. Such development is under hormonal control and has to be in tandem with fetal development. High levels of estrogens and progesterone are necessary for mammogenesis.

The mammary gland undergoes involution at the end of the lactation period during which the young no longer need milk for survival and depend on other sources of food. Many mammals are in an anestrous state (lactational anestrous) during early lactation and the length of this period varies among mammalian species. In some mammals, the anestrous state persists for as long as lactation lasts. In kangaroos and some other mammals, further development of the next zygote is arrested till the young have stopped suckling.

EMBRYO TRANSFER

The transfer of pre-implantation stage for embryos from a donor to a surrogate recipient is known as embryo transfer. Gonadotropins are used to super-ovulate the donor followed by artificial insemination. The zygotes are normally harvested and stored in oxygenated and buffered tissue culture that is supplemented with serum for immediate use or may be frozen in liquid nitrogen at $-196^{\circ}C$ and stored. Approximately a third of the embryos are damaged in such a procedure. The time embryo transfer is done varies in different mammals and depends on the optimal survival and accessibility of the blastocysts. The transfer is normally done before a secondary layer that is secreted by the follicular cells around the cell membrane of the ovum known as zona pellucida disintegrates. Zona pellucida offers some protection from damage to the embryo.

Storage of embryos for transfer purposes is a more costly and specialized process than storing sperm. Embryo transfer has been used in the dairy industry to increase the number of offspring from cows that deliver highly productive offspring. The method has also been practised to introduce new breeds in an area and to investigate cases of recessive alleles in offspring of known carriers. The technique has also been carried out in endangered species to increase their population.

Aging in Vertebrates

Aging is a gradual process that results in loss of physiological functions while increasing the probability of natural death. Cells that are incapable of replication in mature vertebrates such as neurons in the brain, skeletal and cardiac muscle and kidney cells show reduced function with age when compared to those capable of mitosis throughout life including blood and intestinal epithelial cells. In the natural environment, few vertebrates show signs of aging since several conditions including infectious diseases, predation, lack of food and harsh environment will lead to premature death. Deaths that occur before many members of a species have reached sexual maturity are a threat to the survival of the group. The human being and animals that are protected show signs of aging.

Ectothermic vertebrates (fishes, amphibians and reptiles) that survive environmental hazards have long life spans and continue to grow throughout life. Birds and mammals do have a fixed maximal body size and show signs of aging when protected from environmental hazards. From an evolutionary perspective, several mechanisms that cause limitations in somatic maintenance leading to an accumulation of damage have been implicated in the aging process (Kirkwood, 2005). **Reactive oxygen species** (ROS) such as superoxide and hydrogen peroxide that are generated by mitochondria as a result of electron transport during oxidation of foodstuff cause damage to macromolecules in cells including lipids, proteins and DNA. The levels of oxidative damage products are positively correlated with aging (Landis and Tower, 2005). Damage to DNA leads to a decline in cell function with age. The accumulation of damaged macromolecules in cells especially non-dividing types such as neurons produces the aging pigment known as **lipofuscin**. Lipofuscin accumulates at a slow rate in animals on a low calorie diet. Studies on rodents show that calorie restriction increases longevity and delays the onset of certain diseases and also has positive effects on aging (Le Bourg, 2005).

REFERENCES

Abel, M.H., Baker, P.J., Charlton, H.M., Monteiro, A., Verhoeven, G., De Gendt, K., Guillou, F. and O'Shaughnessy, P.J. (2008). Spermatogenesis and Sertoli cell activity in mice lacking Sertoli cell receptors for follicle stimulating hormone and androgen. Endocrinology, 149(7):3279-3285.

Bhandari, R.K., Higa, M., Nakamura, S. and Nakamura, M. (2004a). Aromatase inhibitor induces complete sex change in the protogynous honeycomb grouper (*Epinephelus merra*). Mol. Reprod. Dev., 67(3):303-307.

Bhandari, R.K., Komuro, H., Higa, M. and Nakamura, M. (2004b). Sex inversion of sexually immature honeycomb grouper (*Epinephelus merra*) by aromatase inhibitor. Zoolog. Sci., 21(3):305-310.

Chang, C.F., Lin, B.Y., Lau, E.L., Lee, M.F., Yueh, W.S., Lee, Y.H., Chang, C.N., Huang, J.D., Tacon, P., Lee, F.Y., Du, J.L. and Sun, L.T. (1997). The endocrine mechanism of sex reversal in the protandrous black porgy, *Acanthopagrus schlegeli*: a review. Chin. J. Physiol., 40(4):197-205.

Chen, L. and Li, H. (2004). Progress in the studies of parthenogenetic embryonic stem cells. Zhonghua Nan Ke Xue, 10(1):55-58.

Conti, M., Andersen, C.B., Richard, F.J., Shitsukawa, K. and Tsafriri, A. (1998). Role of cyclic nucleotide phosphodiesterases in resumption of meiosis. Mol. Cell Endocrinol., 1-2(145):9-14.

Cuellar, O. (1977). Animal parthenogenesis. Science, 197(4306):837-843.

de Magalhães, J.P., Costa, J. and Church, G.M. (2007). An analysis of the relationship between metabolism, developmental schedules and longevity using phylogenetic independent contrasts. J. Gerontol. A. Biol. Sci. Med. Sci., 62(2):149-160.

Eddy, E.M. (2002). Male germ cell gene expression. Recent Prog. Horm. Res., 57:103-128.

Enders, A.C. and Carter, A.M. (2004). What can comparative studies of placental structure tell us? A review. Placenta. 25 Suppl. A:S3-9.

Franca, L.R., Avelar, G.F. and Almeida, F.F. (2005). Spermatogenesis and sperm transit through the epididymis in mammals with emphasis on pigs. Theriogenology, 63(2):300-318.

Holdcraft, R.W. and Braun, R.E. (2004). Hormonal regulation of spermatogenesis. Int. J. Androl., 27(6):335-342.

Kirkwood, T.B. (2005). Understanding the odd science of aging. Cell., 120(4):437-447.

Landis, G.N. and Tower, J. (2005). Superoxide dismutase evolution and life span regulation. Mech. Ageing Dev., 126(3):365-379.

Le Bourg, E. (2005). Is caloric restriction a means of delaying ageing and increasing longevity? Presse Med., 34(2 Pt 1):121-127.

Ostermeier, G.C., Miller, D., Huntriss, J.D., Diamond, M.P. and Krawetz, S.A. (2004). Reproductive biology: delivering spermatozoan RNA to the oocyte. Nature, 429(6988):154.

Pierantoni, R., Cobellis, G., Meccariello, R., Palmiero, C., Fiega, G., Minucci, S. and Fasano, S. (2002). The amphibian testis as a model to study germ cell progression during spermatogenesis. Comp. Biochem. Physiol. B Biochem. Mol. Biol., 132(1):131-139.

Tanaka H. and Baba, T. (2005). Gene expression in spermiogenesis. Cell Mol. Life Sci., 62(3):344-354.

Vrana, K.E., Hipp, J.D., Gross, A.M., McCool, B.A., Riddle, D.R., Walker, S.J., Wettstein, P.J., Studer, L.P., Tabar, V., Cunniff, K., Chapman, K., Vilner, L., West, M.D., Grant, K.A. and Cobelli, J.B. (2003). Nonhuman primate parthenogenetic stem cells. Proc. Natl. Acad. Sci. U.S.A., 101(2):693.

Zikopoulos, B., Kentouri, M. and Dermon, C.R. (2000). Proliferation zones in the adult brain of a sequential hermaphrodite teleost species (*Sparus aurata*). Brain Behav. Evol., 56(6):310-322.

Zikopoulos, B., Kentouri, M. and Dermon, C.R. (2001). Cell genesis in the hypothalamus is associated to the sexual phase of a hermaphrodite teleost. Neuroreport., 12(11):2477-2481.

10

Circulatory System

The circulatory system is made up of the cardiovascular or blood vascular and lymphatic systems and performs the major roles of transporting nutrients and oxygen to every cell of the body and removal of metabolic waste products including carbon dioxide from the body. The system also plays a major role in the maintenance of body homeostasis. The circulatory system has been designed according to the metabolic activity and size of vertebrates. As working tissues such as muscle consume large quantities of nutrients and oxygen thus generating corresponding quantities of toxic wastes and heat, the circulatory system has to function in a coordinated manner to meet the metabolic requirements of all tissues in the body. The cardiovascular system consists of a closed circuit system that comprises a muscular pump, blood vessels that include arteries, veins and capillaries and blood. The blood is important in the biochemical and physiological functions of the system. The cardiovascular system has also evolved elaborate feedback controls. The lymphatic system drains lymph that is derived from blood and tissue fluid back to the cardiovascular system through the lymphatic vessels.

The circulatory system has undergone a major change in design with the transition from water to land and also change from ectothermic to endothermic states. From early embryonic life, the survival of an embryo depends on the circulatory system for normal development and maintenance of homeostasis and as a result the circulatory system appears early in development and attains a functional state before the other systems of the body have developed.

SINGLE AND DOUBLE CLOSED CIRCUIT SYSTEMS

Unlike the simple open transport system of circulation that is found in most invertebrates where blood fills spaces between internal organs, the closed circuit system of circulation found in vertebrates comprises a set of branching vessels through which blood circulates as it is pumped by the heart. Blood flows through the same pathway all the time to complete a circuit as it flows to various parts of the body. An important factor in the transition from an open to a closed circulatory system was a change in vessel wall structure and composition that enabled the large arteries to store and release energy during the cardiac cycle as a result of the presence of an elastic fiber network organized by medial smooth muscle (Wagenseil and Mecham, 2009).

A single circuit system of circulation is found in fish and consists of a single heart with one atrium and one ventricle. The heart pumps blood through the lungs or gills first to obtain oxygen and eliminate carbon dioxide. The larger vessels (arteries) supplying these structures branch into the first capillary bed. The capillaries then unite to form arteries that will supply blood to tissues of the body thereby giving rise to the second capillary bed that will eventually lead to the venous system that returns blood to the heart. Since blood flows through two capillary beds in a series before returning to the heart, there is a substantial drop in blood pressure. Cephalopod mollusks such as squid and octopus have a similar system of circulation but have overcome the pressure drop problem by evolving extra **booster** or **gill hearts** that pump deoxygenated blood from body tissues to gills before flowing to the main heart and to the rest of the body.

The double circuit system of all terrestrial vertebrates comprises two circulatory pathways. One pathway leads to the lungs or skin and is known as the pulmonary circulation whereas the systemic circulation leads to body tissues. A single heart pumps blood to these two pathways (circuits) at the same time. There are no booster hearts in terrestrial vertebrates.

The Heart

The vertebrate heart (Fig. 10.1) consists of at least one thin-walled chamber known as an **atrium** that receives blood from the rest of the body and one thick-walled chamber that receives blood from the atrium referred to as the **ventricle** that pumps blood to body tissues. Valves that control a unidirectional flow of blood are found between the two chambers. The heart is surrounded with a sac that is made of fibrous tissue known as **pericardium** (Fig. 10.2). The outer layer of the

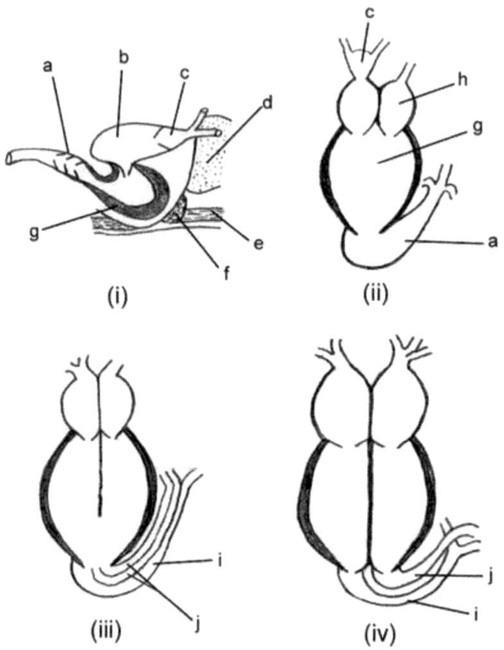

Fig. 10.1 The hearts of (i) fish, (ii) amphibians, (iii) reptiles and (iv) birds and mammals showing the heart compartments and vessels that enter and leave the heart. In life, the initial parts of vessels that originate from the heart of higher vertebrates are part of the heart. (a) conus or bulbus arteriosus, (b) atrium, (c) sinus venosus, (d) liver, (e) hypaxial muscle, (f) pectoral girdle, (g) ventricle, (h) left atrium, (i) pulmonary artery (j) the two aortic trunks of reptiles or the aorta of birds and mammals.

pericardium is tough and fibrous whereas the inner part is a smooth serous membrane that has two layers that surround the pericardial cavity. The innermost serous membrane is also known as the **epicardium** and is the outermost layer of the heart wall. The pericardial cavity contains pericardial fluid that acts as a lubricant and ensures smooth contraction of the heart.

The middle layer of the heart is the **myocardium** and forms the bulk of the heart. It contains the cardiac muscle and is highly vascularized in birds and mammals. The innermost lining of the heart wall is the **endocardium** that comprises a thin elastic connective tissue membrane that has a simple squamous epithelium that covers its innermost surface. The heart valves are formed by the endocardium when it forms a double fold of itself with scanty connective tissue between the folds.

Fish

The hearts of fish vary greatly in relative development and size. Compared to other vertebrates, fish have relatively smaller hearts and

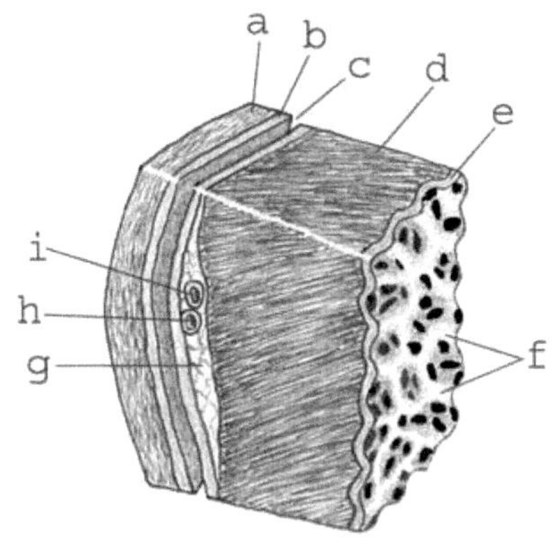

Fig. 10.2 Wall of the heart. (a) fibrous pericardium, (b) visceral pericardium or epicardium, (c) pericardial cavity or space, (d) myocardium, (e) endocardium, (f) trabeculae, (g) adipose tissue, (h) coronary vein and (i) coronary artery. The inside of the heart bears trabeculae that are beam-like projections that are covered by the endocardium. The pericardial cavity contains pericardial fluid that acts as a lubricant and is secreted by the serous pericardium.

lower blood volume and pressure. Hearts are generally smaller in sluggish and sedentary fish species. Fish hearts are composed of typical vertebrate cardiac muscle fibers (Yamauchi and Burnstock, 1968). The diameter of fish cardiac muscle fibers is smaller than that of similar mammalian fibers and in some fish species the fibers are about half the size of mammalian cardiac fibers. Blood flows to the atrium of the heart from the thin walled **sinus venosus**. It then flows to the thick walled ventricle. Valves are found between the atrium and ventricle and ensure unidirectional flow of blood. The ventricle lies in a ventrocaudal position in relation to the atrium. The ventricle leads to a conus or bulbus arteriosus that is located in an anterior position.

The **conus arteriosus** is found in cartilaginous fish and lungfishes and has up to seven transverse rows of valves. The conus wall is contractile and consists of a thin layer of cardiac muscle that covers a fibrous elastic sheath. Teleosts have an elastic and non-contractile **bulbus arteriosus** that lacks cardiac muscle and varies in shape depending on the species. The wall of the bulbus consists of the three layers that are present in arteries and include the outer, middle and inner layers. The middle layer is thick and contains spirally arranged smooth muscle fibers, elastic fibers and occasionally fibrocytes. Internally, the bulbus can be smooth, may contain ridges or trabeculae and is lined with squamous, cuboidal or columnar endothelial cells. A valve or a series of valves are present in the bulbus. The bulbus makes the flow of cardiac output smooth as it acts as an elastic reservoir by expanding during cardiac systole during which it stores a large part of the cardiac stroke volume. At diastole, the bulbus rebounds back to a smaller volume. The bulbus thus ensures constant perfusion of the gills with blood in the process. Knowledge on the bulbus wall is still fragmentary (Icardo et al., 2000). Blood flows anteriorly from the conus or bulbus arteriosus to the thick walled **ventral aorta** that gives off pairs of **afferent branchial arteries** that supply gills on either side of the pharyngeal cavity.

In lampreys and sharks, efferent branchial arteries from the left and right gills carry oxygenated blood to the **median** or **dorsal aorta** that lies dorsal to the gills. In teleosts, efferent branchial arteries of each side lead to the left and right dorsal aortae on either side of the body. The two dorsal aortae then unite behind the gills into a single dorsal aorta that will run caudally below the vertebral column to supply blood to the rest of the body.

Lungfishes have pulmonary and branchial circulatory pathways that represent a transition from water to land breathing (Fig. 10.3). When gills are used for gas exchange, the pulmonary artery is constricted and the ductus arteriosus is open so blood flows through the second, fifth and sixth arches and to the dorsal aorta. The third and fourth aortic arches lack gills. When faced with hypoxia in water, the lungfish revert to pulmonary circulation. The **spiral valve** of the conus arteriosus diverts oxygen-poor blood into the sixth aortic arch. Blood flow to the gills is

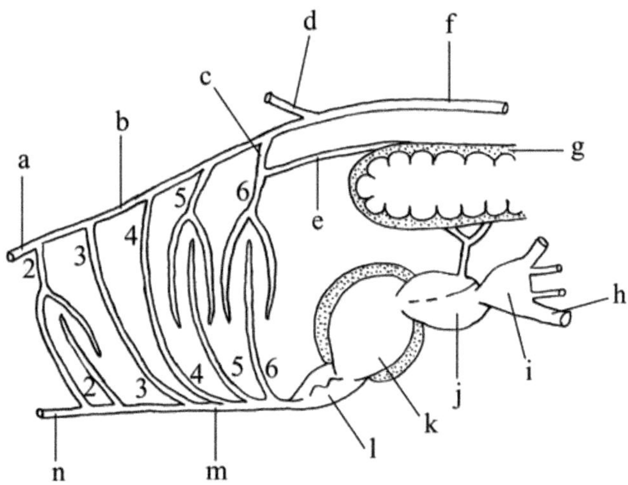

Fig. 10.3 Arrangement of the branchial and pulmonary circulation in the African lungfish, *Protopterus*. (a) internal carotid artery, (b) left dorsal aorta, (c) ductus arteriosus, (d) right dorsal aorta, (e) pulmonary artery, (f) dorsal aorta, (g) lung, (h) vena cava, (i) sinus venosus, (j) partly divided atrium, (k) ventricle, (l) conus arteriosus with a spiral valve, (m) ventral aorta and (n) external carotid artery. The lower and upper level numbers designate afferent and efferent branchial arteries (arches) respectively. The pulmonary vein leads from the lungs to the atrium.

reduced by contraction of the ductus arteriosus so that blood will flow through the pulmonary artery to the lungs. Normal blood flow to gills in hypoxic water can lead to considerable diffusion of oxygen from the gills to water if the branchial circulation is not reduced. The gills still remain important as sites of loss of carbon dioxide and ammonia from the body as well as ion exchange under hypoxic conditions. The pulmonary vein returns oxygenated blood to the heart that has an atrium that is partly divided in lungfishes. The partial division ensures that oxygenated blood from the lungs and deoxygenated blood returning from body tissues mix only partly. The spiral folds of the conus arteriosus direct blood flow to the right direction. From the heart, oxygenated blood flows to the ventral aorta and third and fourth arches that bypass the gills to the dorsal aorta.

Amphibians

The heart of amphibians has two atria and one ventricle. The two atria receive blood from pulmonary and systemic circulatory pathways. There is mixing of oxygenated and deoxygenated blood in the single ventricle. The presence of internal ridges in the ventricle reduces the mixing of blood from the two circuits as the ridges direct most of the blood from the two pathways to the right path. In frogs, about three quarters of the blood that joins the systemic circulation is oxygenated. Since gas exchange takes place in the amphibian skin, this organ also forms part of the pulmonary circulation.

Reptiles

The reptilian heart is four chambered. The sinus venosus is mainly incorporated into the right atrium of most reptiles. The reptilian ventricle is partly divided in non-crocodilian reptiles while it is completely divided in crocodiles (Fig. 10.4). The reptilian conus arteriosus is divided into two aortic trunks and one pulmonary trunk. In crocodilian reptiles, the pulmonary trunk that carries deoxygenated blood leaves the right ventricle then divides into the left and right pulmonary arches that lead to the lungs. One of the aortic trunks runs from the left ventricle after which it gives rise to the right systemic arch and carotid arches. The other aortic trunk leaves the right ventricle before giving rise to the left systemic arch. The two aortic trunks of crocodilians communicate after emerging from the heart ventricles through the **foramen of panizza**. Foramen of panizza is open when the animal is breathing thus allowing oxygenated blood from the left ventricle to enter the left systemic arch since the valve between the right ventricle and the aortic trunk is closed. The right systemic arch also receives oxygenated blood from the left ventricle while the pulmonary trunk will contain deoxygenated blood coming from the right ventricle during this time.

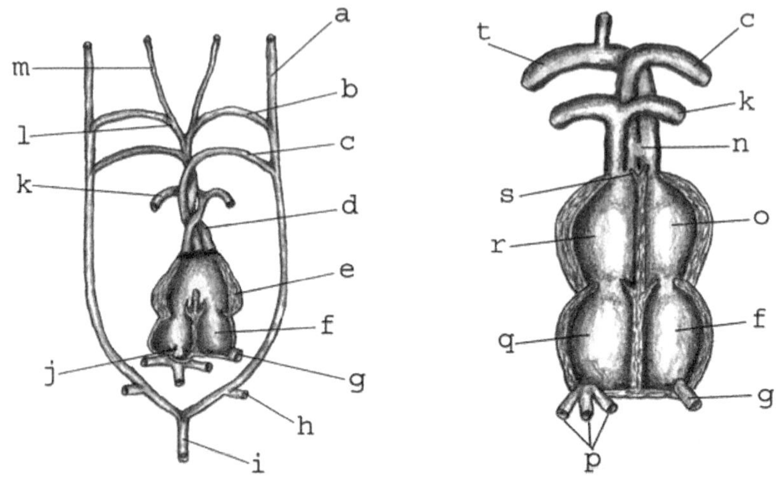

Fig. 10.4 Ventral views of the heart and aortic arches of a reptile (left) and the heart of a crocodile (right). The structure of the reptilian heart shown in the diagram to the left does not apply to crocodiles. (a) internal carotid artery, (b) third arch that contributes to the internal carotid artery, (c) fourth left systemic or aortic arches, (d) the three divisions of the ventral aorta that include the origin of the pulmonary artery and right as well as left systemic arches, (e) ventricle, (f) left atrium, (g) pulmonary vein, (h) subclavian artery, (i) dorsal aorta, (j) sinus venosus (incorporated into the right atrium), (k) pulmonary artery, (l) and (m) common carotid and external carotid arteries respectively, (n) foramen of panizza, (o) left ventricle, (p) systemic veins, (q) right atrium, (r) right ventricle, (s) valve that is closed by back pressure and (t) fourth right systemic arch.

Crocodilian lungs are not used for gaseous exchange when the animal is completely under water since some blood is shunted away from the lungs. Blood is also shunted away from the lungs when the reptile is resting. Valves that control blood flow between the right ventricle and the right aortic trunk open during such times. Some deoxygenated blood with a low pH from the right ventricle flows to the left systemic arch while another fraction will undergo a right to left shunt through the foramen of panizza to flow into the left aortic trunk that leads from the left ventricle. The left systemic arch gives off some branches to the stomach and intestines. The acidic blood from the right ventricle is important in secreting hydrochloric acid into the stomach after feeding and when the crocodilian reptile is resting.

Snakes, lizards and turtles have a partially subdivided ventricle that allows for the mixing of oxygenated and deoxygenated blood to a certain degree. The mixing of the two types of blood can be avoided when necessary. Blood flows from the ventricles to the two aortic arches through the **cavum venosum** where the interventricular septum is incomplete (Fig. 10.5). The cavum venosum communicates with both

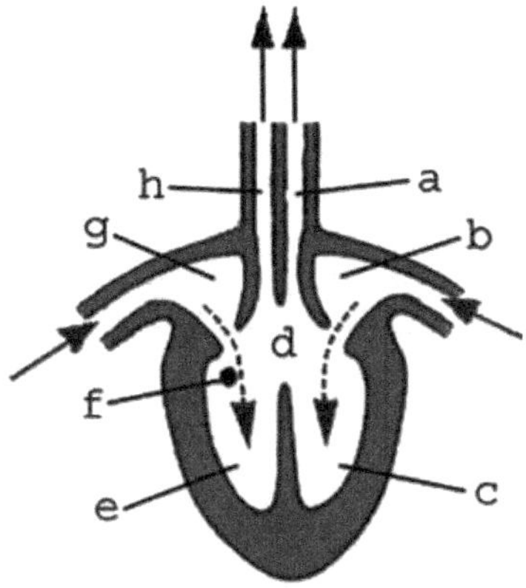

Fig. 10.5 Ventral view of a non-crocodilian heart. (a) Left aortic trunk that leads into the right aortic arch, (b) left atrium, (c) cavum arteriosum, (d) cavum venosum, (e) cavum pulmonale, (f) entrance to pulmonary trunk, (g) right atrium and (h) right aortic trunk that leads to the left aortic arch. Arrows indicate the direction of blood flow.

ventricles. Venous blood flows from the right atrium through the cavum venosum to the right ventricle or **cavum pulmonale** before being pumped into the pulmonary trunk during ventricular systole. The path between the cavum venosum and cavum pulmonale is closed during ventricular systole whereas the passage between the **cavum arteriosum** to the left and the cavum venosum is opened. This action is achieved by the displacement of the septal wall as well as the valve between the cavum venosum and cavum pulmonale. Ventricular systole is also accompanied by flow of blood from the cavum arteriosum to the cavum venosum before moving into the aortic trunks that lead to the right and left aortic arches. There is no mixing of oxygenated and deoxygenated blood during this period when lungs are used for gaseous exchange.

Aquatic reptiles such as turtles and some snakes that do not breathe for a long time when they are under water, employ the right to left shunt. During this process, some of the deoxygenated blood in the right atrium is not pumped into the lungs but flows to the right side of the heart into the two aortic trunks. The deoxygenated blood is recycled throughout the body in the process. The right to left shunt also occurs when the weather is hot. It is believed that the shunting of some of the warmed blood away from the lungs is a thermoregulatory process that reduces heat loss through lungs during exhalation when the weather is cold. When resistance to blood flow is low in pulmonary arteries, there is a left to right shunting (reverse shunting) that returns oxygenated blood to the lungs. The left to right ventricular shunt ensures that blood is saturated with oxygen.

The demands of amphibians and reptiles on their circulatory systems are not as high as those of endothermic birds and mammals. The reptilian circulatory system is able to adjust the amount of blood flowing to the lungs to match the level to which the lungs are in use thus optimizing on energy expenditure. In squamates and turtles, lungs can receive up to 60% of the blood being pumped from the heart when lungs are being used for breathing during which they offer less resistance to blood flow. Blood returning from the lungs can be recycled through the organ in the process.

Birds and Mammals

The hearts of birds and mammals have ventricles that are completely partitioned by the interventricular septum (Fig. 10.6). Since the pulmonary and systemic circuits are completely separated, a heart in these groups of vertebrates is like two hearts that have been mounted side by side. Oxygen rich blood flowing to the left atrium from pulmonary veins is pumped to the left ventricle and will circulate unmixed to the aorta then to systemic circulation. Blood flowing to the right atrium from systemic circulation via veins is deoxygenated. It will circulate to the right ventricle that will pump it to the lungs through the pulmonary artery.

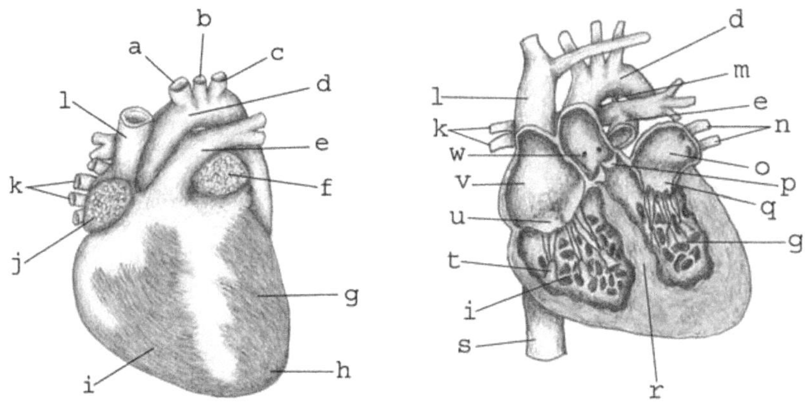

Fig. 10.6 Anterior surface view of a human heart (left) and the interior of the heart (right). (a) brachiocephalic trunk, (b) left common carotid artery, (c) left subclavian artery, (d) arch of aorta, (e) pulmonary trunk, (f) auricle of left atrium, (g) left ventricle, (h) apex, (i) right ventricle, (j) auricle of right atrium, (k) right pulmonary veins,(l) superior vena cava, (m) ligamentum arteriosum, (n) left pulmonary veins, (o) left atrium, (p) aortic semilunar valve, (q) left atrioventricular (mitral) valve, (r) interventricular septum, (s) inferior vena cava, (t) papillary muscle, (u) right atrioventricular (tricuspid) valve, (v) right atrium and (w) opening into coronary artery. Auricles (L. *auricula*, external ears) are flaps that extend from each atrium.

The atria have thin walls and are also referred to as the 'receiving chambers' as they receive blood from veins. The right atrium receives deoxygenated blood from mainly the anterior or superior and caudal or inferior venae cavae whereas the right atrium gets oxygen rich blood from several pulmonary veins. Ventricles are considered as the 'pumping chambers' as they have to pump blood to arteries and body tissues. Since they pump blood for a much longer distance than atria, the ventricular walls have a much thicker myocardium. The myocardium of the left ventricle is thicker than that of the right ventricle as the left ventricle has to pump blood to most of the body whereas the right ventricle pumps blood to lungs. Whereas lower vertebrates can completely regenerate their heart, cardiac injury in higher vertebrates generally leads to progressive failure and the identification of cardiac stem cells suggests that some endogenous repair mechanisms actually exist in the mammalian heart (Germani et al., 2007).

There are four sets of valves in the mammalian heart. Valves act as mechanical devices that allow the flow of blood in one direction only. Two **atrioventricular valves** surround the openings between the two atria and ventricles or **atrioventricular orifices**. The atrioventricular valve guarding the right atrioventricular orifice has three flaps (cusps) and is known as the **tricuspid valve**. The valve that is found between the left atrium and ventricle has two flaps and is known as the **bicuspid** or **mitral valve**. When ventricles contract, ventricular blood is pushed against the valve flaps forcing them to close, as blood is pumped to the pulmonary artery and aorta. **Semilunar valves** are half-moon shaped flaps that are located at the beginning of the pulmonary artery and aorta.

Blood flowing from the head, neck and anterior or upper extremities and thoracic cavity (with the exception of lungs) drains into the **anterior** or **superior vena cava**. Venous blood from the posterior or lower parts of the body and abdomen flows into the **inferior** or **posterior vena cava**. The venae cavae drain blood into the right atrium. Blood from the right atrium then flows through the right atrioventricular valve into the right ventricle from where it will flow through the pulmonary semilunar valve into the pulmonary trunk. The pulmonary trunk divides into the left and right pulmonary arteries that supply the gas exchange tissues of lungs. Blood flows from the gas exchange tissues of lungs to the left atrium via pulmonary veins. Blood flows from the left atrium through the left atrioventricular valve into the left ventricle from where it will flow into the aorta through the aortic semilunar valve.

The two **coronary arteries** are the first branches of the aorta and supply the heart wall with blood. Coronary arteries are located at the beginning of the aorta above the aortic semilunar valves. Most blood returning from the wall of the heart drains into the **coronary sinus** that leads into the right atrium. The aorta makes a 180° turn, forming a curved segment known as the **aortic arch** and follows a posterior or inferior course through the major body cavities. Arteries that supply the body with blood will branch off the aorta.

In birds, the left systemic (aortic) arch has been lost leaving the right one to form the aortic arch. The subclavian arteries that supply the flight and wing muscles are quite large in birds. The left fourth aortic arch is present in mammals and forms the arch of the aorta although six pairs of embryonic aortic arches on each side of the body were present during embryonic development (Fig. 10.7).

FETAL CIRCULATION IN MAMMALS

Fetal circulation (Fig. 10.8) differs from circulation after birth in mammals in various ways. The fetus is totally dependent on the maternal circulatory system for nutrients and gaseous exchange since respiratory and digestive systems are not fully developed. The fetus relies on umbilical blood vessels that are present in the umbilical cord to carry out these vital functions. In most mammalian species, exchange of gases, nutrients and wastes takes place in the placenta without actual mixing of maternal and fetal blood.

There are several structures in the fetus that perform the necessary circulatory functions at this stage but are vestigial or lacking in mammalian life after birth. The **umbilical arteries** branch off the

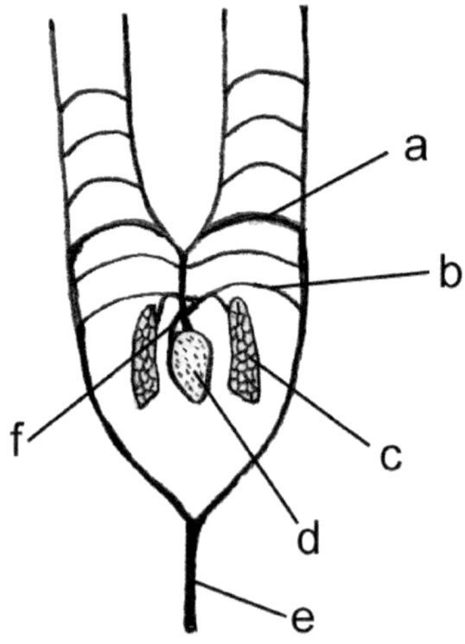

Fig. 10.7 Aortic arches. (a) fourth left arch, (b) ductus arteriosus, (c) left lung, (d) heart, (e) dorsal aorta and (f) pulmonary trunk.

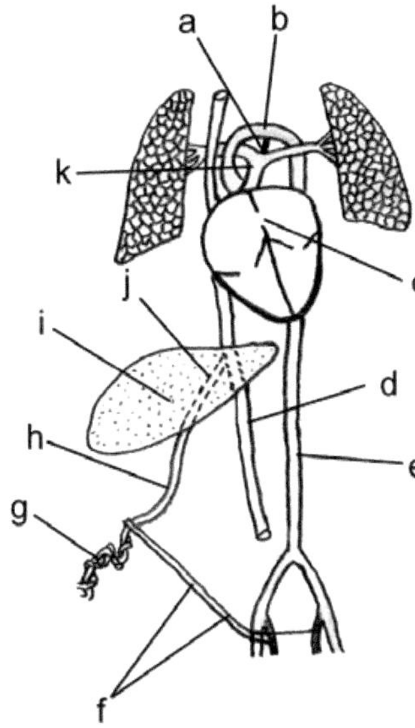

Fig. 10.8 Mammalian fetal circulation. (a) ductus arteriosus, (b) aortic arch, (c) foramen ovale, (d) posterior (inferior) vena cava, (e) abdominal aorta, (f) umbilical arteries, (g) umbilical cord, (h) umbilical vein, (i) liver, (j) ductus venosus and (k) pulmonary trunk.

maternal internal iliac arteries and carry deoxygenated blood and fetal metabolic wastes to the placenta. The **umbilical vein** conveys oxygenated blood from the placenta to the fetus. The vein has some small branches that supply the liver but most of the blood bypasses this organ and flows into the **ductus venosus** that joins the caudal (inferior) vena cava that leads to the heart. The interatrial septum has an opening known as **foramen ovale** that has a valve at the opening of the anterior (superior) vena cava into the right atrium. The valve plays an important role of channeling most of the blood from the right to the left atrium so that it bypasses the non-functional fetal lungs. The small volume of blood that is pumped out of the heart from the right ventricle to the pulmonary trunk is further reduced as it is diverted to the aorta to join systemic circulation by the **ductus arteriosus** that connects the pulmonary artery with the aortic arch. Almost all the blood in fetal circulation is a mixture of oxygenated and deoxygenated blood. Fetal hemoglobin has a higher affinity for oxygen than adult hemoglobin.

After birth, the umbilical cord is severed and the placenta together with the blood vessels of the umbilical cord cease to function. The parts of these vessels that remain in the body become fibrous cords that will remain throughout life. The ductus venosus is transformed into a ligament of the liver. The foramen ovale is closed some time after birth in mammalian species. The ductus arteriosus will contract after birth and will turn into a fibrous cord.

TYPES OF BLOOD VESSELS AND THEIR STRUCTURE

Blood vessels include arteries, veins and capillaries. Arteries transport blood away from the heart. In higher vertebrates, all arteries with the exception of pulmonary artery transport oxygenated blood after birth. Veins return blood back to the heart from systemic circulation. Apart from pulmonary veins of higher vertebrates, all other veins carry deoxygenated blood from body tissues. Small arteries are referred to as arterioles whereas small veins are known as venules. Veins normally contain valves that comprise two flaps and resemble and function as the semi-lunar valves of the heart. Large venous spaces are referred to as sinuses such as the coronary sinus that drains most of the cardiac circulatory blood to the right atrium.

Capillaries are microscopic and are the smallest vessels of the circulatory system that link arterioles to venules. Some structures in the body such as the heart, kidneys, skeletal muscles and endocrine organs have a high blood supply whereas others including tendons, fasciae, ligaments and bones are bradytrophic and have a low blood supply. Cartilage and the cornea of the eye lack a blood supply and depend on diffusion of material from neighboring tissues. The liver and a few other structures have irregular microscopic spaces between cordlike cells that play the role of capillaries and are known as **sinusoids**.

Arteriovenous Anastomoses and Vasa Vasorum

Arteriovenous (AV) anastomoses (Fig. 10.9) are special vascular segments that link arterioles directly to venules thus preceding the capillary bed. These vessels regulate the diameter of their lumina through contraction and relaxation of smooth muscle fibers in their walls. AV anastomoses shunt blood directly from arterioles to venules when they are dilated. Such blood does not flow through the capillary network. Contraction of smooth muscle in AV anastomoses reduces the diameter of their lumina that can be completely closed. Blood will then take its normal course by flowing through capillaries to venules. AV anastomoses are regulated by the autonomic nervous system. These vessels are found in salivary glands, the wall of the gastrointestinal tract, lungs, nasal mucosa, skin, endocrine organs, placenta, uterus, ovary and penis.

Vasa vasorum ('vessels of vessels') are small blood vessels that supply the walls of the larger blood vessels with a diameter that is greater

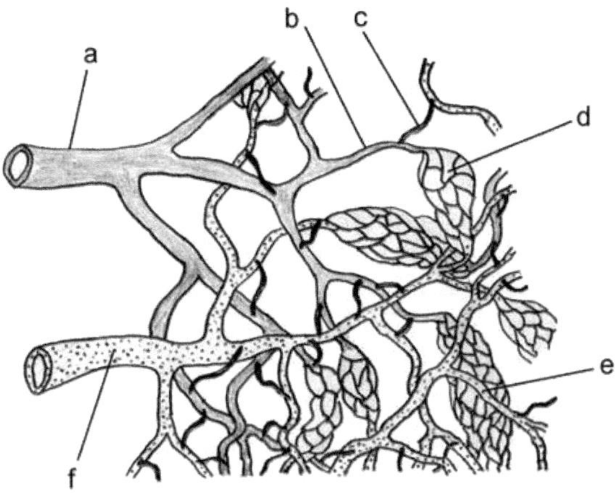

Fig. 10.9 Arteriovenous (AV) anastomoses or shunts in a circulatory system. (a) small artery, (b) arteriole, (c) AV shunt, (d) capillary bed, (e) venule and (f) small vein.

than 1.0 mm. The vessels penetrate the blood vessel wall from the outside and branch into a network of capillaries in the middle and outer layers of the wall.

Wall of Blood Vessels

There are generally three layers (tunics) in the walls of arteries and veins (Fig. 10.10). Corresponding arteries have thicker walls than equivalent veins. The three layers of the wall can be distinguished in arterioles and venules. The outermost layer of a blood vessel is known as the **tunica adventitia** (L. *tunica,* coating; *adventitia,* come first) and contains collagen and elastic fibers that make it strong and flexible. Such a wall keeps blood vessels open and prevents their tear during movement. The tunica adventitia is the thickest of the three layers in veins.

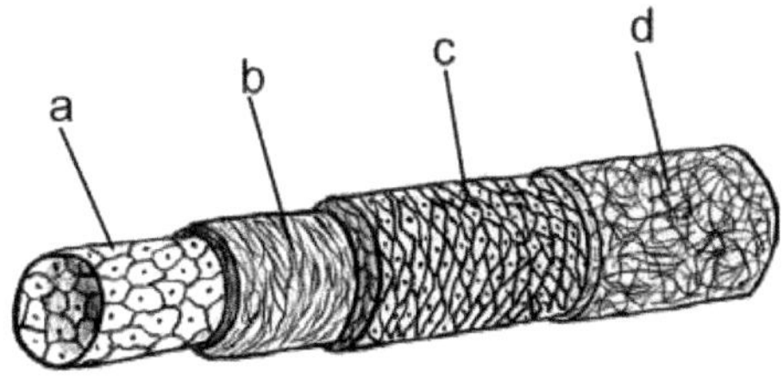

Fig. 10.10 The layers of a blood vessel. (a) internal layer or tunica intima, (b) elastic and (c) muscular layers of the middle layer or tunica media and (d) the outer layer (tunica adventitia).

The middle layer, the **tunica media**, lies between the outer and inner layers of the wall. It is made up of smooth muscle and elastic fibers. Smooth muscle fibers are innervated by the autonomic nervous system and bring about changes in diameter of blood vessels. In the quite distensible elastic (conducting) arteries such as aorta and pulmonary arteries, the tunica media contains a lot of elastic fibers whereas the smaller muscular arteries contain less elastic tissue but more smooth muscle fibers. Smooth muscle fibers in the wall of arterioles play the role of pre-capillary sphincters just before the arterioles give rise to capillaries. The arterioles are able to regulate the amount of blood flowing through capillaries by acting as regulatory valves when they contract and relax.

The innermost layer of a blood vessel wall is the **tunica intima** and consists of an endothelium with a single layer of rhomboid cells that is continuous with the endocardium of the heart. The endothelium rests on an elastic membrane and between these two layers are a few collagen fibers. The endothelium of arteries is a smooth lining whereas that of veins has valves (Fig. 10.11). Valves prevent a backflow of blood in veins towards capillaries since pressure is much lower in these vessels than arteries.

Capillaries

Capillaries are about 1.0 mm long and 3.0 μ to 10 μ in diameter. The internal diameter of capillaries increases from the arteriole end to the beginning of venules. Capillaries have only a thin endothelium that lies on a basal layer in its wall (Fig. 10.12). Diffusion of material between the circulatory system and tissue cells occurs across the thin walls of capillaries. The three types of capillaries are continuous, fenestrated and sinusoidal capillaries. **Continuous capillaries** are the most common

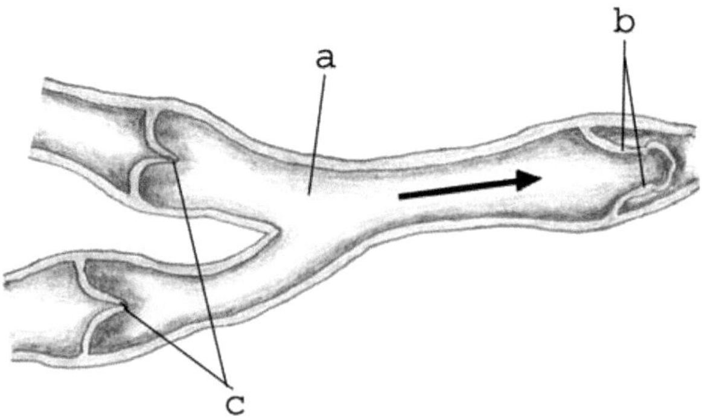

Fig. 10.11 The inside of a vein showing valves. (a) lumen, (b) open valves and (c) closed valves. The arrow indicates the direction of blood flow.

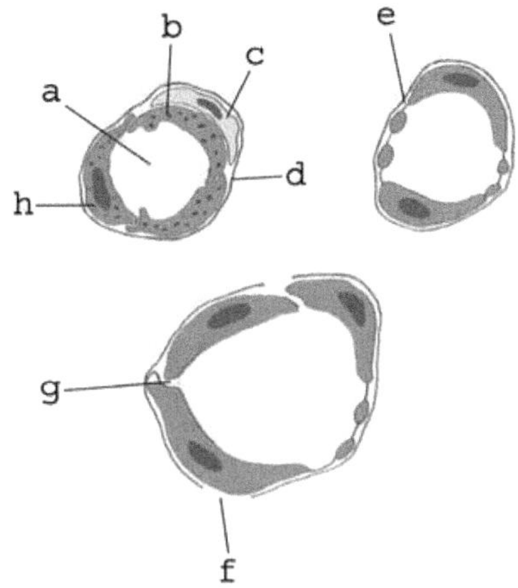

Fig. 10.12 A cross-section of the various types of capillaries. Continuous (top left), fenestrated (top right) and sinusoidal (bottom) capillaries. (a) lumen, (b) pinocytotic vesicle, (c) pericyte or pericapillary cell, (d) basal lamina, (e) fenestration, (f) gap in basal lamina, (g) intercellular cleft and (h) endothelial cell. Pericytes (also known as pericapillary, adventitial, perivascular or perithelial cells) are elongated pluripotential cells that have the ability to contract and wrap themselves to the outside of precapillary arterioles, certain continuous capillaries and postcapillary venules within the basal lamina. Pericytes can differentiate into smooth muscle cells, macrophages and fibroblasts.

capillaries and their endothelial cells are connected by tight junctions that lack large intercellular clefts between the endothelial cells. Vesicles that transport material into and out of capillaries are present within the endothelial cells. Lipid soluble material can be transported through the cells in these vesicles whereas water and ions are transferred in the clefts between the cells. The brain capillaries seem to use vesicles for transport of material within capillary cells.

Fenestrated capillaries (L. *fenestra,* window) have pores within capillary cells. The cells rest on a basement membrane and have vesicles. The pores enhance the flow of fluid and solutes across the capillary wall except large molecules and red blood cells. Fenestrated capillaries are found in kidney glomeruli, the gastrointestinal tract and certain endocrine organs. There are larger gaps between the endothelial cells of **sinusoidal capillaries**. The endothelial cells of sinusoidal capillaries have fenestrations and lack vesicles. Since the basement membrane of such capillaries is incomplete or lacking, a lot of material can be transported through the gaps in sinusoidal capillaries. Sinusoidal capillaries are found in the liver, bone marrow, spleen and some endocrine organs.

Portal System

A portal system is a blood vessel that starts from one capillary bed and terminates in another capillary bed instead of leading to the venous system. There are three portal systems in higher vertebrates. The **hepatic portal vein** transports freshly absorbed nutrients from the gastrointestinal tract to the liver. The **efferent renal arteriole** arises from the glomerular capillaries and leads to the peritubular capillaries that supply the tubular system of kidneys. In the brain, the **hypothalamohypophyseal portal veins** originate from the capillary bed in the hypothalamus to terminate in another capillary bed in the pituitary gland.

Nerve Supply to Blood Vessels

The smooth muscle fibers of blood vessels receive a nerve supply from autonomic vasomotor nerves. The nerves form a coarse network in the outer layer of vessels before dividing into ramifications that penetrate deeper into the vessels. Stimulation of sympathetic and parasympathetic nerves that supply these muscle fibers results in vasoconstriction and vasodilation of blood vessels respectively. Blood flow can be adjusted in the vertebrate body by balancing the extent to which these two nervous pathways are stimulated. The small arteries and arterioles are innervated to a greater level than the larger blood vessels. Some sensory nerves that act as baroreceptors innervate a few blood vessels. Baroreceptors detect stretch in these vessels and are able to monitor blood pressure.

Lymphatic System and Lymphoid Organs

Although the lymphatic system is perhaps secondary to the blood vascular system in mammals, the later system seems to develop on the basis of an ancestral lymphatic system with lymph hearts (Wilting et al., 2004). The chief function of the lymphatic system of teleosts and tetrapods is the maintenance of fluid balance in the internal environment by returning excess fluid and some proteins to the cardiovascular system. The lymphatic system is therefore important in the vital homeostatic role of ensuring a return of plasma proteins to the circulatory system to maintain the correct balance between the blood's osmotic and hydrostatic pressures that are vital in exchange of water between the body's compartments. The system also absorbs the end products of fat digestion in the small intestines. The lymphoid organs of the lymphatic system also play a major role in the body's immunity.

The lymphatic vessels begin blindly in the intercellular spaces of body tissues as lymphatic capillaries (Fig. 10.13). The central nervous

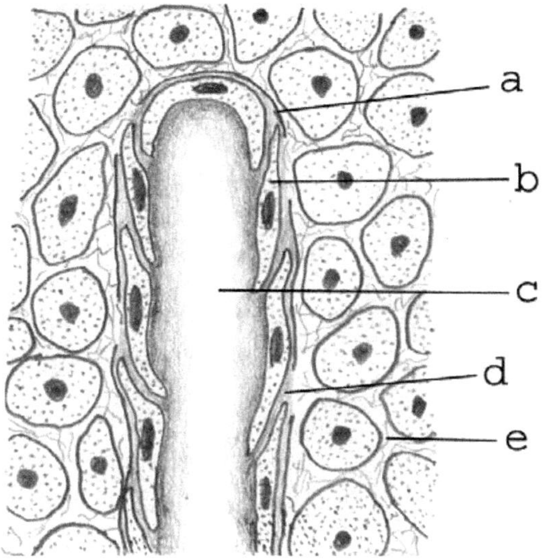

Fig. 10.13 A lymphatic capillary that starts blindly in intercellular spaces. (a) discontinuous basal lamina, (b) endothelial cell, (c) lumen, (d) cleft between neighboring endothelial cells and (e) intercellular space with collagen and reticular fibers. The simple squamous epithelial cells of the wall of a lymph capillary overlap and form valve-like structures that permit fluid to enter vessel but prevent fluid from flowing back to intercellular spaces.

system, eye and inner ear are not drained by the lymphatic system but are drained by cerebrospinal fluid, humors of the eye and endolymph respectively. Lymphatic drainage is also lacking in cartilage, bone marrow, teeth, central parts of the spleen and liver and structures that lack blood vessels such as the epidermis. The lymphatic system is not a closed circuit system of circulation as is the case with the cardiovascular system since it is not continuous throughout.

The wall of a lymphatic capillary is made of a single layer of flattened endothelial cells that are thinner than those of capillaries. Larger clefts than those found in blood capillaries are found between the endothelial cells of lymph capillaries. The clefts together with the underlying thin and discontinuous basal lamina make lymphatic capillaries quite permeable to large molecules that cannot cross directly into the blood capillaries. The large molecules are then able to join the cardiovascular circulation via the lymphatic system.

Lymphatic capillaries branch and anastomose much in intercellular spaces and are attached to tissue cells by filaments. The capillaries will keep joining up to form larger lymphatic vessels of the body that eventually drain into the major veins near the heart where pressure is low. Lymphatic vessels have thinner walls than arteries and veins though the three layers of the blood vessel wall are present. Semilunar valves are more numerous in mammalian lymphatic vessels than veins. Lymph nodes are found along the course of lymphatic vessels, especially in mammals, but are lacking in fish. Movements of the body promote the flow of lymph. Amphibians, reptiles and bird embryos as well as some adult birds including the ostrich, cassowaries, gulls, storks and some passerines have **lymph hearts** that aid in movement of lymphatic fluid. Lymph heart is also responsible for preventing embryonic edema in birds during embryogenesis (Valasek et al., 2007). Lymph hearts comprise segments of smooth muscle fibers in the wall of lymphatic vessels.

Lower fish such as lampreys and hagfishes and cartilaginous fish lack a true lymphatic system but have vessels that drain tissues known as hemolymphatic vessels. **Hemolymphatic vessels** form more numerous connections with veins than is seen in other fish and terrestrial vertebrates. The vessels are similar to lymphatic vessels and sometimes contain some erythrocytes. Hemolymphatic vessels could be the precursors of the true lymphatic system. The lymphatic system of teleosts and tetrapods has evolved independently.

Lymphoid (lymphatic) organs such as spleen, thymus, lymph nodes and tonsils produce lymph and lymphocytes as well as possessing cells that are capable of phagocytosis and are an important component of the body's immune system. The chief function of lymphoid organs in the vertebrate body is defense. Agnathans lack the lymphoid organs that play a role in adaptive immune responses against local infections but have an innate immune system. The other non-mammalian vertebrates lack organized lymph nodes but have a thymus and spleen as major lymphoid organs as well as a reticuloendothelial system. The non-lymphoid/hematopoietic part of lymphoid and hematopoietic organs has played a role in evolution of these vertebrate organs (Zapata et al., 1995). The thymus, spleen and alimentary canal lymphoid organs were the first to appear in vertebrates then lymph nodes and bone marrow followed much later.

Lymphoid organs have some general characteristics in vertebrates. The organs have the resident reticular cells that possess long cytoplasmic processes that run in various directions and make contact with those of other neighboring reticular cells leading to formation of a three-dimensional meshwork that has spaces between the cells. These cells are supported by reticular fibers. Reticular cells engulf foreign material from lymphatic circulation in spaces between these cells and are part of the **reticuloendothelial system** that consists of phagocytic cells that line sinusoids in lymphoid and related organs. Lymphoid organs normally have free cells that include lymphocytes, macrophages, plasma cells, granulocytes and mast cells. Lymphocytes are usually arranged in spherical accumulations known as **lymph follicles** or lymph nodules (Fig. 10.14) except in the thymus where they are arranged in a diffuse manner. Lymphocytes mainly produce antibodies. In some vertebrates, lymphocytes accumulate in the mucous membranes of the digestive, respiratory and urogenital tracts as part of the local immunological defense mechanism. In the mammalian intestines are large accumulations of lymphocytes that are interconnected and known as Peyer's patches.

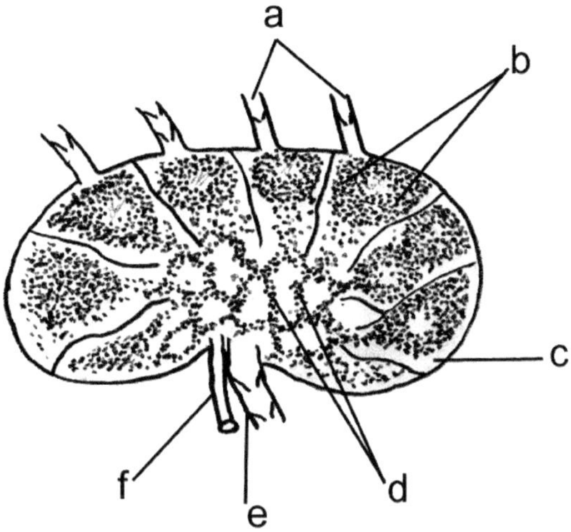

Fig. 10.14 Internal structure of a lymph node. (a) afferent lymph vessels with valves, (b) lymph follicle (nodule) with a less dense germinal center where B-lymphocytes begin their final stages of maturation, (c) sinus through which lymph flows, (d) medullary cords comprising lymphocytes, (e) efferent lymph vessel and (f) blood vessel. Lymph nodes filter lymph by removing wastes and some fluid as well as destroying bacteria and cancerous cells.

Immune System

The immune system of vertebrates has the role of recognizing the normal body antigens from foreign substances that have to be eliminated by the body defense system. The whole process of inflammation is a chemical reaction between an antigen and an antibody. Foreign substances to the body can be microorganisms, their toxins, foreign tissues or abnormal materials from the body itself such as cancerous cells. Each antibody is specific to a certain antigen.

Immunologists study the immune system and normally divide the system of vertebrates into innate and adaptive types. **Innate immune systems** respond rapidly to challenges but subsequent stimulation will not be accompanied with an enhanced response. The cellular components of an innate immune response include granulocytes and macrophages. Such an immune system is said to be amnestic or lacks a memory. An **adaptive immune system** is said to be anamnestic or have a memory since the body is able to adapt to the antigenicity of the environment. The adaptive system is enhanced by the presence of T- and B-lymphocytes that produce cell-mediated immunity and antibodies respectively. T-cells mature in the thymus from precursor cells that migrate to this primary lymphoid organ from haemopoietic tissues. During a cell-mediated response, the T-cells react with foreign antigens on the surface of the invading cell they locate using their surface receptors resulting in the destruction of the invading cell. B-cells that are produced in the bone marrow, fetal liver and bursa of Fabricius located above the cloaca in birds are activated into antibody producing cells, the most mature of which are referred to as **plasma cells**. Plasma cells are large and are rich in rough endoplasmic reticulum.

Agnathans do not produce antibodies (immunoglobulins) and also lack T-cell receptors but have lymphocyte-like cells. Molecular markers that are specific for cellular components of the adaptive immune system in lampreys have been expressed moderately in presumed lymphoid tissues such as the gut epithelium and high levels of expression have been detected in the gonads especially the ovary (Mayer et al., 2002). The jawed vertebrates possess B cells. T cells are present in bony fishes and higher vertebrates and might be present in cartilaginous fish as the molecule that presents part of the antigen to T cells for stimulation known as **major histocompatibility complex** (MHC) is present in cartilaginous fish. Reptiles do not have organized lymph nodes but have lymphoid aggregates in various organs. The reptilian inflammatory response including the synthesis of immunoglobulins is dependent on temperature. Some reptiles have a low immune response in winter as the splenic lymphocytes undergo lymphocytolysis during this season. It has been proposed that evolution of the mammalian immune system has resulted in cellular interactions at sites of injury during inflammation that have brought about tissue defense such as scarring or fibrosis and promoted tissue repair. Such an immune response has been accompanied by loss of regenerative capacity in higher vertebrates (Mescher and Neff, 2005).

BLOOD

The major transport medium of the vertebrate body is blood. The relative volume of blood in the closed circulatory system of vertebrates is much lower than that in the open system of circulation of most invertebrates. The survival of vertebrates depends on the continuous transportation of nutrients and oxygen to tissue cells and removal of metabolic wastes and carbon dioxide to elimination sites. Blood also transports hormones, enzymes and various substances throughout the body for metabolic functions. In homeotherms, blood plays a major role in heat regulation of tissues since it is able to transfer heat from the body to its surface without much increase in body temperature. Blood thus plays an important role in homeostasis of the entire internal environment. Blood is composed of the liquid part known as plasma and a cellular component that has various types of cells.

Blood Cells

The various blood cells are also known as the formed elements of blood and include erythrocytes (red blood cells), leukocytes (white blood cells) and platelets (thrombocytes). Blood cells form a large proportion of blood and this varies in different vertebrates. In the human being, blood cells form about 45% of blood with plasma forming the remaining 55%. Most of the formed elements are red blood cells that form 99% of the blood cell volume with white blood cells and platelets forming about 1% of the other volume.

Red Blood Cells (Erythrocytes)

Mammalian red blood cells or erythrocytes (Gr. *erythros,* red; *kytos,* cell) lack a nucleus and are biconcave round disks whereas those of other non-mammalian vertebrates are nucleated and oval shaped (Fig. 10.15). The red blood cells of the camel and related species are oval. An immature mammalian red blood cell contains a nucleus that is extruded just before the cell joins the circulatory system from the bone marrow. The mature mammalian erythrocyte also lacks organelles that are typical of most body cells. The main component of a red blood cell is the red pigment hemoglobin that is important in transport of gases and accounts for about a third of the red blood cell volume in mammals. The number of erythrocytes present in the blood is far more than that of white blood cells. In the human being, the number of erythrocytes is about 5,000,000 per cubic millimeter of blood. These cells together provide a very large surface area for exchange of gases that in the human being could be as large as a football field.

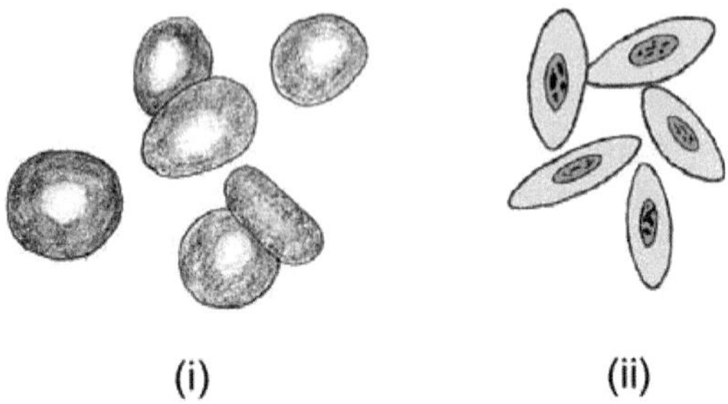

Fig. 10.15 (i) Mammalian and (ii) non-mammalian erythrocytes. The anucleate mammalian erythrocytes have thinner centers and their rims are thicker.

The lower mammals show greater variation in size, shape and staining of erythrocytes. The blood cells of lower mammals also become distorted more easily if removed from the body as happens with the blood of younger mammals when compared to older individuals of the same species. The appearance of red blood cells of lower mammals and also younger individuals in a group is affected more seriously than in higher and older groups by minor nutritional variations.

Most fish have round or oval and nucleated yellowish erythrocytes. The nuclei vary from round to rod-like in shape. Some fish, including some Antarctic and deep-sea species and larvae of eels, lack erythrocytes. The Antarctic species that lack erythrocytes inhabit highly oxygenated water and have low metabolic rates. The species of fish without red blood cells depend on diffusion for gaseous exchange. The hemoglobinless icefish which are about 16 species, such as the blackfin icefish (*Chaenocephalus aceratus*) (Fig. 10.16) lack respiratory pigments including hemoglobin and have had structural adjustments to meet their oxygen requirements. The radius of capillaries is wider and the fish has a relatively strong heart that is comparable in weight to that of mammals of equivalent weight. The cardiac fibers have high mitochondrial densities and short oxygen diffusion distances between the ventricular

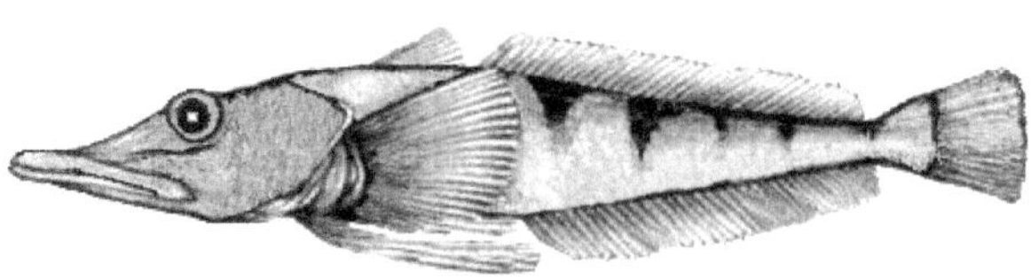

Fig. 10.16 The blackfin icefish, *Chaenocephalus aceratus*, a hemoglobinless Antarctic icefish from the Southern Ocean. Icefish are perciformes that look pale in color and live in waters at subzero temperatures. The large and strong hearts of icefish pump large quantities of blood at low pressures. Icefish can attain weights of 3.5 kg and lengths of 60 cm.

lumen and the mitochondrial membrane and also contain a lot of lipid material that enhances oxygen diffusion since oxygen is more soluble in fat when compared to the cytoplasm (O'Brien and Sidell, 2000). Such a structure ensures that the heart is able to receive enough oxygen to maintain its functions. The icefish normally feed on smaller fish and rest on the seabed waiting for their prey to swim closer thus expending minimal energy in looking for food.

The long diameter of fish erythrocytes ranges from 7.0 m in many fishes to 36.0 m in the African lungfish and this is several times the diameter of mammalian erythrocytes. The proportion of hemoglobin in the blood also varies from 37 to 79% of the dry weight of erythrocytes in many teleosts. Fish red blood cells counts are comparatively low in relation to other vertebrates. Quite often hemoglobin is present in the protoplasm of fish red blood cells in the form of granules.

Amphibian erythrocytes are oval in most species as are the nuclei that can also be irregular in shape. Amphibian erythrocytes are the largest of all vertebrate red blood cells. The cells can be as long as 70.0 μ by 41.0 μ in the three-toed salamander *Amphiuma tridactylum*. Some amphibian species have round erythrocytes. Occasionally, non-nucleated erythrocytes may be present in amphibian blood. The nucleus of the amphibian erythrocyte has a net-like arrangement known as the **network of linin** that comprises viscous filamentous material that interconnects chromatin granules. Amphibians show greater seasonal variation in red blood cell counts than any other vertebrate group. The seasonal variation is greater within the same animal than in blood from animals belonging to different genera at the same time.

The red blood cells of reptiles are smaller than those of amphibians and are oval shaped. The oval shaped reptilian erythrocyte contains a nucleus that also has a linin network. Hemoglobin is located mainly around the edges of the reptilian erythrocyte. Seasonal variation in blood cell counts are less marked than is the case in amphibians.

The oval and nucleated avian erythrocytes are much larger that those of reptiles. The nuclei are of the same general form as the cells. Round or non-nucleated erythrocytes are rare in birds. Avian red blood cells are of more uniform size and form than is the case in lower vertebrates. Hemoglobin is located mainly at the periphery of the avian red blood cell and the protoplasm is more efficient as an oxygen-carrier than that of reptilian erythrocytes.

White Blood Cells (Leukocytes)

The total number of leukocytes per certain volume of blood varies more between individuals and the same individual at different times in lower vertebrates than in the human being. The white blood cells form about 1% of the total blood volume. Leukocytes (Fig. 10.17) are classified according to the presence or absence of granules in their cytoplasm into granulocytes and agronulocytes.

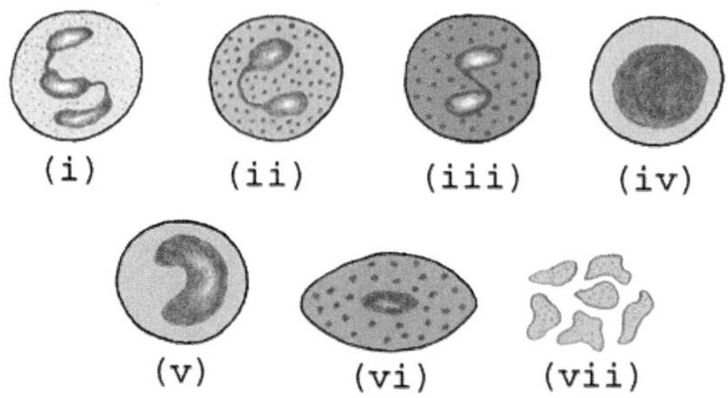

Fig. 10.17 Other blood cells of vertebrates. (i) neutropil, (ii) eosinophil, (iii) basophil, (iv) lymphocyte, (v) monocyte (macrophage), (vi) mast cell and (vii) platelets.

GRANULOCYTES

Granulocytes have granules in their cytoplasm. The granules vary in size and staining characteristics depending on the type of granular leucocytes. The three types of granulocytes found in mammals are neutrophils, eosinophils and basophils. Some of the non-mammalian cells granular leukocytes differ from the mammalian granulocytes but still have comparable functions.

Mammalian **neutrophils** have fine and numerous granules in their cytoplasm that stain light purple with neutral dyes. The equivalent cell to the neutrophil in other vertebrates is the **heterophil**. In non-mammalian vertebrate groups, the granules can stain weakly basophilic or weakly eosinophilic. The nucleus of neutrophils has several lobes and as a result neutrophils are also referred to as **polymorphonuclear leucocytes**. Neutrophils are the most numerous granulocytes and account for about 65% of the total leukocyte count in the human being. Neutrophils are highly phagocytic and can leave the vascular system to enter intercellular spaces. The granules in the cytoplasm of neutrophils contain powerful lysosomes. Fish heterophils are spindle-shaped or round and contain acidophilic granules that seem to participate in coagulation of blood. The heterophils of fish often have vacuoles and the granules can fill the cell and push the nucleus to one corner of the cell. The amphibian heterophils have granules that show various staining characteristics. The avian heterophil is the most common white blood cell in some avian species. The cytoplasm has rod shaped granules that stain red-orange and partly obscure the nucleus that has two to three lobes.

Eosinophils or **acidophils** of mammals contain large basic oval cytoplasmic granules that stain orange with acid dyes. The granules are round in human eosinophils. Generally the nuclei of eosinophils have two lobes. Eosinophils in mammals vary in size considerably and account for 2% to 5% of leukocytes. They are also abundant in the wall of digestive and respiratory tracts. Mammalian eosinophils are weak phagocytes and ingest material that is associated with antigen-antibody reaction complexes. Eosinophils play a major role in protecting the mammalian body from parasitic worms. Fish eosinophils vary in size and have large acidophilic granules. Eosinophils are abundant in reptilian blood and their granules vary in size considerably depending on the species. The nuclei of reptilian eosinophils are quite large though some species have small nuclei. Some reptilian eosinophils contain rod-like granules. Avian blood does not have as many eosinophils as reptilian blood. Avian blood contains two types of eosinophils and the more common eosinophil has fine granules and is capable of mitosis while in circulation. The less common avian eosinophil has densely packed rod-shaped granules in the cytoplasm and the shape of the nucleus could be roundish or lobated. The eosinophils of birds are phagocytic and tend to be related to mammalian neutrophils.

The **basophils** of higher mammals contain large cytoplasmic granules that stain dark purple with basic dyes. They are the least common leukocytes and form about 0.5% to 1.0% of the total white blood cell count. The cytoplasmic granules contain the inflammatory chemical histamine and heparin (an anticoagulant) that mediate a hypersensitive reaction. The nucleus tends to be S-shaped. Basophils are rare in fish blood. The type of basophil present in amphibian blood has abundant and fine granules that are sometimes weakly basophilic. The nucleus is normally round or has two lobes. The reptilian basophil has a large nucleus that occupies about two-thirds of the cell and deeply basophilic granules that are of moderate size.

Lymphocytes

Mammalian lymphocytes are the smallest leukocytes and measure about 6.0 μ to 15 m in diameter. They are found in the circulatory system, lymphoid tissues and lymphatic vessels. Most lymphocytes in circulation are in the resting stage. Lymphocytes of lymphoid organs are activated by antigenic stimulation. Lymphocytes normally have large spherical nuclei that are surrounded by little cytoplasm and are the second most numerous white blood cells, forming about 25% of the entire leukocyte count. Lymphocytes are the main constituents of the immune system and defend the body against microbial attack. Mammals have T- and B-lymphocytes. T-lymphocytes mature in the thymus and attack infected or cancerous cells of the body, causing lysis of such cells. B-lymphocytes breed quickly on stimulation into plasma cells that produce antibodies against specific antigens. Production of antibodies is the humoral response of the body's immune system. Mammalian lymphocytes are also divided into two types according to size. The small lymphocytes comprise about 90% of all lymphocytes and measure about 6.5 m in diameter. Large lymphocytes measure about 10-15 m in diameter. The life span of mammalian lymphocytes ranges from several days to years. Fish possess mammalian T- and B-like lymphocytes and differences in various fish species are likely to be evident (Ellis, 1986). Amphibians, reptiles and birds also have T- and B-lymphocytes. In birds, the most numerous white blood cells are lymphocytes.

Monocytes

Monocytes are also known as macrophages and are the largest leukocytes in most mammals, measuring 10-15 m in diameter. Their nuclei are kidney-shaped and the cytoplasm is abundant. Monocytes are phagocytes that engulf microorganisms including cells that have been infected by viruses. They form 2% to 10% of mammalian leukocytes. The avian monocytes are not easily distinguished from the large lymphocytes since they are similar in shape and size. The avian monocytes show irregular indentations.

Platelets (Thrombocytes)

Platelets are small irregular, round or spindle shaped disks of mammalian blood that measure about 2.0 μ to 4.0 m in diameter. Mammalian platelets lack nuclei. They are necessary for the clotting of blood in vertebrates. Platelets adhere to each other and to surfaces they contact as soon as blood is outside the vascular system. The main functions of platelets are haemostasis (Gr. *haima,* blood; *stasis,* a 'standing') and coagulation or blood clotting. Immediately after injury, the blood is exposed to air, platelets stick to the damaged lining of the blood vessel and also to each other. Platelets release factors that promote blood coagulation such as serotonin that causes a reduction in the diameter of the injured vessel, slowing down blood loss in the process. The threadlike fibrin is formed from fibrinogen traps cells. If the injury is not too large and blood pressure not too high, a haemostatic plug will form to stop flow of blood into tissues or to the outside.

Non-mammalian vertebrates have platelets that function in a similar manner to mammalian platelets. The platelets are nucleated and are quite abundant in avian blood. The cells are oval in shape and measure 4.0 μ to 5.0 m in length. The cytoplasm of avian platelets has fibrils that are extruded when blood is exposed to air after an injury. These fibrin threads form net-like masses. Meseguer et al. (2002) have reviewed the possible extra role played by platelets of non-mammalian vertebrates and mammalian platelets in inflammation.

Blood Cell Formation

The process of blood formation is known as **hemopoiesis** or **hematopoiesis**. Since blood cells have a relatively short life span and are incapable of mitosis once formed, a continuous supply of these cells to the circulatory system is necessary. In warm-blooded vertebrates, hemopoiesis takes place in the red bone marrow, the spleen and lymph nodes. In the fetuses of these vertebrates, hemopoiesis also takes place in the yolk sac, mesenchyme, blood vessels, spleen, liver and bone marrow. The red bone marrow is responsible for production of most types of blood cells. The lymphoid organs produce lymphocytes on stimulation by antigens. Many more organs are involved in the manufacture of blood cells in lower vertebrates.

In fish, blood cells are formed from the inner wall of blood vessels and other organs of the body. The diffused agnathan spleen that is located in the submucosa of the digestive tract manufactures blood cells. Other diffused sites in several blood vessels could also be involved in hemopoiesis. The distinct spleen of jawed fish manufactures erythrocytes and thrombocytes in its cortex while lymphocytes and some granulocytes are formed in the medulla. Granulocytes are also formed in the mesonephric kidney, liver, gonads and submucosa of the digestive tract. The mesonephric kidney also manufactures thrombocytes. In teleosts, the spleen is involved in the destruction of red blood cells.

The initial stem cell from which blood cells will form is the **hemocytoblast**. This cell is pluripotential and is capable of differentiating into several different lines of specialized cells. Hemocytoblasts give rise to restricted progenitor cells that have limited ability to develop into various cell lines. Restricted progenitor cells are still relatively undifferentiated. Later on, these cells become differentiated into blood cell precursors and adopt specific lines that lead to maturation of particular functional blood cells. Conserved genetic programs regulate vertebrate hematopoiesis and vasculogenesis (Song et al., 2004). A group of some small proteins known as cytokines have been identified as colony-stimulating factors that are the key to hematopoiesis, modulation of blood cell functional responses, maintenance of homeostasis and overall immune competence (Barreda et al., 2004).

BLOOD PLASMA

The liquid part of blood is known as plasma and lacks blood cells. Plasma is composed of water and solutes. In the human being, 90% of plasma is water and 10% are solutes. The largest proportion of solutes includes proteins that form 6% to 8% of plasma. Other solutes include glucose, amino acids, lipids, hormones, enzymes, urea, uric acid, creatinine, lactic acid and the gases—oxygen and carbon dioxide. There are three main proteins in plasma which include albumins, globulins and fibrinogen. Albumins form most of the proteins (up to 60%) whereas fibrinogen, the chief clotting protein, forms about 7% of the plasma proteins. At the time receptors of adrenal and sex steroid hormones were evolving in protochordates and lower vertebrates, albumins that normally bind to a variety of lipophylic compounds could have regulated access of steroids to their receptors as well as protecting these animals from endocrine disruptors such as phytochemicals, fungal chemicals as well as other chemicals formed by geo-chemical processes (Baker, 2002). Such a process could have favored the evolution of vertebrates with such albumins.

The blood proteins are important in the maintenance of the right osmotic pressure and viscosity of blood. Globulins play a role in the body's immunity. Antibodies are modified gamma globulins. Many blood proteins are manufactured in the liver except antibodies that synthesized in plasma cells.

The solute content of fish plasma varies in freshwater and marine fishes. The freezing point depression has been used to indicate the quantity of solutes and osmotic pressure of fish plasma. The freezing point depression of fish plasma is lowest in freshwater bony fishes where it is about 0.5°C and highest in marine elasmobranchs with a value of 2.17°C. The protein content of fish plasma is relatively low when compared to other vertebrates. The iodine binding protein iodurophorine is unique to fish plasma. The levels of fibrinogen are relatively low in fish blood. The Arctic and Antarctic fishes such as the ice fish living in water below 0°C do not freeze as a result of the presence of antifreeze glycoproteins that were discovered in 1960s in their plasma. Antifreeze glycoproteins lower the freezing point of fish blood. It is believed that these proteins bind to the tiny ice crystals and inhibit their growth. Ice fish look pale as they lack hemoglobin and grow up to 60 centimeters long and weigh as much as 3.5 kilograms. Each antifreeze glycoprotein is made up of the amino acids—alanine and threonine—in a ratio of 2:1. The proteins also contain the disaccharide beta-D-galactosyl(1->3)-alpha-N-acetyl-D-galactosamine that is joined as a glycoside to the hydroxyl oxygen of the threonine residues (Harding et al., 2003).

BLOOD CLOTTING

The process of blood clotting or coagulation begins as soon as normal blood leaves a ruptured or severed blood vessel. The clotting mechanism involves a series of complex activities that occur fast to form a meshwork of fibers that trap blood cells. Several factors (about 30) are necessary before coagulation of blood can occur in mammals. Prothrombin, thrombin, fibrinogen, fibrin, calcium ions and various factors have been identified to be necessary in clotting of mammalian blood.

Injured tissue normally produces chemicals that start the clotting mechanism. Phospholipids and lipoproteins are normally produced and in the presence of calcium ions and other clotting factors lead to production of prothrombin activator known as **prothrombinase**. Exposure of collagen fibers in the blood vessel wall activates other clotting factors that will eventually lead to production of factor X or Stuart factor that is important in production of prothrombinase. Platelets stick together when outside blood vessels and produce the platelet factor that is a phospholipid and activates factor X to also produce prothrombinase. This activator, in the presence of calcium ions converts prothrombin into thrombin. Thrombin catalyzes the conversion of the soluble **fibrinogen** into insoluble **fibrin** in the presence of calcium ions. Fibrin strands will then undergo polymerization to form a fibrin clot in the presence of a fibrin-stabilizing factor. Blood cells will be trapped in the fibrin meshwork of the clot.

Fibrinogen and prothrombin are synthesized in the liver. The fat-soluble vitamin K is vital for the synthesis of prothrombin. The vitamin is acquired through dietary intake or can be synthesized by bacteria in the intestine. A deficiency of vitamin K in the body is followed by incidences of bleeding.

Throughout vertebrate evolution, proteins have been added to the blood clotting mechanism to improve its efficiency from the much simpler systems as are found in living agnathans. The extrinsic pathway that comprises chemicals released from damaged tissues that lead to formation of the prothrombin factor evolved to be part of the initial blood clotting mechanism about 500 million years ago. The jawless vertebrates have a coagulation network that involves a few factors including fibrinogen (Davidson et al., 2003). About 50 million years ago, the intrinsic pathway that involves contact of platelets with the rough exposed surfaces of vessels was added to the coagulation process. The processes involved in clotting increased in sensitivity and steps with such additions. The net effect of this evolution is creation of many pathways that are totally dependent on each other and clotting will not take place in the absence of any of the pathways. There is also a close relationship between clotting and inflammation that has been preserved throughout vertebrate evolution and is readily demonstrable in the human being on exposure to potentially injurious stimuli (Opal, 2000).

FLUID COMPARTMENTS OF THE BODY

The total amount of water present in the vertebrate body varies depending on the species, age, individual sex and body condition. In the adult human being, the amount varies from 40-60%. Vertebrates with more fat have less water per kilogram of body weight than those with less fat as adipose tissue contains less water than other soft tissues. Extra-cellular fluid is found outside body cells and includes interstitial fluid that is found around body cells, blood plasma, lymph and transcellular fluid that comprises cerebrospinal and synovial fluids.

Intracellular fluid is found within body cells and forms about 63% of body water in the human being with the rest of the body water being extra-cellular fluid. Extra-cellular fluid forms the internal environment as it bathes the tissues cells and provides a stable environment in which the body cells are able to carry out their metabolic activities. The composition of body electrolytes including sodium, potassium, chloride and bicarbonate ions plays an important role in controlling the movement of electrolytes between the body compartments as well as maintaining the acid-base balance of the body. Whereas plasma and intercellular fluids are almost similar in composition, the two body fluids differ substantially from intracellular fluid.

As capillary walls are impermeable to almost all plasma proteins, the amount of these proteins in interstitial fluid is quite low when compared to plasma. The protein anions have a negative charge and as a result there are generally more electrolytes in plasma than interstitial fluid. The level of protein anions in intracellular fluid is quite high when compared to extra-cellular fluid. There is also a high level of potassium ions in intracellular fluid whereas the extra-cellular fluid has high quantities of sodium and chloride ions.

Tissue cells tend to be within 100 m from capillaries for effective exchange of material between the two structures. Interstitial fluid is replaced on a continuous basis by fresh fluid from blood. Material normally diffuses between interstitial fluid and intracellular fluid of body cells. Maintenance of a stable pH and level of ions in blood also stabilizes the pH and ionic composition of interstitial fluid. Fluid homeostasis of the body compartments thus depends on chemical composition of body fluid and an abnormality in the proper composition will cause a fluid imbalance.

Fluid balance in the body compartments is best understood in mammals. The balance is normally maintained if input of fluid equals output. There are mechanisms in the body that adjust the levels of fluid loss from the body primarily to equal intake and also some for regulating intake secondarily to meet the output from the body. During decreased fluid intake, the output is normally decreased under hormonal control. **Aldosterone** hormone is produced by the adrenal gland cortex under the influence of the juxtaglomerular cells of the kidney when blood pressure falls below a certain level. Aldosterone causes increased reabsorption of sodium salts in the kidney. This action is followed by increased water retention. **Antidiuretic hormone** (ADH) that is produced by the hypothalamus and stored in the posterior part of the pituitary gland also plays a role in water retention by the body. The increased osmotic pressure of blood that results from reduced water intake is detected by the osmoreceptors of the brain leading to release of ADH. ADH acts on the distal and collecting tubules of the kidneys causing them to be more permeable to water. Water then flows osmotically out of the tubules into the interstitial fluid.

Intracellular and Interstitial Fluid

The control of the intracellular fluid composition is by the plasma membrane of tissue cells. The hydrostatic and colloid osmotic pressures of fluids in the intracellular and interstitial compartments determine fluid exchange between these two compartments. The colloid pressure determines most of the transfer of fluid between the two compartments since it varies more than the hydrostatic pressure. Electrolyte concentrations in the two body fluids also vary with sodium being the main electrolyte of interstitial fluid while potassium is the chief electrolyte of intracellular fluid. Variation in the concentration of these electrolytes which are the main determinants of osmotic pressure in the two fluid compartments causes an imbalance in exchange of water between the compartments. Sodium and potassium ions can diffuse or be selectively transported through the selectively permeable pores in the cell membrane. The large molecules of the intracellular fluid such as proteins cannot pass through these pores.

Movement of fluid across the plasma membrane by osmosis will depend on the concentration of solutes in each compartment. The concentration is determined by the quantity of solutes and fluid present as water determines the level of dilution of the solutes and similarly the colloid osmotic pressure of the compartments. A decrease of the major electrolytes in each compartment is followed by a decrease in colloid osmotic pressure resulting in the establishment of a colloid osmotic pressure gradient and vice versa. Osmosis normally occurs with movement of water to the compartment with a higher colloid osmotic pressure.

CARDIAC CYCLE

A complete heart beat that involves contraction and relaxation of both atria and ventricles is known as a cardiac cycle (Fig. 10.18). Contraction and relaxation of the two atria in higher vertebrates occurs simultaneously and the same action applies to the two ventricles. Atria fill with blood then contract to pump blood through the atrioventricular valves. As the atria contract the ventricles will be expanding and filling with blood during this phase of diastole (Gr. *diastole,* dilation). Diastole is followed by systole (Gr. *systole,* drawing together) when the ventricles contract and the atrioventricular valves close. Aortic and pulmonary semilunar valves open at the same time for blood to flow into the aorta and pulmonary artery respectively. Blood from the heart flows to the pulmonary and systemic circuits at the same time in this double circuit system of circulation, as opposed to the single circuit system of fish where blood has to flow to the gills first before circulating to tissues then returning to the heart. As the atria fill with blood from systemic circulation, the atrioventricular valves are closed since the ventricles will be contracting at the same time. During most of the time the atrioventricular valves are closed, the aortic and pulmonary artery

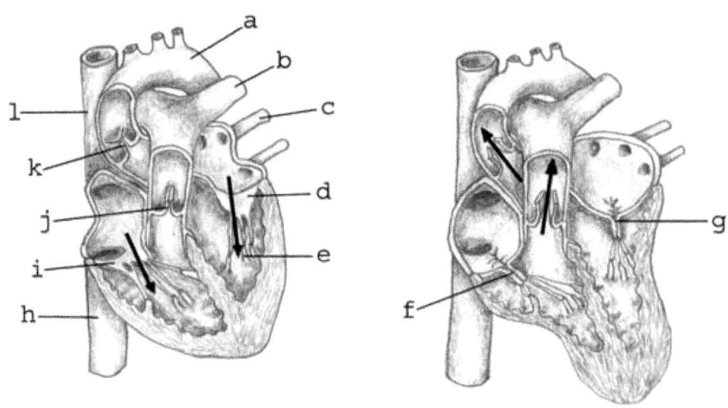

Fig. 10.18 The cardiac cycle showing the state of the heart chambers and valves when atria contract and ventricles relax (left) and when the ventricles contract and atria relax (right). (a) aortic arch, (b) left pulmonary artery, (c) one of the pulmonary veins, (d) open bicuspid or mitral valve, (e) papillary muscle, (f) closed tricuspid valves, (g) closed bicuspid valves, (h) posterior or inferior vena cava, (i) open tricuspid valves, (j) closed pulmonary semilunar valves, (k) closed aortic semilunar valves and (l) anterior or superior vena cava. Arrows indicate the direction of blood flow.

semilunar valves will be open and vice versa. The opening of atrioventricular and semilunar of valves occurs when pressure in the pumping chambers exceeds pressure in the ventricles and the main arteries leaving the heart respectively. Reverse pressure in the pumping chambers leads to closure of these heart valves.

Contraction and relaxation of the heart is a coordinated activity under the control of the conducting system of the heart (see Chapter 3). The impulses that initiate the contraction of the heart originate in the **sinoatrial node** or '**pacemaker**'. The sinoatrial node is a remnant of the sinus venosus. The node has specialized cells that possess an intrinsic rhythm at regular intervals. The evolution and development of the earliest components of cardiac pacemaking and conduction system and persistence of such cells into old age in vertebrate hearts has been reviewed (Thompson et al., 2003). The rate at which the sinoatrial node generates impulses is under the control of the autonomic nervous system and hormones. The ratio of sympathetic and parasympathetic impulses to the sinoatrial node per unit time determines the rate at which the pacemaker generates impulses. Should the sinoatrial node lose its ability to start an impulse, other **ectopic pacemakers** that form the excitable parts of the conducting system such as the atriaoventricular node or purkinje fibers will generate impulses though at a slower pace.

BLOOD CIRCULATION THROUGH BLOOD VESSELS

Since the cross-sectional area of the branches of a blood vessel exceed that of the parent vessel, the total cross-sectional area of capillaries is greater than that of any other group of vessels in the circulatory system. As the rate of blood flow varies inversely with the total cross sectional area of blood vessels, blood flow in capillaries will be slow when compared to other vessels of the blood vascular system since blood flowing through corresponding segments of the system is equal. The pressure in arteries that supply blood to body tissues from the heart should be high enough to maintain circulation in the cardiovascular system. Arterial blood pressure depends on the volume of blood that is present in arteries and the force of the heartbeat. These factors operate under the control of the autonomic nervous system and hormones.

Sympathetic nervous stimulation together with its associated hormones epinephrine and norepinephrine increase the rate at which the sinoatrial node fires and also the force with which the heart contracts. Stimulation by the parasympathetic nervous system has a negative effect. During the cardiac cycle, blood pressure increases and falls in series. Ventricular contraction is accompanied by a rise in systolic pressure of blood being pumped from the heart. This pressure spreads through the arterial tree with time. Between two ventricular contractions is the minimum diastolic pressure. Systolic pressure decreases with increase in distance from the heart due to loss of energy resulting from an increase in friction with a decrease in diameter of arteries and an increase in the overall surface area down the arterial tree. The elasticity nature of the arterial wall also contributes to the fall in pressure. The stretch that occurs in arteries during systole temporarily reduces arterial pressure as well as resistance. Part of the energy is restored during diastole when the arterial walls return to their normal size.

Air-breathing diving animals can sometimes stay under water for a long time. Since they cannot breathe at such times, they depend on the oxygen stores in the blood to supply oxygen to organs that need a continuous supply of the gas such as the brain, heart and some endocrine organs. When the levels of oxygen drop and carbon dioxide levels increase, arterial chemoreceptors are stimulated resulting in constriction of peripheral blood vessels, bradycardia and decreased cardiac output. Such effects of hypoxia are different from the effects that result from low levels of oxygen due to low breathing on land. Blood flow to many tissues including muscle is diverted to organs that cannot withstand lack of oxygen at times of hypoxia.

Blood Flow in Capillaries

The rate of blood flow in the cardiovascular system is slowest in capillaries due to the very low pressure that results from the large surface area of these vessels. Such a rate of blood flow facilitates diffusion of gases, nutrients and metabolic wastes between the circulatory system and tissue cells. The arteriole end that leads to capillaries has relatively more smooth muscle than the preceding segment and forms pre-capillary sphincters. These sphincters contract and relax to regulate blood flow through capillaries since not all capillaries of the body tissues are filled with blood at the same time. Most cells in the body are less than 100 μ from capillaries thus shortening the diffusion distances for exchange of material. Krogh (1919) and Paff (1930) stated that the average distance between capillaries varies inversely with the rate of gas exchange.

The higher hydrostatic pressure at the arteriole end of capillaries forces plasma fluid into interstitial spaces (Fig. 10.19). At the venous end, the reduced volume of blood plasma in capillaries has a higher osmotic pressure (due to plasma proteins) than at the arteriole end. This higher pressure is able to counteract the effect of the hydrostatic pressure so that some fluid flows back to blood plasma by osmosis. Small molecules and ions are exchanged in capillaries by osmosis. The higher oxygen diffusion

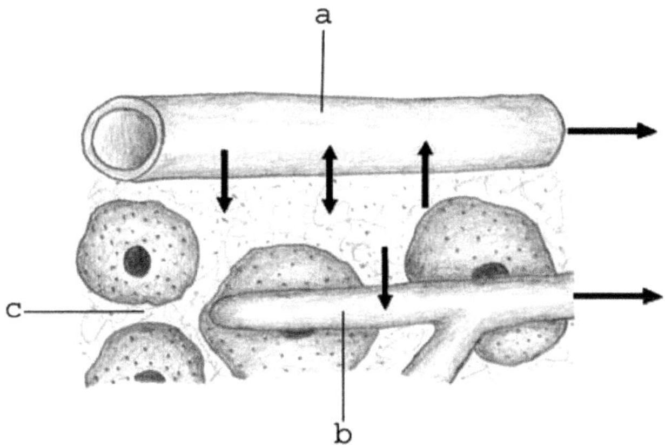

Fig. 10.19 Exchange of material between capillaries and intercellular spaces. The arterial end of the capillary lies to the left whereas the venous end is to the right. (a) blood capillary, (b) lymphatic capillary that begins blindly in intercellular space and (c) intercellular space. Arrows indicate the direction of flow in capillaries and movement of material at the tissue level. Lymphatic capillaries drain off excess interstitial fluid and some proteins as lymph. Apart from temporary changes, all the body's fluid compartments are in osmotic equilibrium.

gradient at the arteriole end results in greater diffusion of oxygen into tissue than at the venous end. The same principle applies to carbon dioxide.

Blood cells normally have a diameter that is greater than that of capillaries and have to be deformed as they circulate through these vessels since capillaries do not change their diameters much.

Exchange of Water and Electrolytes in Capillaries

At the capillary level, there is exchange of water and electrolytes between plasma and the interstitial fluid. The blood hydrostatic pressure forces fluid out of capillaries into interstitial fluid whereas colloid osmotic pressure tends to draw back the fluid into capillaries. Interstitial fluid hydrostatic pressure pushes fluid in this compartment into capillaries while interstitial colloid pressure draws fluid back from capillaries. Capillary hydrostatic and interstitial colloid pressures act in one direction by drawing fluid towards the interstitial compartment whereas capillary colloid and interstitial hydrostatic pressures act in an opposite direction towards capillaries. The net filtration pressure is the difference between the two opposing pressures and will determine the net fluid movement between interstitial fluid and blood in capillaries. Excess interstitial fluid and some proteins are drained as lymph by the blind ending lymphatic capillaries into lymphatic circulation and eventually to the cardiovascular circulation. Should production of lymph exceed lymph uptake by lymphatic capillaries then **edema** (swelling due to excess fluid) will result. Lymph is transparent and yellowish in color. The movement of interstitial fluid into lymphatic capillaries is by osmotic and diffusion forces. The low pressure in lymph capillaries enables fluids to flow easily into these vessels from interstitial spaces.

Venous Blood Flow

Veins return blood to the heart from capillaries. In the resting condition, veins are blood reservoirs as they store more than 50% of the total blood volume. Veins have larger diameters than corresponding arteries and can collapse when not filled with blood. Venous return to the heart is enhanced by contraction of surrounding skeletal muscle, inhalation and heart rate that 'sucks' blood towards the heart. Contraction of smooth muscle in the wall of veins also helps. The presence of valves in veins prevents a back flow of blood. The only vein that lacks valves is the vena cava. There are more veins than arteries in vertebrates since they have to drain large areas when necessary. Peripheral veins tend to follow a parallel path with corresponding arteries.

BLOOD VOLUME AND VISCOSITY

Blood volume and viscosity is controlled by several factors. The mechanisms that control thirst and urine production also regulate the volume of water within the body. The kidneys play a key role in adjusting the proportion of sodium and potassium in the body and this affects the volume of fluids in extracellular and intracellular compartments. Plasma albumin is a major protein in regulating osmotic pressure in blood and leads to movement of fluid into blood vessels. Hormones such as aldosterone, antidiuretic and atrial natriuretic from the right atrium of the heart contribute to the overall blood volume of the body. **Atrial natriuretic hormone** is produced when the right atrium is distended and promotes loss of sodium and water in kidneys.

The correct viscosity of blood ensures the right blood pressure and flow within vessels. The quantity of blood cells and albumin present determine the viscosity of blood. Blood with a high viscosity requires high pressure to be pumped within the circulatory system. Mechanisms that adjust the production or distribution of red blood cells such as the spleen that can store some blood adjust some of the changes in viscosity.

CONTROL OF BLOOD FLOW

The body has the ability to shunt blood to tissues that are active at a particular time from the less active ones. This conserves the body's energy since blood does not have to be pumped to all tissues at the same time. The autonomic nervous system together with signals from tissues in demand of higher circulation enable the cardiovascular system to adjust its blood flow in a short time to meet the metabolic requirements of body tissues. Blood is then channeled to the right areas of the body. During exercise or when an animal is excited, the sympathetic system is stimulated. Blood flow to the body viscera is reduced and shunted to skeletal muscle. Specific organs also produce their own signals when there is a demand for an increased blood supply. The endothelium of blood vessels produces nitric oxide in tissues that experience low oxygen levels. This free radical with a half-life of a few seconds is produced in response to acetylcholine and other vasodilators. Nitric oxide diffuses across biological membranes and causes relaxation of smooth muscle fibers in walls of arterioles and pre-capillary sphincters leading to their dilation. As the blood and oxygen supply to the tissue increase, the rate of synthesis of nitric oxide decreases. Such a negative feedback between oxygen and nitric oxide can apply to small areas of a tissue. Nitric oxide also plays other roles in the body including mitochondrial respiration that results in oxidative stress as reviewed by Stuart-Smith (2002). Other roles of nitric oxide include neurotransmission and a messenger in the response of macrophages to bacteria and cancer cells.

When the body is relaxed during resting the parasympathetic system is activated and blood mainly flows towards visceral organs. The visceral organs are actively processing digested food and maintaining the normal state of the body at such times.

A normal circulatory system is well coordinated in its functions ensuring that all tissues of the body are well nourished and supplied with oxygen that meets their metabolic requirements as well as removal of metabolic waste and excess heat from the body. There are receptors in the cardiovascular system that are sensitive to changes in blood pressure and the chemical composition of blood and transmit the information to the brain for the necessary homeostatic adjustments when an imbalance occurs.

REGULATION OF BLOOD PRESSURE

In higher vertebrates, blood pressure is detected by stretch receptors known as **baroreceptors** that are located near the heart in the wall of the aorta and carotid arteries as the rise of the blood pressure stretches these receptors. Afferent nerve fibers originate from these receptors and lead to the autonomic cardiac control center in the medulla oblongata of the brain. Secondary signals also inhibit the vasoconstrictor center of the medulla and stimulate the vagal center. The integrators in these brain centers have a negative feedback effect on the heart whenever the blood pressure is outside the normal range through the sympathetic and parasympathetic autonomic nervous system that regulates the pacemaker.

Blood pressure and blood distribution in the body can also be controlled by factors that alter the diameter of arterioles. The vasomotor or vasoconstrictor center in the medulla oblongata on stimulation will send impulses through sympathetic nerves to arterioles, venules and veins causing them to undergo vasoconstriction. Blood will then circulate back to the heart to be channeled to areas with increased activity.

The chemoreceptor reflex is sensitive to low oxygen and excess carbon dioxide or hydrogen ions. Chemoreceptors are located in aortic and carotid bodies (Fig. 10.20). The **aortic body** is located between the aortic arch and pulmonary artery whereas the **carotid body** (also known as **carotid glomus** or **glomus caroticum**) is found at the point of bifurcation of the common carotid artery in mammals. These bodies have a rich blood supply. A fall in arterial blood pressure below a certain critical level stimulates chemoreceptors as the low blood supply is accompanied with low oxygen levels and increased levels of carbon dioxide and hydrogen ions. Signals are sent to the vasomotor center that will then stimulate the sympathetic nervous system to increase blood pressure. The low pressure thus prevents a further drop in blood pressure.

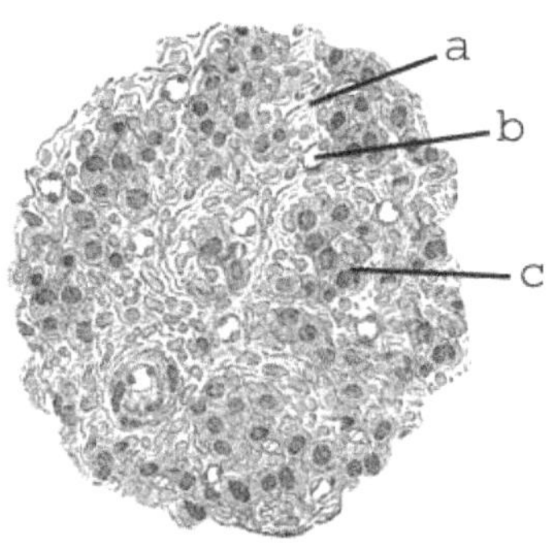

Fig. 10.20 Diagram of a carotid body. (a) connective tissue with supportive or sustentacular cells, (b) capillary and (c) glomus cells. A carotid body is a small cluster of chemoreceptors and supportive cells (about 5 mm in diameter in the human being) that is richly supplied by fenestrated capillaries and is located in the tunica adventitia at the angle of branching of the common carotid artery. The aortic body also comprises chemoreceptors and supportive cells as well as baroreceptors.

REFERENCES

Baker, M.E. (2002). Albumin, steroid hormones and the origin of vertebrates. J. Endocrinol. 175:121-127.

Barreda, D.R., Hanington, P.C. and Belosevic, M. (2004). Regulation of myeloid development and function by colony stimulating factors. Dev. Comp. Immunol., 28(5):509-554.

Ellis, A.E. (1986). The function of teleost fish lymphocytes in relation to inflammation. Int. J. Tissue React., 8(4):263-270.

Davidson, C.J., Tuddenham, F.G. and McVey, J.H. (2003). 450 million years of homeostasis. J. Thromb. Haemost., 1(7):1487-1494.

Germani, A., Di Rocco, G., Limana, F., Martelli, F. and Capogrossi, M.C. (2007). Molecular mechanisms of cardiomyocyte regeneration and therapeutic outlook. Trends Mol. Med., 13(3):125-133.

Harding, M.M., Anderberg, P.I. and Haymet, A.D. (2003). 'Antifreeze' glycoproteins from polar fish. Eur. J. Biochem., 270(7):1381-1392.

Icardo, J.M., Colvea, E., Cerra, M.C. and Tota, B. (2000). Light and electron microscopy of the bulbus arteriosus of the European eel (*Anguilla anguilla*). Cell Tiss. Org., 167:184-198.

Krogh, A. (1919). The number and distribution of capillaries in muscles with calculations of the oxygen pressure head necessary for supplying the tissues. J. Physiol. (London), 52:409-415.

Mayer, W.E., O'Huigin, C., Tichy, H., Terzic, J. and Saraga-Babic, M. (2002). Regenerative capacity and the developing immune system. Adv. Biochem. Eng. Biotechnol., 93:39-66.

Mescher, A.L. and Neff, A.W. (2005). Regenerative capacity and the developing immune system. Adv. Biochem. Eng. Biotechnol., 93:39-66.

Meseguer, J., Esteban, M.A. and Rodriguez, A. (2002). Are thrombocytes and platelets true phagocytes? J. Microsc. Res. Tech., 57(6):491-497.

O'Brien, K.M. and Sidell, B.D.(2000). The interplay among cardiac ultrastructure, metabolism and the expression of oxygen-binding proteins in Antarctic fishes. J. Exp. Biol., 203(pt.8):1287-1297.

Opal, S.M. (2000). Phylogentic and functional relationships between coagulation and the innate immune response. Crit. Care Med., 28(9 suppl.):S77-80.

Paff, G.H. (1930). A quantitative study of the capillary supply in certain mammalian skeletal muscles. Anat. Record, 46:401-406.

Song, H.D., Sun, X.J., Deng, M., Zhang, G.W., Zhou, Y., Wu, X.Y., Sheng, Y., Ruan, Z., Jiang, C.L., Fan, H.Y., Zon, L.I., Kanki, J.P., Liu, T.X., Look, A.T. and Chen, Z. (2004). Haematopoietic gene expression profile in zebrafish kidney marrow. Proc. Natl. Acad. Sci. U.S.A., 101(46):16240-16245.

Stuart-Smith, K. (2002). Demystified. Nitric Oxide. Mol. Pathol., 55(6):360-366.

Thompson, R.P., Reckova, M., deAlmeida, A., Bigelow, M.R., Stanley, C.P., Spruill, J.B., Trusk, T.T. and Sedmera, D. (2003). The oldest, toughest cells in the heart. Novartis Found. Symp., 250:157-174.

Valasek, P., Macharia, R., Neuhuber, W.L., Wilting, J., Becker, D.L. and Patel, K. (2007). Lymph heart in chick--somitic origin, development and embryonic edema. Development, 134(24):4427-4436.

Wagenseil, J.E. and Mecham, R.P. (2009). Vascular extracellular matrix and arterial mechanics. Physiol. Rev., 89(3):957-989.

Wilting, J., Papoutsi, M. and Becker, J. (2004). The lymphatic vascular system: secondary or primary? Lymphology. 37(3):98-106.

Yamauchi, A. and Burnstock, G. (1968). An electron microscopic study on the innervation of the trout heart. J. Comp. Neurol., 132:567-588.

Zapata, A.G., Torroba, M., Vicente, A., Varas, A., Secedon, R. and Jimenez, E. (1995). The relevance of cell microenvironments for the appearance of lympho-haemopoietic tissues in primitive vertebrates. Histol. Histopathol., 10(3):761-778.

11

Nervous System and Endocrine Organs

Although the nervous and endocrine systems are independent, the two systems are highly dependent on each other in performing the vital roles of communication, integration and overall control of all systems in the vertebrate body. The rapid responses to stimuli in vertebrates are mediated by the nervous system on stimulation of nerves by sensory cells or other nerve cells. The endocrine system produces hormones that are transported by the blood to specific target organs that have receptors for the hormones all over the body. The effects of hormones are slower but last for a longer time when compared to those of the neurotransmitters of the nervous system. The overall homeostasis of the vertebrate body is dependent on the coordination of body functions by the nervous and endocrine systems.

ORGANIZATION OF THE NERVOUS SYSTEM

The nervous system is divided into the central and peripheral nervous systems. The central nervous system comprises the brain and spinal cord whereas the peripheral nervous system consists of nerve tissues that lie to the outer region of the median system such as the cranial nerves that originate from the brain and the spinal nerves that arise from the spinal cord. Signals are conveyed to the central nervous system by **afferent** or **sensory nerves** to be processed so that the right combination of effectors can be activated. Signals are then transmitted from the central nervous system to effectors by **efferent** or **motor nerves**. The cells of the nervous system that transmit signals are known as **neurons**. Neurons are supported in their function by glial cells.

Action Potential

Neurons are excitable and are capable of initiating and conducting impulses that relay specific signals throughout the body. In a neuron that is not transmitting an impulse or is 'resting', there is a high concentration of sodium ions to the outside of the plasma membrane making it have a positive charge relative to the inside of the membrane. The difference in electrical charge across the membrane is known as the membrane potential and at rest the resting membrane potential is about -70 m volts. A stimulus that exceeds the threshold value causes opening of ion channels in the plasma membrane. This changes the permeability of the membrane and causes an axon to develop an action potential. There is an inward influx of sodium ions into the axon as the sodium channels are opened. This action results in momentary reversal of the membrane polarity and is known as **depolarization**. Depolarization affects the next adjacent area along the axon membrane resulting in more depolarization. This action is repeated along the axon and is known as a nerve impulse. After depolarization, sodium ions are pumped out of the axon actively by the sodium-potassium pump thus restoring the original resting potential of the membrane.

The terminal parts of an axon, the telodendria (Gr. *telos*, end; *dendria*, trees) has several branches that make contact with the next neuron at the synapse (Gr. *synapsis*, union) or the effectors at the neuromuscular or axoglandular junctions. The terminal synaptic knobs of telodendria contain neurotransmitter substances including acetylcholine, noradrenalin, dopamine and serotonin that are stored in synaptic vesicles. The arrival of a nerve impulse causes release of the neurotransmitter substance that will cross into the synaptic cleft and bind onto receptor sites of the postsynaptic membrane of the next neuron or effector organ. A new action potential will follow or the membrane will be inhibited from generating one.

Neurotransmitters

Neurotransmitters are chemicals that relay information between neurons or neurons and effector organs in one direction by crossing the synapse. The chemicals are released at the terminal end of neurons known as pre-synaptic nerve terminals. Neurotransmitters bind to specific receptor sites of the next nerve or receptor organ where they either elicit an excitatory postsynaptic potential by causing depolarization or may cause an inhibitory postsynaptic potential through hyperpolarization. After release from the neuron, neurotransmitters are either inactivated by enzymes or are taken back by the neurons through a reuptake mechanism. The first neurotransmitter was discovered in 1921 by Otto Loewi, an Austrian scientist, and is acetylcholine.

Neurotransmitters may be small or large molecules. Based on their chemical structure, there are four types of neurotransmitters that include **acetylcholine**; **biogenic amines** such as serotonin, histamine and the catecholamines epinephrine and norepinephrine; **excitatory amino acids** such as glutamate, aspartate, gamma-aminobutyric acid (GABA) and glycine and **neuropeptides**. The most numerous neurotransmitters are neuropeptides and comprise about 50 neurotransmitter substances that include bradykinin, cholecystokinin, oxytocin, secretin, gastrin, beta-endorphin and many other hormones. Acetylcholine is classified on its own as it has a chemical structure that is unique from that of other neurotransmitters and is formed by the combination of choline and acetyl-coenzyme-A. Amine transmitters are synthesized from amino acids.

Neurotransmitters can also be classified based on their function as **excitatory** or **inhibitory neurotransmitters**. Some neurotransmitters can have an excitatory effect on some postsynaptic membranes and an inhibitory effect on others. For example, acetylcholine has an excitatory effect on skeletal muscle and an inhibitory effect on cardiac muscle. The postsynaptic receptors determine the action of neurotransmitters. Other neurotransmitters bind to their postsynaptic receptor sites and cause the opening or closing of ion channels whereas others activate chemical messengers within the postsynaptic cell. The most common inhibitory neurotransmitter in the brain is GABA.

Whereas neuropeptides are synthesized in the nerve cell body and transported in vesicles, other neurotransmitters are synthesized at the axon terminals and stored in synaptic vesicles. The neurotransmitter substance is released from a vesicle into a synaptic cleft when the vesicle fuses with the cell membrane at the terminal end of an axon. Nitric oxide (NO) is not stored in synaptic vesicles but diffuses out of a neuron after production.

CENTRAL NERVOUS SYSTEM

The central nervous system is major to the regulatory control of the entire nervous system and the entire vertebrate body and comprises the brain and spinal cord. The delicate central nervous system is covered by the skull and vertebrae and the protective layers of connective tissue known as **meninges** (Gr. *meninx*, membrane). The meninges develop from the mesoderm and neural crest cells. The number of layers that compose the meninges varies in different vertebrate groups. In fish, a single layer known as the **primitive meninx** closely covers the central nervous system leaving space between the system and the surrounding skeletal system that is filled with semi-liquid material. The amphibian and reptilian meninges comprise a **dura mater** (L. *dura mater,* hard mother) to the outside and **a secondary meninx** to the inside covering the central nervous system.

In birds and mammals, the secondary meninx has given rise to the **arachnoid membrane** (Gr. *arachne,* spider; *eidos,* form) and **pia mater** (L. *pia,* tender) thus giving rise to three meninges in these species. The pia mater closely conforms to the contours of the brain and spinal cord and is the innermost layer of the meninges. The arachnoid membrane resembles a cobweb and lies between the dura mater and pia mater. The dura mater is made of tough fibrous material and is the outermost layer of meninges or the inner periosteum of cranial bones. Below the arachnoid membrane is the subarachnoid space through which cerebrospinal fluid circulates in the cavities of the central nervous system.

Cerebrospinal fluid cushions the central nervous system against concussion to the body. The fluid also nourishes the brain and drains off excess extra-cellular fluid since the brain lacks a lymphatic system. The composition of cerebrospinal fluid is regulated by the brain capillaries that are less permeable than other capillaries in the body and the tight sheaths of astrocytes around the capillaries that form the blood-brain barrier. The blood-brain barrier is selective in material that crosses from capillaries into the intercellular fluid and is also an active transport site for many substances. Lipid soluble molecules including oxygen and carbon dioxide easily cross the blood-brain barrier.

BRAIN

Evolution of the vertebrate nervous system may have started in planaria (free-living flatworms), the first animals to possess a brain, before the divergence of metazoans into invertabrate and chordate branches (Sarnat and Netsky, 2002). Neurons of the planarian brain resemble those of vertebrates more closely than those of advanced invertebrates. Planarian neurons have a single axon and are multipolar in shape, possess dendrites with synaptic knobs, express vertebrate-like neural proteins and generate relatively slow spontaneous impulses.

The vertebrate brain is divided into three major regions or vesicles that include the forebrain or prosencephalon, midbrain or mesencephalon and hindbrain or rhombencephalon (Fig. 11.1). The **prosencephalon** (Gr. *pro,* before, *enkephalos,* brain) will develop into the diencephalon and telencephalon later on during neurogenesis. Anterior to the midbrain and above the brainstem lies the **diencephalon** (Gr. *dia,*

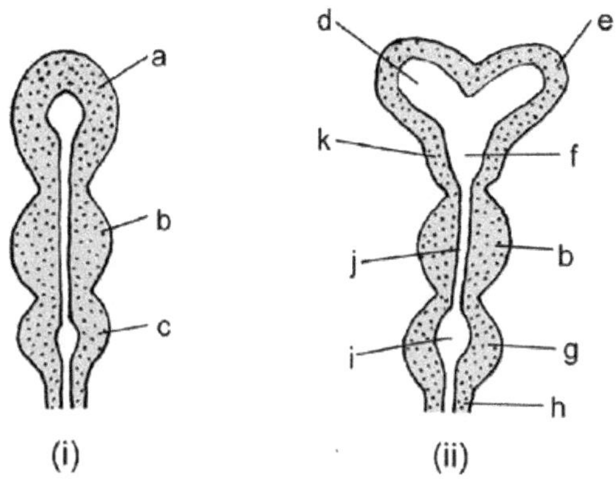

Fig. 11.1 The major divisions of the brain during (i) early and (ii) later embryogenesis. (a) prosencephalon, (b) mesencephalon, (c) rhombencephalon, (d) lateral ventricle, (e) telencephalon, (f) third ventricle, (g) metencephalon, (h) myelencephalon, (i) fourth ventricle, (j) cerebral aqueduct formally known as aqueduct of Sylvius and (k) diencephalon. The cerebral aqueduct never thickens with development as is the case with other parts of the anterior neural canal. The two lateral ventricles are the first and second ventricles.

through or across) that comprises the thalamus, hypothalamus, pineal gland and pituitary gland. Above the diencephalon is the **telencephalon** (Gr. *telos,* end) or endbrain that gives rise to the cerebral hemispheres. The **mesencephalon** (Gr. *mesos,* middle) is located between the diencephalon (rostrally) and the rhombencephalon (caudally) and forms the optic lobes or corpora quadrigemina dorsally. The **rhombencephalon** (Gr. *rhombos,* shaped like a rhombus) develops into the metencephalon anteriorly, the myelencephalon posteriorly and reticular formation in between the two divisions. **Metencephalon** (Gr. *meta,* after) gives rise to the cerebellum and pons (in birds and mammals) while the **myelencephalon** (Gr. *myelos,* core, marrow) develops into the medulla oblongata. The hindbrain is the oldest part of the brain and formed the major part of the brain in early life.

The **brain stem** (also known as the **reptilian** or **primitive brain**) is a transitional area of the central nervous system between the brain and spinal cord and evolved about 500 million years ago. The brain stem comprises the midbrain, pons and medulla oblonga as is seen in the brain of living reptiles. This part of the brain deals with basic life support processes such as breathing, heart contraction and digestion. With evolution of amphibians, the processing of sensory information shifted to the forebrain. Simple reflex pathways predominated in early vertebrates. The hindbrain of living vertebrates deals with reflex coordination whereas the parts that evolved later in the forebrain are concerned with memory, learning and reasoning.

Reticular formation (L. *reticulum*, small net) is found in the middle of the brainstem and is composed of short branching inter-neurons that form an ancient integrating system of vertebrates. The inter-neurons run from sensory to motor nuclei of cranial nerves attached to the brainstem. The neurons also have connections with other parts of the brain. In mammals, reticular formation filters incoming stimuli and passes only important information to the cerebrum. Reticular formation also transmits inputs from the cerebrum to motor neurons. Sleep centers are also located in reticular formation. The **reticular activating system** (RAS) is a network of neurons that is responsible for maintenance of consciousness since impulses from this network continuously excite neurons in the cerebral cortex. Serotonin is produced from another part of reticular formation and inhibits RAS impulses causing drowsiness or sleep. Drugs that depress RAS such as barbiturates decrease alertness and induce sleep.

Since the brain size has tended to increase with evolution, vertebrates show considerable variation in the size of this organ. The ability of vertebrates to show varied behavioral responses is related to the quantity of nerve cells and the way they relate to each other in the brain. The varying levels of intelligence observed in different vertebrates is related to the absolute or relative size of the brain. The different parts of the brain also vary much in size depending on the mode of life adapted by the various vertebrate groups. The human being has the largest brain relative to body weight and displays the most complex forebrain. Intelligence has evolved many times independently in vertebrates. The human being has more cortical neurons than any other mammal although only marginally more than whales and elephants and human intelligence has drawn from a combination and enhancement of various factors including language, imitation and theory of mind (Roth and Dicke, 2005). Agnathans have relatively the smallest brain of all vertebrates groups.

Fish brain (Fig. 11.2) is less complex than that of other vertebrates groups. The parts that are highly developed are related to survival instincts in an aquatic environment such as feeding, reproduction and defense. Since smell plays a vital role in the life of fishes, the **olfactory bulbs** are large and the cerebrum is of moderate size as it integrates signals related to senses of smell. Other parts of the brain that are well developed include the **hypothalamus** that integrates visceral activity, the **cerebellum** that controls locomotor activity and **medulla oblongata** that controls feeding, cardiac and respiratory movements.

The brain of amphibians (Fig. 11.3) has not changed much from that of fish or ancient amphibian fossils. The typical amphibian brain plan is found in the brain of the tiger salamander (*Amblystoma*). Most of the cell

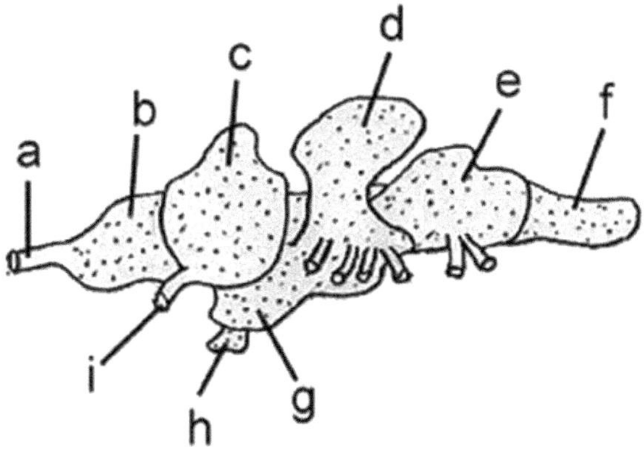

Fig. 11.2 Brain of a teleost fish. (a) olfactory tract, (b) cerebrum, (c) optic lobe, (d) cerebellum, (e) facial and vagal lobes of the medulla, (f) medulla oblongata, (g) hypothalamus, (h) pituitary gland and (i) optic nerve.

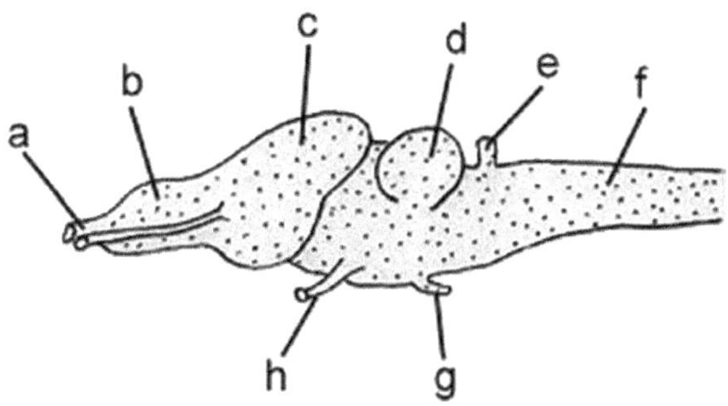

Fig. 11.3 The amphibian brain. (a) olfactory tract and (b) bulb, (c) cerebrum, (d) optic lobe, (e) cerebellum, (f) medulla oblongata, (g) pituitary gland and (h) optic nerve.

bodies of the amphibian neurons in the telencephalon lie near the ventricles whereas the axons are located superficially. Since locomotion is not as complex as is the case in many fish, the amphibian cerebellum is relatively smaller than that of most fishes. The brain of salamanders is even less complex than that of some vertebrates which are lower in the phylogenetic tree including cartilaginous and bony fishes as many brain features are less differentiated (Roth et al., 1993). Differences in amphibian brain morphology have largely been attributed to desynchronization of the ontogenic processes (heterochrony) and is related to changes in genome sizes and life histories (Schmidt and Roth, 1996).

Reptiles are the only amniotes with a central nervous system that is known to be capable of regeneration on its own. Reptiles have a brain that has greatly expanded cerebral hemispheres (Fig. 11.4) when compared to that of amphibians. The cerebrum evolved about 200 million years ago and is the most modern part of the brain. The cerebral hemispheres have expanded caudally and cover part of the diencephalon in reptiles. The reptilian cerebrum integrates more information than the amphibian cerebrum as it receives, in addition to olfactory signals, visual, auditory and sensory sensation from all over the body. The cerebellum in reptiles is also much larger than is the case in amphibians. The cerebellum evolved about 400 million years ago and is sometimes known as the little or hindbrain. The cerebellum was the main brain before evolution of the cerebrum. The brain of the turtle is relatively unspecialized.

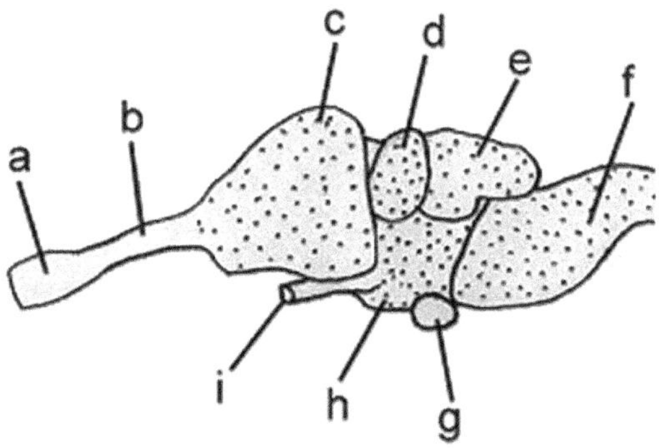

Fig. 11.4 The brain of an alligator. (a) olfactory bulb, (b) olfactory tract, (c) cerebrum, (d) optic lobe, (e) cerebellum, (f) medulla oblongata, (g) pituitary gland, (h) hypothalamus and (i) optic nerve.

The avian brain (Fig. 11.5) is comparatively much larger than that of reptiles and is as large as that of many mammals in relation to body weight. Evolution of ectothermy and high metabolic rates had to be accompanied with complex neuronal circuits that transmit impulses at high rates. The avian brain has distinct cerebral hemispheres that obscure entirely the diencephalon and the greater part of the midbrain. The cerebrum makes contact with the expanded cerebellum. The good vision of birds needs an accompanying larger cerebrum to be able to process much more information than happens in reptiles. Birds perform complex motor activities such as flight and feeding and the large cerebellum is able to receive an increase in tracts from the spinal cord and send others to the cerebrum for proper action in coordinating these activities.

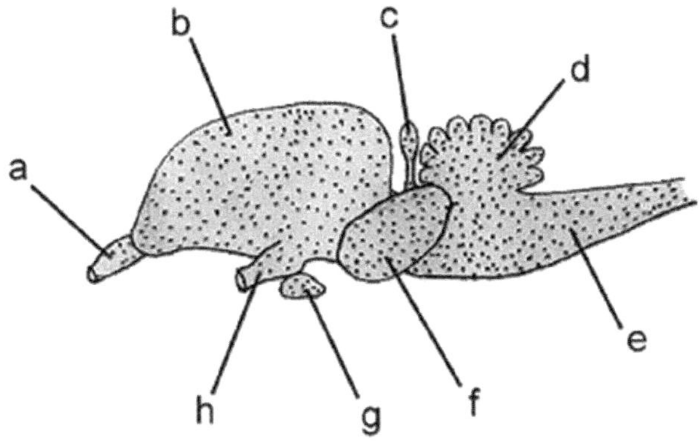

Fig. 11.5 The avian brain. (a) olfactory bulb, (b) cerebrum, (c) pineal gland, (d) cerebellum, (e) medulla oblongata, (f) optic lobe, (g) pituitary gland and (h) optic nerve.

Mammalian Brain

The mammalian cerebrum and cerebellum are greatly enlarged (Fig. 11.6) in comparison to other vertebrates. The cerebrum has increased in size as a result of expansion of the **neopallium** (Gr. *neos*, new; *pallium*, mantle). The surface of the neopallium in many mammals is convoluted and has ridges known as **gyri** (Gr. *gyros*, circle) that are separated by grooves referred to as **sulci** (L. *sulcus*, furrow). Such a surface increases the surface area for more neurons that process greater sensory input and motor output. The surface of the neopallium in some mammals that include many rodents, rabbits, duckbilled platypus and opposum is smooth. The mammalian thalamus is also larger than in the other vertebrates as it has to process more sensory inputs before relaying some of it to the neopallium.

The second largest part of the mammalian brain is the cerebellum. In the human being, the cerebellum is partly covered by the cerebrum above. There are more pathways that link the cerebellum and cerebrum in the mammalian brain when compared to other vertebrates and these pathways coordinate motor activity. The gray matter of the cerebrum and cerebellum is located to the outer part of the brain while the white matter dominates the internal part. The internal part of the cerebellum has a pattern that is similar to a branching tree known as **arbor vitae** (L., *arbor*, tree; *vita*, life). The surface of the cerebellum bears gyri and sulci though these are smaller and less distinct than those of the cerebrum. The cerebellum and the cerebrum coordinate the activities of muscle groups bringing about accuracy of movement. The cerebellum also controls posture and skeletal muscle in maintaining balance.

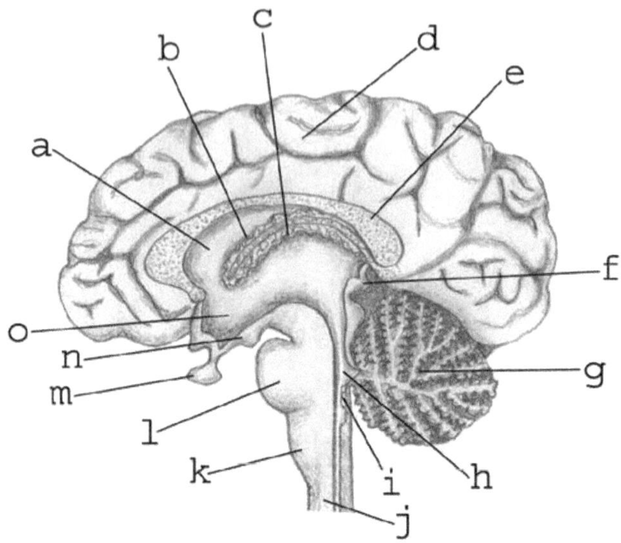

Fig. 11.6 The human brain. (a) lateral ventricle, (b) choroid plexus of (a), (c) choroid plexus of third ventricle, (d) cerebrum, (e) corpus callosum, (f) pineal gland, (g) cerebellum, (h) fourth ventricle, (i) choroid plexus of (h), (j) spinal cord, (k) medulla oblongata, (l) pons, (m) pituitary gland, (n) hypothalamus and (o) third ventricle.

Blood Supply to the Mammalian Brain

The mammalian brain receives its blood supply from the left and right internal carotid and left and right vertebral arteries (Fig. 11.7). **Internal carotid arteries** are branches of the common carotid arteries while **vertebral arteries** branch off subclavian arteries. Vertebral arteries pass through transverse foramina of cervical vertebrae before entering the cranial cavity through the foramen magnum. The left and right vertebral arteries unite at the base of the brain stem into the **basilar artery** that runs in an anterior manner while giving off branches including the **anterior inferior cerebellar, pontine, superior cerebellar** and **posterior cerebral arteries**.

Internal carotid arteries enter the cranial cavity in the middle part of the cranial floor where they are referred to as **arterial cerebral arteries** before branching into **anterior** and **middle cerebral arteries**. The anterior cerebral arteries are joined together by the **anterior communicating artery** while the **posterior communicating arteries** join internal carotid arteries to posterior communicating arteries leading to formation of an arterial circle at the base of the brain that is known as the **circle of Willis** or **cerebral arterial circle**. The circle of Willis is an example of arterial anastomosis.

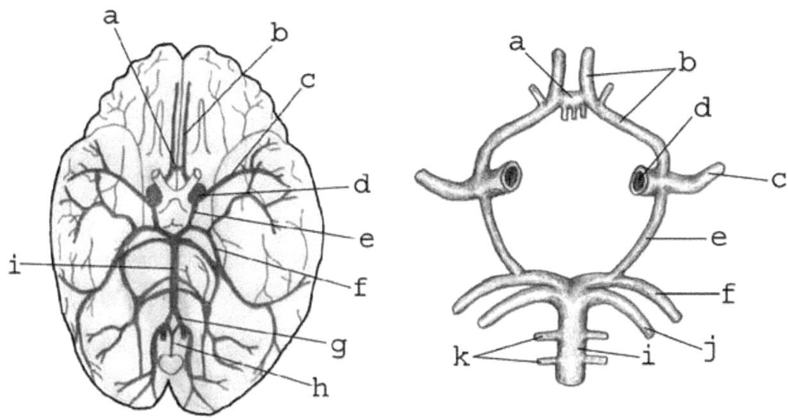

Fig. 11.7 The base of the brain showing blood supply to the organ (left) and circle of Willis (right). (a) anterior communicating artery, (b) anterior cerebral artery, (c) middle cerebral artery, (d) internal carotid artery, (e) posterior communicating artery, (f) posterior cerebral artery, (g) vertebral artery, (h) anterior spinal artery, (i) basilar artery, (j) superior cerebellar artery and (k) pontine arteries.

The Spinal Cord

The spinal cord runs from the hindbrain to the posterior or inferior part of the body trunk in the neural canal of vertebrae. In some teleosts, frogs and mammals, the length of the spinal cord is shorter than that of the spinal canal as the cord grows at a lower rate than the rest of the body. The unequal rates of growth in the two structures result in the corresponding vertebral segments lying in a more posterior or inferior position to the respective spinal nerves. The spinal nerves have to travel caudally or inferiorly within the spinal canal for sometime before exiting the canal thus forming a bundle of nerves within the canal known as **cauda equina** (L. *cauda*, tail, *equinus*, horse) (Fig. 11.8). In most other vertebrates, the spinal cord extends the entire length of the spinal canal and vertebral column. The cord also varies in shape in different vertebrates. The extent of brain control over the spinal cord has increased with the level of vertebrate evolution.

In agnathans with a less complex nervous system, the spinal cord has non-myelinated axons and as a result a distinction cannot be made between gray and white matter. The thin spinal cord in these jawless fish lacks a blood supply and is flattened dorsoventrally to facilitate diffusion of gases and nutrients. The spinal cord of other vertebrates is round in cross-section and has a blood supply. The number of ascending and descending myelinated tracts has increased with evolution resulting in expansion of the white matter that bulges outwards. Such an expansion

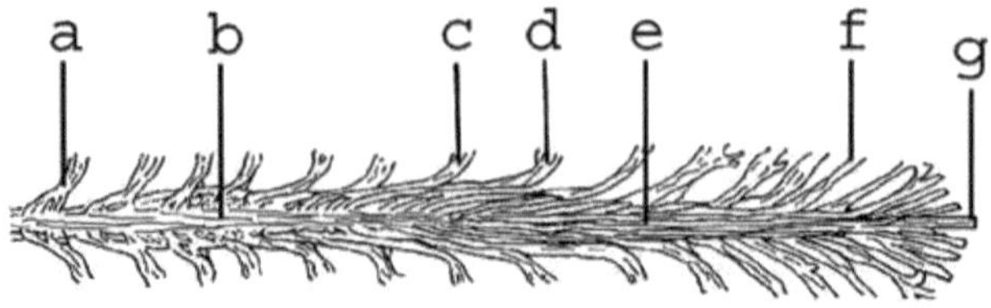

Fig. 11.8 The middle and terminal parts of the spinal cord. (a) spinal ganglion, (b) lumbar enlargement of the cord, (c) last thoracic spinal nerve, (d) first lumbar nerve, (e) cauda equina that resembles a horse's tail, (f) first sacral nerve and (g) terminal filament.

has resulted in the formation of a **ventromedian** or **anterior median fissure** and a **dorsomedian** or **posterior median sulcus**.

In amniotes, the white matter does not contain the synapses formed by sensory, interneurons and motor neurons but only ascending and descending fiber tracts. The gray matter of the cord has a butterfly shape. In birds and mammals, the number of ascending and descending tracts is more than in other vertebrates as more signals to the brain and back are required since the brain has more control of body activities in these groups than other vertebrates. The number of neuroglia and blood supply to the spinal cord is also higher in birds and mammals. In vertebrates with limbs, the spinal cord is thicker in cervical and lumbar regions as nerves that supply the limbs originate in these areas. Such enlargement of the spinal cord is more pronounced in the cervical region of vertebrates with more developed and active pectoral appendages such as bats and apes and in the lumbar region where pelvic limbs are appendages of locomotion as occurs in bipedals including the ostrich.

Apart from providing routes for conducting impulses to and from the brain, the spinal cord performs considerable integration for all spinal nerves. The tracts of the spinal cord comprise several axons that originate from neuron cell bodies of a particular area of the central nervous system that will terminate in another specific area of the system. The axons of a tract perform a particular function and as a result there are many tracts in the white matter of the spinal cord. In performing the role of integration, the spinal cord acts as a reflex center for all spinal nerves. The reflex centers are located in the gray matter of the cord and are areas of a reflex arc where sensory impulses to the cord turn into motor impulses. Reflex centers can be inter-neurons between sensory and motor neurons or synapses in two-neuron arcs.

Spinal Nerves

Spinal nerves are attached to the spinal cord by dorsal and ventral nerve roots (Fig. 11.9) save for lampreys where they form individual dorsal and

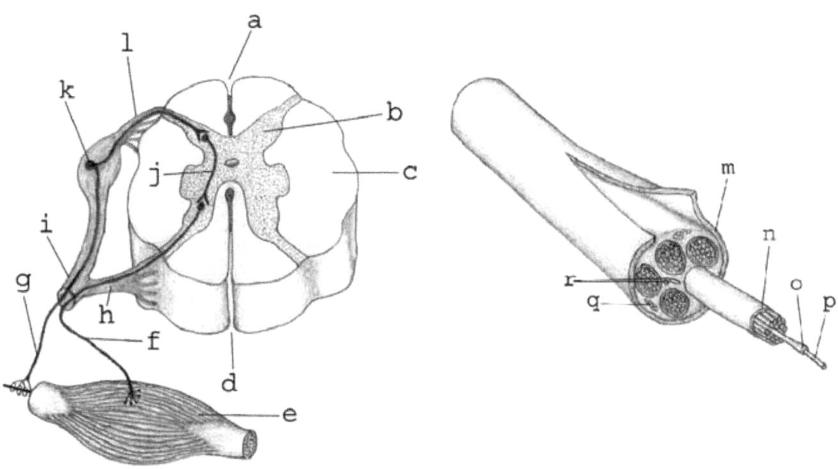

Fig. 11.9 Spinal nerves (left) and a cross-section through a nerve (right). (a) dorsomedian or posterior median sulcus, (b) gray matter, (c) white matter, (d) ventromedian or anterior median fissure, (e) muscle, (f) motor and (g) sensory neurones, (h) ventral root of spinal nerve, (i) spinal nerve, (j) interneurone, (k) cell body in dorsal root ganglion, (l) dorsal root of spinal nerve, (m) epineurium, (n) perineurium surrounding a fascicle, (o) endoneurium, (p) axon, (q) fat and (r) artery and vein.

ventral nerves. The dorsal nerve root comprises sensory nerves that transmit impulses to the spinal cord. Sensory nerves are unipolar and their cell bodies are located in the dorsal root ganglion that is a region of gray matter in the dorsal nerve root. Motor nerves are located in the ventral nerve root and relay impulses away from the spinal cord. Motor nerves are multipolar neurons and their cell bodies are located in the gray matter within the spinal cord. Interneurons that link sensory to motor nerves are located within the gray matter of the spinal cord. The dorsal and ventral nerve roots on each side of the body unite to give rise to a single spinal nerve that contains both motor and sensory fibers. In lampreys, the ventral nerves contain motor neurons whereas the dorsal nerves carry sensory and some motor neurons. In most fishes and amphibians, motor nerves leave the spinal cord through the dorsal and ventral roots. The number of spinal nerves present in each vertebrate depends on the number of segments present in the body. Long vertebrates such as snakes have hundreds of paired spinal nerves whereas those with short trunks have fewer spinal nerves. In animals with defined body regions, spinal nerves bear specific names such as cervical and thoracic spinal nerves.

Spinal nerves divide into large branches soon after emerging from the spinal cavity known as rami. The **dorsal ramus** subdivides into smaller nerves that bear sensory and motor fibers that innervate muscles and skin that are located on the dorsal aspect of the animal. Many **ventral rami** take a more complex course before supplying the target tissues. Such rami unite first to form complex networks known as **plexuses**. Nerves that innervate hypaxial musculature will emerge from such plexuses. Plexuses are well developed in mammals and include cervical, brachial, lumbar and sacral plexuses.

CRANIAL NERVES

Cranial nerves (Fig. 11.10) do not show uniformity in terms of location, branches and composition of fiber neurons when compared to spinal nerves. Cranial nerves are named according to their functions or structures they supply. The nerves are also numbered using Roman numerals in a superior to inferior or an anterioposterior manner as the sequence occurs in the human being with 12 cranial nerves. Some cranial nerves bear sensory, motor or both types of nerves. The 13 cranial nerves found in vertebrates are terminal (0), olfactory (I), optic (II), oculomotor (III), trochlear (IV), trigeminal (V), abducens (VI), facial (VII), vestibulocochlear (VIII), glossopharyngeal (IX), vagus (X), accessory (XI) and hypoglossal (XII).

Cranial nerves can be divided into three broad groups based on their points of exit from the undersurface of the brain in comparison with the

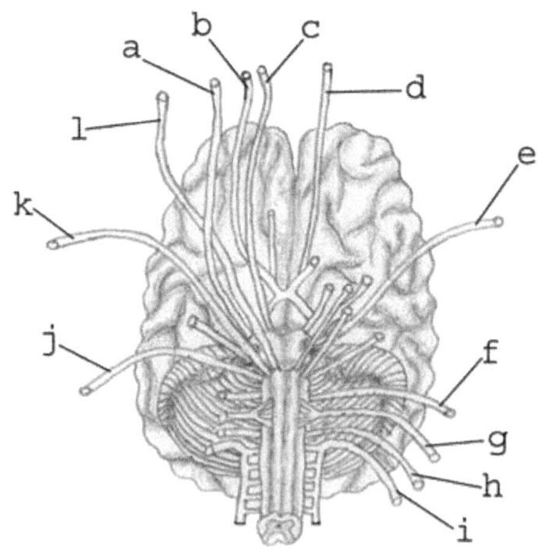

Fig. 11.10 A diagram of the ventral surface of the brain of a mammal showing attachment of the cranial nerves. (a) Trochlear (IV), optic (II), (c) oculomotor (III), (d) olfactory (I), (e) trigeminal (V), (f) glossopharyngeal (IX), (g) hypoglossal (XII), (h) vagus (X), (i) accessory (XI), (j) vestibulocochlear (VIII), (k) facial (VII) and (l) abducens (VI).

relationship of spinal nerves to the spinal cord. The nerves in series with dorsal roots of spinal nerves include V, VII, IX and X. These nerves arise from the lateral part of the brainstem. In early life, these nerves could have served branchial and pharyngeal areas. Nerves that emerge from the ventral part of the brainstem are in series with the ventral spinal nerves and include cranial nerves III, IV, VI and XII. The trochlear nerve is considered ventral since its nucleus lies in the ventral part of the brainstem. These nerves innervate hypobranchial muscles and derivatives of the head somites. The other nerves have no series with spinal nerves and are the sensory cranial nerves 0, I, II, VII, VIII and IX. The nerves innervate the eye, nose, ear and lateral line system. The nerves that appear in more than one series have roles that fit such functions. Cranial **nerve 0** or **terminal nerve** is closely apposed to the olfactory nerve and is present in most mammals apart from cyclostomes. Cranial nerve 0 is numbered so since it was discovered after the others had been assigned their current numbers. The terminal nerve appears as a nerve or in some cases a complex of nerves that most likely derive from the olfactory placodes and has been observed in at least some stages of development in all major groups of vertebrates except the hagfish (Demski, 1993). The nerve is part of the chemosensory system that regulates reproduction in relation to pheromones.

Fish lack cranial nerves XI (accessory) and XII (hypoglossal). The accessory branch of the vagus that supplies the cucullaris muscle is present in fish but in amniotes has separated off the vagus to become an independent cranial nerve that has somatic motor fibers that supply derivatives of the cucullaris that include the trapezius and sternocleidomastoid muscles. The hypoglossal nerve (Gr. *hypo*, under, *glossa*, tongue) contains both sensory and motor fibers and supplies the muscles of the tongue. Tongue movements in fish are brought about by the contraction of hypobranchial muscles.

The longest cranial nerve is the vagus (L. *vagus*, wandering). The four branches of the vagus nerve in fish supply motor branches to branchiomeric muscles of visceral arches four to seven. Sensory nerve branches arising from the last four posterior gill pouches are also contained in the branches of the vagus nerve. The vagus nerve could have evolved as a result of the union of four dorsal cranial nerves. In mammals, the sensory fibers of the vagus supply the pharynx, larynx, trachea, bronchi, lungs, carotid body, esophagus, stomach, intestines and gall bladder whereas the motor fibers supply the muscles of the pharynx, vocal ligaments in the larynx and thoracic and abdominal organs. Most motor fibers of the vagus nerve belong to the autonomic nervous system.

AUTONOMIC NERVOUS SYSTEM

Involuntary effectors are regulated by the autonomic nervous system (Gr. *autos*, self; *nomos*, law) (Fig. 11.11). The system supplies cardiac muscle, smooth muscle of viscera and glands. Coordination of cardiac and smooth muscle contraction as well as glandular secretion by this nervous system maintains homeostasis in the vertebrate body. The sensory autonomic pathway is essential for the necessary feedback that leads to regulation of autonomic effectors. The most studied autonomic nervous system is that of mammals. The two parts of the efferent autonomic nervous system are sympathetic and parasympathetic divisions. Many autonomic effectors receive a nerve supply from these two autonomic divisions. The effects of these two systems are antagonistic to each other as they release different transmitters at the neuron-effector junctions. The effectors are stimulated by the

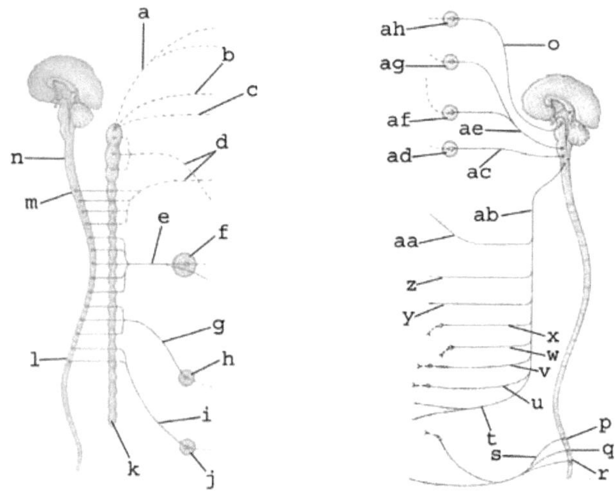

Fig. 11.11 The main sympathetic (left) and parasympathetic (right) pathways leading to various structures in the body. The **sympathetic pathways** lead to (a) eye and nasal mucosa, (b) sublingual and submandibular salivary glands, (c) parotid salivary gland, (d) heart and respiratory system, (e) liver, stomach, spleen, kidney and adrenal gland by the greater splanchnic nerve, (g) intestines by the lesser splanchnic nerve and (i) colon and urinary bladder via the lumbar splanchnic nerve, (f) celiac ganglion, (h) superior mesenteric ganglion, (j) inferior mesenteric ganglion, (k) sympathetic chain, (l) 2nd lumbar and (m) 1st thoracic vertebrae and (n) spinal cord. The **parasympathetic pathways** comprise (o) 3rd cranial nerve to eye, (p) 2nd, (q) 3rd and (r) 4th sacral nerves and (s) pelvic nerve to colon and urinary bladder, (ab) vagus nerve with various branches leading to (t) large intestine, (u) small intestines, (v) pancreas, (w) spleen, (x) stomach, (y) liver, (z) heart (aa) and lungs, (ac) 9th cranial nerve to parotid salivary gland and (ae) 7th cranial nerve to sublingual and submandibular salivary and lacrimal glands as well as the nasal mucosa. (ad) otic ganglion, (af) submandibular ganglion, (ag) pterygopalatine ganglion and (ah) ciliary ganglion.

neurotransmitters produced by the nerves of the sympathetic division whereas substances produced by the parasympathetic nerves at the neuron-effector junctions inhibit effectors.

The reflex arc of an autonomic nervous system includes information sent to the central nervous system by sensory nerves and impulses that are conducted to autonomic effectors from the brainstem and spinal cord by two autonomic neurons (Fig. 11.12). The first of the two autonomic motor neurons is known as a **preganglionic neuron** and transmits impulses from the brainstem and spinal cord to an **autonomic ganglion** that has cell bodies of some of the **postganglionic neurons** and lies outside the central nervous system. The postganglionic neuron then transmits impulses to the autonomic effectors.

The design of sympathetic and parasympathetic pathways is different in terms of location in the body. Sympathetic preganglionic neurons of mammals originate from the spinal cord in the thoracic and lumbar segments and are said to be of thoracolumbar origin. Most ganglia of the sympathetic division are located on both sides of the ventral or anterior surface of the spinal cord. Neighboring ganglia on the same side of the vertebral column are joined by sympathetic axons forming sympathetic chain ganglia that resemble beads in a chain.

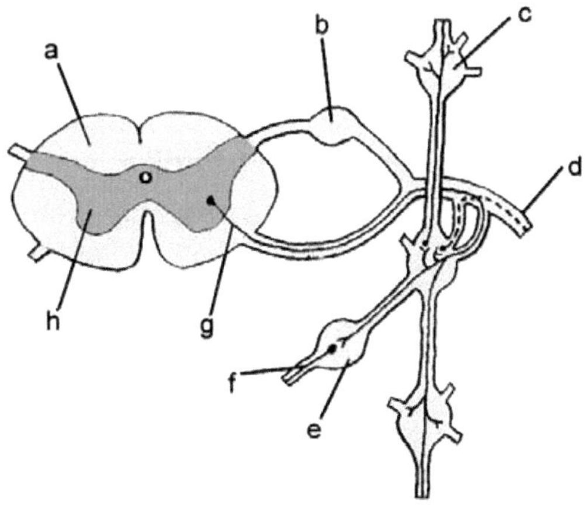

Fig. 11.12 Diagram showing the pathway of a sympathetic reflex arc. (a) white matter, (b) dorsal root ganglion, (c) sympathetic ganglion, (d) spinal nerve, (e) collateral ganglion, (f) a postganglionic neuron, (g) preganglionic neuron and (h) gray matter. The dotted line is a postganglionic neuron that forms a synapse with a shorter branch of the preganglionic fiber and joins the spinal nerve via the white ramus. Sympathetic ganglia are joined together by sympathetic trunks that contain axons that run in opposite directions. Adjacent ganglia together with sympathetic trunks on either side of the vertebral column give rise to structures that resemble chains of beads that are sometimes referred to as 'sympathetic chain ganglia'.

Sympathetic preganglionic neurons exit the spinal cord via the ventral roots then break from other spinal nerves through the white ramus to the sympathetic chain ganglia. In the sympathetic ganglia, the preganglionic neurone may synapse with a sympathetic postganglionic neurone, may give off branches to synapse with other postganglionic neurons in other sympathetic ganglia or may pass through several ganglia to splanchnic nerves that lead to prevertebral ganglia. The cell bodies of most sympathetic postganglionic neurons are located in the sympathetic chain ganglia or prevertebral ganglia. The larger postganglionic nerves will either return to the spinal nerve or continue separately with autonomic nerves to various sympathetic effectors in the body. Sympathetic responses affect many synapses since one preganglionic neuron synapses with many postganglionic neurons that lead to several organs.

The cell bodies of preganglionic neurons of the parasympathetic or craniosacral autonomic division are located in the nuclei of the brain stem or in the lateral gray columns of the sacral part of the spinal cord. The long axons of the parasympathetic preganglionic neurons are located in cranial nerves III, VII, IX and X and some of the pelvic spinal nerves. They synapse with postganglionic nerves in terminal ganglia that are located near or are embedded in the effectors in the chest and abdomen. A preganglionic parasympathetic neuron synapses with several short postganglionic neurons that supply a single autonomic effector and as a result parasympathetic stimulation often evokes response by a single organ.

Role of the Autonomic Nervous System

The fast responses that maintain homeostasis are brought about by the autonomic nervous system that regulates autonomic effectors. The type of response that occurs depends on the antagonistic roles played by the sympathetic and parasympathetic systems that continuously transmit impulses to autonomic effectors. Axons of preganglionic sympathetic neurons and both preganglionic and postganglionic parasympathetic neurons are **cholinergic fibers** since they release acetylchloline at their terminal ends. Fibers of postganglionic sympathetic neurons are **adrenergic nerves** and produce norepinephrine as a neurotransmitter substance. Sympathetic impulses stimulate an effector whereas the same effector is inhibited by parasympathetic impulses. In effectors that receive both sympathetic and parasympathetic nerve supplies, the action taken by the effector depends on the dominating type of autonomic nerve impulse.

The function of the autonomic nervous system is influenced by autonomic centers in the brain located in areas including the cerebral cortex and hypothalamus that send impulses indirectly to autonomic preganglionic neurons. The sympathetic nervous system is important in emergency cases and stimulation by this system brings about changes in various parts of the body in readiness for aggression in what is sometimes referred to as the 'fight-or-flight' reaction. Physiological changes that accompany such a phenomenon include a faster heart beat, increased blood supply to skeletal muscle due to dilation of blood vessels, dilation of bronchi and increased glycogenolysis that results in elevated blood glucose levels. Norepinephrine that is produced by the sympathetic postganglionic neurons is further enhanced by production of epinephrine and norepinephrine by the adrenal medulla on stimulation by the sympathetic neurons. Most activities of the body under normal conditions are under the control of the parasympathetic nervous system. The system stimulates digestive system processes and slows the heartbeat.

The **enteric system** (Gr. *enteros*, intestine) forms a network of nerves and is also considered to be part of the autonomic nervous system. The system comprises the **submucosal** or **Meissner's plexus** that lies in the submucosa of the gastrointestinal system and the **myenteric plexus** or **Auerbach's plexus** that is located between the circular and longitudinal smooth muscle layers of the same system. Although these intramural plexuses can function autonomously by detecting changes in the gut lumen, sympathetic and parasympathetic neurons that supply these plexuses also have some control on their function. The intramural plexuses are interconnected and comprise inter-neurons and motor neurons that control peristaltic movements, gastrointestinal gland secretions and blood vessel tone.

Autonomic Nervous System of Non-mammalian Vertebrates

Amniotes seem to have an autonomic nervous system that is similar to that of mammals. The autonomic nervous system of anamniotes has not been studied as much as in higher vertebrates and studies that have been conducted on anamniotes show that the sympathetic and parasympathetic divisions of the system are not as distinct as is found in amniotes. Not many organs in anamniotes have dual autonomic innervation.

The lampreys and hagfish lack ganglia in their sympathetic pathways that supply some blood vessels. The vagus nerve supplies the heart and digestive system in these agnathans. Some sympathetic ganglia appear in cartilaginous fish and some sympathetic nerves travel in the vagus to organs such as the stomach where they cause gastric secretion on stimulation. Parasympathetic neurons found in the vagus supply the heart and stomach and those in the oculomotor nerve supply the eye. The role of the parasympathetic nerves to the stomach is not understood.

The autonomic nervous system becomes better defined in teleosts and amphibians where sympathetic ganglia are found by the long axis of the spinal cord. More parasympathetic nerves leave the central nervous system via cranial nerves in these groups and through the sacral nerves in some amphibians such as anurans.

ENDOCRINE SYSTEM

The endocrine system is present in vertebrates and many animals with well-developed nervous and circulatory systems. The endocrine glands of birds and mammals are almost similar and are more complex than those of low vertebrates. The study of the endocrine system is most advanced in mammals with less being known about the system in most low vertebrates. Although parts of the endocrine system are located in various parts of the body, the system does not form an organ system. Both endocrine and nervous systems stimulate or inhibit target cells via transmission chemicals. The two systems can act independently or together as a neuroendocrine system. Whereas neural transmitters are produced by neurons, hormones are secreted by endocrine glands. In a few cases, hormones and transmitters are the same substance. Whereas neurons are activated into action by other neurons or receptor cells, endocrine organs are stimulated by several factors including changes in blood levels of their own hormone, other endocrine glands, nervous impulses and presence of certain substrates within the circulatory system. The transit time of hormones is longer than that of a nervous impulse.

Hormones

A hormone (Gr. *hormaein*, to excite) is a regulatory substance that is produced by endocrine glands but has an effect on target organs some distance from the site of production. About 50 different hormones are present in vertebrates. Hormones have been involved in the control of various processes in the vertebrate body including metabolism, growth, reproduction, sexual development, water and electrolyte balance, sugar and calcium balance, slow color changes, metamorphosis, and development of the nervous system. Hormones thus play a major role in the maintenance of homeostasis in the entire vertebrate body. Although some hormones are species specific as far as their functions are concerned, many of the hormones produced by vertebrates have a similar action in other vertebrates and show little or no species specific action. Hormones are normally effective in small quantities.

Hormones are transported passively to various parts of the body either on their own or in bound form using carrier plasma proteins. Once in the target organ, a hormone has to move out of capillaries to target cells with receptor molecules that are specific to the particular hormone. Hormones control biochemical reactions in target cells in various ways. A hormone can stimulate the synthesis of certain substances such as proteins, suppress existing cellular enzymes and affect the rate at which some substances enter or leave the target cell. Although integration by the endocrine system is slower than that of the nervous system, its effect is much longer lasting.

Hormones can be classified according to their chemical structure or general function. The chemical classification of hormones is based on whether the hormone is of steroid or non-steroid origin. Nonsteroid hormones are normally proteins, peptides, amino acid derivatives or glycoproteins. Steroid hormones are produced by endocrine glands of mesodermal origin such as the adrenal cortex and gonads whereas nonsteroid hormones are synthesized by glands of ectodermal or endodermal origin. Based on general function, hormones are known as tropic hormones if they influence the secretion of hormones by other glands.

Steroid hormones are synthesized from the lipid choresterol and are very similar to this lipid structurally (Fig. 11.13). Steroid hormones are not stored in vesicles of cells that synthesize the hormones. The hormones

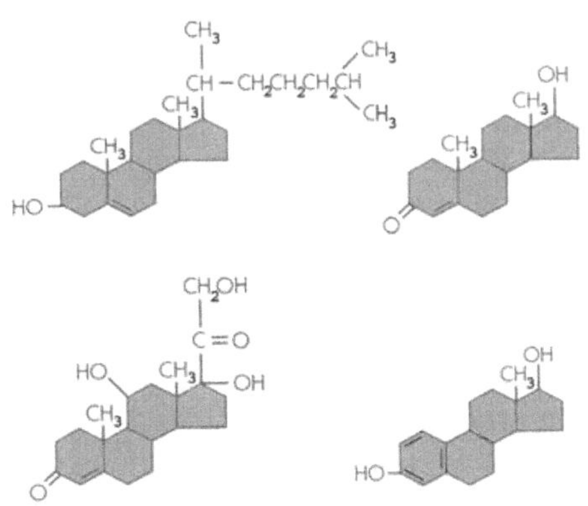

Fig. 11.13 Cholesterol (top left) and the steroid hormones testosterone (top right), cortisol (bottom left) and the estrogen estradiol (bottom right). Cholesterol and its derived hormones all have a steroid nucleus that is highlighted in gray.

normally diffuse out of the endocrine cells soon after they are synthesized to enter the bloodstream. Since steroid hormones are lipid soluble, they easily pass through the plasma membrane of target cells. Some of the steroid hormones include testosterone, estrogen, progesterone, aldosterone and cortisol (hydrocortisone).

Nonsteroid hormones are of amino acid origin. Protein nonsteroid hormones comprise long chains of amino acids that are folded and include insulin, glucagon, growth hormone, prolactin, parathyroid hormone, adrenocorticotropic hormone, and calcitonin. Glycoprotein hormones such as follicle-stimulating, luteinizing, thyroid-stimulating and chorionic gonadotropin hormones have a carbohydrate group attached to their amino acid chains. Peptide hormones are smaller than protein hormones as their amino acid chains are shorter. Examples of peptide hormones include oxytocin, antidiuretic, gonadotropin-releasing, thyrotropin-releasing, somatostatin and melanocyte-stimulating hormones. Hormones that are derived from a single amino acid such as epinephrine, norepinephrine, melatonin, thyroxine (T_4) and triidothyronine (T_3) are known as amino acid derived hormones. Amino acid hormones are smaller than other hormones structurally but that does not reflect on their effect in the body. Nonsteroid hormones can be strored in vesicles of endocrine cells before release.

Hormonal Action

Hormonal action obeys the 'lock-and-key' mechanism whereby hormones bind to receptor sites that conform to their structure and such action ensures that hormones are specific in their functions as they can only influence the action of specific receptors or target organs. Normally, cells have several different receptors for hormones. Hormone-receptor interaction regulates various chemical processes within the target cell. The type of regulation brought about by hormones varies depending on such interaction since it can activate or inactivate certain enzymes thus altering the biochemical pathway and products that are eventually synthesized by the target cells.

Sometimes different hormones complement each other during synergism in integrating the activities of a target cell and the combined effect of such hormones is more than that of the effect of either hormone acting on its own. In other cases, the combined effect of hormones on a target cell results in antagonism whereby one hormone acts in an opposite manner to another hormone. The combined regulation of cellular activity by hormones ensures that the desired effect in target cells is as precise as possible. Hormones that are able to lock onto target cells are either degraded or excreted by the kidneys.

Steroid hormones are carried in the blood attached to soluble plasma proteins since they are not soluble in water. The lipid soluble steroid hormones cross the plasma membrane of target cells easily to attach to their receptors that are mainly found inside cells. The steroid hormone then moves to the nuclear membrane where it binds to a mobile receptor molecule to produce an activated hormone-receptor complex that binds to DNA. The complex activates specific genes leading to an increase in production of proteins. Nonsteriod hormones attach to fixed receptors on the plasma membrane of target cells where they activate the desired cellular processes via a 'second messenger'. Five different second messenger molecules including cyclic AMP are known. Intracellular chemicals are activated by second messenger chemicals to produce the target cell response.

Hormonal secretion is regulated mainly by negative feedback mechanisms known as an **endocrine reflexes** and **cycles of secretion**. The cycles of secretion can last from hours to months in duration and sustain both physiological and homeostatic control. Under normal circumstances endocrine glands are sensitive to levels of substances that are produced by target cells or their own hormone levels in blood. When levels of a certain substance fall, endocrine cells will increase their production of the desired hormone and vice versa in a process that is normally gradual. Several hormones including insulin show dramatic changes in their levels in plasma in a short period of time depending on rapid changes that are associated with changes of certain substances in the blood.

PROSTAGLANDINS

Prostaglandins are lipid molecules that are synthesized from 20-carbon unsaturated fatty acids and play several integrative roles in the vertebrate body. Prostanglandins derive their name from the prostate gland as they were first discovered in semen in 1957. About 16 prostanglandins were known by 1993. The levels of prostaglandins in circulation are rather low as they are rapidly metabolized. These molecules generally have an effect in tissues in which they are produced and have been referred to as tissue hormones. Prostaglandins play a role in regulation of endocrine glands by binding to cell membrane receptors and use cyclic adenylate monophosphate (AMP) as a second messenger within the target cells.

In birds and mammals, prostaglandins have been implicated in reproductive functions including parturition, ovulation, oviposition and luteal function. Although the molecules also stimulate contraction of the oviducts in various ways, there is neural control over such contractions.

The roles played by prostaglandins and the nervous system in the evolution of viviparity in vertebrates have been reviewed by Guillette Jr. et al. (1991). These local hormones could have coordinated oviductal contractions after ovulation in ancestral amphibians. Lower vertebrates have a greater diversity of prostaglandins and their production is regulated by environmental temperature (Herman, 1990). Prostaglandins also play a role in inflammation and induce a fever by causing local vasodilation and an increase in blood flow.

ENDOCRINE GLANDS

Endocrine glands (Gr. *endon,* within; *krino,* to separate) are ductless glands or glands of internal secretion that have lost their original connection with the epithelium or a free surface. Endocrine glands are distributed in various parts of the body (Fig. 11. 14). The glands are small and since they lack ducts when compared to exocrine glands (Gr. *ex*, out), the hormones they produce are secreted into the rich blood supply through the fenestrated or sinusoidal endocrine capillaries.

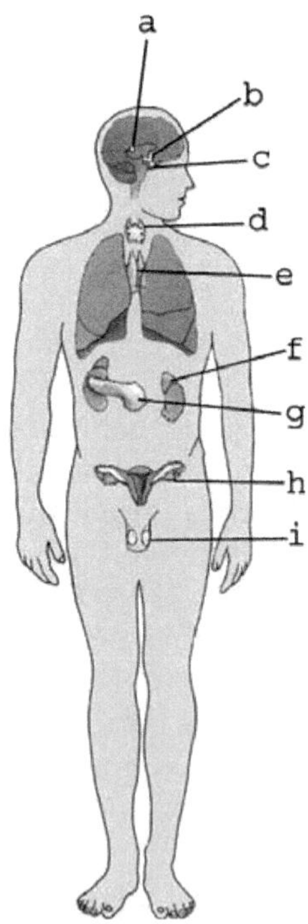

Fig. 11.14 Location of endocrine organs in the human body. (a) pineal gland; (b) hypothalamus, (c) pituitary, (d) thyroid, (e) thymus and (f) adrenal glands; (g) pancreas, (h) ovary of females and (i) testis of males.

Endocrine glands appear in various forms and can be distinct, may be associated with exocrine glands, or be part of other endocrine glands or may be composed of diffusely distributed cells such as the argentaffin cells of the digestive system. In lower vertebrates, endocrine glands are often diffuse in nature although they are mainly discrete in tetrapods. Endocrine glands show diverse but distinct structural and functional organization. The glands normally store their secretory products within their cells or in follicles or sacs in readiness for release after specific stimulation.

The Pituitary Gland

The pituitary gland (L. *pituitarius*, slime) or hypophysis (Gr. *hypo*, under; *physis*, growth) (Fig. 11.15) is the major endocrine gland of vertebrates and is sometimes referred to as the master gland since hormones that are produced by the gland have an effect on most of the other endocrine glands in the body. The gland is attached to the ventral part of the hypothalamus in the brain by a stem-like stalk known as the **infundibulum**. The pituitary gland is made of two separate glands both of which are of different embryological origin. The anterior pituitary or **adenohypophysis** (Gr. *aden*, gland) develops from the roof of a developing mouth as a pouch known as **Rathke's pouch** that grows dorsally and the posterior part of the gland or **neurohypophysis**

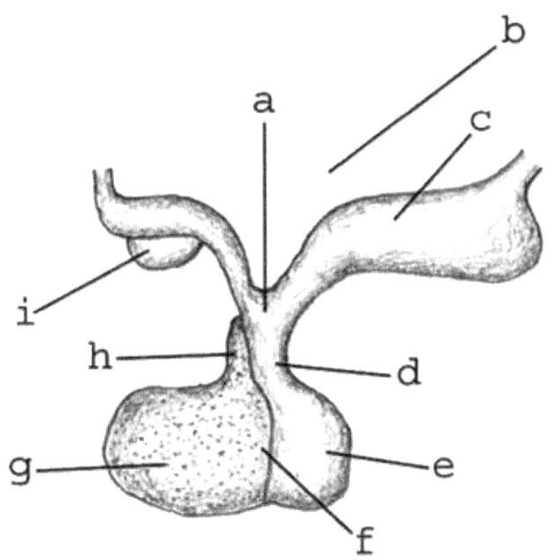

Fig. 11.15 Structure of the mammalian pituitary gland or hypophysis. (a) medial eminence, (b) 3rd ventricle, (c) hypothalamus, (d) infundibulum, (e) neurohypophysis or posterior pituitary, (f) pars intermedia, (g) adenohypophysis or anterior pituitary, (h) pars tuberalis and (i) optic chiasma. The pars distalis is the part of the adenohypophysis that lies anterior to the pars intermedia. The pars distalis, intermedia and tuberalis form the adenohypophysis while pars nervosa, infundibular stalk and medial eminence are parts of the neurohypophysis.

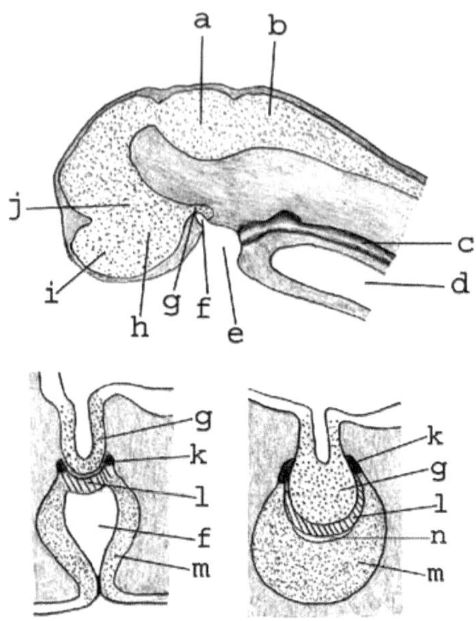

Fig. 11.16 The anterior part of an embryo showing the location of the developing pituitary gland (above) and the relationship between the developing neurohypophysis and adenohypophysis (below) at an earlier stage (left) and later on (right). (a) metencephalon, (b) rhombencephalon, (c) archenteron, (d) coelom, (e) stomodeum, (f) Rathke's pouch, (g) neurohypophysis, (h) telencephalon, (i) diencephalon, (j) mesencephalon, pars (k) tuberalis, (l) intermedia and (m) distalis and (n) cleft that is a remnant of Rathke's pouch. (k), (l) and (m) form the adenohypophysis. The stomodeum develops into the anterior part of the oral cavity. The apposition of the stomodeal ectoderm and the foregut endoderm forms the buccopharyngeal membrane.

develops from a ventral growth from the diencephalon (Fig. 11.16). The two growths meet and will develop into a pituitary gland. A cleft that represents the remnant of the Rathke's pouch lumen may be present and divides the adenohypophysis into the thin **pars intermedia** posteriorly beside the **pars nervosa** and a thicker **pars distalis** anteriorly. The pars intermedia normally fuses with the pars nervosa. Morphogenesis of the pituitary primodium is induced by signals from the diencephalon (Takuma et al., 1998). In some vertebrates, especially mammals, the **pars tuberalis** is present as an upgrowth around the stalk of the infundibulum.

The adenohypophysis is composed of normal endocrine tissue that is made of masses and cords of secretory epithelial cells that are separated by sinusoidal capillaries in loose connective tissue. The cells of the pars tuberalis show signs of little secretory function. The neurohypophysis is a neurosecretory tissue that serves as a storage depot and consists of mainly long parallel nerve fibers that originate from the hypothalamus with a ramification of many small blood vessels. Supportive and branched cells that play a role similar to that of neuroglia are also present in the neurohypophysis. The pituitary glands of various vertebrate groups are shown in Fig. 11.17.

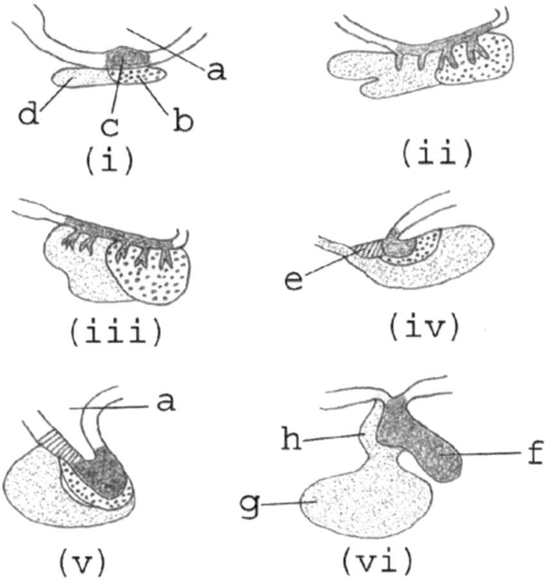

Fig. 11.17 Pituitary glands of the lamprey (i), elasmobranch (ii), teleost (iii), urodele (iv), reptile (v) and bird (vi). (a) 3rd ventricle, (b) pars intermedia and (d) pars distalis of the adenohypophysis, (c) pars nervosa of the neurohypophysis, (e) medial eminence, (f) neurohypophysis and (g) adenohypophysis with pars tuberalis (h). The pars intermedia is lacking in birds. A distinct median eminence lacks in fishes apart from lungfishes.

The anterior pituitary gland is functionally connected to the hypothalamus by the **hypothalamohypophyseal portal system** (Fig. 11.18) that is lacking in fish. The pituitary gland in many fish species has finger-like projections that extend from the neurohyposphysis into the adenohypophysis and might play the role of this portal system. The hypothalamohypophyseal portal vein carries hypothalamic hormones that control the synthesis and discharge of hormones from the anterior pituitary gland. The function of the anterior pituitary gland can also be influenced by hormones in the cerebrospinal fluid. The mammalian anterior pituitary gland produces the following hormones: growth, thyroid-stimulating, adrenocorticotropic, follicle-stimulating, luteinizing, melanocyte-stimulating and prolactin. Several of these hormones are also present in other vertebrates.

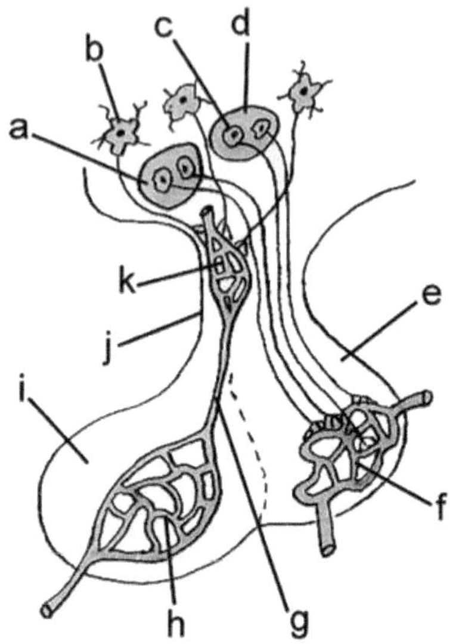

Fig. 11.18 Diagram showing the relationship of the mammalian pituitary gland with the hypothalamus. (a) supraoptic nucleus, (b) and (c) neurosecretory cells of the hypothalamus, (d) paraventricular nucleus, (e) neurohypophysis, (f) and (h) capillary beds, (g) hypophyseal portal vein, (i) adenohypophysis, (j) infundibular stalk (L. *infundibulum*, little funnel) and (k) capillaries in median eminence.

The **growth hormone** (**GH**) promotes growth in various ways including promoting protein synthesis by stimulating the liver to synthesize factors known as somatomedins that promote entry of amino acids into cells. This action is followed by protein anabolism in bone, muscle, connective tissue and other body tissues. Such an effect of GH has been seen in some fish and most terrestrial vertebrates. Lower than normal levels of GH in the human being have been associated with dwarfism and excess levels with gigantism. The growth hormone has been identified in the sea lamprey and seems to be the only member of the family of growth hormones that also comprise prolactin and somatolactin in this fish that is a representative of the most ancient vertebrates (Kawauchi et al., 2002). The hormone could be the ancestral hormone in the molecular evolution of the growth hormone family.

The thyroid gland is located ventral or anterior to the larynx and is stimulated by the thyroid stimulating hormone (TSH) or thyrotropin to synthesize the hormones **thyroxine** and **triiodothyronine**. These hormones promote the rate of oxidative metabolism in vertebrates. **Adrenocorticotropic hormone** (**ACTH**) or **adrenocorticotropin** sustains the growth of the adrenal cortex and also stimulates the production of its hormones that have an effect on metabolism. The adrenal gland lies cranial to or above the kidney. **Follicle stimulating hormone** (FSH) and **luteinizing hormone** (**LH**) stimulate the development of ovarian follicle and the formation of corpus luteum after ovulation. In males, FSH plays a role in the development of seminiferous tubules of the testis that produce sperm and together with LH (also known as interstitial cell stimulating hormone in males) promotes the synthesis of testosterone by the interstitial cells of the testis.

Melanocyte (**Melanophore**) **stimulating hormone** (**MSH**) is produced by cells of the anterior and middle part of the pituitary gland. This hormone stimulates melanocytes in the skin to synthesize melanin. In birds and mammals, melanin is deposited outside melanocytes in epidermal cells whereas in fishes, amphibians and reptiles the pigment remains in the pigment producing cells. The dispersal of melanin within melanocytes is subject to physiological changes in fishes, amphibians and reptiles and is important in adjusting the color of the skin to the background.

The anterior pituitary gland also produces the lactogenic hormone **prolactin** that stimulates the synthesis of milk by the mammary glands. This hormone is found in most vertebrates with the exception of hagfish and lampreys and has the widest range of action of all the anterior pituitary gland hormones. Prolactin plays a role in the production of somatomedins by liver cells as well as production of testosterone by the testis and progesterone by the corpus luteum. In fish, prolactin plays an important role in freshwater osmoregulation by preventing both the loss of ions and the uptake of water (Manzon, 2002). Prolactin is also associated with the water-seeking behavior of adult amphibians during the breeding season and the production of crop or pigeon 'milk' in the crop of pigeons and discus 'milk' in the discus fish. Discus 'milk' is a skin secretion that is rich in protein and is used to nourish the young. Prolactin shares a common tertiary structure with the polypeptide hormones such as growth hormone, somatolactin of fish and placental lactogen of mammals. The hormones are thought to have arisen as a result of gene duplication and subsequent divergence early in vertebrate evolution (Forsyth and Wallis, 2002).

The anterior pituitary also produces **endorphins** that control pain receptors in the brain. The three types of endorphins are beta-endorphins that are found primarily in the pituitary gland and enkephalins and dynorphin that are found throughout the nervous system. Endorphins or 'endogenous morphine' are neurotransmitters that were discovered in 1975 and are morphine-like substances or opioids. These polypeptides contain 30 amino acids that consist of tyrosine mainly and are naturally produced in the brain during deep breathing, meditation, eating spicy food, during acupuncture, chiropractic manipulation or even hearty laughter. Endorphins are produced in cases of severe stress and block pain signals arising from the nervous system thus reducing stress and relieving pain. The 20 or so different endorphins have also been implicated in enhancing the body's immune system and slowing down the aging process.

Control of Hormonal Secretion in the Anterior Pituitary

The hypothalamus controls the entire endocrine system. Synthesis and secretion of hormones by the adenohypophysis is under the control of the hypothalamus. The hypothalamus also receives information from nerves about the internal and external environment and in response sends out appropriate nerve signals to endocrine glands. The cells bodies of neurons in the hypothalamus are able to synthesize **releasing** and **inhibiting hormones** that are small neuropeptide hormones that stimulate or inhibit the release of each of the hormones of the adenohypophysis under negative feedback mechanisms respectively. Releasing hormones are secreted into the blood by axons and will be carried by the hypothalamohypophyseal portal system to the anterior pituitary gland where they will stimulate the target cells to synthesize the relevant hormones. As the portal system transports blood directly from the hypothalamus to the adenohypophysis, only small amounts of releasing hormones are necessary to bring about the desired effect. The pituitary gland will then be able to adjust the activity of target tissues in many parts of the body.

Neurohypophysis

The posterior pituitary gland or neurohypophysis secretes the related peptide hormones antidiuretic hormone (ADH) and oxytocin (OT) that are synthesized in cell bodies of neurons in the supraoptic and paraventricular nuclei of the hypothalamus. The hormones are then secreted by axons into the neurohypophysis. The release of the two hormones into blood is dependent on nervous stimulation.

ADH or **vasopressin** is produced when the osmotic pressure of blood increases as a result of dehydration. This change is detected by osmoreceptors near the supraoptic nucleus of the hypothalamus. This triggers the neurohypophysis to release ADH into circulation. ADH causes an increase in reabsorption of water in the kidney tubules by opening of more water channels in the wall of the tubules. An increase in water content of blood is followed by a fall in the blood osmolality and restoration of the normal osmotic pressure.

Oxytocin (OT) (Gr. *okytokos*, swift birth) is capable of stimulating contraction of smooth muscle in the uterus and mammary glands. OT and related hormones play a role in the reproduction of all vertebrates at several levels (Gimpl and Fahrenholz, 2001). The release of OT from the posterior pituitary depends on several factors including suckling, parturition and stress. Oxytocin secretion is controlled by a positive feedback mechanism from the target cells. An increase in signals from the target cells is followed by an increase in the production of OT. Various tissues of the body also synthesize oxytocin. The placenta, uterus, amnion, corpus luteum, heart and testis have been implicated in the process of OT production. Oxytocin plays a central role in social approach behavior in nonhuman mammals and improves the ability to infer the mental state of people from social cues of the eye region (Domes et al., 2007).

Pineal Gland

The pineal gland (L. *pineus*, relating to pine) (Fig. 11.6), also known as the pineal body, epiphysis or 'third eye' is a small organ located in the brain. The gland is called the 'third eye' due to its role of sensing light and dark cycles in lower vertebrates where it is better developed than in mammals. The gland is lacking in cetaceans. In fish, amphibians, reptiles and birds, the pineal gland is located near the skin where it is able to detect light. In the human being, the pineal gland does not receive light transmission directly since it is located above the thalamus between the two cerebral hemispheres. Light to the mammalian pineal gland is transmitted through the eyes and the optic nerve.

The pineal gland is sensitive to different intensities of light detected by the eye and plays a role in the proper functioning of an animal's biological clock. The gland also produces **melatonin**, a hormone, during darkness. Light inhibits the synthesis of this hormone that has been implicated in various roles in the body including sleep, sexual development, seasonal breeding, hibernation, metabolism, growth and migration. In lower vertebrates the pineal gland has a structure that is similar to that of the eye and is a light receptor that is able to register day and night cycles without the use of eyes. Sensing of such cycles by the pineal gland in mammals has been superseded by the eyes and the nervous system.

The pineal gland of anamniotes has **pinealocytes** (main parenchymal cells) with retinal cone photoreceptor-like characteristics. Sauropsids had less developed photoreceptors with secretory characteristics that have been thought to represent a transition between the anamniote pineal photoreceptors and the mammalian non-sensory and secretory pinealocytes (Ekstrom and Meissl, 2003). The avian pineal gland also represents a transition between a photosensory organ of lower vertebrates and the endocrine gland of mammals and shows great changes in its innervation and structure during development (Sato, 2001). Some photoreceptive cells are present in the avian pineal gland.

Urophysis

The urophysis (Gr. *oura*, tail; *physis*, growth) or neurohemal contact site is a neurosecretory organ that is located on the ventral aspect of the caudal end of the spinal cord in teleost fish. Secretory neurons from the caudal end of the spinal cord send unmyelinated axons into the urophysis where they terminate close to capillaries. The urophysis sysnthesizes the hormones **urotensin I** and **urotensin II.** Urotensin I has a molecular structure that is similar to somatostatin while urotensin II belongs to the family of peptides that includes corticotropin-releasing hormone of mammals. Urotensins cause an increase in blood pressure of teleosts and also play a role in osmoregulation. Urotensin II is the most potent vasoconstrictor hormone known and may have an aetiological role in hypertension and its complications (Cheung et al., 2004).

In most vertebrate species, urotensin II is synthesized in neurons of the central nervous system. The heart, liver and kidney are also sources of urotensin II (Christopher et al., 2005).

Thyroid Gland

The thyroid gland (Gr. *thyroides*, resembling an oblong shield) in jawed vertebrates is a bi-lobed structure that is found caudal or inferior and ventral or anterior to the pharynx. In fish, the gland remains close to its original position and consists of follicles that are scattered throughout the connective tissue that is ventral to the middle part of the pharynx. The thyroid follicles can be widely distributed in fish and can extend to the ventral aorta, hepatic veins and anterior part of the kidney. In a few fishes and tetrapods, the gland becomes surrounded by a connective tissue capsule. The gland is a paired structure in the floor of the pharyngeal cavity in amphibians. In reptiles and birds, the gland has migrated much more caudally to lie close to the posterior end of the trachea and can be either a single median organ or is paired. The thyroid gland of mammals is bi-lobed and is located close to the posterior end of the larynx. In the human being, the gland is located on the anterior and lateral surfaces of the thyroid cartilage of the larynx and part of the trachea. The thyroid gland is composed of nearly spherical follicles that lie in connective tissue that is well vascularized (Fig. 11.19). Each follicle is hollow and has a wall composed of a simple cuboidal glandular epithelium that synthesizes the thyroid hormones. The hormones are stored in the thick fluid known as **thyroid colloid** that occupies the interior part of the follicle.

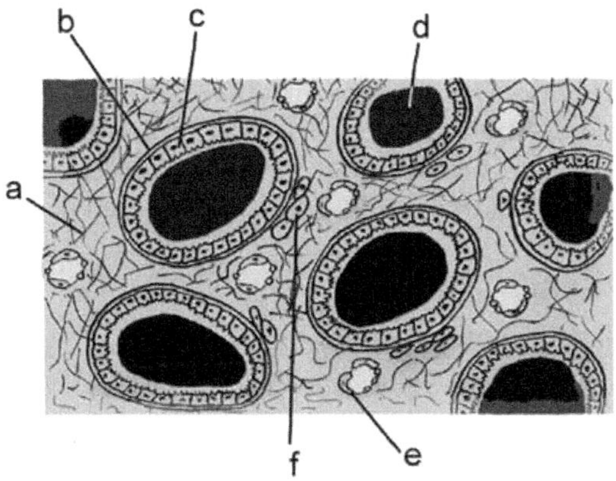

Fig. 11.19 Thyroid gland. (a) interfollicular tissue, (b) basement membrane surrounding a follicle, (c) cuboidal epithelium with microvilli on the luminar surface, (d) colloid in follicle, (e) fenestrated capillary and (f) parafollicular or C cells.

The thyroid gland produces **tetraiodothyronine** (T_4) or **thyroxine** and **triidothyronine** (T_3). Most of the hormone produced is T_4. The thyroid gland is stimulated to produce these hormones by the thyrotropic hormone of the anterior pituitary that is influenced by the thyrotropic releasing hormone of the hypothalamus. The hormones of thyroid gland have a wide range of effects in vertebrates and generally stimulate and increase in the metabolic rate of body cells. The thyroid hormones are also important during development and maturation. The hormones are necessary in the metamorphosis of larvae of lampreys, bony fishes and amphibians, bone and nerve development in mammals and maintenance of heart rate, blood pressure, muscle tissue and digestion. Many vertebrates require the thyroid gland hormones for the normal development of the skin and its derivatives. Skin molting in amphibians and reptiles is abnormal in cases of low levels of these hormones.

The hormone **calcitonin** was discovered much later than T_3 and T_4 and is produced by **parafollicular** or **C cells** that are located between thyroid follicles in the thyroid gland of mammals. In other vertebrates, calcitonin is produced by ultimobranchial bodies. The hormone is also synthesized in other tissues including lungs and intestines. In fish, amphibians, birds and mammals, calcitonin is important in calcium and phosphorus metabolism. Calcitonin normally lowers blood calcium levels by inhibiting the tubular re-absorption of calcium and phosphorus in kidneys.

The thyroid gland evolved from the **endostyle** of protochordates that is located in the lower part of the pharyngeal region. The endostyle still persists in larvae of lampreys. The endostyle is an iodide-trapping and glycoprotein-secreting gland that has the ability to bind iodide to a glycoprotein. Swallowing of the iodinated glycoproteins is followed by enzymatic degradation resulting in production of iodinated amino acids that are referred to as thyroid hormones. In lampreys, the endostyle fragments and the resulting cells rearrange themselves into follicles of the thyroid gland when a larva undergoes metamorphosis into an adult.

Parathyroid Gland

Parathyroid glands develop from the third and fourth pharyngeal pouches and are lacking in fish. The parathyroid gland of tetrapods and gills of fish are related structures in terms of evolution and the gland could have emerged as a result of the transformation of the gills during tetrapod evolution (Okabe and Graham, 2004). Most tetrapods have four parathyroid glands that lie close to the thyroid gland or are embedded in it as is found in most mammals. The glands comprise irregular rows of cells and a network of blood capillaries. Parathyroid glands secrete **parathyroid hormone** (PTH) or **parathormone**. In bony fish (lack cellular bone), **corpuscles of Stannius** that form islands of eosinophilic granular cells which are found in paired organs on the ventral surface of the kidney secrete **hypocalcin** or **teleocalcin** that lowers blood calcium levels.

Parathormone is antagonist to calcitonin and so plays a role in calcium homeostasis. The cells of the gland monitor calcium levels in blood. A fall in blood calcium levels is followed with an increase in the release of parathormone. The hormone restores the normal blood calcium levels by increasing the resorption of the mineral from bone, decreasing the loss of calcium by the kidneys and promoting the uptake of the mineral in the intestines by activating vitamin D (cholecalciferol). Vitamin D is normally obtained from food or is synthesized from cholesterol in the skin under ultraviolet light and increases the uptake of calcium in the intestines into blood.

Ultimobranchial Bodies

The paired ultimobranchial bodies or glands (L. *ultimus*, farthest; Gr. *branchia*, gills) or postbranchial bodies develop from the epithelium of the posterior surface of the last (fifth or sixth) pair of pharyngeal pouches. Ultimobranchial bodies are distinct in fishes, amphibians, reptiles and birds but are incorporated into the thyroid gland of mammals as parafollicular cells during embryological development. Ultimobranchial bodies lie ventral to the esophagus in the transverse septum that separates the heart from the abdominal cavity and are made of cells of various shapes which form follicles, cellular masses or cell strands. The bodies secrete the hormone calcitonin.

Adrenal Gland

In mammals, adrenal glands (L. *ad*, prefix for toward; *rene*, kidney) or suprarenal glands are paired. The glands are located in front or on top of the kidneys. Each gland has an outer cortex and an inner medulla (Fig. 11.20). The two parts of the gland are of different embryological origin and are structurally and functionally so different that they are like two separate glands in one gland. The cortex is made of ordinary endocrine tissue that comprises cords of cells whereas the medulla is a neurosecretory tissue.

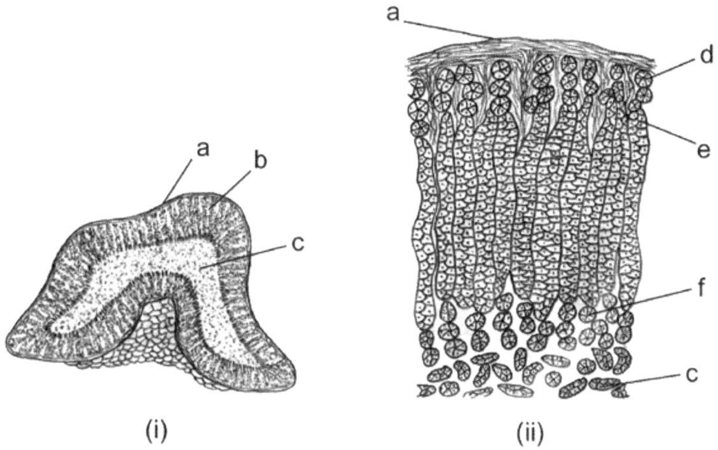

Fig. 11.20 Structure of the adrenal gland (i) and detailed arrangement of cells of the gland (ii). (a) capsule, (b) cortex, (c) medulla and zonae (d) glomerulosa, (e) fasciculata and (f) reticularis.

The adrenal cortex develops from the epithelial cells of the coelom and secretes groups of steroid hormones (corticosteroids) known as mineralocorticoids, glucocorticoids and gonadocorticoids. The most important **mineralocorticoid** (hormones regulating the processing of minerals in the body) is **aldosterone** and is secreted by **zona glomerulosa**. Aldosterone plays a role in sodium homeostasis in blood. It promotes reabsorption of sodium in the kidney and excretion of potassium and hydrogen ions thus increasing retention of water in the body. It also influences the pH of blood in the process. **Glucocorticoids** are synthesized by cells of **zona fasciculata** and affect carbohydrate metabolism. The three main glucocorticoids are **cortisol** (**hydrocortisone**), **cortisone** and **corticosterone**. Glucocorticoids stimulate pathways that lead to an increase in levels of blood sugar. The hormones therefore affect virtually every cell in the body. High levels of glucocorticoids decrease the body's immunological responses. **Gonadocorticoids** (cortical androgens) are sex hormones produced by **zona reticularis** of the adrenal gland. Small quantities of male hormones (androgens) are normally produced. Normal levels of androgens promote protein synthesis and muscle growth in male and female vertebrates. Cells of zona reticularis also produce small quantities of glucocorticoids.

The **chromaffin cells** (Gr. *chromo*, color; *affinis*, affinity) are of neural crest origin. The cells are modified sympathetic postganglionic fibers of the autonomic nervous system that occupy the adrenal medulla. These cells secrete their products into the venous sinuses they surround. Chromaffin cells sythesize **catecholamines** known as **epinephrine** and **norepinephrine**. Chromaffin cells are innervated by preganglionic fibers of the sympathetic nervous system. Stimulation by sympathetic fibers leads to secretion of epinephrine and norepinephrine into capillaries. The two hormones can bind to receptors of sympathetic effectors such as smooth muscle, the heart and glands to prolong sympathetic effect. Norepinephrine is also a neurotransmitter that is produced by sympathetic postganglionic neurons. Epinephrine is synthesized from norepinephrine in mammals and forms the largest quantity of hormones produced by the adrenal medulla. Epinephrine is also more potent than norepinephrine.

Non-mammalian vertebrates have cells that are similar to those of the adrenal cortex known as **interrenal cells** while those that correspond to the medulla are chromaffin cells. In low vertebrates (fish and amphibians), adrenal glands are sometimes referred to as **interrenal glands**. The certex and medulla of non-mammalian vertebrates are not as discretely arranged as those of mammals since chromaffin cells of non-mammalian vertebrates are not completely surrounded by cortical cells and show various degrees of structural association (Fig. 11.21). Apart from a few groups of fish such as sculpins, no complete adrenal gland as a unit is present in fish. In fishes, interrenal cells are found in the anterior part of the kidney where they are associated major blood vessels. Chromaffin cells sometimes vary in location but are normally found in clumps together with sympathetic ganglia in the interrenal tissue or between the anterior part of the kidney and spine. The adrenal glands of amphibians are diffused and elongated structures that are embedded in the ventral part of kidneys. Although reptiles and birds have complete adrenal glands, the cortex and medulla are not distinct structures. In non-mammalian vertebrates and young mammals, norepinephrine is the main secretion of the chromaffin cells.

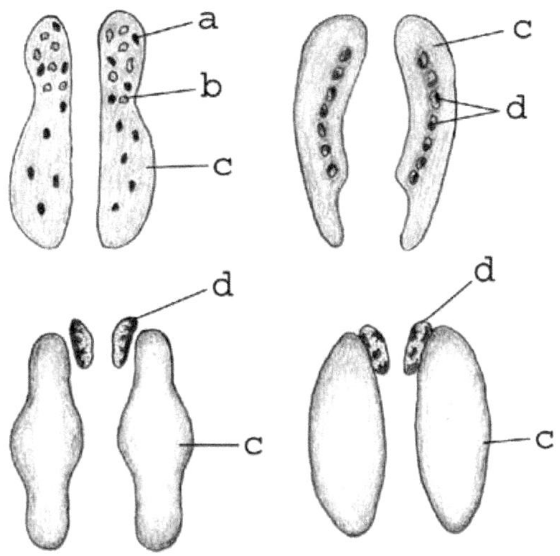

Fig. 11.21 The adrenal glands of various vertebrate groups. Fish (top left), anurans (top right), reptiles (bottom left) and birds (bottom right). (a) clumps of chromaffin cells, (b) interrenal cells, (c) kidney and (d) adrenal glands.

Though the human genome contains two to three times (35,000 genes) the number of genes in invertebrates, the human being shows quite some complex development and differentiation when compared to lower forms of life. Only about 7,000 of human genes seem to be unique to vertebrates. The unique vertebrate ligand based mechanism that entails the actions of adrenal gland and sex gland hormones through receptors that evolved from ancestral nuclear receptors in protochordates influences most form of vertebrate differentiation and development (Baker, 2003).

Pancreas

As stated in Chapter 7, the pancreas (Fig. 11.22) has both exocrine and endocrine functions. The exocrine and endocrine pancreas of tetrapods is well developed and the endocrine cells are organized into distinct islets of Langerhans that consist of clusters of hormone producing cells whose abundance of different cell types varies. In the human being and various mammals, the islets number about 1 million and constitute about 1-2% of the pancreatic mass that receives about 15% of the blood supply to the pancreas. The islets are also innervated by sympathetic and parasympathetic nervous systems that modulate the secretion of their hormones. Most fishes lack a discrete pancreas and the location and level of development of the organ varies greatly in this group of vertebrates.

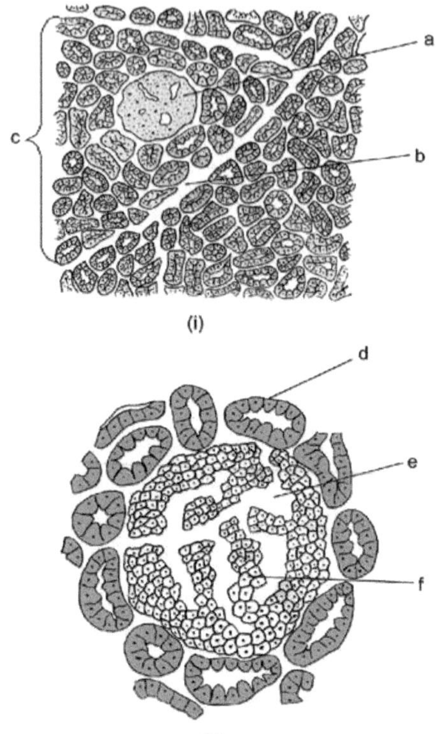

Fig. 11.22 Portion of the pancreas (i) and islet of Langerhans (ii). (a) Islet of Langerhans, (b) interlobular connective tissue, (c) part of a lobule, (d) exocrine acinus, (e) capillary and (f) cords of cells of the endocrine pancreas.

In low fishes such as hagfish, the endocrine pancreas comprises a few somatostatin producing cells that are scattered within the intestinal mucosa and insulin-producing lobular cell nests that are closely associated with the bile duct. The endocrine part of the pancreas lacks an exocrine component in these fish. The islets of Langerhans are associated with the exocrine pancreas in most fish. Holocephalans have a simple pancreas that is made of exocrine cells with a duct that opens directly into the lumen of the intestines and islets that comprise three different types of cells that produce insulin, somatostatin and glucagon. The pancreas of sharks, rays and related fish has an exocrine portion and islets have a fourth cell type that produces pancreatic polypeptide. Discrete islets known as **Brockmann bodies** or **principal islets** that contain four types of endocrine cells are found in the pancreas of teleosts. In some teleosts, Brockmann bodies can be quite large that can be seen grossly while in other species, the size and number of islets will increase during the spawning season. Apart from insulin, cells that produce other characteristic pancreatic endocrine polypeptides are also scattered throughout the wall of the gastrointestinal tract.

Glucagon is produced by alpha cells that form about 15-20% of the islets in mammals. Glucagon increases the blood glucose levels by stimulating the enzymes that convert glycogen to glucose in the liver cells. The enzyme also stimulates the conversion of fatty acids and amino acids to glucose during gluconeogenesis. The beta cells constitute 65-80% of the mammalian islets and produce **insulin** that increases movements of glucose, fatty acids and amino acids into mainly muscle and adipose tissue, by increasing the permeability of the plasma membrane thus lowering blood glucose levels. Insulin also activates enzymes that convert glucose to glycogen in liver and muscle cells and to fat in adipose tissue. Insulin also stimulates the conversion of amino acids into protein. The overall antagonistic effect of glucagon and insulin determines the blood glucose levels. Abnormally low levels of insulin lead to a disease known as diabetes mellitus in which there are elevated levels of glucose in blood, a condition that is accompanied with various complications. The quantities of glucose in blood stimulate the secretion of either glucogen or insulin.

Pancreatic polypeptide is produced by the pp or F cells that constitute about 1% of the mammalian pancreatic islets and plays a role in digestion and distribution of food in the body. With the exception of rodents, the primary structure of pancreatic polypetide has been quite strongly conserved in mammals and the extreme variability in the sequences of amphibian and reptilian pancreatic polypeptides means that the peptide is a useful molecular marker to study the branching order in early tetrapod evolution (Conlon, 2002). Delta cells form 3-10% of the mammalian islets and produce **somatostatin** that inhibits alpha, beta and pp cells. Somatostatin also inhibits the production of the growth hormone from the anterior pituitary. Somatostatin is also produced by other endocrine cells in the body.

Gonads

Although gonads are the primary sex organs of vertebrates, each organ produces its own hormones. The male gonad, the testis, contains interstitial cells of Leydig that are located between the coiled seminiferous tubules (Fig. 11.23) and produce steroid hormones known as **androgens** or male sex hormones. The main androgen produced is testosterone that is vital for growth and maintenance of male sexual features and sperm production. Androgens have been involved in

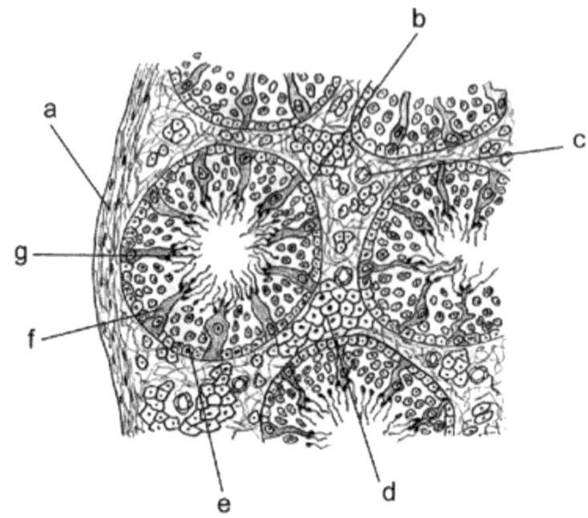

Fig. 11.23 Part of the testis. (a) the outer covering or tunica albuginea, (b) seminiferous tubule, (c) capillary, (d) interstitial cells of Leydig, (e) spermatogonium, (f) Sertoli cell and (g) spermatozoa.

mechanisms that are not sex specific including enhanced neural survival, stimulation of muscle satellite cell proliferation and release of somatostatin and as a result these hormones could play a major role in normal female development (Staub and De beer, 1997). In many amphibian and reptilian species, the levels of androgens in females are as high as those in males and such high levels could be important in the normal functioning of the female reproductive system. Steroids have been implicated in hormone-dependent cancers such as prostate cancer. Secretion of testosterone is under the control of the hypothalamus that produces luteinzing hormone releasing hormone (LHRH). The effect of the latter releasing factor on the anterior pituitary is stimulation of production of follicle stimulating hormone (FSH) and luteinizing hormone (LH). FSH promotes spermatogenesis while LH causes sperm release. Both gonadotropins stimulate the secretion of androgens such as testosterone by the interstitial cells of the testis. Secretion of testosterone shows diurnal and seasonal variation in some vertebrate species.

Estrogens, including estradiol and estrone, are produced by cells of ovarian follicles (Fig. 11.24). The development of follicles is enhanced by FSH while LH causes ovulation. FSH and LH also stimulate the secretion of estrogens. Estrogens are necessary for the normal development and maintenance of female characteristics and the proper development of body organs and tissues including bones and the skin. Synthetic compounds or plant products that have estrogens are used to promote growth in the livestock industry whereas others are metabolites from pesticides and are known as **ecoestrogens**. Ecoestrogens bind to estrogen

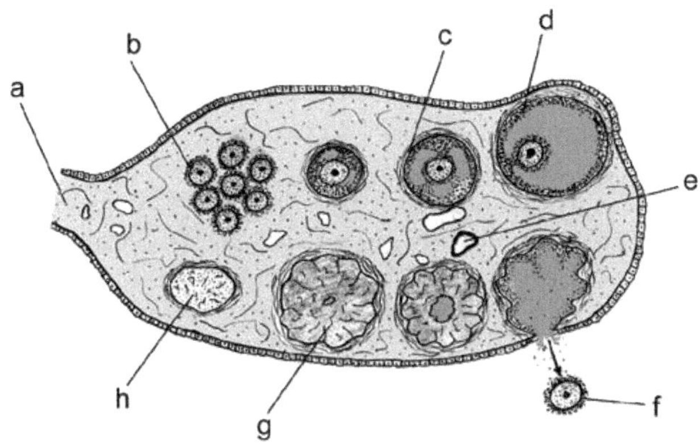

Fig. 11.24 A mammalian ovary showing various stages of the ovarian cycle. (a) mesovarium (ovarian ligament), (b) primary follicle, (c) maturing follicle, (d) mature or graafian follicle, (e) blood vessel, (f) discharged ovum, (g) corpus luteum and (h) corpus albicans.

receptors and influence estrogen-signaling pathways thus influencing growth, development and behavior. There have been various effects of ecoestrogens in fish including feminization of male fish and development of hermaphroditism in male birds feeding on such fish. Ecoestrogens are normally concentrated in animals higher up in the food chain.

Progesterone (Gr. *pro,* before; L. *gestatus,* to bear) is secreted by the corpus luteum after ovulation and together with estrogen the two hormones promote the development of the lining of the mammalian uterus for normal gestation. Progesterone is also produced by the adrenal gland in low quantities. Progesterone is one of the earliest steroid hormones to evolve and in lower vertebrates is associated with glucose metabolism, bone formation and development of intelligence. In the male human being, progesterone receptors are found in the endothelial lining of blood vessels. The hormone is responsible for keeping the endothelial lining of the vessels smooth.

Chorionic Gonadotropin

In primates and the horse family, chorionic gonadotropin is produced by the placental chorion in pregnant animals and is the first hormone to be produced by the fetus in these species. The placenta therefore acts as a temporary endocrine gland. Chorionic gonadotropin stimulates the synthesis and secretion of estrogens and progesterone by the **corpus luteum** (L. *corpus,* body; *luteus,* yellow). Secretion of chorionic gonadotropin is high early in pregnancy as it sustains the corpus luteum and the hormones it secretes that are vital for maintenance of the uterine lining and complete formation of the placenta. Since chorionic gonadotropin is excreted in the kidney, urine samples have been used to detect early pregnancies in the horse and primates including the human being. In some mammalian species, secretion of prolactin by the anterior pituitary is responsible for the sustenance of the corpus luteum. In some mammals, later in pregnancy, the placenta synthesizes its own estrogens and progesterone and the production of these hormones by corpus luteum diminishes. The corpus luteum persists till late in pregnancy in some species and gradually regresses to turn into **corpus albicans** (L. *albus*, white).

The presence of high levels of progesterone in circulation of placental mammals inhibits the estrous cycle and ovulation as this hormone inhibits the secretion of follicle stimulating hormone and luteinizing hormone by the anterior pituitary gland. Towards the end of gestation, the levels of estrogens in material blood rise to maximal levels. The estrogens stimulate the growth of the myometrium and counteract the myometrial relaxing effect of progesterone thus inducing the oxytocin receptors of the myometrium and facilitating parturition. Estrogens together with other hormones are necessary for mammary gland development. The placenta and ovaries also produce the hormone relaxin in many mammals as pregnancy advances. **Relaxin** causes relaxation of the pelvic ligaments and pelvic symphysis at the end of gestation thus facilitating parturition. In some species, the corpus luteum is the main source of relaxin. **Placental lactogens** have molecular structures that are close to those of prolactin and growth hormone. They are found in primates, ruminants and rodents and seem to modulate fetal and maternal metabolism as well as stimulating the growth of mammary glands.

Thymus Gland Hormones

The thymus gland (Fig. 11.25) is found in all vertebrates and develops from the epithelium of some of the pharyngeal pouches. During development, the thymic epithelial cells of the pouches are invaded by stem cells from the bone marrow and spleen. The invading stem cells develop into T lymphocytes that will later on circulate to other lymphoid organs where they will participate in cell-mediated immunity among other roles.

The thymus is subdivided into small lobules that measure 1-2 mm in diameter by connective tissue septae that extend from the covering fibrous capsule into the organ. Each lobule is composed of an outer cortex that has a higher density of lymphocytes and a less dense cellular medulla. Unlike other lymphoid organs where lymphocytes are arranged

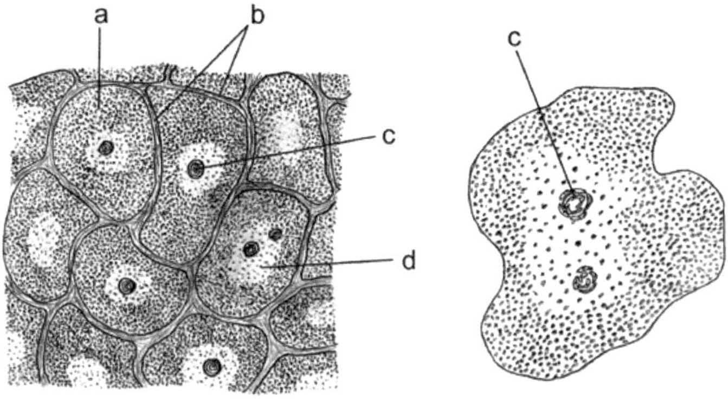

Fig. 11.25 Part of the thymus showing various lobules (i) and a thymic lobule (ii). (a) cortex, (b) lobule, (c) thymic or Hassall's corpuscle and (d) medulla.

in discrete lymph nodules, lymphocytes of the thymus are diffusely arranged in the cortex and medulla of lobules. The medulla of the thymus contains large spherical structures that consist of concentric layers of keratinized epithelial cells known as **thymic** or **Hassall's corpuscles**.

Apart from being a lymphoid organ, the thymus gland produces the thymic hormones—**thymosin** and **thymopoietin**. Several other hormones are produced within the thymus gland in the human being. These polypeptides play a role in the body's immune system. The hormones stimulate rapid white blood cell proliferation, reduce autoimmune reactions and increase antibody production. The hormones can also stimulate the pituitary gland to produce growth hormone. Thymus gland extracts obtained from young calves have been used in the human being to treat various conditions such as cancer and hepatitis B.

Gastrointestinal Mucosa

The mucous membrane of the gastrointestinal tract has cells that produce both endocrine and exocrine secretions. The hormones produced by this mucous membrane include gastrin, secretin, cholecystokinin-pancreozymin and gastric inhibitory peptide. These hormones have a regulatory role of coordinating the secretions and motor activities of the digestive system. **Gastrin** is produced by part of the lining of the stomach and stimulates the gastric glands to secrete pepsinogen and hydrochloric acid and it also stimulates contractions of the stomach wall. Secretion of gastrin is stimulated by the tasting of food or by arrival of food in the stomach and distension of the pyloric antrum. Secretion of gastrin is inhibited by low pH and somatostatin. **Secretin** is produced in the duodenum and stimulates the secretion of sodium bicarbonate and digestive enzymes by the pancreas and bile secretion in the liver. Secretin also has an inhibitory role on cells that produce acid in the stomach. Secretion of secretin is triggered off by the acidity in the chyme as well as presence of fats and partly digested proteins. Secretin is also thought to play a neurocrine role and studies using rats show that this polypeptide hormone functions as a messenger that facilitates transmission of GABA which is an indicator that it can modulate motor and other functions (Ng et al., 2002). **Cholecystokinin-pancreozymin** (CCK) is secreted in the duodenum in response to the presence of fat in chyme. The hormone stimulates the pancreas to secrete its digestive enzymes and the gall bladder to release bile. The hormone also decreases gastric emptying. **Gastric inhibitory peptide** (GIP) is produced in the duodenum and slows the emptying of the stomach by lowering stomach contractions.

Heart

The heart atria secrete **atrial natriuretic hormone factor** (L. *natrium,* sodium) in response to an increase in stretch of the atrial wall as a result of high blood volume or blood pressure. An increase in the level of atrial natriuretic hormone (discovered in 1981) in circulation results in the loss of more sodium from the body through urine (natriuresis). Loss of sodium is accompanied with loss of more water from the internal environment, an action that results in a fall in blood volume and pressure. The action of atrial natriuretic is antagonistic to that of aldosterone. The pharmaceutical industry is trying to manufacture drugs that mimic the effects of atrial natriuretic hormone. Teleosts and tetrapods generally have atrial natriuretic hormone and other related natriuretic peptides that include brain natriuretic peptide and ventricular natriuretic peptide that are secreted by the heart. C-type natriuretic peptide is found in the brain and is the only natriuretic peptide that is found in the heart and brain of elasmobranchs.

A new peptide known as Ebu natriuretic peptide has been cloned from the heart and brain of a hagfish (Kawakoshi et al., 2003). Some teleosts lack the brain natriuretic peptide. Natriuretic peptides play a role not just in the cardiovascular system but in osmoregulation as well and these hormones diverged during fish evolution that could have reflected changes in osmoregulatory systems (Inoue et al., 2003). Baralis et al. (1997) have done a literature survey and analysis on physiology, hemodynamic and humoral effects on the central nervous system as well as clinical application of natriuretic peptides.

EVOLUTION OF THE ENDOCRINE SYSTEM

The earliest endocrine system could have evolved from unicellular organisms that produced chemicals that were used in perceiving intracellular signals and in communicating with neighboring cells. With evolution of higher multicellular animals, true endocrine glands evolved. Initially, the glands comprised diffuse endocrine tissues in various parts of the body. Some of the cells in the endocrine tissues could have migrated from the nervous system leading to the development of a neurosecretory type of system whereby the nervous system either secretes neurohormones directly into circulation or stores them in neurohemal organs. Neurohormones are hormones that act on or are secreted by nervous tissue while neurohemal organs are neurons that make contact with blood vessels. Neurohormones are secreted directly into the circulatory system by neurohemal organs. The main type of chemical endocrine messenger in arthropods is the neurohormone. Endocrine glands are best developed in vertebrates that have normal glandular endocrine tissues but still have the neurohypophysis and adrenal medulla as neurosecretory tissues.

The control that is exerted by the hypothalamus over the pituitary and the effect of the latter on other organs of the body is found in all vertebrates. Hagfish have a poorly developed hypothalamic neurosecretory system. All the basic rudiments of the hypothalamic neurosecretory system are present in the closely related lampreys. Several well-developed nuclei that play the role of neurosecretory centers are found in the hypothalamus of more advanced jawed fishes. The nuclei become more distinct and increase in numbers in amphibians to the more advanced birds and mammals.

The tropic hormones produced by pituitary glands under the control of neurohormones from the hypothalamus are basically the same in vertebrates. Teleosts depend on not only neurohormones to produce their tropic hormones but also neurotransmitters as well since the pituitary gland cells are innervated directly. The nonmammalian hypothalamus of other vertebrates also secretes hormones that have biological properties similar to oxytocin and vassopressin. **Isotocin** and **mesotocin** are oxytocin-like peptides that are secreted by many fishes and amphibians, reptiles and birds respectively. Arginine vasotocin is produced in nonmammalian vertebrates and fetal mammals and has biological properties of both oxtyocin and vasopressin.

REFERENCES

Baker, M.E. (2003). Evolution of adrenal and sex steroid action in vertebrates; a ligand-based mechanism for complexity. BioEssays, 25:396-400.

Baralis, G., Antonielli, E., Pizzuti, A., Tanga, M., Rovere, M.E., Leonardi, G., Doronzo, B., Bassignana, A., Riva, G., Mantovani, M., Correndo, L. and DiLeo, M. (1997). Natriuretic peptides and the heart. Critical review and application. Minerva Cardioangiol., 45(12):605-613.

Cheung, B.M., Leung, R., Man, Y.B. and Wong, L.Y. (2004). Plasma concentration of urotensin II is raised in hypertension. J. Hypertens., 22(7):1341-1344.

Christopher, J.C., Miriam, T.R., Mark, R. and Timothy, G.Y. (2005). Urotensin II: Evidence for cardiac, hepatic and renal production. Peptides, 26(11):2211-2214.

Conlon, J.M. (2002). The origin and evolution of peptide YY(PYY) and pancreatic polypeptide (PP). Peptides, 23(2):269-278.

Demski, L.S. (1993). Terminal nerve complex. Acta Anat. (Basel). 148(2-3):81-95.

Domes, G., Heinrichs, M., Michel, A., Berger, C. and Herpertz, S.C. (2007). Oxytocin improves 'mind-reading' in humans. Biol. Psychiatry, 61(6):731-733.

Ekstrom, P. and Meissl, H. (2003). Evolution of photosensory pineal organs in new light: the fate of neuroendocrine photoreceptors. Philos. Trans. R. Soc. London B Biol. Sci., 358(1438):1679-1700.

Forsyth, I.A. and Wallis, M. (2002). Growth hormone and prolactin–molecular and functional evolution. J. Mammary Gland Biol. Neoplasia, 7(3):291-312.

Gimpl, G. and Fahrenholz, F. (2001). The oxytocin receptor system: structure, function and regulation. Physiol. Rev., 81(2):629-683.

Guillette Jr., L.J., Dubois, D.H. and Cree, A. (1991). Prostaglandins, oviductal function and parturient behavior in nonmammalian vertebrates. Am. J. Physiol. Regul. Integr. Comp. Physiol. 260:R854-R861.

Herman, C.A. (1990). Prostaglandins in lower vertebrates. Prog. Clin. Biol. Res., 342:608-613.

Inoue, K., Naruse, K., Yamagami, S., Mitani, H., Suzuki, N. and Takei, Y. (2003). Four functionally distinct C-type natriuretic peptides found in fish reveal evolutionary history of the natriuretic peptide system. Proc. Natl. Acad. Sci. U.S.A., 100(17):10079-10084.

Kawakoshi, A., Hyodo, S., Yasuda, A. and Takei, Y. (2003). A single and novel natriuretic peptide is expressed in the heart and brain of the most primitive vertebrate, the hagfish (*Eptatretus burgeri*). J. Mol. Endocrinol., 31(1):209-220.

Kawauchi, H., Suzuki, K., Yamazaki, T., Moriyama, S., Nazaki, M., Yamaguchi, K., Takahashi, A., Youson, J. and Sower, S.A. (2002). Identification of Growth Hormone in the Sea Lamprey, an Extanct Representative of a Group of the Most Ancient Vertebrates. Endocrinol., 143(12):4916-4921.

Manzon, L.A. (2002). The role of prolactin in fish osmoregulation: a review. Gen. Comp. Endocrinol., 125(2):291-310.

Ng, S.S., Yung, W.H. and Chow, B.K. (2002). Secretin a neuropeptide. Mol. Neurobiol., 26(1):97-107.

Okabe, M. and Graham, A. (2004). The origin of the parathyroid gland. Proc. Natl. Acad. Sci. U.S.A., 101(51):17716-17719.

Roth, G. and Dicke, U. (2005). Evolution of the brain and intelligence. Trends Cogn. Sci., 9(5):250-257.

Roth, G., Nishikawa, K.C., Naujoks-Manteuffel, C., Schmidt, A. and Wake, D.B. (1993).

Paedomorphosis and simplification in the nervous system of salamanders. Brain, Behavior and Evolution, 42(3):137-170.

Sarnat, H.B. and Netsky, M.G. (2002). When does a ganglion become a brain? Evolutionary origin of the central nervous system. Semin. Pediatr. Neurol., 9(4):240-253.

Sato, T. (2001). Sensory and endocrine characteristics of the avian pineal organ. Microsc. Res. Tech., 53(1):2-11.

Schmidt, A. and Roth, G. (1996). Differentiation processes in the amphibian brain with special emphasis on heterochronies. Int. Rev. Cytol., 169:83-150.

Staub, N.L. and DeBeer, N. (1997). The role of androgens in female vertebrates. Gen. Comp. Endocrinol., 108(1):1-24.

Takuma, N., Sheng, H.Z., Furuta, Y., Ward, J.M., Sharma, K., Hogan, B.L., Pfaff, S.L., Westphal, H., Kimura, S. and Mohan, K.A. (1998). Formation of Rathke's pouch requires dual induction from the diencephalon. Development, 125(23):4835-4840.

12

Sense Organs

Sense organs and sensory receptors detect information about the environment such as light, sound, smell, taste and touch and cause sensory transduction whereby cells convert extra-cellular signals into electrical signals (energy). Usually, the plasma membrane of a sensory receptor is stimulated to cause depolarization that will result in an action potential once the threshold limit has been exceeded. The sense organs are vital to vertebrates as they detect and provide information that enables these animals to make appropriate responses. The sense organs of vertebrates are complex in comparison to those of invertebrates and have contributed to the complex behaviors seen in vertebrates that also have a complex nervous system that is highly cephalized or shows a high degree of brain development. Since the anterior or superior end of bilaterally symmetrical animals moves through the environment first or is the highest from the ground depending on species, this end contains most of the sense organs. With evolution, the anterior or superior end of the nervous system become larger with more complex pathways to be able to process information received from the sense organs.

Receptors are able to sense general senses such as touch, pain and temperature after which they initiate signals that are transmitted to the relevant part of the central nervous system for appropriate action that contributes to the maintenance of homeostasis. **Special sense organs** are organs where sensory receptors have been organized into specific organs in particular parts of the body and detect information related to light, sound, smell, taste and balance and also play a vital role in maintenance of homeostasis.

SENSORY RECEPTORS

Although vertebrates still retain free nerve endings in some parts of the body that act as sensory receptors as is the case in invertebrates, unique receptor cells and nerves that are specific to various types of sensation are present in many cases. Receptors in vertebrates receive stimuli from external and internal environments. The receptors respond to different stimuli and are capable of detecting and absorbing minor energy changes in the environment and will start nerve impulses that are transmitted to elicit the appropriate response to stimulation. On stimulation, the ion channels in the plasma membrane of the receptor cell will open, leading to depolarization and development of a receptor potential that is a graded response to the intensity of the stimulus. The receptor potential has to attain a particular threshold before an action potential is initiated in the axon of the sensory neuron. The impulse is then transmitted in coded form to the central nervous system where it will be interpreted (uncoded) to belong to a particular sensation and the relevant effectors activated.

During adaptation, continuous stimulation of a receptor cell at the same level results in a decrease in the strength of receptor potential. Such an action results in a decrease in the rate at which impulses are transmitted by the sensory neurons and greater stimulation will be necessary to bring about the initial rate of impulse transmission. The rate at which receptors adapt to stimulation varies depending on types of receptors. For example, touch receptors adapt fast whereas proprioceptors in the muscular system do so slowly.

CLASSIFICATION OF RECEPTORS

Receptors are classified based on their location, stimulus detected and structure. According to their location or source of stimulus, there are three general types of receptors that include exteroreceptors, interoreceptores (visceroreceptors) and proprioceptors. Depending on stimulus detected or energy transduced are mechanoreceptors, chemoreceptors, photoreceptors, thermoreceptors, electroreceptors and nociceptors. The two types of sensory receptors based on structure are free nerve and encapsulated nerve endings.

Classification by Location

Sensory receptors are located in most parts of the body and detect various types of stimuli. The receptors include exteroreceptors, interoreceptors and proprioceptors.

Exteroreceptors lie close to or on the body surface and detect stimuli that are mainly external to the body. Such receptors are located mainly in the skin and are sometimes referred to as **cutaneous receptors**. Such receptors detect changes in the external environment including temperature, pressure as well as touch and pain. The vertebrate responds to detection of such sensation by taking the appropriate action including moving to a more suitable environment.

Interoreceptors or **visceroreceptors** are located in the deeper part of the body such as many organs and detect various changes in the internal environment. Changes in blood pressure, chemical composition of body fluids and stretching of organs as happens with the urinary bladder are detected by interoreceptors.

The **proprioceptors** (L. *proprius*, one's own; *ceptus*, taken) are mechanoreceptors and are limited to skeletal muscle, tendons and joint capsules. These mechanoreceptors detect changes in contraction and relaxation of skeletal muscle as well as movement. Proprioceptors coordinate movement and maintain muscle tone as part of a reflex arc. The receptors are also able to detect the position of various body parts and enable the vertebrate orient its body in space. Proprioceptors are not as numerous as and are more specialized than interoreceptors. **Muscles spindles** are able to detect stretching in skeletal muscle while **Golgi tendon organs** detect a similar action in tendons (Chapter 4). The stretching of ligaments is sensed by joint receptors. Proprioceptors evolved with the transition of life from water to land.

Classification according to Energy Transduced

Sensory receptors are capable of detecting various types of stimuli since each type of receptor is capable of discriminating a sensation based on the energy that can elicit a nervous impulse. There are a variety of receptors that are placed into broad groups that detect related stimuli.

Mechanoreceptors are located superficially and in various parts of the vertebrate body and respond to mechanical stimuli that distort or alter the position of a receptor leading to generation of an action potential. Mechanoreceptors in the skin of vertebrates are touch receptors that sense the surfaces of other animals, plants and objects in order to obtain signals for coordination of muscle activity. Internally, mechanoreceptors detect stretch to smooth muscle fibers in various tissues, movement and gravity. Receptors for posture, texture and changes in organs of the vertebrate body are mechanoreceptors.

Chemoreceptors are stimulated by certain chemical changes in the external and internal environment such as taste and smell that are detected by neuron endings in their mucous membranes. Chemoreception of external chemicals is important in locating food, communication among members of a species or group and in detection of predators. Although the sense of smell is very sensitive, it adapts fast. Internal chemoreceptors monitor the concentration of certain chemicals in the body. The carotid and aortic bodies (Chapter 10) that are located near the heart detect levels of oxygen and carbon dioxide in the blood. Information received from these receptors will lead to the appropriate adjustment in the rates of breathing and circulation. Internal chemoreceptors also detect changes in levels of blood glucose.

Thermoreceptors are sensitive to changes in temperature. Pit vipers and boas have thermoreceptors which they use to locate their endothermic prey undetected. In pit vipers, very sensitive infrared detectors are located in **facial pit organs**. An increase in temperature by 0.002°C inside the pit organs stimulates an increase in the rate of action potentials by sensory fibers that supply the organs. A viper can detect the body heat of a small rodent about 40 cm away if the body temperature of the rodent is above 10°C that of the surrounding air temperature. The two pit organs on the head enable the snake to tell the direction of the source of the heat. Mammals possess warm and cold receptors that will send the right signals to the hypothalamus for body temperature regulation.

Nociceptors are sensitive receptors that detect pain as a result of tissue damage. Many stimuli cause pain including pressure, intense light, toxic chemicals and heat. Nociceptors represent the first line of defense against any potentially threatening or damaging environmental inputs (Woolf and Ma, 2007).

Photoreceptors possess photoreceptive cells that have the ability to absorb light and initiate a nerve impulse due to the presence of pigments known as rhodopsins. Photoreceptors are found in the eyes of vertebrates. Vertebrates are able to detect visible light whose wavelength varies between 380nm (violet) and 760nm (red). The longer wavelength light has lower energy whereas the shorter wavelength light has a lot of energy and could be destructive to body tissues. Vertebrates with photoreceptive cells that are specific to various wavelengths can receive a lot of information about the external environment. In many vertebrates, light plays an important role in regulation of reproductive cycles, feeding and behavior in general.

Electroreceptors sense electrical currents in water and are present in many fish species and urodeles. In many fishes, electroreceptors are **ampullary organs** (L. *ampulla*, flask) that are located in the head region and comprise modified hair cells (Fig. 12.1). The weak electric currents that are generated by muscular contraction of animals of prey are

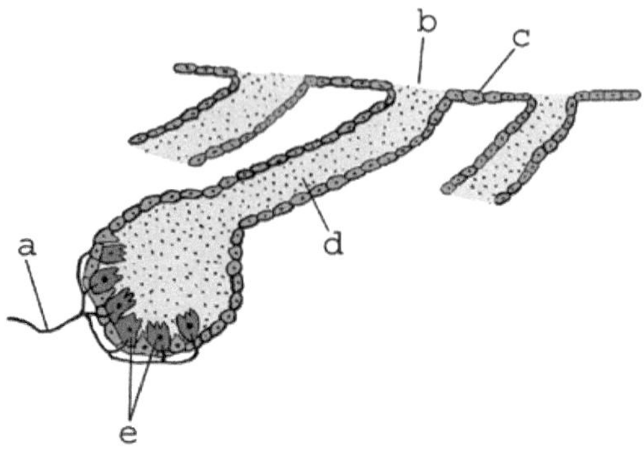

Fig. 12.1 The ampullary organ that is found in chondrichthyans, some teleosts and urodeles. (a) nerve, (b) pore opening into the tuberous organ, (c) skin surface, (d) jelly and (e) sensory hair cells. Groups of ampullary organs are also known as **ampullae of Lorenzini**.

detected by electroreceptors even when the prey animals are buried under sand. The low voltage that results from the magnetic field of the earth in water can be used by many fish including sharks to orient themselves in certain positions. Some fish use such voltage sensation in migration. Some bony fishes have tuberous organs that detect electric currents that are generated by the same fish. The electric catfish, lungfishes and sturgeons use such currents to detect objects in murky water since the objects distort the electric fields of these fish. Electroreceptors were lost with evolution of life on land as air is not a good conductor of electricity.

Structural Classification of Sense Organs

Sense organs can be classified anatomically into free or encapsulated nerve endings. Organs in either of these groups can be activated by various stimuli. Free or encapsulated sensory nerve endings are located in various parts of the body. Free nerve endings or those that have been encapsulated are responsible for detecting various sensations including pain, temperature changes, touch, itching and stretching.

Free nerve endings are the most numerous and widely distributed sensory nerve endings in the vertebrate body. They also display the simplest form of sensory receptors. Free nerve endings start off as small swellings known as **dendritic knobs**. They are also known as nociceptors as they are the primary receptors for pain. Whereas free nerve endings are widespread in many visceral organs, these nerve endings are lacking in the brain. Free nerve endings can be of acute or chronic types. **Acute pain fibers** relay localized pain feelings that are sharp whereas the

chronic pain fibers transmit impulses that are related to mild but longer lasting pain.

Free nerve endings can adopt various patterns (Fig. 12.2). **Root hair plexuses** form a net-like pattern of free nerve endings around the root of hair follicles in the dermis of the mammalian skin. Root hair plexuses are sensitive to hair movement. The non-laminated and slow adapting **Merkel's disks** are superficially located in the dermis. The disk-shaped or flattened Merkel's disks sense light and touch. These sensory receptors are quite precise on the location of touch.

Encapsulated nerve endings have connective tissue capsules around the dendritic end of sensory nerves. The non-laminated **Krause's end bulbs** or **Krause's corpuscles** that detect cold and **Ruffini's corpuscles** that sense warmth are slow adapting modified free nerve endings that are located in the dermis of the skin. Krause's corpuscles are also found in mucous membranes of many body systems. Krause's end bulbs are egg shaped and have coiled dendritic endings within their capsules. The corpuscles are also sensitive to touch. Ruffini's corpuscles are more flattened than Krause's corpuscles and are located deeper in the dermis. The low adapting Ruffin's corpuscles are also sensitive to touch.

Pacinian corpuscles are large and laminated, fast adapting sensory receptors that are located deep in the dermis of the skin. These receptors

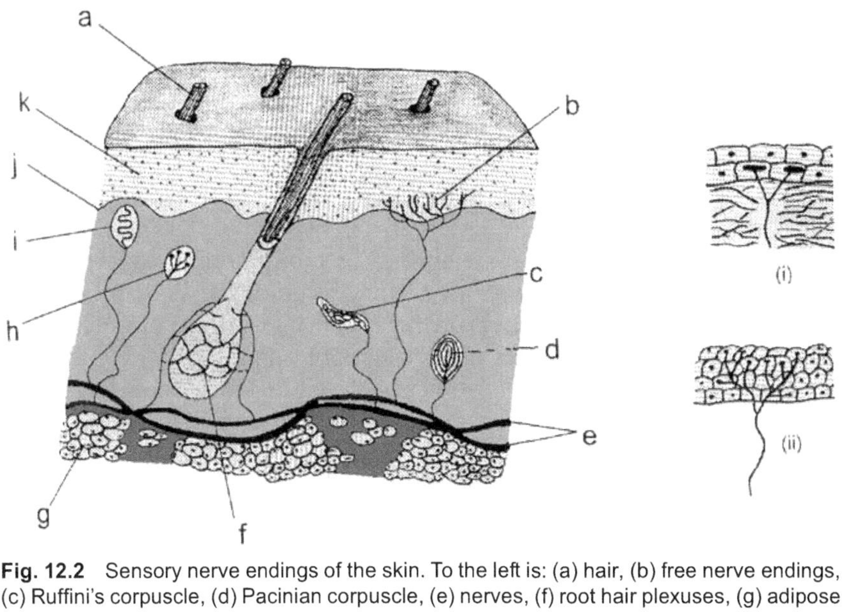

Fig. 12.2 Sensory nerve endings of the skin. To the left is: (a) hair, (b) free nerve endings, (c) Ruffini's corpuscle, (d) Pacinian corpuscle, (e) nerves, (f) root hair plexuses, (g) adipose tissue, (h) Krauses's end bulb, (i) Meissner's corpuscle, (j) basement membrane and (k) epidermis. To the right is: (i) Merkel's disks and (ii) nociceptors.

are activated by deep pressure, stretch and high frequency vibrations. **Meissner's corpuscles** are located superficially in the dermis of the skin. The ovoid-shaped, large and laminated Meissner's (tactile) corpuscles are sensitive to touch and low frequency vibrations and are rapidly adapting receptors. In mammals, Meissner's corpuscles are numerous in areas that lack hair. The other capsulated nerve endings are muscle spindles and Golgi tendon organs that are stretch receptors (see Chapter 5).

SPECIAL SENSE ORGANS

Special senses are a group of sensory cells that are grouped together and organized into organs that sense particular stimuli. Sensory organs have a much wider range of sensitivity when compared to a single receptor cell since such organs have several receptors that show various degrees of sensitivity. Individual receptors show a certain level of sensitivity of the overall range of a receptor system. Very sensitive receptor cells show a response to quite low levels of stimuli and vice versa. The receptors of a sensory organ show different levels of graded response to varying intensities of stimuli until all receptors are fully active and there will be no more response to a further increase in the level of the stimulus. The special sense organs in the vertebrate body are the olfactory receptors (smell), taste buds (taste), ears (hearing and equilibrium), eyes (vision) and the lateral line receptors of fishes and amphibians (water disturbances).

OLFACTORY RECEPTORS

Olfactory receptors (L. *olfacio*, to smell) are chemoreceptors that smell and detect odorants. Olfaction in vertebrates is used to detect food, enemies as well as members of the same species or group and in migration and homing. In fish, the sensitivity of olfactory receptors is about the same as that of gustatory (taste) receptors while in terrestrial vertebrates the olfactory receptors are several times more sensitive than gustatory receptors and detect volatile chemicals from a distance. In jawed fishes, smelling is controlled by about 100 genes. In mammalian genomes the olfactory receptor genes form the largest multi-gene family and in the human being comprise 800 olfactory receptor genes (Niimura and Nei, 2005). In fishes, the olfactory organs comprise sacs that have folds of sensory epithelium (Fig. 12.3). Water circulates over the olfactory epithelium that is pleated by entering through the incurrent nasal openings (nares) and flows out through the excurrent nasal opening that is located on the roof of the head. Pharyngeal contractions that occur

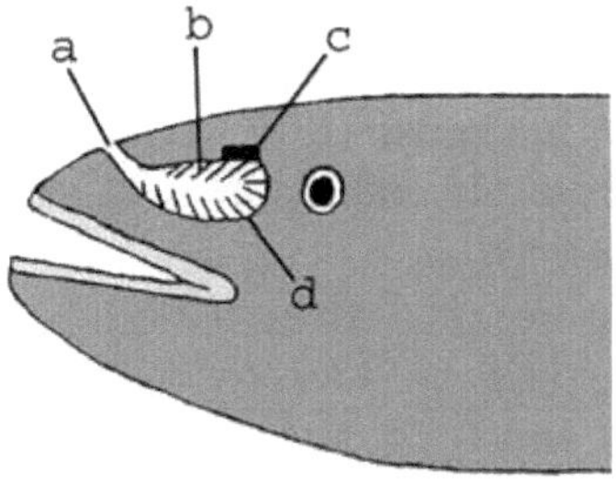

Fig. 12.3 The nostrils of fish showing: (a) incurrent naris, (b) fold of sensory olfactory epithelium, (c) excurrent naris and (d) the nasal capsule (sac). Folds (pleats) increase the olfactory surface area. Fish nasal openings do not lead to the oral cavity.

during respiration pump water over the olfactory epithelium. The nasal openings of fish do not open into the oral cavity. Lampreys and hagfish have single nostrils and changes in pressure in the olfactory sac resulting from respiratory movements ventilate the olfactory epithelia through the single opening. Fish are able to sense chemical substances that are mainly of organic nature using their olfactory sensory system though others are sensed by the taste buds.

In tetrapods, the olfactory epithelium is located to the caudal or dorsal part of the nasal cavity (Fig. 12.4). The nasal cavities of tetrapods play the dual role of olfaction as well as serving as passageways of air to lungs and as a result the olfactory epithelium of tetrapods is ventilated during respiratory movements. As amphibians and reptiles do not inhale air frequently to expose their olfactory epithelium to new odors, extra pumping movements by the floor of the buccopharyngeal cavity

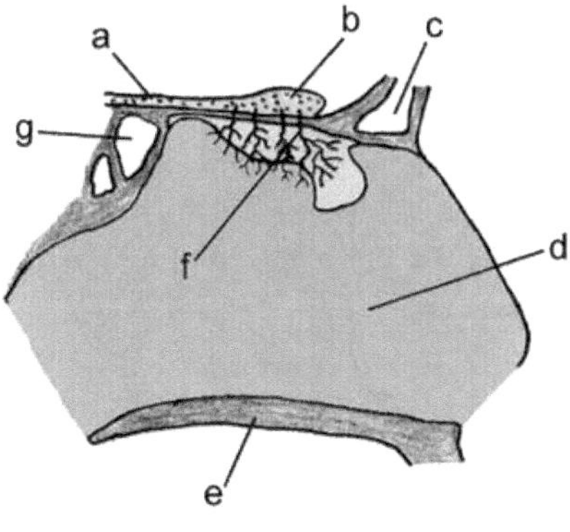

Fig. 12.4 The nasal cavity of a human being showing the olfactory region. (a) olfactory tract, (b) olfactory bulb, (c) frontal sinus, (d) nasal cavity, (e) palate, (f) olfactory nerve fiber and (g) sphenoidal sinus. Olfactory nerves enter the cranial cavity through foramen in the cribriform plate of the ethmoid bone in mammals.

ventilates the olfactory epithelium. The presence of the scroll-like ethmoturbinates at the back or roof of the nasal cavity in amniotes especially mammals amplifies the surface area of the olfactory epithelium. A good sense of smell is quite helpful when vision is not of much use during darkness.

The olfactory neurons (Fig. 12.5) are quite similar in the different vertebrate groups. The olfactory receptor neurons are of the bipolar type and have non-motile olfactory cilia that extend from the end of the dendrites. The olfactory neurons are replaced on a regular basis by basal cells that are found in the olfactory epithelium. In between the neurons are supportive or sustentacular cells and goblet cells. Mucus is produced by goblet cells and sustentacular cells in fishes while multi-cellular glands (Bowman's glands) of other vertebrates secrete the product. The axons of the olfactory neurons extend as a group known as the olfactory nerve to the olfactory bulb of the brain. Chemical or gas molecules dissolve in mucus that covers the olfactory epithelium and bind to receptor sites (odor-binding proteins) on the olfactory nerve cilial membrane, generating receptor potentials in the process. Each olfactory neuron is able to express a single receptor gene (Kratz et al., 2002). The olfactory receptors are quite sensitive. Vertebrates can distinguish several thousand different odors.

The olfactory neurons are sensitive to new odor but show rapid adaptation when stimulated continuously by the same odor. The rapid adaptation results from inhibition of action potentials by granule cells in the olfactory bulb. Olfaction is important on land and in water but not in flying and arboreal vertebrates as there are gaps between trails of material that stimulate olfactory neurons. Scavengers and the kiwi have

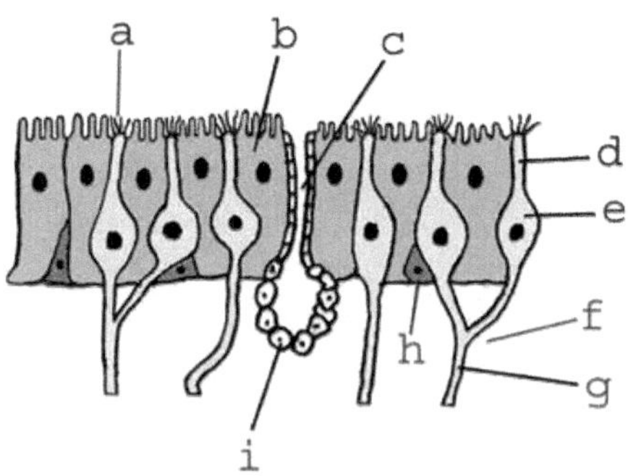

Fig. 12.5 Olfactory epithelium showing (a) cilia, (b) supporting cell, (c) duct of mucous or Bowman's gland, (d) dendrite, (e) olfactory neuron, (f) lamina propria, (g) axon, (h) basal cell and (i) mucous or Bowman's gland. The free surface of the supporting cells bears microvilli.

good senses of smell whereas the human being has a sense of smell that is lower than that of many animals.

Many vertebrates produce **pheromones** (Gr. *pherein*, to bear; *hormaein*, to excite) into the external environment that act as specific chemical messengers within species. Pheromones are used for purposes of communication and the sense of smell is part of this process. Most pheromones are small molecules that are easily synthesized and can be applied to objects. Pheromones have been used by vertebrates to mark territories, as a warning sign in cases of danger and in reproduction as a sign of sexual preparedness. Pheromones are detected by the terminal nerve (cranial nerve O) that lies close to the olfactory nerve in most vertebrates except cyclostomes. The terminal nerve might regulate some aspects of reproduction as it contains luteinizing hormone releasing factor in jawed fishes, amphibians and some mammals.

THE VOMERONASAL ORGAN

The vomeronasal organs or organs of Jacobson (Fig. 12.6) are paired and first appeared in amphibians during vertebrate evolution and so are lacking in fish. The aquatic salamanders (amphiuma and sirens) possess both vomeronasal and olfactory systems and this shows that vomeronasal organs evolved in aquatic tetrapods and not as an adaptation to terrestrial life (Eisthen, 2000). Vomeronasal organs are located below the olfactory epithelium in the middle of the nasal cavity. These organs are not present in most aquatic vertebrates, birds, and many mammals though they may appear as vestiges during embryonic development. Recent findings show that vomeronasal organs are present in the human being and might be functional (Halpern and Martinez-

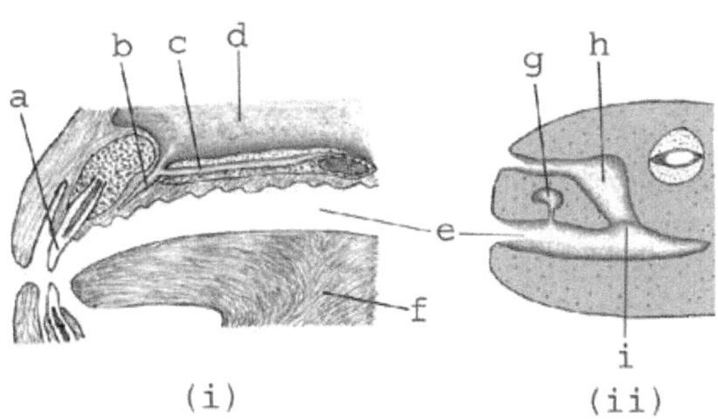

Fig. 12.6 Drawings showing vomeronasal (Jacobson's) organs of (i) mammals and (ii) snakes. (a) upper incisor tooth, (b) nasopalatine duct, (c) vomeronasal duct that is partly supported by cartilage of vomeronasal organ, (d) nasal and (e) oral cavities, (f) tongue, (g) vomeronasal organ, (h) nasal cavity and (i) choana.

Marcos, 2003). The sensory neurons of vomeronasal organs are like olfactory neurons but have microvilli instead of cilia. Axons lead from vomeronasal organs to the accessory olfactory bulb. Vomeronasal organs can detect trace quantities of chemicals.

In amphibians and lower reptiles, the vomeronasal organs are close to the olfactory epithelium and are located in the floor of the nasal passage while in snakes and lizards these organs are sac-like structures in the roof of the oral cavity. The vomeronasal organ of amphibians and lower reptiles monitors chemicals in air as it passes over the organ in the nasal cavity. Snakes and lizards flick out their tongues to pick up molecules in the air or nearby objects then retract the tips of the tongue close to the opening of vomeronasal organs. Snakes use a similar method to trace prey they have bitten but escaped by trailing the path of their own venom left behind by the escaping prey.

In mammals, vomeronasal organs comprise two ducts that are located at the floor of the nasal cavity on either side of the nasal septum. The ducts are partly supported by thin cartilages. The epithelium of vomeronasal organs like that of the nasal cavity has both respiratory and olfactory components. The ducts are open at their anterior end into the dorsal part of the mouth through the **nasopalatine duct** or both the nasal and oral cavities but end blindly caudally. Vomeronasal organs of mammals are also used in detecting pheromones that are produced in relation to reproduction at the time of breeding or for social behavior. Several mammals exhibit the **flehmen reaction** during which they appear to sneer and curl the upper lips thus exposing the vomeronasal organs for sensing chemical substances. Some male mammals display the flehmen reaction at the time of breeding when they smell the urine of females. Recent studies show that both the main olfactory and the vomeronasal systems are actively involved in pheromonal communication (Tirindelli et al., 2009).

TASTE BUDS

Taste buds are sensory receptors that detect taste or gustatory stimuli (L. *gusto*, to taste). In fishes and amphibians, taste buds are found in the oral cavity, the pharynx and the skin. In fishes that do not depend on sight for feeding or live in murky water such as catfishes, cod, minnows and loaches, taste buds are located over the entire body surface and are supplied by the facial nerve. The taste buds of the body surface develop from endodermal cells that have migrated to these areas. In other vertebrates, taste buds are located mainly in the oral cavity and pharynx. The mammalian taste buds are located on the chemical papillae that project from the tongue. The taste buds in the oral cavity are supplied by the facial nerve while the glossopharyngeal and vagus nerves supply taste buds of the pharynx.

A taste bud resembles the segments of a peeled and intact orange (Fig. 12.7) and is made of gustatory cells that have gustatory hairs that project from the apical end of each cell into an opening known as a taste pore. Supporting cells are found between the gustatory cells. There are undifferentiated cells in taste buds that are capable of developing into either sensory or supporting cells since the buds are exposed to wear and tear and have a lifespan that lasts for a few to several days. The entire taste bud is surrounded with an epithelial cell capsule. Taste buds are chemoreceptors and are activated by chemicals dissolved in liquid such as saliva. Taste-producing chemicals bind to receptor sites on the cell membrane of the gustatory hairs to initiate a receptor potential. The action potential generated depends on the nature and concentration of the chemical. The number of substances taste buds can detect is much less than those that olfactory neurons are able to sense in higher vertebrates. Taste buds in the human being respond most effectively to substances that are sweet, salty, sour and bitter. The combination of these four tastes gives the varied tastes that are detected.

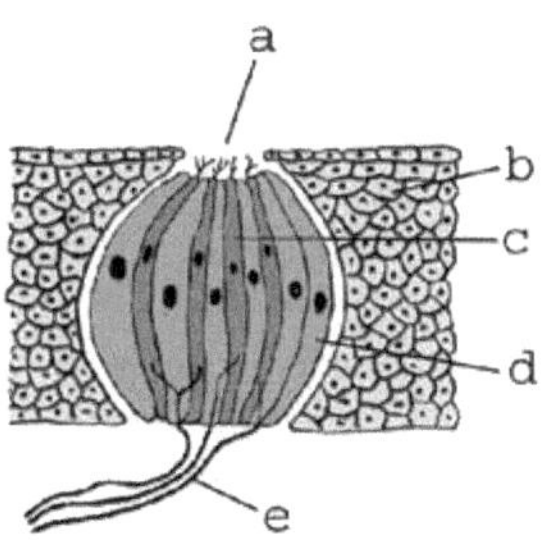

Fig. 12.7 A taste bud. (a) taste pore, (b) epithelium of surrounding structure, (c) gustatory cell, (d) supporting cell and (e) nerve fiber to olfactory nerve.

EAR

The higher vertebrate ear is related to the lateral line system of fishes and amphibians in terms of sensation as the two systems possess hair cells that are sensitive to forces that cause movement of their cilia. The lateral line system is exposed to the external forces of water while the equivalent part in the ear is encased in the skull where it is in contact with the liquid material of the inner ear. The three parts of an ear are external, middle and inner ears (Fig. 12.8) and are present in many vertebrates except fishes that have only an inner ear. The vertebrate ear plays the two roles

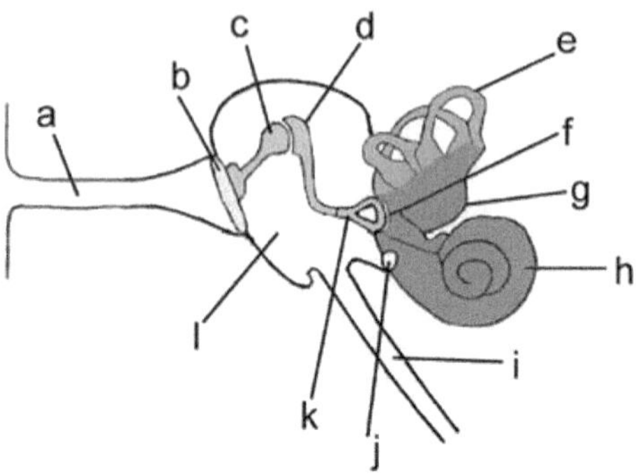

Fig. 12.8 The mammalian ear. (a) external auditory meatus, (b) tympanic membrane or ear drum, (c) malleus or hammer, (d) incus or anvil, (e) semicircular canal, (f) oval window, (g) vestibule, (h) cochlea, (i) auditory or Eustachian tube, (j) round window, (k) stapes or stirrup and (l) middle ear or tympanic cavity. Each semicircular canal forms an enlargement at the junction with the vestibule known as an ampulla.

of hearing and balance or equilibrium. Both these roles are detected by hair cells that are mechanoreceptors.

Hair cells are specialized columnar epithelial cells with several modified microvilli or stereocilia on their apical surface that evolved in early vertebrates and have a similar form in all classes of vertebrates (Pickles and Corey, 1992). Sound waves and movement stimulate the hair cells by causing deflection of the hairs and stretching of linkages between the cells leading to the opening of mechanotransducer channels to generate action potentials to the brain that will be interpreted as sound or balance. Hair cells maintain optimum sensitivity by adapting and keeping the resting tension on the transducer channels constant.

The **external ear** comprises the flap-like **pinna** or **auricle** in mammals and a tube that leads to the middle ear known as the **external auditory meatus** or **ear canal**. The wall of the canal has modified sweat glands that secrete the wax-like cerumen. The length of the canal determines its ability to amplify sound frequencies. The auditory meatus is separated from the middle ear by a **tympanic membrane** (L. *tympanum*, drum) or **eardrum**.

The **middle ear** or **tympanic cavity** of mammals contains three small bones or auditory ossicles that are (from the outside to inside) **malleus** or **hammer**, **incus** or **anvil** and **stapes** or **stirrup**. The mammalian middle ear contains the tensor tympani and stapedius muscles. The **tensor tympani** muscle is attached to the malleus and on contraction pulls the eardrum inwards, thus adding tension to the tympanic membrane. The **stapedius muscle** is attached mostly to the head of the stapes. The two muscles of the middle ear contract when there is intense sound and thus resist the transmission of vibrations to the inner ear that is protected in the process. Apart from the external auditory meatus, several openings lead from the middle ear. The **oval window** and **round window** lead to the internal ear while the **eustachian** or **auditory tube** leads from the middle ear to the nasopharynx. Eustachian tubes maintain equal pressure on both sides of the eardrum, an action that enables the tympanic membrane to vibrate freely. There are many other air spaces in the temporal bone that open into the middle ear.

The two main parts of the inner ear are the **bony** or **cartilaginous labyrinth** to the outside and a **membranous labyrinth** that lies within the bony labyrinth. These structures are referred to as labyrinth as a result of their complicated structure. The lymph-like fluid that fills the membranous labyrinth is known as **endolymph** while **perilymph** that resembles cerebrospinal fluid occupies the space between the bony and membranous labyrinths and contains strands of connective tissue fibers. The bony labyrinth consists of the **vestibule**, **cochlea** and **semicircular canals** whereas the membranous labyrinth comprises the **utricle** and **saccule** inside the vestibule, the **cochlear duct** within the cochlea and the **membranous semicircular canals** that lie within the bony semicircular canals. The organs of balance include the vestibule together with the semicircular canals while the cochlea is concerned with hearing.

Cochlea and Hearing

There have been strong and selective forces in the evolution of the ear in most vertebrate groups for the kinds of sound encoding and processing abilities that has resulted in efficient detection, localization and identification of sound sources in noisy environments (Popper and Fay, 1997). Vertebrates show great variation in the range of sounds they hear. In mammals, the evolution of the malleus and incus together with the adaptations of the cochlea have enabled the mammalian ear to hear sounds with frequencies greater than 10 kHz. Whereas the human being can hear sounds up to 20 kHz, mammals that depend on sound to locate prey by echolocation such as bats, dolphins and whales hear sounds in the range of 100 to 140 kHz.

The mammalian cochlea (L. snail shell) appears from the outside as the shell of a snail. To the inside of this bony part is the membranous **cochlear duct** (Fig. 12.9) that is concerned with hearing. Throughout its winding course, the cochlear duct divides the bony cochlea into upper and lower sections. The upper and lower sections are filled with perilymph that has a high concentration of sodium while the cochlear duct contains endolymph that is rich in potassium. The **organ of Corti** is the sense organ for hearing and lies on the entire **basilar membrane** that

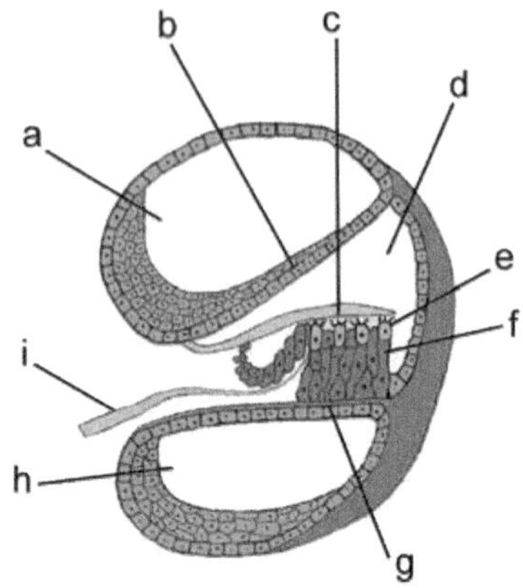

Fig. 12.9 A cross section through the membranous cochlea. (a) scala vestibuli, (b) vestibular membrane, (c) tectorial membrane, (d) cochlear duct, (e) hair cell, (f) supporting cell, (g) basilar membrane, (h) scala tympani and (i) cochlear nerve. The perilymph occupies the scala vestibuli and scala tympani while the endolymph is found in the cochlear duct. Sound waves cause vibrations that move hair cells against the tectorial membrane leading to displacement of stereocilia that are embedded in the gelatinous tectorial membrane and generation of an action.

forms the floor of the cochlear duct. The organ of Corti contains hair cells that make contact with the endolymph and supporting cells. The **tectorial membrane** lies on top of the organ of Corti. The tectorial membrane is the evolutionary homologue of the cupula of the fish lateral line system. The dendrites of sensory neurons lie at the base of hair cells. The hearing sensation in transmitted by neurons that eventually form the cochlear nerve that is a branch of the vestibular-cochlear (8th cranial) nerve.

Sound waves produce vibrations in various media. In mammals, the pinna aids in trapping the sound waves. Sound waves have to exceed a certain amplitude (height) value before they can cause vibration of the tympanic membrane. Vibrations of the tympanic membrane are transmitted to the ossicles of the middle ear that in turn will move the stapes against the oval window in a piston-like manner. Sound waves are amplified by up to 20 times in the ear before reaching the cochlea. The movement of the stapes towards the inner ear increases pressure of the perilymph since liquids are incompressible. Increase in pressure of the perilymph starts a wave that will eventually be transmitted to the endolymph to cause vibration of the organ of Corti with its basilar membrane and the tectorial membrane. The frequency of sound waves per time unit determines the pitch of the sound.

The basilar membrane shows varying levels of thickness along its length and as a result different frequencies of sound waves will cause its vibration and bulging at specific points. The narrow portion of the basilar membrane that is located close to the oval window is activated by high-frequency sound waves while the thicker part of the membrane that is located at the apical end of the cochlea responds to low frequency waves. When the basilar membrane vibrates and bulges upwards and downwards as a result of sound waves, the cilia of hair cells in the affected region are stimulated and register sound of a specific pitch. The magnitude of the bulge in the basilar membrane corresponds to the level of sound and enables vertebrates hear various intensities of sound. The vibrations of the basilar membrane are then transmitted to the perilymph below and will die out towards the round window. Since the cochlea is encased in bone, pressure applied at the oval window is only relieved at the round window. An inward movement of the oval window is followed by an outward movement of the round window and vice versa.

The movement of hair cells against the tectorial membrane that is located above the sensory cells initiates an action potential in the bipolar sensory neurons whose dendrites are attached to the sensory hair cells. The impulses are transmitted to the brainstem by the cochlear nerve before eventually ending up in the auditory area of the cerebral cortex. Sensory cells are produced in the internal ears of birds and mammals mainly during early development and this makes these groups of vertebrates vulnerable to deafness and balance disorders later on in life should these hairs be lost as a result of various causes (Corwin, 1985). Some birds and other nonmammalian vertebrates will suffer from temporary deafness after the hair cells have been damaged since the cells will regenerate later from the division and subsequent differentiation of supporting cells that surround the hair cells into these sensory cells (Kwan et al., 2009). Studies show that some early mammalian postnatal supporting cells that retain a latent capacity to divide and differentiate into hair cells could be important therapeutic targets for continued efforts to induce hair growth regeneration (Kwen et al., 2009).

Hearing in Fishes

Fishes lack external and middle ears so only the internal ear is present. The sensitivity of fishes to sound in water varies greatly. Hearing in fishes is in the form of detection of disturbances in the water that cause sound waves. Sound waves travel at a much higher speed in water than air since water has a higher density than air. Although most fishes can detect sound waves between 1 and 3 kHz, some fish species of the herring group can detect sounds up to 150 kHz. The herring group could have evolved the hearing of such ultrasounds so as to detect a major predator, the echolocating dolphin (Popper, 2000). The ultricle in herrings is also highly sensitive and could be involved in detection of the ultrasounds.

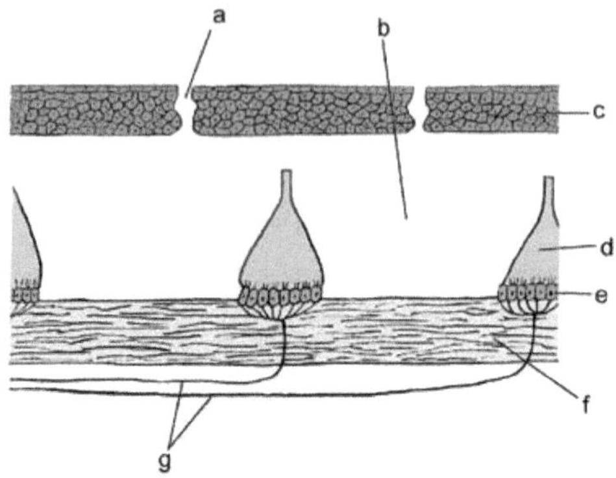

Fig. 12.10 Lateral line system. (a) pore, (b) lateral line canal, (c) epidermis, (d) cupula, (e) sensory cell, (f) skeletal muscle and (g) nerves. Cupula and sensory cells constitute a neuromast.

The low-frequency sound waves that are closer to the fish are detected by the **lateral line system** that contains **neuromasts** (Fig. 12.10). The lateral line system is also found in amphibians. The lateral line system is closely related to the inner ear in evolution, basic structure and function. The relative motion between fish and the surrounding water is detected by the hair cells of neuromasts. Neuromasts can be located on the skin or under the skin in fluid-filled lateral canals. Neuromasts are deposited by migrating primordia that originate from pre- and postotic placodes (thickening of embryonic ectoderm behind the brain) and follow defined pathways on the head and body (Pichon and Ghysen, 2004). The vertebrate inner ear develops from the otic placode. During development of the lateral line system, the migrating sensory precursors guide axons that in turn guide glia that eventually control the formation of sense organs (Ghysen and Dambly-Chaudiere, 2005).

Lateral line canals are small canals that are located on the head and both sides of the body trunk. The canals have several openings or pores to the outside. Neuromasts that comprise hair cells with cilia embedded in a gel-like cupula are located within the canals with a neuromast organ lying between two neighboring lateral line pores. Vibrations in water cause pressure changes around the fish and flow of fluid within the canals. Flow of water through the canals causes the cupula to move thus bending the hair cells and initiating generation of impulses to the brain in a similar manner to the auditory inner ears in other vertebrates. Fish are then alerted of some nearby movements around their bodies.

The lateral line system and eyesight are used to coordinate the movement of fish in a school. The lateral line system is able to detect water flow from other fish in the school. The water movements that are produced by fish in a particular direction are detected and enable the other fish to adjust their movements in a similar direction. This enables a school of fish to swim in a particular direction and almost at the same speed. Some fish species are able to detect low frequency sound waves using the inner ear.

High frequency sounds will travel through water, especially seawater, and the fish body at almost the same amplitude as the two bodies have a density that is about the same. Such waves are detected by the hair cells of the inner ear. Ostariophysians such as carps, catfishes and minnows use their swim bladders to amplify sound waves as the bladder has a different density from the rest of the fish. The gas in the swim bladder is compressed in the process by sound pressure waves making the bladder pulsate and cause tissues associated with it to move. The amplified waves are transmitted to the back of the skull then to the inner ear by **Weberian ossicles** that are located between and connect the anterior end of the swim bladder and the back of the skull (Fig. 12.11). Weberian ossicles play an equivalent role to that of the tympanic membrane and auditory ossicles in tetrapods. Ostariophysians have a more acute sense of hearing than any other fish as they can detect the

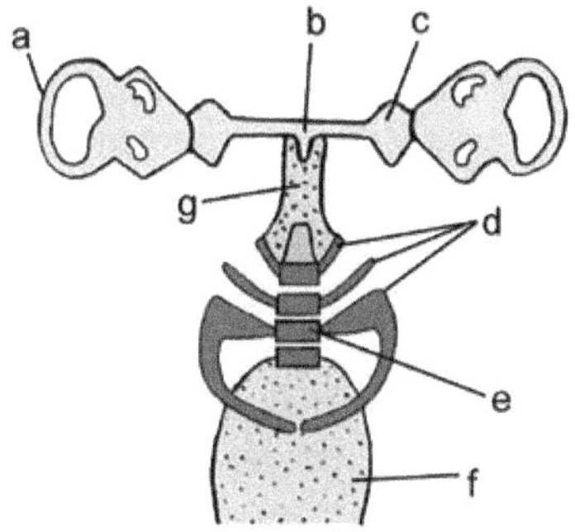

Fig. 12.11 Weberian apparatus. (a) semicircular canal, (b) endolymphatic sac, (c) sacculus, (d) weberian ossicles, (e) vertebra, (f) swim bladder and (g) perilymph.

highest frequencies of any fish group. Weberian ossicles are modified parts of anterior vertebrae including vertebral bodies, neural arches, pleural ribs and neural spines. In the herring family (clupeids) a pair of long gas ducts link the swim bladder to the inner ear. Fish with a well developed sense of hearing tend to make a lot of sounds.

Otoliths (Gr. *ot,* ear; *lithos,* stone) are saccular stone-like structures made of calcium carbonate that are found in the saccules of the inner ear (Fig. 12.12). Since otoliths are much more dense than the rest of the fish, they move at a slower rate than when they receive sound vibrations. The difference in the rate of movement of the fish and the otoliths causes the

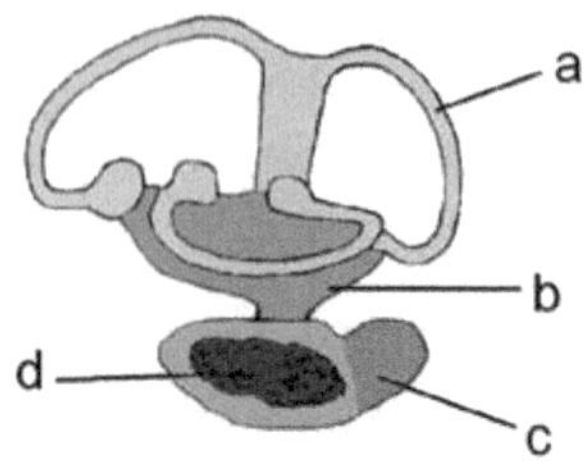

Fig. 12.12 The inner ear of fish. (a) semicircular canal, (b) utriculus, (c) lagena that is modified into the coiled cochlea of mammals and (d) otolith within a sacculus.

bending and stimulation of cilia on hair cells of the vestibule. Sensory neurons with dendrites on hair cells generate impulses that will be interpreted as sound. Otoliths can discriminate between sound waves of different frequencies. Different acoustic signals cause variation in otolith movements. Otoliths are also used in the ageing of fish based on the layers they contain that represent seasonal deposition of salts.

Hearing in Terrestrial Vertebrates

The sense of hearing has evolved independently in tetrapods. In the less dense air that has 3600 times greater acoustic impedance than water resulting in a slower rate of transmission of sound waves when compared to water, the waves have to be amplified several times to be able to reach the more dense medium of the inner ear to be able to cause movement of sensory hair cells before the generation of an action potential. Terrestrial vertebrates can detect high frequency sound waves since the ancestral ear underwent some changes with transition from water to land. Air is also better at transmitting high frequency sounds than water that is better at transmission of high amplitude and low frequency sounds. Hearing airborne sounds entailed change in articulation of the jaws. The hyomandibular (Fig. 4.6) previously suspended the lower jaw from the cranium. Changes that brought about direct articulation of the lower jaw with the cranium were accompanied by separation of the upper part of the hyomandibular from the rest of the bone. The upper part of the hyomandibular moved to the middle ear as columella to participate in transmission of higher frequency sound. The columella was the precursor to the stapes. The lower part of the hyomandibular continued to suspend the jaw muscles as the hyoid bone.

The outer ear has a direct connection to the outside through a tube in many terrestrial vertebrates. The middle ear contains air and has an auditory or eustachian tube that leads to the pharynx. The tympanic membrane is responsible for the conversion of pressure waves into displacement waves. Low frequency waves that are quite intense are sensed in snakes, some lizards, terrestrial salamanders and caecilians as they travel through the ground and air. Such sound waves cause a minor displacement of superficial bones of the skull that will in turn transmit the waves to the middle ear.

Terrestrial vertebrates possess at least one auditory ossicle in the middle ear. Amphibians, reptiles and birds have one ossicle, the stapes. The stapes is located between the tympanic membrane or superficial skull bones to the outside and the oval window medially. The superficial bones are slightly moved by sound waves that are transmitted to the stapes. Mammals possess three ossicles in their middle ear. Waves received by the stapes are passed to a perilymphatic duct and to a sensory region with receptor hair cells that are covered at their tips by a tectorial membrane in place of otoliths. The waves are then transmitted to the cranial cavity or round window that opens between the inner and middle ears.

Amphibian Ear

When compared to the ears of other terrestrial vertebrates, the amphibian ear is unique in basic structure and mechanics of operation and could have evolved from the labyrinthine structures of primitive fishes since it is quite independent of the general principle of ear design that is found in reptiles, birds and mammals (Wever, 1981). Frogs lack pinnae and other sound collecting structures and possess a tympanic membrane and an extra small plate-like ossicle known as the **operculum** in the oval window that is attached to the caudal part of the stapes or **columella** (L. *columella,* small column). Other amphibians lack a tympanic membrane. The two types of sensory hair cells in the inner ear of amphibians are **basilar papilla** and **amphibian papilla**. Frogs are capable of detecting low-frequency sound vibrations and high frequency vibrations. The amphibian papilla is overlain with a tectorial membrane and is more elaborate than the basilar papilla and contains about 1,000 to 1,500 hair cells. The basilar papilla contains 60 to 90 hair cells.

The **opercularis muscle** joins the operculum to the suprascapula of the frog pectoral girdle (Fig. 12.13). Contraction of this muscle brings together the stapes and the operculum and relaxion of the muscle causes separation of the two ossicles. When the two ossicles are coupled, they have a greater mass and transfer lower frequency sound waves to the perilymph to stimulate the amphibian papilla. When the ossicles are disengaged, the stapes (with a lesser mass) transmits higher frequency waves to the perilymph. High frequency waves stimulate the basilar papilla. The amphibian ear is thus a dual-ear system (two ears in one). The social signals such as the mating sound with higher frequencies are

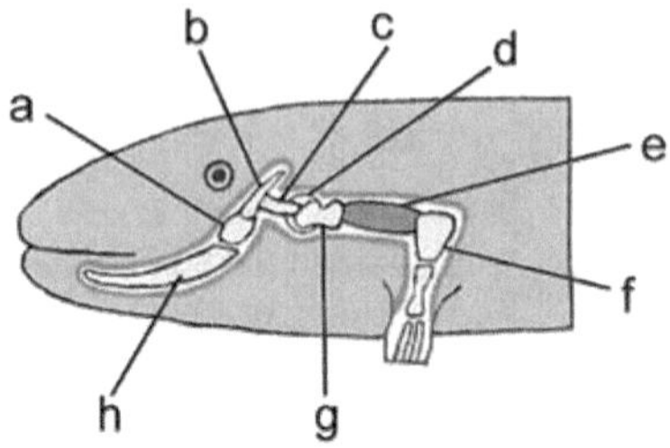

Fig. 12.13 Amphibian ear. (a) quadrate, (b) squamosal, (c) stapes, (d) oval window,(e) opercularis muscle, (f) scapula, (g) operculum and (h) lower jaw.

detected by the basilar papilla while the low sound frequencies including those made by predators are sensed by the amphibian papilla.

Reptilian Ear

External ear structures are present in some reptiles (Fig. 12.14). Reptiles with external ear structures such as iguanids have tympanic membranes that are normally superficially located close to the body surface and are visible from the outside. The ear drum is located deeper in the head region if an external acoustic meatus is present as is found in lizards. Reptiles with external ears lack pinnae to trap sound waves though some have scales that stick out from the head and are located cranial to the opening into the ear canal while some species have canals that are angled. The stapes is the only ear ossicle in the middle ear and runs from the inner part of the tympanic membrane to the oval window, crossing the middle ear cavity in the process. The outer end of the stapes has a cartilage known as **extrastapes** that makes contact with the tympanic membrane. In reptiles that lack eardrums, the extrastapes is attached to the quadrate that supports the lower jaw to the upper jaw.

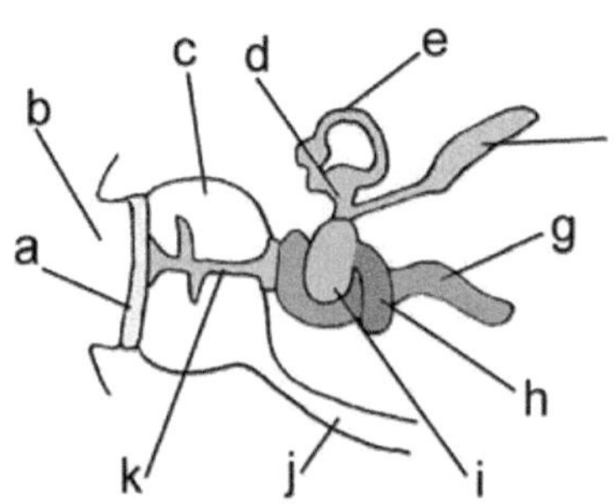

Fig. 12.14 The ear of a lizard. (a) tympanic membrane, (b) external auditory meatus, (c) tympanic cavity, (d) utriculus, (e) semicircular canal, (f) endolymphatic sac, (g) perilymphatic sac, (h) lagena, (i) sacculus, (j) auditory tube and (k) stapes.

Crocodilians and geckos have the stapedius muscle that attaches to the stapes and may serve a similar function to that of the mammalian stapedius of reducing oscillations from strong vibrations. The tympanic membrane normally senses vibrations in the air while the quadrate detects vibrations from the surface the reptile is in contact with. The vibrations cause the tympanic membrane or quadrate to vibrate and they transit this sensation through the stapes to the cochlear duct and to the hair cells. The tympanic membrane lacks in snakes amphisbaenians, the tuatara and other fossorial lizards. The tympanic membrane could have been lost as an adaptation to burrowing in reptiles. Though most snakes live above the ground, they could have evolved from a burrowing ancestor. Snakes can also sense low-frequency vibrations using their ears. Reptiles that lack tympanic membranes can also conduct vibrations from mechanoreceptors in the skin of their body trunks to the quadrate through spinal nerves. The vibrations caused by the quadrate can then be transmitted to the inner ear.

Avian Ear

The bird's ear is similar to that of reptiles in many ways (Fig. 12.15) though the sense of hearing is more sensitive in birds. The avian ear lacks an ear lobe and the external opening into the short external auditory meatus is covered with the specialized auricular feathers. These feathers protect the opening from objects and also funnel sound waves towards the opening. The wall of the auditory meatus has ceruminous glands that secrete wax-like material. The meatus is separated from the middle ear by a tympanic membrane. The middle ear has a single rod shaped ossicle known as columella that transmits sound vibrations from the tympanic membrane to the oval window and eventually to the hair cells of the long and uncoiled lagena.

The **paratympanic** or **vitali organ** that measures about 1.0 mm in length is found on the medial wall of the middle ear. The banana-shaped

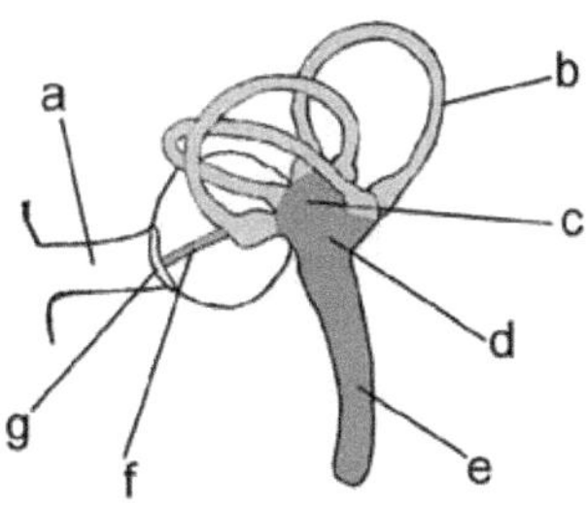

Fig. 12.15 The avian ear. (a) external auditory meatus, (b) semicircular canal, (c) utriculus, (d) sacculus, (e) lagena, (f) columella and (g) tympanic membrane.

paratympanic organ is embedded in the connective tissue and has a lumen that is surrounded by an epithelium which contains sensory and non-sensory cells. The sensory hair cells are similar to type II receptor cells that are present in neuroepithelia of the vestibule of the inner ear and lateral line systems and may represent a sense organ which is derived phylogenetically and ontogenetically from the lateral line system (von Bartheld, 1990). There is also evidence that the paratympanic organ is also present in the middle ear of juvenile alligators (Neeser and von Bartheld, 2002) as well as bats. Paratympanic organs may be part of a neural circuit that controls the position of the tympanic membrane and may mediate barometric perception in birds (von Bartheld, 1994).

Various birds are capable of hearing different sound intensities from members of their species and some sounds inaudible to the human being and other vertebrates. Birds are able to discriminate sounds of different pitches and many are capable of perceiving frequencies that are far above the range of the human being. The hair cells of birds are found in the **basilar papilla** of the uncoiled cochlear duct and include stereocilia that are hexagonally arranged relative to kinocilia. The individual hair cells vary in their length, width, number and distribution of their cell extensions. The length of the stereocilia also shows a gradient in its numbers, length distribution and width and may play a major role in frequency discrimination in the cochlea (Tilney and Saunders, 1983).

Echolocation

Some vertebrates use high frequency sound such as sonar and radar that bounce off objects as reflected vibrations known as echos. The echo is then interpreted including the time it takes to return to the vertebrate to determine the direction, distance and size of objects in the neighboring environment. The animal is then able to orient, navigate and locate food. Echolocation or biosonar has evolved independently in various vertebrates including bats, porpoises, whales, shrews, oilbirds and swiftlets. Events leading to echolocation are well preserved in the fossil remains of ancestors of porpoises and whales. Early fossil remains of bats show that echolocation was already present.

The ultrasonic chirps of bats (over 15 Kz) are produced by the larynx through the mouth or nostrils. After sensing the echos that bounce off the prey, the bat increases the chirp rate as it flies in the direction of the object. Most bats alternate between emitting sound and listening to the echo. Echolocation enables many bats to catch insects for food as well as navigate in poorly lit caves. Evolution of echolocation has led to poor visual abilities in most microchiropterans in comparison to macrochiropterans.

Dolphins produce 'clicks' and 'whistles' of echolocation probably in the nasal air sacs and valves along the opening into the blowhole. The sounds are then directed towards the 'melon' that contains fats and forms a bulge on the forehead by the inwardly curved surface of the head. The melon focuses the vibrations and transduces the air-borne vibrations into water-borne vibrations. The ensuing echo passes through the lower jaw to the middle ear.

Sense of Balance

The sense of balance or equilibrium is detected by hair cells within the ampullae of semi-circular canal and the vestibule that comprises the utriculus and sacculus in vertebrates. The vestibule is located in the middle of the inner ear. The **utriculus** (L. *utricules,* small bag) runs into the larger **sacculus** (L. *sacculus,* small sac) ventrally. The utriculus and sacculus contain endolymph. The size of the sacculus varies in different vertebrates. The utriculus and sacculus are used in sound detection in some lower vertebrates. The sacculus forms an extension on its caudoventral part known as **lagena** (L. *lagena,* flask) in fishes, amphibians and most reptiles. In some reptiles and birds, the lagena is larger and is mainly coiled into a cochlea in mammals except monotremes. Lagena is used in sound detection. In some birds including pigeons, the lagena is a main element in the magnetic sensory system that is the key to their homing abilities (Harada, 2002). In crocodiles, birds and monotremes, the basilar papillae become incorporated in the organ of Corti that is the sole receptor for sound. Otoliths are found in the utriculus, sacculus and lagena and their specific names are **lapillus**, **sagitta** and **asteriscus** respectively. Sagittae are the largest otoliths in many fish (Yamauchi et al., 2008) and are used for ageing these vertebrates.

Three semi-circular canals are connected to the utriculus and are arranged in planes that are approximately at right angles to each other. Two of the canals are arranged in a vertical plane but perpendicular to each other while the third canal is horizontally arranged. The horizontal canal is not present in lampreys while only one canal is present in the hagfish. The membranous semi-circular canals are located within the bony semi-circular canals and contain endolymph. The perilymph occupies the space between the membranous and bony semi-circular canals. Each semi-circular canal forms an enlargement at the junction with the utriculus known as an **ampulla** that contains receptive cells.

There are various types of hair cells within the membranous labyrinth of the inner ear of vertebrates. The ampullae of semi-circular canals contain cluster of hair cells known as **cristae** (L. *crist,* crest) (Fig. 12.16). The **maculae** (L. *macula,* spot) (Fig. 12.17) are hair cells found

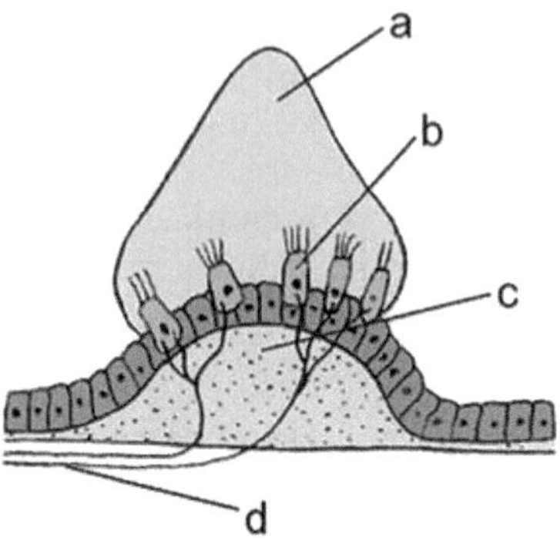

Fig. 12.16 Part of the inside of an ampulla. (a) cupula, (b) sensory cell, (c) crista ampullaris and (d) vestibular nerve fiber.

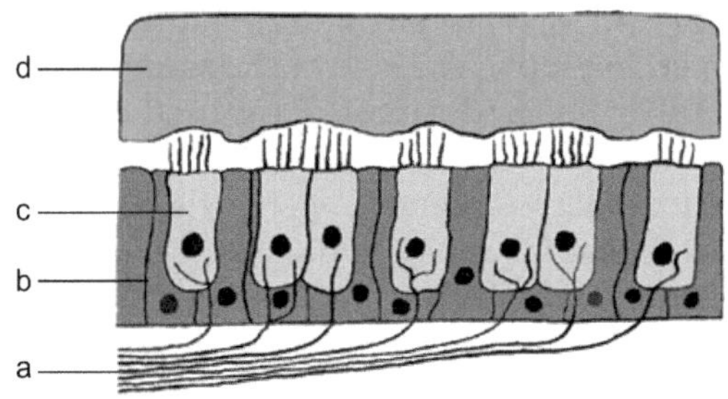

Fig. 12.17 Part of the macula of utriculus and sacculus. (a) sensory nerve fibers to the vestibular nerve branch, (b) supporting cell, (c) sensory cell and (d) otolith.

in the utriculus and saccules of all vertebrates and the lagena of many other groups. In the utriculus of fishes is also found a type of hair cells known as the **crista neglecta**. A group of hair cells and supporting cells known as **sensory papillae** is found in the sacculus, lagena or cochlea of tetrapods and is sensitive to sound. Above sensory papillae lies a tectorial membrane. In gnathostome vertebrates, otoliths (in fish) or **otoconia** (in other vertebrates) are found in the utriculus, sacculus and lagena. Whereas the mineral crystals are held together as a single solid mass in otoliths, the crystals are found in a pasty mass in otochonia. Endogenous otoliths are thought to develop in the macula and endolymphatic duct (Marmo, 1982). The development of the otolioth and the role it performs in vertebrates has been reviewed extensively by Fermin et al. (1998). The utricules, saccules and lagena of teleosts are used in sound detection. Otoliths have a higher density than these fluid-filled chambers and are affected by sound waves differently resulting in independent movement of these organic crystals as they intercept sound vibrations. The hair cells are stimulated mechanically by the movement of otoliths, an action that triggers off the auditory impulse.

Sensory cells of the utriculus, sacculus and the ampullae of semi-circular canals sense balance by detecting changes in position and movement. **Static equilibrium** senses the orientation of the body in space. The position of the head relative to gravity when the body is motionless is detected by sensory receptors in the utriculus and sacculus. The sensory apparatus in the ampullae detect **dynamic equilibrium** that is associated with rotational or sudden movements of the head or body.

The maculae of the utriculus and saccules comprise hair cells and supporting cells that are covered by a gelatinous matrix. Otoliths are located in and above the matrix of the macula. As the otoliths are of greater density than the macula, the pull of gravity on these structures that results from head movement during static equilibrium moves the cilia of the hair cells resulting in a generation of action potentials to the brain. The position of the head is then detected as well as a change in the pull of gravity-like during acceleration. The **righting reflex** involves muscular action and related structures and ensures that the body is back to its normal anatomical position after displacement and is initiated after stimulation of the macula. Proprioceptors as well as the eyes play a role in triggering off the righting reflex.

The crystae of the ampullae sense the dynamic equilibrium state. The processes of the hairs cells of the crista are embedded in a gelatinous substance known as cupula. Movement of endolymph in the semi-circular canals with movements of the body in various directions move the cupula that then bends the hairs of the sensory cells. Since semi-circular canals are arranged in a perpendicular manner to each other, movements of the body in various directions can be sensed by the hairs of the cristae. The cupula normally moves in a direction that is opposite to that of the direction in which the body moves as inertia prevents the endolymph in the semi-circular canals from moving at the same rate as the body.

EVOLUTION OF THE MAMMALIAN MIDDLE EAR

Mechanoreceptors are homologous structures across animal phyla. The ear is a mixture of highly conserved developmental components that is combined with co-opted genes of newly evolved developing parts of the body including the forebrain, limbs and kidneys (Fritzsch and Beisel, 2001). As mammals evolved from synapsids, the evolution of the mammalian ear can be traced from these extinct mammal-like reptiles. The early synapsids had lower jaws that were made up of several bones which were attached to each other by strong fibrous joints. The tooth-bearing part of the lower jaw was the dentary and the other more posterior components of the jaw comprised the articular, quadrate and

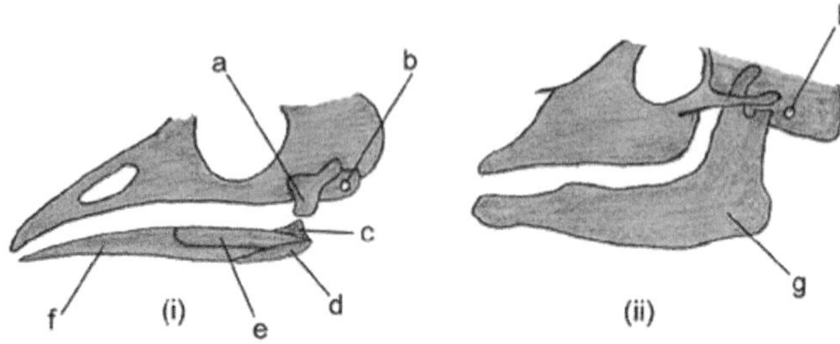

Fig. 12.18 A diagram comparing the differences between the jaws of (i) a bird and (b) a mammal. (a) quadrate, (b) opening of external auditory (ear) canal, (c) articular, (d) angular, (e) supra-angular, and (f) dental bones and (g) mandible. The quadrate and articular bones have evolved into the incus and malleus respectively whereas the dental, angular and supra-angular bones have fused into the mandible in mammals.

angular bones (Fig. 12.18). The synapsids could detect low frequency sound vibrations and pass them onto the articular, quadrate, stapes and the inner ear. With further evolution, the postdentary bones and the stapes decreased in size and weight as their efficiency to transmit sound improved. The articulation of the lower jaw with the upper jaw shifted from the quadrate and articular bones to the dentary and squamosal bone. The quadrate and articular bones were now free to act in sound transmission and were acted on by natural selection to act as ear ossicles. The early mammals could then use their ears to detect high-frequency sounds. Forces that brought about changes in articulation of jaws could also have favored the evolution of the articular into the malleus and the quadrate into the incus. Evidence gathered from the dentary bone of an ancient toothed monotreme shows that the middle ear bones evolved independently in the monotremes and therian mammals (Martin and Luo, 2005; Rich et al., 2005).

EYE

The eye is the vertebrate's sense organ for vision. Some vertebrates that burrow or live at the bottom of seas have secondarily reduced or lost their sense of sight. The eyes of hagfish lack lens and those of the genera *Myxine* and *Neomyxine* are smaller than those of *Eptatretus* and *Paramyxine* and are partly covered by skin and body trunk musculature. In most vertebrates, the two eyes are located on either side of the head so that there is little or no overlap in the visual fields. Such vertebrates have high periscopy as each eye has a wide field of view. In those vertebrates where both eyes are located in front of the head (binocular vision) such as primates and owls, there is a considerable overlap in the field of vision resulting in greater visual acuity.

Most of the eyeball in vertebrates is located in the eye socket and only a small part is exposed to the outside. The basic structure of the eyeball is the same in various vertebrate groups. The eyeball is surrounded with three layers (tunics) of tissue (Fig. 12.19). The outermost layer is the sclera, the middle layer is the choroid and the innermost layer is the retina. The **sclera** (Gr. *skleros,* hard) is the white part of the eye that is visible from the outside and is made of tough fibrous tissue. The **cornea** (L. *corneus,* horny) is the part of the sclera that forms the central part of the exposed eye. The cornea lacks a blood supply and light enters the eye through this transparent structure. The epithelial cells of the cornea rest on a thick basal lamina known as **Bowman's membrane**. The thickness of the corneal epithelium varies in the different vertebrate groups. Most of the underlying tissue of the cornea is made up of collagen fibers. The outer surface of the cornea is covered by the conjuctiva that is an epithelial tissue. The **conjuctiva** (L. *conjunctus,* joined together) also covers the inner surface of the eyelids, when present. Apart from the cornea, the rest of the sclera is opaque.

The **choroid layer** (Gr. *chorioeides,* like a membrane) has a rich blood supply and is pigmented. The choroid layer nourishes the underlying retina. The choroid is also highly pigmented with the melanin. The pigmentation reduces internal reflections by absorbing any stray light

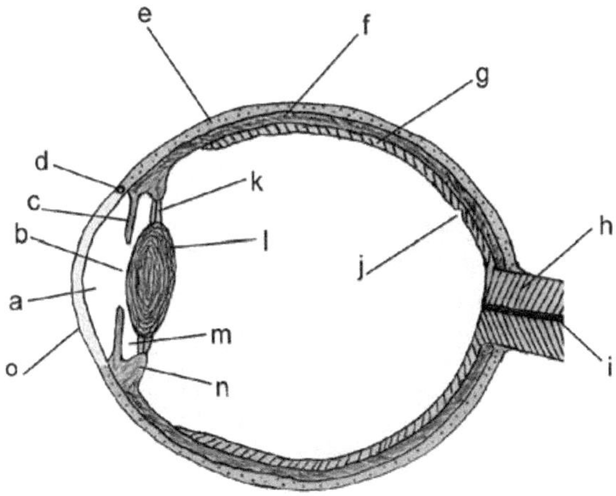

Fig. 12.19 A mammalian eye. (a) anterior chamber, (b) pupil, (c) iris, (d) canal of Schlemm, (e) sclera, (f) choroid, (g) retina, (h) optic nerve, (i) central retinal artery and vein, (j) fovea, (k) suspensory ligaments, (l) lens, (m) posterior chamber, (n) ciliary body and (o) cornea. The optic disk or blind spot is the central part at the beginning of the optic nerve and borders the vitreous humor. The anterior cavity lies between the cornea and the lens together with the suspensory ligaments while the posterior cavity extends backwards up to the retina. Aqueous and vitreus humor occupy the anterior and posterior cavities respectively.

penetrating the retina thus increasing contrast and visual acuity. At the margin of the cornea, the choroid forms the ciliary body, the suspensory ligaments and the iris. The choroid also forms folds that extend into the vitreous body known as the falciform process of most teleosts, the papillary cone of some squamates and the pecten of birds (Fig. 12.20). The **falciform process** is curved like a scythe and is a ridge-like process on the floor of the eyeball. The **retractor lentis** muscle attaches to the falciform process and to the lens. Contraction of the muscle pulls the lens backwards. The falciform process is thought to have a nutritive function. The **pecten** is a thin and highly folded tissue that extends from the retina towards the lens. The pecten is highly developed in predatory birds that also have a good sense of sight such as eagles and hawks. The structure supplies oxygen and nutrients to the vitreous humor of the eye and as a result the number of blood vessels supplying the retina is lessened resulting in a reduction in scattering of light and a sharp vision in these raptors.

The **ciliary body** contains the ciliary muscle in its anterior part. The ciliary muscle contains smooth muscle fibers. The **suspensory ligaments** hold the lens in place. The **iris** is the colored visible part of the eye that is surrounded by the sclera and contains circular and radial smooth

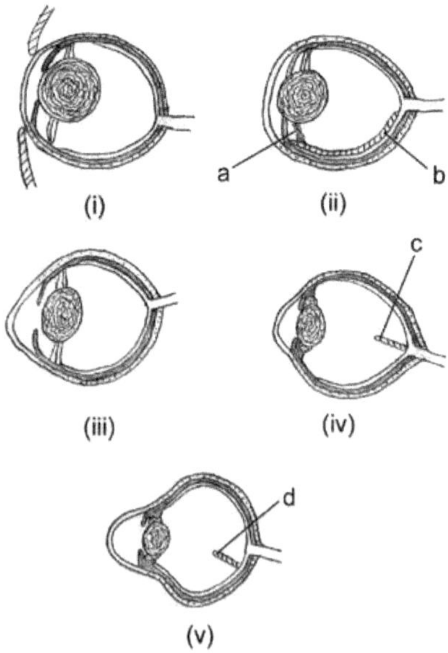

Fig. 12.20 The eyes of various vertebrates: (i) lamprey, (ii) teleost, (iii) amphibian, (iv) lizard and (v) bird. (a) retractor lentis, (b) falciform process, (c) papillary cone and (d) pecten. Eyes of lampreys lack intrinsic muscles and their extrinsic muscles are almost like those of gnathostomes. The eyes of hagfish are small, lack lens and are covered by skin.

muscles. The iris is attached to the ciliary body and surrounds an opening in the middle of the eye known as the **pupil**. The smooth muscles of the iris contract and relax to change the size of the pupil thus controlling the amount of light that enters the eye. The lens is located deeper to the iris and is covered by an elastic capsule and an epithelium that comprises cuboidal cells. The cuboidal cells are the source of most of the lens. The lens are capable of growth throughout life. The lens varies in shape from a flattened structure when the suspensory ligaments are under tension to almost a spherical shape when the ligaments relax as a result of contraction of ciliary muscles. Changes in the shape of the lens enable the eye to change its focus when viewing objects at various distances. The lens is in contact with the aqueous humor and vitreous humor.

The **retina** (Fig. 12.21) is the innermost layer of the eye. It contains neurons and glial cells and is partly an extension of the brain. The retina

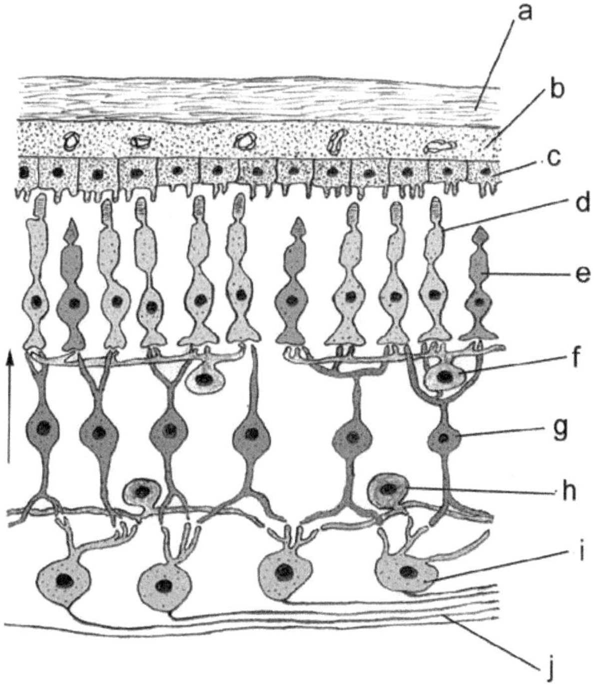

Fig. 12.21 Structure of the retina. (a) sclera, (b) choroid, (c) pigmented layer of retina, (d) rod, (e) cone, (f) horizontal cell as part of the outer plexiform layer, (g) bipolar cell, (h) amacrine cell of the inner plexiform layer, (i) ganglion cell and (j) fibers to optic nerve. The arrow indicates the direction in which light travels. There are about 40 different types of amacrine cells and most lack axons. Amacrine cells play a role in the complex processing of the retinal image by adjusting the brightness of the image as well as integrating activation of neurons detecting motion in a sequence.

is the photoreceptive part of the eye and is supported by the other ocular structures. The retina contains pigmented cells next to the choroid coat. The pigmented layer of the retina is not part of the retina proper and plays a role in phagocytosis of worn out photoreceptor cells that are continuously being replaced and is also involved in the formation of visual pigments. Photoreceptor neurons are found in the retina and have dendrites that differ in the shape of their distal ends. The distal ends that are rod-shaped belong to the **rods** while those that resemble the shape of a cone are known as **cones**.

The eyes of animals have two types of photoreceptors that differ in the way their surface areas are increased, the site the photoreceptor molecules are located and the path taken during photoreceptor transduction. A larger surface area of the cell membrane is able to accommodate more photoreceptor molecules. **Rhabdomeric photoreceptors** (Fig. 12.22) are found in the compound eyes of arthropods and are closely related to the ganglion cells of the vertebrate retina. Rhabdomeric photoreceptors increase their surface area by forming folds at their apical surfaces. **Ciliary photoreceptors** of the vertebrate eye are found in the image forming eyes as well as the pineal eye. A cilium extends from a ciliary photoreceptor cell and increases its surface area by forming several membranous projections. The polycheate marine rag-worm *Platynereis dumerilii* has rhabdomeric photoreceptor cells in the eyes and ciliary photoreceptor cells in the brain where the latter uses a photopigment that closely resembles vertebrate rod and cone opsins and that in *Urbilateria,* the last common ancestor of insects and vertebrates, both types of photoreceptors had distinct opsins that coexisted (Arendt et al., 2004).

Rods and cones contain light sensitive pigments that are derived from vitamin A and are attached to proteins known as **opsins**. Light activates opsin leading to a change in the conformation of the photopigment. Opsin then binds to a G-protein. Since the G-protein is common in photosensitive cells and plays a role in most photoreceptor

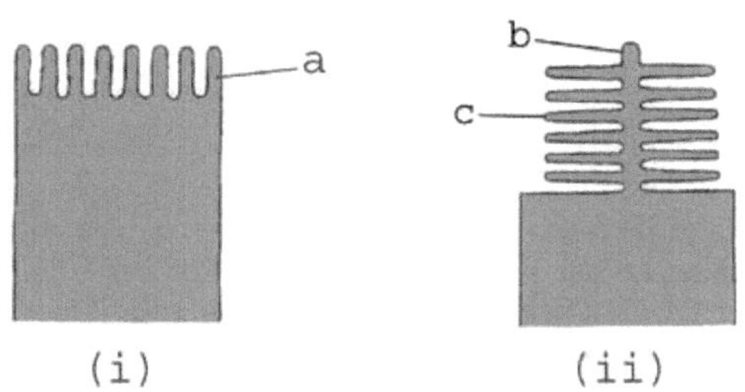

Fig. 12.22 The apical surfaces of (i) rhabdomeric and (ii) ciliary photoreceptor cells. (a) fold at the apical surface, (b) cilium and (c) membranous projection on (b).

signal transduction cases, all eyes could have a common evolutionary ancestor. The photosensitive pigments vary in the light sensitive cells and determine the type of light that is absorbed. Cones contain the pigments **iodopsin** and rods contain **rhodopsin**. Cones are sensitive to bright light and colors while rods are sensitive to dim light. The difference in sensitivity to light between rods and cones is likely due to the difference in the molecular properties of the phototransduction protein isoforms in these pigmented cells (Hisatomi and Tokunaga, 2002). Cones are concentrated in a small depression of the retina known as the **fovea**. Since cones are sensitive to color, vertebrates with color vision have various types of this photoreceptor cell and pigments. Multiple opsin genes originated early in vertebrate evolution, prior to the separation of the jawed and jawless vertebrate lineages and this provided the genetic basis for color vision in all vertebrates (Collin and Trezise, 2004).

Colored oil droplets are useful for color vision and are found in the retina of most diurnal reptiles and birds. The droplets are found in the cones and rarely in rods. Oil droplets consist of lipids that contain dissolved carotenoid pigments. The droplets can appear transparent, pale yellow, green, orange or red. Oil droplets cover the entire distal inner parts of cones. Light has to traverse the droplets before reaching the photosensitive parts of cones. Oil droplets absorb light with a lower wavelength than the color of the droplets. Light with longer wavelengths passes through the droplets to reach the photopigments. The droplets also increase the number of object colors that can be discriminated by filtering light in the retina thus reducing spectral overlap between spectrally adjacent cones (Vorobyev, 2003).

Vertebrates that are active at low intensities of light have a light reflecting structure known as **tapetum lucidum** (L. *tapete*, carpet; *lucidus*, shining) in the back part of the retina. Such action by the tapetum lucidum maximizes on the low light level as the reflected light is detected by the photoreceptor cells of the retina. Some of the reflected light is lost out through the pupil and will cause the eye to glow when light is directed at the eyes. The tapetum is lacking in birds, red kangaroos, pigs, squirrels and primates but is common in fish, some reptiles especially crocodiles and carnivores. The tapetum varies in composition of the reflective layer and location as well as physical and chemical properties (Ollivier et al., 2004) and this will affect the color of the glow. In fish and crocodiles, the tapetum is made up of guanine crystals whereas in most mammals it is composed of a layer of reflective connective tissues including specially arranged collagen fibers.

Photoreceptor neurons synapse with bipolar neurons that form connections with ganglion neurons. The axons of ganglion neurons lead to the **optic disk** in the deeper part of the eye from where they will give rise to the optic nerve. The optic disk lacks rods and cones and so cannot visualize light rays and is sometimes referred to as the **blind spot**. Studies in fish, mice and the human being have shown that there are other photopigments in the retina apart from cones and rods and their role in photoreception is being investigated (Foster and Hankins, 2002). Studies using the naked mole rat, a fossorial rodent with a poor sense of vision, have shown that although image formation on the retina is not important in these rodents, circuits beyond those required for circadian entrainment remain in place (Mills and Catania, 2004). The vertebrate retina including that of birds and mammals contains stem cells and this has stimulated research in transplantation of such stem cells into damaged retinas (Amato et al., 2004).

The inside of the eye contains liquid that is found in the anterior and posterior cavities. The **anterior cavity** is found in front of the lens and is filled with the watery **aqueous humor**. The aqueous humor nourishes the avascular lens. The anterior cavity is divided into the **anterior chamber** that is located anterior to the iris and posterior to the cornea and the **posterior chamber** that lies between the iris and lens. The **posterior cavity** is much larger than the anterior cavity and is posterior to the lens, ciliary body and suspensory ligaments. The posterior cavity is filled with a gelatinous fluid known as **vitreous humor** (L. *vitreus*, glassy). The aqueous humor and vitreous humor maintain the right intraocular pressure and prevent the eyeball from collapsing.

Most of the aqueous humor is formed from blood vessels of the ciliary body. Aqueous humor is actively and passively secreted into the posterior chamber. Aqueous humor will then circulate from the posterior chamber through the pupil into the anterior chamber. The liquid flows back to the venous system via the **canal of Schlemm** that is similar to a lymphatic vessel and is located at the junction between the sclera and the cornea. Aqueous humor has to be continuously produced and drained to maintain the normal intraocular pressure. When this pressure is too high or too low, there is distortion of the shape of the eyeball and impaired vision. An increase in production or a decrease in the drainage of aqueous humor can result in an increase in the intraocular pressure known as **glaucoma**. If unattended to, glaucoma can lead to blurred vision, retinal detachment, atrophy of the optic nerve and permanent blindness.

FORMATION OF IMAGES ON THE RETINA

Light has to pass through the cornea, aqueous humor, lens and vitreous humor before reaching the retina and crossing this structure to the photoreceptive receptors (rods and cones) at the back of the retina. When light rays pass through media of different optical densities, they undergo refraction or bending. In such cases light passes obliquely from one medium to another. Refraction is greater when the surface of the medium is more convex. Light undergoes refraction in the various structures it crosses from the outside to the retina. The parallel light rays that are reflected from an object have to undergo proper deflection in the four media that include the cornea, aqueous humor, lens and vitreous humor before the rays are properly focused on the retina.

Several changes occur in some structures of the eyes as light passes through them. The pupil adjusts its size and thus controls the amount of light passing through it to the retina. When the light is intense, the circular muscle fibers of the iris contract thus reducing the size of the pupil and amount of light passing through. Constriction protects the retina from light that is too intense. The pupil is also constricted to prevent divergent rays of an object from entering the eye through the side of the cornea and lens. The side rays will cause a blurred image on the retina since they are not properly refracted. When the light intensity is low, the size of the pupil increases due to relaxation of circular muscles and contraction of longitudinal muscles of the iris.

The lens adjusts its convexity when objects at various distances are viewed. During **accommodation**, the lens increases its convexity when near objects are viewed (Fig. 12.23). The ciliary body muscles contract thus easing tension on the suspensory ligaments leading to an increase in the curvature of the lens. The divergent light rays from near objects are bent more acutely by the more convex lens for the rays to be focused on the retina. The continued viewing of closer objects in the human being strains the eyes as a result of contraction of ciliary muscles for a long time. When viewing distant objects with almost parallel light rays, the lens is relatively flat as the rays do not have to be refracted much. A less curved lens is achieved by an increase in the tension of suspensory ligaments as a result of relaxation of ciliary muscles. About two thirds of refraction of light rays occurs in the cornea with most of the remaining percentage occurring in the lens. Lenses lose their elasticity and their ability to curve for use in viewing closer objects with age. In the human being, the condition is known as **presbyopia** and people with such a condition are long sighted.

The incoming parallel light rays are brought together by the converging double convex lens after passing through this structure. The distance from the lens to the covergence point of the rays is the focal length. The eye normally varies its focal length to accommodate objects at various distances. When an object is nearer the eye, the focal length is

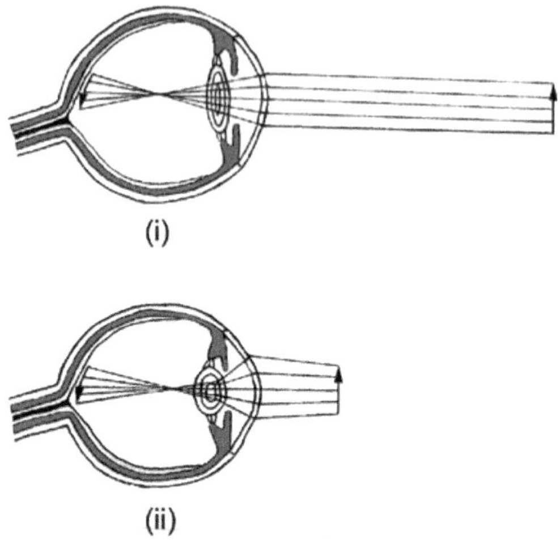

Fig. 12.23 Accommodation of the lens. When viewing an object at a long distance (i) the ciliary muscles are relaxed and the suspensory ligaments are stretched, leading to the flattening of the lens. When viewing objects at close range, the ciliary muscles contract and this leads to a reduction of tension on the suspensory ligaments and rounding of the lens (ii). In both cases, images are focused on the retina.

shorter and vice versa. Variation in focal length ensures that objects that are closer to the vertebrate appear larger than those of similar size at a distance. The center of the object is focused to the most sensitive part of the retina, the fovea that has the highest concentration of photoreceptor cells, mainly cones. The image formed on the retina is inverted and reduced in size. The brain interprets the image as originating from a right-side-up object.

Once light reaches the photoreceptor rods and cones, the photopigments within these receptors undergo structural changes that will elicit nerve impulses to the brain. When the highly light sensitive pigment rhodopsin found in rods is exposed to even low intensities of light, it disintegrates into the protein opsin and a vitamin A derivative known as **retinal** that absorbs light. When light is absorbed by the retina, the proteins of the cell membrane of rods are stimulated altering the membrane potential and leading to the generation of an action potential. Three types of cones are present in the human being and contain different photopigments that include **erythrolabe**, **chlorolabe** and **cyanolabe** for the colors red, green, and blue respectively. Each color reflects light of a specific wavelength that is sensed by a particular cone. The breakdown of a specific pigment by light of a particular wavelength leads to generation of an action potential to the brain. Cones produce vision in intense light since they are less sensitive to light than rods and their photopigments require more intense light before they can break down.

MUSCLES OF THE EYE

The two types of ocular muscles are extrinsic and intrinsic muscles. Extrinsic muscles of the eye (Fig. 5.10) are skeletal muscles that attach to the outer part of the eyeball and to the bone of the orbit and are also known as extraocular muscles. Extrinsic ocular muscles move the eyeball in any direction. There are six extraocular muscles: four straight and two oblique muscles. Intrinsic muscles are smooth muscles found in the ciliary body and the iris. The contraction and relaxation of these muscles regulates the convexity of the lens and size of the pupil respectively.

ACCESSORY STRUCTURES OF THE EYE

The accessory structures associated with the eye include eyelids, lachrymal glands, eyelashes and eyebrows. Tetrapods have **eyelids** or **palpebrae** that are lacking in fish and are lined to the inside by a conjuctival epithelium. The tarsal conjuctiva that lines the eyelids continues onto the outer surface of the cornea as the bulbar conjuctiva. The rare type of conjuctival epithelium is stratified columnar and contains goblet cells. The rest of the eyelid is made of skeletal muscle and skin. The eyelids can move over the corneal surface to protect and cleanse it. At the free edge of an eyelid is thick connective tissue known as the **tarsal plate**. The upper and lower eyelids join at either side of the eye forming angles known as **medial canthus** and **lateral canthus**. Modified sebaceous glands known as **tarsal** or **Meibomian glands** are found in the eyelids and open at their margins. There are also modified sweat glands known as **glands of Moll** that release their secretions into the follicle of eyelashes in mammals.

The **third eyelid** or **nictitating membrane** (L. *nictare,* to wink) is found in many vertebrates but is lacking in fishes apart from many sharks and the human being. The nictitating membrane, also known as **haw**, is normally a transparent and thin conjuctival fold that contains a layer of connective tissue that is supported by a T-shaped cartilage. The stratified epithelium of the membrane also contains goblet cells. When at rest, the nictitating membrane is folded at the medial canthus of the eye. In many amniotes, the membrane moves across the surface of the cornea, spreading lachrymal secretions and cleaning the corneal surface in the process. Most animals can control movement of the nictitating membrane. Some vertebrates draw the membrane across the cornea to spread moisture over the eyeball while maintaining visibility at the same time. Such action reduces the rate of blinking and enables some vertebrates sustain continuous visibility. In the polar bear, the nictitating membrane is closed to filter out ultraviolet light and to lower cases of snow blindness while in sharks it protects the eyes when these cartilaginous fish attack their prey. The remnant of the nictitating membrane in the human being is a vestigial semilunar fold or the visible pink nub or lump in the inner corner of the eye.

Eyelashes are only present in mammals and are associated with eyelids. Eyelashes together with the eyebrows of the human being offer some protection against the entrance of small objects into eyes. These hairs also shade the eyes to a certain extent.

Lacrimal glands (L. *lacrima*, tear) secrete tears that keep the surface of the eyeball moist. Tears protect the cornea against damage such as abrasion. The structure of lacrimal glands and nature of their secretions resembles that of serous salivary glands. The secretions of lacrimal glands leave through various ducts to the upper surface of the eyeball. Tears also contain the anti-microbial agent lysozyme. In mammals, tears are drained into the nasal cavity through the lacrimal duct. There are other tear glands whose ducts open the inside of the eyelids.

The **Harderian gland** or **Harder's lacrimal gland** was first described in a red deer by a Swiss physician known as Johann Harder in 1694. Harderian glands are found in most vertebrates with a nictitating membrane and are quite large in ungulates and rodents. The tubulo-alveolar Harderian glands are located within the orbit on the posterior part of the eyeball at the base of the nictitating membrane around the optic nerve and contain large quantities of lipid. Harderian glands play various roles depending on vertebrate species including lubricating the eye and nictitating membrane, synthesis of pheromones, photoprotection, osmoprotection, thermoregulation, immunity and endocrine functions. Photoreceptors that can sense light and help regulate the pineal gland are present in the Harderian gland. In rodents, the Harderian gland contains the reddish brown porphyrin pigment that makes the tears of these mammals appear red in color.

VISION IN WATER

Visibility in water is not as good as on land. Turbidity diminishes visibility in an aquatic environment further as light rays are scattered in various directions. Aquatic vertebrates have to rely on other sense organs to sense their surrounding. Water with poor visibility is generally inhabited by vertebrates with dull colors such as benthic fish whereas colorful vertebrates tend to inhabit clear waters that are well lit including coral reef water that is inhabited by the brightly colored coral reef fish. As water and the cornea have a refractive index that is close to each other, light moving from water to the cornea will not be deflected much.

Vertebrates that live on land and have to spend many hours in water such as seals have eyes that have great ability to focus.

Fish lack tear glands as water cleans the cornea externally continuously. The iris of most fish lacks longitudinal muscle fibers and as a result enlargement of the pupil after the circular fibers have relaxed is slow. This should not be a problem to fish as levels of light undergo slower variations in water than on land. The lens of fish is more spherical in shape and less flexible than that of terrestrial vertebrates and is the main refractive structure in a fish's eye. Such a lens provides great refraction.

The fish lens is located close to the cornea and further away from the retina and this makes most fish short sighted. Fish focus by changing the relative distance between the lens and retina through the action of the retractor lentis muscle. In order to focus on a more distant object, the muscle pulls the lens closer to the retina and vice versa. The choroid of fish contains the reflective guanine crystals in the tapetum lucidum. The guanine crystals are useful in dim conditions as they reflect the light back to the photoreceptor cells. Most deep-sea fish have rods as photoreceptor cells since cones are lacking. Other deep-sea fish possess tubular eyes that concentrate the low rays of light. Fish living in illuminated waters have four types of cones that detect red, green, blue and ultraviolet light. Others have two or three of these cell types. Many fish are most sensitive to light in the green-yellow wavelength range. Ultraviolet light cones might be important in navigation, communication and avoidance of ultraviolet radiation as this high-energy short wavelength radiation can be damaging to biological life.

THIRD EYE

The non-image forming third or median eye (Fig. 12.24) develops from the diencephalon and is connected to the pineal gland. The eye is found in some fishes, amphibians and most reptiles and comprises one or two light sensitive eyes in the parietal foramen on top of the head. Detection of light by these eyes is important in these vertebrates as it enables them to adjust their physiological processes in response to the intensity of light. Median eyes are superficially located and are covered by skin and sometimes a thin plate of the skull.

The retina of the parietal eye has photoreceptive and ganglion cells but lacks bipolar, horizontal and amacrine cells that are found in the image forming eyes of vertebrates (Rieke, 1998). The photoreceptors have sensory nerve endings that lead to the brain and depolarize when exposed to light unlike other photoreceptors of vertebrates that hyperpolarize under similar conditions. In darkness, the median eye

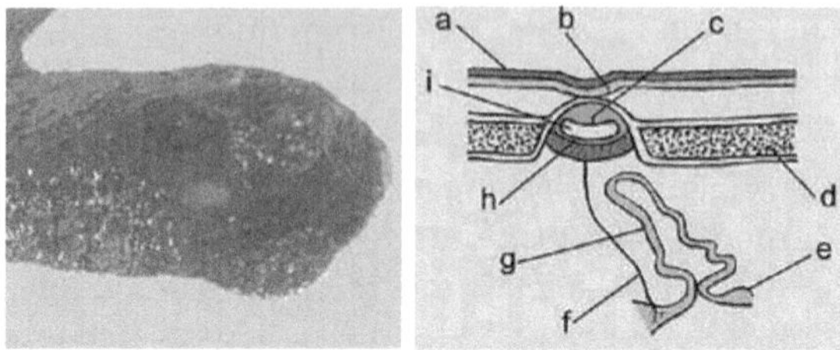

Fig. 12.24 The non-image forming third or median (parietal) eye that appears as an oval light yellowish-brown structure on top of the head of an iguana (left) and a drawing of a section through the eye (right). The median eye is found in many vertebrates. (a) skin, (b) cornea, (c) lens, (d) skull, (e) midbrain, (f) parietal nerve, (g) pineal organ, (h) light sensitive cells of the retina and (i) parietal eye. The third eye appears sometimes as an opalescent spot.

produces an enzyme that converts serotonin into melatonin. Melatonin acts on melanocytes and causes melanin to concentrate making the animal look lighter. Light inhibits the production of the enzyme leading to dispersal of the pigments in melanocytes resulting in the animal looking darker. The median eye has been shown to be essential for the development of sexual maturity.

The level of development of the median eye varies in different vertebrates. The median eye complex was present in extinct ostracoderms and placoderms. In cartilaginous fish and many higher bony fishes, the median eye is represented by a smaller pineal organ or epiphysis. In some reptiles such as lizards, the pineal complex is a sophisticated structure with a median pineal eye, a paraphysis and a pineal sac. In the tuatara, the pineal eye has a cornea-like structure, rudimentary lens and a simple retina and the pineal sac is a very large saccular organ with a poorly differentiated retina (Ung and Molteno, 2004). In young tuataras, the median eye can be seen clearly through the translucent covering scale before the skin thickens in adults. In birds and mammals, the median eye is represented by the pineal gland that is an endocrine organ and its functions are affected by light.

REFERENCES

Amato, M.A., Arnault, E. and Perron, M. (2004). Retinal stem cells in vertebrates: parallels and divergencies. Int. J. Dev. Biol., 48(8-9):993-1001.

Arendt, D., Tessmar-Raible, K., Snyman, H., Dorresteijn, A.W. and Wittbrodt, J. (2004). Ciliary photoreceptors with a vertebrate-type opsin in an invertebrate brain. Science, 306(5697):869-871.

Collin, S.P. and Trezise, A.E. (2004). The origins of colour vision in vertebrates. Clin. Exp. Optom., 87(4-5):217-223.

Corwin, J.T. (1985). Perpetual production of hair cells and maturational changes in hair cell ultrastructure accompany postembryonic growth in an amphibian ear. PNAS, 82(11): 3911-3915.

Eisthen, H.L. (2000). Presence of vomeronasal system in aquatic salamanders. Philos. Trans. R. Soc. Lond. B Biol. Sci., 355(1401):1209-1213.

Fermin, C.D., Lychakov, D., Campos, A., Hara, H., Sondag, E., Jones, T., Jones, S., Taylor, M., Meza-Ruiz, G. and Martin, D.S. (1998). Otoconia biogenesis, phylogeny, composition and functional attributes. Histol. Histopathol., 13(4):1103-1154.

Foster, R.G. and Hankins, M.W. (2002). Non-rod, non-cone photoreception in vertebrates. Prog. Retin. Eye Res., 21(6):507-527.

Fritzsch, B. and Beisel, K.W. (2001). Evolution and development of the vertebrate ear. Brain Res. Bull., 55(6):711-721.

Ghysen, A. and Dambly-Chaudiere, C. (2005). The three-sided romance of the lateral line: glia love axons love precursors love glia. Bioessays, 27(5):488-494.

Halpern, M. and Martinez-Marcos, A. (2003). Structure and function of the vomeronasal system: An Update. Prog. Neurobiol., 70(3):245-318.

Harada, Y. (2002). Experimental analysis of behavior of homing pigeons as a result of functional disorders in their lagena. Acta Oto-Laryngol., 122(2):132-137.

Hisatomi, O. and Tokunaga, F. (2002). Molecular evolution of proteins involved in vertebrate phototransduction. Comp. Biochem. Physiol. B Biochem. Mol. Biol., 133(4):509-522.

Kratz, E., Dugas, J.C. and Ngai, J. (2002). Odorant receptor gene regulation: implications from genomic organization. Trends Genet., 18(1):29-34.

Kwan, T., White, P.M. and Segil, N. (2009). Development and regeneration of the inner ear. Ann. N.Y. Acad. Sci., 1170:28-33.

Marmo, F. (1982). Development, structure and composition of the otoliths in vertebrates. Basic Appl. Histochem., 26(2):117-130.

Martin, T. and Luo, Z.X. (2005). Paleontology: Homoplasy in the Mammalian Ear. Science, 307 (5711):861-862.

Mills, S.L. and Catania, K.C. (2004). Identification of retinal neurons in a regressive rodent eye (the naked mole rat). Vis. Neurosci., 21(2):107-117.

Neeser, J.A. and von Bartheld, C.S. (2002). Comparative anatomy of the paratympanic organ (Vitali organ) in the middle ear of birds and non-avian vertebrates: focus on alligators, parakeets and armadillos. Brain Behav. Evol., 60(2):65-79.

Niimura, Y. and Nei, M. (2005). Comparative evolutionary analysis of olfactory receptor gene clusters between humans and mice. Gene, 346:13-21.

Ollivier, F.J., Samuelson, D.A., Brooks, D.E., Lewis, P.A., Kallberg, M.E. and Komaromy, A.M. (2004). Comparative morphology of the tapetum lucidum (among selected species). Vet. Ophthalmol., 7(1):11-22.

Pichon, F. and Ghysen, A. (2004). Evolution of posterior lateral line development in fish and amphibians. Evol. Dev., 6(3):187-193.

Pickles, J.O. and Corey, D.P. (1992). Mechanoelectrical tranduction by hair cells. Trends Neurosci., 15(7):254-259.

Popper, A.N. (2000). Hair cell heterogeneity and ultrasound hearing: recent advances in understanding fish hearing. Philos. Trans. R. Soc. Lond. B Biol. Sci., 355(1401):1277-1280.

Popper, A.N. and Fay, R.R. (1997). Evolution of the ear and hearing: issues and questions. Brain Behav. Evol., 50(4):213-221.

Rich, T.H., Hopson, J.A., Musser, A.M., Flannery, T.F. and Vickers-Rich, P. (2005). Independent origins of middle ear bones in monotremes and therians. Science, 307 (5711):910-914.

Rieke, F. (1998). Opening the third eye. Nature Neuroscience, 1:339-441.

Tilney, L.G. and Sauders, J.C. (1983). Actin filaments, stereocilia, and hair cells of the bird cochlea. I. Length, number, width and distribution of sterocilia of each hair cell are related to the position of the hair cell on the cochlea. J. Cell Biol., 96(3):807-821.

Tirindelli, R., Dibattista, M., Pifferi, S. and Menini, A. (2009). From pheromones to behavior. Physiol. Rev., 89(3):1921-1956.

Ung, C.Y. and Molteno, A.C. (2004). An enigmatic eye: the histology of the tuatara pineal complex. Clin. Experiment Ophthalmol., 32(6):614-618.

von Bartheld, C.S. (1990). Development and innervation of the paratympanic organ (Vitali Organ) in chick embryos. Brain Behav. Evol., 35(1):1-15.

von Bartheld, C.S. (1994). Functional morphology of the paratympanic organ in the middle ear of birds. Brain Behav. Evol., 44(2):61-73.

Vorobyev, M. (2003). Colored oil droplets enhance color discrimination. Proc. Biol. Sci., 270(1521):1255-1261.

Wever, E.G. (1981). The role of the amphibians in the evolution of the vertebrate ear. Am. J. Otolaryngol., 2(2):145-152.

Woolf, C.J. and Ma, Q. (2007). Nociceptors - noxious stimulus detectors. Neuron., 55(3):353-364.

Yamauchi, M., Tanaka, J. and Harada, Y. (2008). Comparative study on the morphology and composition of the otoliths in the teleosts. Acta Otolaryngol., 128(8):846-855.

13

Excretion and Osmoregulation

Excretion (L. *ex*, out; *cretus*, separated) is a process that is concerned with elimination of nitrogenous metabolic wastes from the body whereas osmoregulation is the active maintenance of the normal osmotic pressure of body fluids so that the fluids are neither too concentrated nor too dilute. Whereas the extra-cellular fluid of invertebrates has an osmotic composition that is similar to that of seawater, the same fluid in vertebrates is about a third the concentration of some ions in seawater. Marine invertebrates are thus **osmoconformers**. Terrestrial vertebrates have a lower concentration of ions in their extra-cellular fluid than marine vertebrates, a fact that could be attributed to the freshwater evolution of many vertebrates. The maintenance of the proper ionic composition in the body is vital to the maintenance of the homeostasis of the body's water levels.

Aquatic vertebrates have to cope with the problem of having gills or oral mucous membranes that are permeable to water and salts. Freshwater and seawater have osmotic pressures that differ from those of aquatic vertebrates inhabiting these bodies of water. Seawater is hypertonic to most aquatic vertebrates since it contains a higher solute concentration or osmotic pressure. A **hypotonic solution** (Gr. *hypo*, under; *tonos*, tension) (Fig. 13.1) such as freshwater has a lower solute concentration than bodies of aquatic vertebrates inhabiting such a body of water. Water moves into bodies of vertebrates inhabiting freshwater through permeable membranes by osmosis. Seawater is **hypertonic** (Gr. *hyper*, above) in relation to most vertebrates since it has a higher osmotic pressure resulting from the high salt levels and so water leaves bodies of

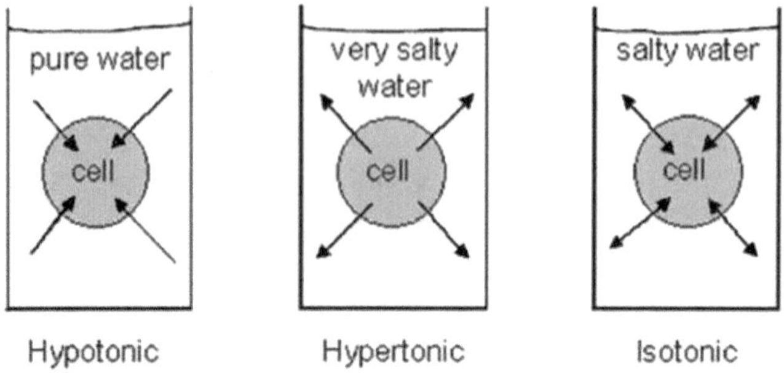

Fig. 13.1 A diagram showing solutions of various osmolalities in relation to a cell. Arrows indicate the direction of water movement by osmosis. Movement of water in an isotonic solution in relation to a cell is equal in both directions.

such vertebrates to the surrounding seawater. An **isotonic solution** (Gr. *isos,* equal) has a similar solute concentration in relation to another solution. Ideally water moves from either isotonic solution to another in equal quantities.

The vertebrate body excretes ammonia as a toxic by-product of protein metabolism that is normally converted into less toxic products such as urea or uric acid before elimination from the body. Excretion is part of osmoregulation as the excretory products and other substances are eliminated from the body through the kidney, especially in terrestrial vertebrates, together with water. Renal function thus plays a major role in osmoregulation.

The first vertebrates to move from a marine to a freshwater environment over 500 million years ago had to undergo reduction of body salt content even though such a salt concentration remained higher than that of the freshwater they had inhabited. Later on, most fish groups returned to seawater where they live today but the group that gave rise to amphibians continued to live in freshwater. Although most vertebrates that evolved from amphibians are terrestrial, some returned to aquatic life and have had to cope with osmoregulation in that environment. During evolution of osmoregulation in vertebrates, there was development of a water-impermeable epithelium that resulted from the development of water-impermeable intercellular spaces with a high electrical resistance (Natochin, 1999).

Aquatic Environment

About 71% of the Earth's surface is covered in water, with seawater covering approximately 70.8% of the water area. Oceans contain 97% of the 287,131,678 cubic miles of water on Earth with the remaining 3%

comprising freshwater mainly (Table 13.1). Both freshwater and seawater contain dissolved salts, organic compounds and gases. Seawater is quite salty and has an average of 35 parts of salt per thousand parts of water (ppt) or 35g/l with a density of 1020 to 1030 kg/m^2. Freshwater normally has a salinity of less than 0.5 ppt whereas brackish water such as estuaries where freshwater rivers meet salty oceans has 0.5 to 17 ppt of salt. The Caspian Sea, a remnant of the ancient Tethys Sea or Thetis Ocean of the supercontinent Pangaea (Fig. 14.58) and the largest lake in the world, is brackish and contains 10 to 13 g of salt per liter of water. The lower level of salinity in the Caspian Sea when compared to other oceans has been attributed to the supply of huge amounts of thawed freshwater from glaciers after the sea had become isolated. The Caspian Sea is home to 90% of the world's sturgeons. With a salt concentration of 120 to 250 g/l of water, The Great Salt Lake of Utah is more than three times saltier than ocean water. The Great Salt Lake lacks an outlet to the sea. The Dead Sea, with a salt content of around 340 g/l of water (Oren and Ventosa, 1999), does not have any aquatic vertebrates.

Table 13.1 Location of the world's freshwater reserves. Approximately 0.336% of the Earth's freshwaters are in liquid form as surface water.

Freshwater Compartment	*Volume (mi^3)*	*Percentage of freshwater*
Polar ice	4,976,900	68.6
Ground water	2,181,479	30.1
Snow Pack	70,561	0.98
Lakes	18,852	0.25
Atmosphere	2,672	0.035
Marshes/Wetlands	2,376	0.026
Rivers/Streams	439	0.006
Living organisms	232	0.003
TOTAL	**7,253,511**	**100**

Natural seawater (in parts per thousand) is mainly made up of chlorine (19 parts) mainly as chloride, sodium (10.7 parts) mainly as NaCl, NaF, and $NaHCO_3$, sulfur (2.7 parts) as sulfate or sulfide magnesium (1.3 parts) as $MgCl_2$ or $MgSO_4$, calcium (0.4 parts) as $CaCO_3$ mainly and potassium (0.39 parts) as KCl or KBr. The percentages of these chemical constituents in relation to the total salt content of seawater are shown in Table 13.2. The osmotic pressure of seawater is about 1000 mOsm/l and this value is far above the osmotic pressure of body fluids of most aquatic vertebrates. Freshwater has an osmotic pressure of less

Table 13.2 The main salt constituents of seawater. The various open oceans of the world contain these constituents in almost constant proportions.

Chemical Constituent	*Content (parts per thousand)*	*Percentage of Total Salt Content*
Chloride (Cl)	19.350	55.16
Sodium (Na)	10.710	30.53
Sulfur (S)	2.690	7.67
Magnesium (Mg)	1.304	3.72
Calcium (Ca)	0.419	1.19
Potassium (K)	0.390	1.11
Bicarbonate (HCO_3)	0.146	0.42
Bromide (Br)	0.070	0.20
Total dissolved solids (salinity)	35.079	Total 100.00

than 10 mOsm/l resulting from sodium and chloride ions mainly. The osmolality of body fluids in seawater teleosts is about 350 mOsm/l and is higher than that of freshwater teleosts of 300 mOs/l.

FRESHWATER FISHES

Freshwater fishes include freshwater lampreys and bony fishes and are hyperosmotic in relation to surrounding water. There is an influx of water across the gills and mucous membranes of the oral cavity and intestines into the bodies of freshwater fishes by osmosis as well as loss of salts from the gills to water through diffusion. The influx of water is counter balanced by loss of a lot of diluted urine that is hypotonic in relation to body fluids. Such urine minimizes loss of salt from the body. The kidney tubules reabsorb little of the water that has been filtered by the glomerulus in freshwater fish. Sodium and chloride ions are obtained through dietary intake and active transport across the gills from the water. The active loss of hydrogen and bicarbonate ions from the gills to water creates an electrochemical gradient that favors the uptake of sodium and chloride ions. Hydrogen ions also combine with ammonia that results from the deamination of amino acids to form ammonium ions (NH_4^+). Ammonium ions diffuse from the gills into water. About six to nine times as much nitrogen is excreted through gills in form of the more diffusible compounds that include ammonia, urea, amines and trimethylamine oxide when compared to the remaining amount that is lost through urine as uric acid, creatine and creatinine. Other important ions in osmoregulation of fish are potassium and calcium. Studies using a tilapia, *Oreochromis mossambicus*, have shown that the mitochondria-rich cells of the yolk-sac epithelium of the developing larvae are responsible for a major role in the osmoregulatory mechanism before the function of gills is fully developed (Hwang et al., 1999).

Fishes show varying degrees of tolerance to changes in salt concentration in water. **Stenohaline fish** (Gr. *stenos*, narrow or small; *hal*, salt) are relatively intolerant to changes in water salinity and are strictly freshwater or seawater species whereas **euryhaline species** (Gr. *eurus*, wide) are fairly tolerant to salinity change. Euryhaline fish include true or widely euryhaline species, anadromous and catadromous fish. **True** or **widely euryhaline species** occupy waters that vary greatly in salinity such as the red drum off the coast of Texas on which a commercial harvesting ban of the fish was introduced in 1981. **Anadromous species** (Gr. *ana*, up; *dromos*, running) such as the Atlantic salmon and sea lampreys will migrate from salt water to freshwater where they will spawn. **Catadromous fish** (Gr. *cata*, down) including the European eel (*Anguilla anguilla*) live as adults in freshwater but migrate to seawater to spawn. The physiology of osmoregulation in anadromous and catadromous fish (diadromous groups) is similar to that of freshwater and marine teleosts when they live in freshwater and seawater respectively.

The hormone prolactin that is produced by the anterior pituitary gland plays a role in freshwater osmoregulation in diadromous fish by promoting the loss of large quantities of urine in the kidneys. Prolactin levels are quite low during life in seawater. Studies using the Japanese flounder have shown that prolactin may play an important role in the control of water and electrolyte balance through its receptor which is expressed in the osmoregulatory organs (Higashimoto et al., 2001). An increase in the number of prolactin cells of the anterior pituitary has been associated with entry of the anadromous three-spine stickleback to freshwater (Honma et al., 1976). Most of the effects of prolactin on hydro-mineral balance are specific to euryhaline fish (Manzon, 2002). Studies using a tilapia, *Oreochromis mossambicus* have shown that osmoreception is linked to changes in cell volume rather than to extra-cellular osmolality (Weber et al., 2004). An increase in cell size is followed by release of prolactin. Prolactin is also essential in maintaining low osmotic water permeability of the transport epithelia in fishes in hypotonic water (Nishimura, 1985).

Studies using the Japanese eel (*Anguilla japonica*) have shown that the natriuretic peptide system plays a key endocrine role in the adaptation of this euryhaline fish to various osmotic environments especially at the beginning of the adaptation (Takei and Hirose, 2002). Natriuretic peptides in fish act both directly on ion-transporting cells of osmoregulatory tissues and indirectly through increased vascular flow to osmoregulatory tissues, inhibition of drinking as well as affecting other endocrine systems (Loretz and Pollina, 2000). Atrial natriuretic peptide is secreted in response to an increase in the plasma osmolality of eels leading to excretion of sodium ions specifically thus promoting adaptation to a marine environment in the process. The family of natriuretic peptides could have played a role in the extrusion of sodium ions originally in fish but evolved to be volume depleting hormones promoting the excretion of both sodium ions and water from the body in tetrapods (Takei, 2000).

Freshwater amphibians such as the tadpoles that possess gills have osmoregulatory mechanisms that are generally similar to those of freshwater fishes. Adult freshwater amphibians including frogs and salamanders use their skin in salt movement in a similar manner to the gills of freshwater fishes. In some **aquatic birds** and **mammals**, there is little uptake of water through the skin. Such uptake is accompanied by an increased flow of urine and loss of some salts whose demand is met by dietary intake.

SEAWATER VERTEBRATES

Seawater fishes evolved from freshwater ancestors that returned to salty water from freshwater and radiated into teleosts and cartilaginous fish mainly. Some terrestrial vertebrates returned to the sea secondarily in the course of evolution. The marine vertebrates have adjusted to the high concentration of salt in seawater as they maintain a stable internal ionic balance. Fishes in seawater lose water from their bodies by osmosis and gain salts from the water.

Hagfish

Hagfish are osmoconformers and have an osmotic concentration that is similar to that of the surrounding marine environment. Although the fish are isoionic with seawater, their bodies contain a lower content of chlorine in relation to seawater. The level of sodium ions in the body fluids of hagfish is high while levels of magnesium, calcium and sulfate are low in relation to marine water. The concentration of urea in hagfish varies depending on the food they eat. Divalent ions are secreted in fish actively. Hagfish have been gradually adapted to water with osmolalities of 590 mOsm and 1500 mOsm from the average 1000 mOsm of marine water they inhabit. Under such osmolalities, the fish were iso-osmotic with the surrounding water. In the hagfish *Myxine glutinosa*, muscle fibers have trimethylamine oxide, betaine and free amino acids (mainly proline) that are similar to those of cephalopods and some crustaceans. In *Myxine*, inorganic salts are mainly responsible for the osmolality of body fluids making up to 99% of osmolytes while urea plays a similar role by about 0.5%.

Marine Teleosts

Marine lampreys and marine teleosts lose water from their bodies by osmosis as well as gaining salts through gills and the pharyngeal cavity by diffusion. As a result of such water loss, such marine fish ingest seawater together with the high salt content. The amount of water consumed varies from 0.2% to 0.5% of the fish body weight per hour and 70% to 80% of this water is absorbed into the body system. The salt in the ingested water diffuses across the digestive system into the circulatory system. A solute concentration gradient is created that results in the osmotic withdrawal of water from the digestive tract. Most of the monovalent ions and some divalent ions are absorbed from the digestive tract whereas magnesium, sulfate and carbonate residues are eliminated in feces. The presence of bicarbonate ions in the intestinal fluids of most marine teleosts creates an alkaline environment that precipitates calcium and magnesium as carbonate complexes resulting in a reduction in the osmolality of intestinal fluids and plasma (Wilson et al., 2005). It is thought that bicarbonate ions are secreted into the intestinal lumen by the presence of calcium and play a role in water absorption and osmoregulation. Osmoregulation in sea water requires from 6% to 10% of the total energy budget in fish (Kidder et al., 2006). Extrusion of salt from the gills of marine fish is enhanced by cortisol (Nishimura, 1985).

Water loss through the kidneys is minimal in marine teleosts as the glomerular filtration rate is quite low and some teleosts in the salty water have evolved tubules that lack glomeruli. In such aglomerular fish, sodium chloride and water secretion across the proximal tubules from capillaries and intercellular spaces into the tubular lumen replaces glomerular filtration and this is an early function of the proximal tubule that has been retained throughout evolution (Beyebnach, 2004). The low quantity of urine excreted by marine teleosts need not be a problem to the fish as ammonia is excreted across the gills. Although marine teleosts lose a reasonable amount of salt in their urine, the amount per unit volume cannot exceed that which is present in the body fluids as teleosts lack the loop of Henle that is involved in salt secretion and is present in birds and mammals only.

Most of the salt in marine teleosts is eliminated from the body through the gills. The salt transport proteins in the gills of marine teleosts are similar to those found in the kidneys of higher vertebrates. The monovalent ions are eliminated by the mitochondria-rich **chloride cells** or **ionocytes** that are located above the basal cells of gill filaments. Sodium-sodium and sodium-potassium pumps are responsible for the removal of sodium from plasma. The pumps exchange external potassium with sodium in the gills. Chloride ions are also secreted into the water. The two types of chloride cells are the seawater type and freshwater type and have various ion pumps, transporters and channels (Sakamoto et al., 2001; Hirose et al., 2003). A novel glucose transporter has been found to specifically be expressed and function in gill ionocytes since a sufficient and timely energy supply is a prerequisite for the operation of iono- and osmoregulatory mechanisms in fish (Tseng and Hwang, 2008).

Marine Elasmobranchs

Marine elasmobranchs have evolved a mechanism that maintains a high level of osmotic solutes in body fluids. The presence of high levels of nitrogenous compounds (mainly urea) that form 40% to 55% of solutes which have an osmotic pressure and high levels of chloride ions in elsmobranch blood makes the osmotic pressure of the body fluids in these fish be equal to or slightly higher than that of their marine environment. A freezing point depression of more than -2.0°C of elasmobranch tissue fluids is slightly higher than that of marine water. There is little or no water that passes through the water permeable membranes in elasmobranch fish. The little water that enters the elasmobranch body through the mucous membranes of the gills and the pharyngeal cavity is lost from the body through the kidney as hypotonic urine. The rest of the elasmobranch body is relatively impervious to water.

Inorganic ions make up about half of the total osmolytes in elasmobranch body fluids. Marine elasmobranchs maintain levels of urea at a concentration of about 300 to 400 mM/l in the body fluids. Loss of urea through the kidney is minimized by reabsorption of the osmolyte and gills of elasmobranchs are quite impermeable to this organic solute. Elasmobranch kidneys have a complex renal countercurrent system for urea retention and the epithelium lining the renal tubules in these cartilaginous fish is highly specialized and diverse when compared to that of other fish (Lacy and Reale, 1991). Studies using the marine blue-spotted fantail ray, *Taeniura lymma* and the freshwater white-edge whip ray, *Himantura signifer* have shown that the primary action of the ornithine-urea cycle in ureogenic marine and freshwater elasmobranchs could be to synthesize urea for osmotic water retention and not for ammonia detoxification (Ip et al., 2005).

Trimethylamine oxide (TMO) is another organic solute that contributes 7% to 12% of elasmobranch osmotic pressure. TMO also has effects that counter those of urea at a ratio of approximately 2:1 as urea to TMO (Somero, 1983). TMO counters the denaturing effects of high levels of urea on elasmobranch enzymes. TMO could also be of physiological importance in deep-sea fish as various tissues in these fish show an increase in accumulation of this osmolyte with increasing depth in water when compared to the decreasing but lower levels of urea under similar circumstances (Treberg and Driedzic, 2002). TMO has a considerably lower density than an equimolar solution of urea and solutions of these two osmolytes are considerably less dense than equimolar solutions of most other body fluid solutes suggesting the other adaptive role of TMO and urea in buoyancy of chondrichthyan fishes (Withers et al., 1994).

Since elasmobranch blood contains a lower level of sodium and chloride ions relative to the surrounding water, these ions diffuse across the gill epithelium into the body of the fish. The sodium and chloride ions are excreted by the rectal salt gland and to some extent the gills and kidneys. The **rectal salt gland** (Fig. 13.2) can secrete a solution that contains a higher concentration of sodium chloride than either elasmobranch blood or seawater. The rectal salt gland is a diverticulum at the caudal end of the intestines. A salt solution that has a concentration of 500 mM/l of sodium chloride is secreted by the salt gland that also has a selective permeability barrier to urea (Zeidel et al., 2005). The rectal salt

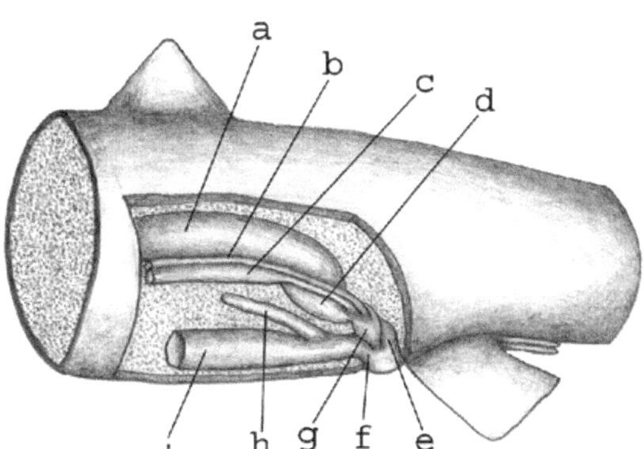

Fig. 13.2 Part of an elasmobranch showing the location of the rectal salt gland. (a) opisthonephros, (b) accessory urinary duct, (c) archinephric duct or seminal vesicle, (d) sperm sac, (e) urodeum, (f) coprodaeum, (g) urogenital sinus, (h) rectal salt or digitiform gland and (i) intestine. The coprodaeum and urodeum form the cloaca.

gland achieves efficient salt secretion as is found in salt secreting glands of other vertebrates including sea birds and is a result of the presence of a counter-current system of flow whereby liquid in the secretory tubules moves in an opposite direction to that of blood flow in capillaries (Newbound and O'Shea, 2001). Secretion of salt by the rectal salt gland is stimulated by C-type natriuretic peptide (Silva et al., 1999). The role played by the rectal gland, gill, kidney, liver in relation to the drinking process and the endocrine control of urea, sodium and chloride ion levels as elasmobranchs acclimate to various levels of salinity has been discussed by Hazon et al. (2003).

Non-piscine Marine Vertebrates

Non-piscine marine vertebrates (Fig. 13.3) have adapted to various methods of osmoregulation. There are not many marine amphibians. The crab-eating frog of Asia is an example of such an amphibian that changes its osmoregulation from that of a marine teleost at the tadpole stage to that of an elasmobranch as an adult amphibian where it accumulates a lot of urea in the body fluids. Marine reptiles lose little water to the sea due to their thick and highly keratinized skin. Salts ingested through food and water consumption is eliminated in various ways as the reptilian kidney lacks the loop of Henle. In crocodilian reptiles, **supra-lingual glands** play the role of salt secretion while **sublingual glands** play a

Fig. 13.3 Some non-piscine marine vertebrates that secrete salt using various mechanisms. The crab-eating frog of Asia (top left), turtle (top right), seagull (bottom left) and Dall's porpoise (bottom right). Marine birds generally have webbed feet. Porpoises are the smallest cetaceans and are smaller but stouter than dolphins. Porpoises have small and rounded heads with blunt jaws and can reach body lengths of up to 2.5 m. Dall's porpoise is one of the fastest swimming cetaceans and can attain a speed of 55 km/h.

similar role in sea snakes. Turtles and sea lizards possess an **orbital salt gland** that secretes salt. Some terrestrial reptiles possess **salt glands** and include the desert iguana of North America, green or Linnean iguana, false iguana and chuckwallas. Marine birds eliminate salts using the **orbital secretory gland** that opens into the nasal cavity and is also known as the **nasal gland**.

Salt glands are also found in some terrestrial birds such as the ostrich, partridge of North Africa and the Middle East deserts, the roadrunner of North America and many birds of prey. Salt gland activity and release of the antidiuretic hormone vasotocin is controlled by nitric oxide of the hypothalamus in marine birds (Hubschle et al., 1999). The salt glands of terrestrial reptiles and birds are specialized for secreting potassium unlike their marine counterparts. Salt glands do not function continuously as the kidney but become active only when there is an osmotic load to secrete mainly sodium and chloride ions with little quantities of potassium and bicarbonate ions (Jouanin, 2000). Birds also have the loop of Henle that is important in salt secretion. Mammals living in seawater lose excess salts obtained from marine food in their kidneys that have the loop of Henle that can concentrate salts in urine two to three times as much as is present in blood.

Excretion

Nitrogenous wastes in the vertebrate body arise from the deamination of amino acids mainly. **Deamination** occurs in the liver before further breakdown of amino acids can occur and involves the removal of the amino (NH_2) group from an amino acid molecule. The NH_2 group combines with a hydrogen ion to form ammonia (NH_3) that is toxic to tissues. Ammonia is highly soluble in water and is a small molecule so it diffuses quickly from the bodies of freshwater fish through the gills into water. Freshwater fish and permanently aquatic amphibians are **ammonotelic** since they excrete excess nitrogen from their bodies in the form of ammonia. Ammonia diffuses from the bodies of amphibians through the skin and kidneys and also gills in larval forms. Ammonia can be converted into the larger molecule urea that has two amino groups and is less toxic or uric acid, an even a larger molecule than urea. Urea diffuses across the gills of some fish but is eliminated from the body through the kidneys in semi-terrestrial adult amphibians and mammals and as a result these groups are referred to as being **Ureotelic**. Uric acid is not soluble in water and does not require much water for transportation. Uric acid is also not toxic and precipitates to form a semisolid mass. Reptiles and birds eliminate their nitrogenous wastes through the kidney in the form of uric acid with relatively little water and are said to be **uricotelic**. Elimination of nitrogenous wastes in vertebrates requires water and occurs together with the process of osmoregulation.

Urinary System

The vertebrate urinary system comprises kidneys that process blood to form urine that contains water and wastes to be excreted from the body and accessory organs through which urine passes from the kidneys to the outside. Accessory organs present depend on the vertebrate species and include ureters, urinary bladder, cloaca and the mammalian urethra.

Evolution of the Kidney

The different kidneys that appear at various times in the developing vertebrate embryo are thought to represent the evolution of the vertebrate kidney from an ancestral kidney known as the **archinephros** (Gr. *arche*, origin or beginning) or **holonephros** (Gr. *holos*, whole; *nephros*, kidney) (Fig. 13.4). The holonephros consisted of segmental tubules that developed from the whole of the **nephric ridge**. Segmental tubules are thought to have led to an **archinephric duct** that opened to the outside at the urinary papilla. The holonephros kidney is thought to have resembled the kidneys of larval hagfish and caecilians.

Evolution of the vertebrate kidney is represented by three different kidneys in succession and whose evolution is separated from each other

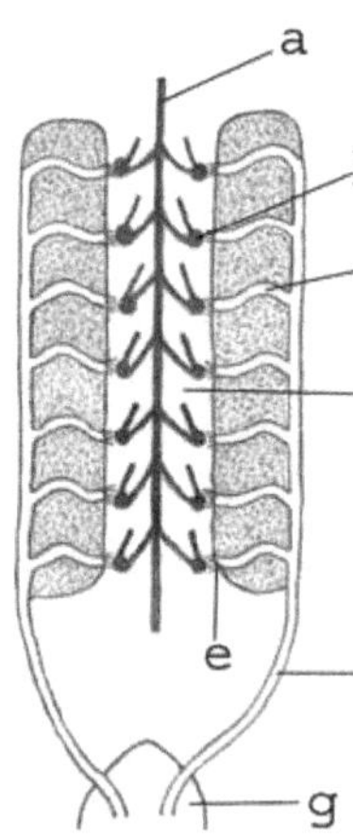

Fig. 13.4 The archinephros. (a) aorta, (b) external glomerulus, (c) segmental tubule, (d) body cavity, (e) nephrostome, (f) archinephric duct and (g) cloaca. The archinephros was the earliest vertebrate kidney that could have extended the entire length of the body cavity. The glomeruli in the archinephros were external and drained the fluids of the body cavity. Archinephros are found in embryos of hagfish.

by about 50 million years. The kidneys develop in the nephric ridge from an anterior to caudal location and represent three different kidneys in succession. The first kidney to evolve in living vertebrates was the pronephros. **Pronephric kidneys** (Gr. *pro,* before) are also the first kidneys to appear during embryonic development and they are formed at the anterior end of the nephric ridge above the pericardial cavity (Fig. 13.5). A few pronephric tubules are formed in the kidney that unite at their distal ends into an archinephric duct. The archinephric duct leads from the kidneys to the allantois. The pronephros is the first excretory organ to appear in the embryo of a vertebrate and is functional as a kidney in the embryos and larvae of cyclostomes, most bony fishes and amphibians. Pronephric kidneys are not functional in cartilaginous fish and amniotes. In most vertebrates, the pronephros regresses later on with further embryonic development. Hagfish and some teleosts still have a pronephros in adult life known as the **head kidney** that is located over the pericardial cavity with a gap separating it from the more caudal functional kidney.

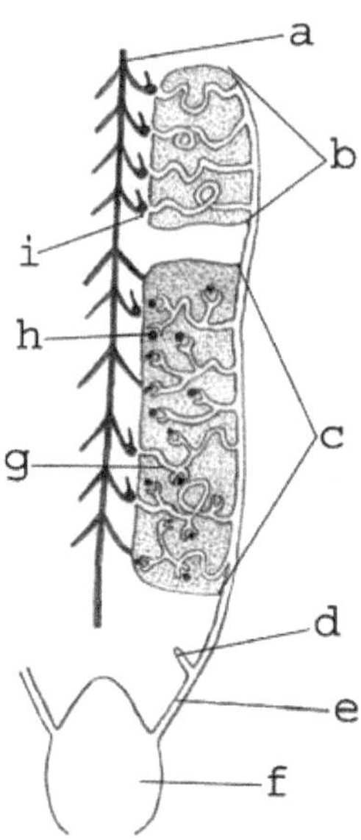

Fig. 13.5 A longitudinal section of pronephric and mesonephric kidneys. The metanephric kidney is the functional kidney of mature amniotes. (a) aorta, (b) pronephric and (c) mesonephric kidneys, (d) metanephric bud and (e) duct, (f) cloaca, (g) renal tubule, (h) internal and (i) external glomeruli. Although the pronephric kidney is segmentally arranged, the mesonephros begins to lose the segmental character due to the branching of renal rubules. Nephrostomes are lost in mesonephric kidneys although they are still retained in the mesonephros of some sharks.

The regression of the pronephric kidney is followed by the development of the mesonephric kidney that was also the second kidney to evolve. A **mesonephric kidney** (Gr. *mesos*, middle) develops caudal to the pronephric kidney and there is a gap between the two kidneys since kidney tubules do not develop in a few body segments between the two renal structures. The tubules of the mesonephros grow towards the archinephric duct with which they unite. An archinephric duct to which mesonephric tubules are attached is also known as a **mesonephric duct**. Mesonephric kidneys are functional in vertebrate larvae or embryos and will be the functional kidneys of adult fishes and amphibians where they are known as opisthonephros. The **opisthonephros** (Gr. *opisthen*, behind or at the back) is enlarged at its caudal part and most of the urine is produced here in most anamniotes. Mesonephric kidneys degenerate in developing embryos of amniotes. Cranial mesonephric tubules are connected to the gonads of many vertebrates with development with the exception of teleosts. The mesonephric duct develops into the ductus deferens of males whereas it will regress in the female with time.

The adult functional kidney of amniotes is the metanephric kidney that develops with the regression of the mesonephric kidney. **Metanephric kidneys** (Gr. *meta*, after) are the last kidneys to develop during embryonic development and are located caudal to mesonephric kidneys. A small diverticulum grows from the archinephric or mesonephric duct toward the developing metanephros or **nephrogenic cord** and divides into many tubules in a dichotomous manner. The tubules penetrate the cord and divide further eventually giving rise to the kidney tubules. The terminal end of each tubule will give rise to the Bowman's capsule. Although the metanephric kidney develops in the caudal part of the nephric ridge, it migrates cranially or superiorly as a result of local differential growth mainly. The metanephric kidney lies caudal or inferior to the liver in fully developed amniotes.

Structure of the Kidney

Vertebrate kidneys vary in shape and are generally located in the sub-lumbar region of the body. In mammals, kidneys can be bean shaped (Fig. 13.6) or are lobated (composite kidneys) in some species including cattle, whales, seals, polar bears and river otters. In fishes, amphibians, reptiles and birds, the kidneys are closely applied to the vertebrae. Kidneys are **retroperitoneal** (L. *retro*, backward) as their dorsal surfaces are not covered by the coelomic epithelium or peritoneum (in mammals). In mammals, the dorsal surfaces of kidneys are in contact with back muscles.

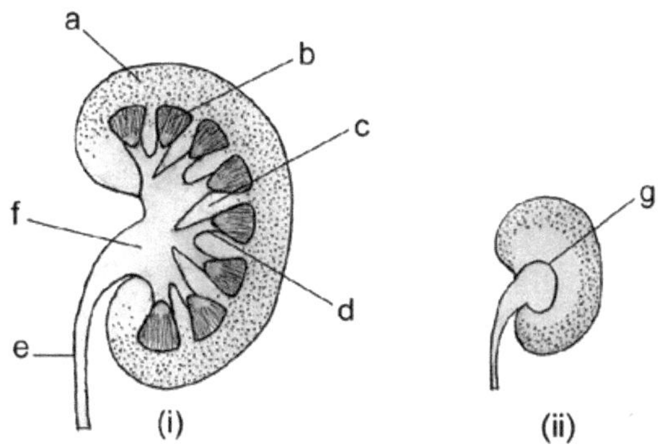

Fig. 13.6 (i) Kidney with incompletely fused lobes at the medulla and (ii) one with completely fused cortical and medullary lobes. (a) cortex, (b) renal pyramid, (c) medulla, (d) calyx, (e) ureter, (f) renal pelvis and (g) renal crest.

The mammalian kidney has an outer cortex and an inner medulla. Kidney lobes are visible externally in some mammalian species with lobated or composite kidneys. The cortical and medullary parts of kidneys show varying degrees of fusion in different mammals. Complete fusion of cortical tissue of neighboring lobes results in a kidney with a smooth external surface. When fusion of cortical tissue is incomplete the kidney shows surface fissures that vary in depth between the lobes. Fusion of lobes can also occur in the medulla. Kidneys that show complete fusion of the medulla have a ridge-like structure at their innermost part known as **a renal crest** while those that have incompletely fused lobes at the medulla have **renal papillae**. All urine leaving the kidney collects at the **renal pelvis** that is an expanded reservoir that leads to a ureter. Lobated kidneys lack a renal pelvis. Kidneys are highly vascular organs and in the human being about 1200 ml of blood or a fifth of all blood pumped from the heart per minute circulates through these organs.

The functional unit of the kidney is the microscopic nephron (Fig. 13.7). **Nephrons** number in millions and make up most of the kidney. The nephron is involved with the process of filtration and formation of urine. The nephron has several segments that play specific roles in urine formation. From the beginning to the terminal end of a nephron are the renal corpuscle (Bowman's capsule and glomerulus), proximal convoluted tubule, loop of Henle and distal convoluted tubule. Distal convoluted tubules open into collecting ducts. Kidney function is under the control of hormones that act via extra-renal and intra-renal mechanisms. Extra-renal mechanisms involve control of kidney function by the nervous and circulatory systems. Internally, kidney function is influenced by dynamics of intra-renal circulation and transport across tubular epithelia. In lower vertebrates, anti-diuretic hormone and the

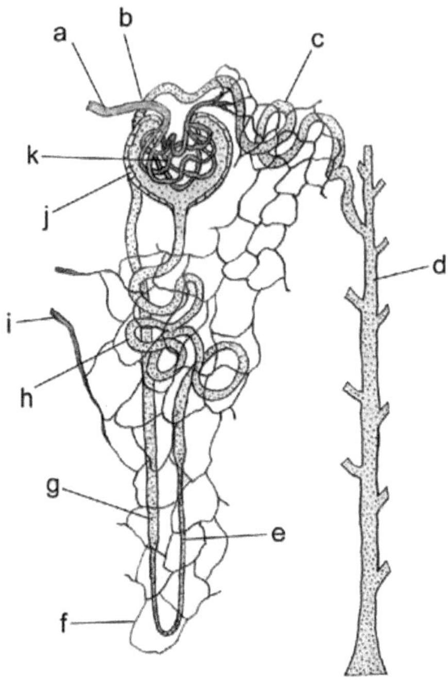

Fig. 13.7 A nephron. (a) renal afferent arteriole, (b) juxtaglomerular apparatus, (c) distal tubule, (d) collecting tubule, (e) descending limb of loop of Henle, (f) vasa recta, (g) ascending limb of the loop, (h) proximal tubule, (i) venule, (j) Bowman's capsule and (k) glomerulus.

renin-angiotensin system appear to act mainly on systemic and preglomerular vasculature whereas action by such hormones is on kidney tubules in higher vertebrates and this shows that both function and site of hormone action seem to have altered during vertebrate evolution with changing environmental conditions (Nishimura, 1985).

A **renal** or **Malpighian corpuscle** (L. *corpusculum*, a minute particle) comprises the Bowman's capsule and glomerulus and is located in the cortex of the kidney. The **Bowman's capsule** is the cup-shaped part of a nephron that surrounds a glomerulus. There is space in a Bowman's capsule that is surrounded to the outside by a simple squamous epithelium and to the inside by cells known as podocytes. **Podocytes** possess feet-like processes that branch and surround capillaries of the glomerulus. Between the podocyte processes are spaces known as **filtration slits**. **Glomeruli** (L. *glomus*, ball) are a cluster of capillary loops that are located between afferent and efferent renal arteries and have a single layer of thin endothelial cells with numerous fenestrations. The fenestrations enhance the level of filtration that meets the excretory and osmoregulatory demands of the body. A basement membrane lies between the glomerulus and the Bowman's capsule. The general structural features and functional properties of the glomerulus appear to be largely similar among different vertebrates but there is a lot of variation in the level of the rate of filtration that is primarily influenced by the rates of water influx and metabolism (Yokota et al., 1985). Birds and mammals have more glomeruli and higher filtration pressures and filtration rates than ectothermic vertebrates with lower rates of metabolism.

After the renal corpuscle is the **proximal tubule** that is located in the cortex of the kidney and also forms the first part of the renal tubule. Since this segment takes a winding course, it is also known as the **proximal convoluted tubule**. This tubule has a single layer of pyramidal cells that have numerous microvilii on their luminal surface. Microvilli form the brush-border that greatly increases the internal surface area of the proximal convoluted tubule. A large surface area by the cells of this segment is important in the role played by the cells of the proximal tubule.

The **loop of Henle** connects the proximal tubule to the distal tubule. The **descending limb** of the loop can run deep into the medulla where it narrows into a thin segment then loops around into an **ascending limb** that thickens and runs back to the cortex. Mammals have two types of nephrons. The type with a loop of Henle that penetrates into the medulla is known as a **juxtamedullary nephron** (L. *juxta,* beside or near) while those that are confined to the cortex are known as **cortical nephrons**. The pyramidal cells of the proximal tubule merge abruptly with those of the thin segment of the descending limb of the loop. The thin segment of the loop of Henle contains squamous cells in its wall. Low cuboidal cells form the wall of the ascending limb of the loop.

The ascending limb of the loop of Henle merges with the **distal tubule** or **distal convoluted tubule** in the juxtaglomerular apparatus region. **Juxtaglomerular apparatus** is found at the point of contact between the afferent arteriole and the distal tubule (Fig. 13.8) and is made up of modified cells in the walls of the two tubular structures and mesangial cells. The juxtaglomerular apparatus controls blood pressure through the renin-angiotensin-aldosterone system. The afferent arterioles contain juxtaglomerular granular cells at this point that secrete renin reflexively from the granules when blood pressure in the afferent arteriole drops and such action is important in the maintenance of homeostasis of blood flow. The drop in blood pressure is sensed by the stretch receptors of the afferent arterioles. Granular cells normally acquire smooth muscle markers that persist throughout life. Secretion of

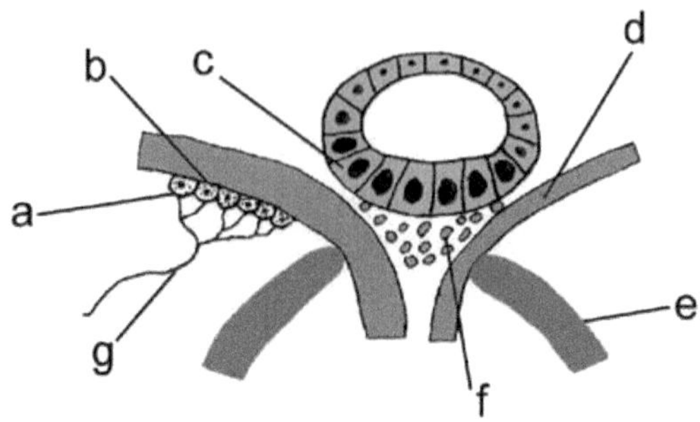

Fig. 13.8 Juxtaglomerular apparatus. (a) granular cell, (b) stretch receptors in afferent arteriole, (c) macula densa of distal tubule, (d) efferent arteriole, (e) Bowman's capsule, (f) mesangial cell and (g) sympathetic nerve. Mesangial cells are phagocytic and engulf macromolecules that escape from capillaries.

renin is inhibited by increasing intracellular Ca^{2+} whereas cyclic AMP stimulates its release (Persson, 2003). The modified cells of the distal tubule at the juxtaglomerular apparatus are known as **macula densa** (L. *macula*, spot; *densa*, crowded or thickly packed) and comprise numerous cells with closely packed nuclei. The cells of the macula densa act as chemoreceptors that detect the concentration of solutes in tubular fluid. The distal tubule is located in the cortex and has low cuboidal cells in its wall.

Mesangial cells (Gr. *mesos*, middle; *angis*, capillary) are modified smooth muscle cells or specialized pericytes that are located around glomerular capillaries within a renal corpuscle and outside the glomerulus near the macula densa where they are known as **lecis cells**. Mesangial cells provide structural support to glomerular capillaries as well as regulating blood flow in the capillaries by their contractile activity in response to vasoactive agents. Mesangial cells organize the glomerular capillaries by adhering to the G domain of laminin $\alpha5$ in the glomerular basement membrane (Kikkawa et al., 2003). Other roles played by mesangial cells include production and breakdown of the glomerular basement membrane and other material of the biomatrix, mediation in inflammation and production of prostaglandins.

The **collecting ducts** are joined by several distal tubules of many nephrons. Collecting ducts are straight and several of the ducts unite with larger ducts that will open at the renal papillae or renal crest. The collecting ducts run into the medulla. Smaller collecting ducts have a single layer of high cuboidal cells in their walls whereas walls of large collecting ducts contain columnar cells.

Blood supply to the glomerular capillaries is by the **afferent renal arterioles**. After circulating through the glomerular capillaries, blood flows to the efferent renal arterioles. The **efferent renal arterioles** branch into **peritubular capillaries** that are found around kidney tubules. Efferent arterioles are thus a portal system as they do not join larger vessels that will eventually end up leading to the heart but rather run from one capillary bed to another. Peritubular capillaries branch and anastomose with each other. Some of the peritubular capillaries known as **vasa recta** (L. *vas*, duct; *recta*, straight or correct) lead to the medulla of the kidney to supply the loop of Henle. Peritubular capillaries eventually unite into venules that return blood to veins and ultimately the heart.

Urine Formation

Terrestrial vertebrates rely mainly on kidneys for osmoregulation and excretion. All vertebrates possess kidneys that show various levels of development of the nephron. Glomeruli are present in hagfish and loops of nephrons are found in the kidneys of lampreys. Further evolution of kidneys has been accompanied with mechanisms that support concentration of urine and this is associated with the subdivision of the kidney into the renal cortex and medulla (Natochin, 1996).

The primary functions of kidneys in higher vertebrates are processing of blood and excretion of urine as well as the maintenance of fluid-electrolyte balance and acid-base balance. Kidneys adjust their functions to ensure that material leaving the body in the form of urine is equal to that entering the circulatory system so that excess or a deficit of such material in the entire body is avoided and homeostais is maintained. Urine is formed by the nephron and involves filtration, reabsorption and secretion.

Filtration is a process that involves the passage of material from the glomerulus into Bowman's capsule. The passage depends on a hydrostatic pressure gradient that is established by the hydrostatic pressure of glomerular blood that causes the passage of most of the water in the plasma and small solutes into Bowman's capsule. Blood cells and proteins remain within the glomeruli and do not enter the Bowman's capsule. The effective or net filtration pressure is the difference between glomerular hydrostatic pressure and the osmotic pressure of glomerular blood together with the hydrostatic pressure of the filtrate. Since the efferent renal arteriole has a smaller diameter than an afferent renal arteriole, the former offers more resistance to blood flow than the latter resulting in a positive net filtration pressure. Changes in systemic pressure as well as diameters of afferent and efferent renal arterioles bring about changes in glomerular filtration rate.

Reabsorption takes place in all parts of the renal tubule and involves the passive or active movement of molecules out of the tubule into the surrounding interstitial spaces and peritubular blood vessels. Reabsorption is necessary since the filtrate contains a lot of material including water, electrolytes and nutrients that the body still requires. Most of the reabsorption takes place in the proximal tubules. Little reabsorption occurs in the remaining parts of renal tubules. In aquatic amphibians that have to excrete water which is gained through the skin by osmosis, the initial parts of the distal nephron have highly developed basolateral interdigitations and absorb 40% of filtered sodium, potassium and chloride ions but are impermeable to water resulting in formation of hypotonic renal tubular fluid (Uchiyama et al., 2000).

The **proximal tubules** reabsorb sodium ions from the tubular lumen back to the interstitial fluid and peritubular capillaries through active transport. Microvilli present on the inner border of the tubules increase the surface area for the transport. Accumulation of sodium ions in the interstitial space creates a temporary positive charge that creates an electrical gradient that is accompanied by the passive movement of the negatively charged chloride and phosphate ions from the tubules to the interstitial space. Accumulation of ions in peritubular capillaries makes the blood hyperosmotic in relation to tubular fluid, an action that is accompanied by movement of water from the tubules to the blood by osmosis. Glucose and amino acids are also reabsorbed back to the blood passively by transport mechanisms using carrier proteins on the cell membranes of tubular cells. Urea remains in tubular fluid so its concentration per unit volume of luminal fluid increases as water and other material are taken out of the proximal tubule. A diffusion gradient is created that results in diffusion of about half of the urea back to the blood.

The **loop of Henle** shows different mechanisms of reabsorption depending on its location in the kidney. Juxtaglomerular nephrons, located in the medulla of the kidney, and the surrounding capillary network known as vasa recta employ the **countercurrent mechanism** of reabsorption. Fluid in the ascending and descending limbs of the loop of Henle flow in opposite directions as is the case with arterial blood and venous blood in the vasa recta that flow towards the medulla and cortex respectively. The countercurrent flow makes the solute concentration of the medulla quite high. Although water and urea are able to diffuse into and out of the thin descending limb of the loop depending on their concentration gradients, the thicker walled ascending limb is selective in the molecules it transports out of the tubule. The ascending limb actively pumps sodium and chloride ions into the interstitial fluid. Since the wall of the limb is impermeable to water, there is an increase in the osmotic pressure of the interstitial fluid while there is a decrease in the osmolality of the tubular fluid. The continued pumping of salts into the intercellular fluid by the ascending limb sustains the high solute concentration of the intercellular fluid in what is known as the **countercurrent multiplier mechanism**.

A high solute concentration in the intercellular fluid causes movement of water to the fluid from the descending limb. Urea also diffuses into the descending limb from the interstitial fluid. After tubular fluid has flown through the loop of Henle, it is hypo-osmotic in relation to interstitial fluid. Blood flow through the vasa recta is slow and because it forms a countercurrent loop in the medulla, it is only able to remove some of the solutes from the interstitial fluid. Blood flowing away from the medulla in the vasa recta contains a slightly higher solute concentration than blood flowing into the medulla since it loses some of the solutes it had gained to incoming blood through diffusion.

Reabsorption of solutes and fluids into interstitial fluid takes place in **distal tubules** and **collecting ducts**. The distal tubule reabsorbs some sodium ions through active transport. Under normal circumstances, the walls of the distal tubules and collecting ducts are to a certain degree impermeable to water even though the collecting ducts are located in the medulla that is hyperosmotic in relation to fluid in the ducts. Such water-impermeable walls of the distal tubules and collecting ducts would lead to production of a lot of hypotonic urine were other regulatory mechanisms not in place.

Antidiuretic hormone (ADH) that is produced in the hypothalamus and secreted into the posterior pituitary gland is a major hormone that regulates the amount of urine produced. The hormone acts on the cells of the distal tubules and collecting ducts by making them more permeable to water. The presence of water channels known as **aquaporins** is responsible for the movement of water across cell membranes. Aquaporins form pores that are composed of identical subunit proteins in membranes of cells. These water channels allow selective passage of water across the cell membranes while preventing the passage of ions and other small molecules. There are various types of aquaporins and they are numbered according to the order in which they were discovered. Aquaporin-1 was discovered in 1988. Antidiuretic hormone regulates water reabsorption in the cells of the collecting duct by largely regulating aquaporin-2 water channel whose synthesis is induced by the hormone and is quickly degraded later in the absence of the hormone (Hasler et al., 2002). The antidiuretic hormone arginine-vasopressin in mammals causes the exocytotic insertion of the water channel aquaporin-2 from intracellular vesicles into the apical membranes of collecting duct cells through the action of cyclic adenylate monophosphate and protein kinase A (Klussmann et al., 2000). In the absence of arginine-vasopressin hormone, aquaporin-2 is retrieved from the cell membrane.

Aquaporins 3 and 4 are located on the basolateral membranes of collecting duct cells and are used by water to exit the cells. Water is then able to flow from the tubules and ducts to the interstitial fluid by osmosis. The level of permeability of the cells of tubules and ducts to water depends on the level of ADH present. ADH thus determines the amount and osmolality of urine produced. As water is reabsorbed under the influence of ADH, the concentration of urea in tubular fluid increases per unit volume resulting in diffusion of urea into interstitial fluid. Urea contributes to the high solute concentration of the medulla. Some of this urea diffuses into the vasa recta.

Secretion of material from the vascular system into tubular liquid also occurs across the wall of kidney tubules. Urea diffuses into the descending limb of the loop of Henle as the distal tubules and collecting duct transport potassium and hydrogen ions actively into tubules in exchange for sodium ions. Ammonium ions are synthesized by the cells of the distal tubules and collecting ducts and will also diffuse into the tubules. The exchange of sodium and potassium ions by sodium-potassium pumps is enhanced by **aldosterone**. Aldosterone may also regulate the expression of aquaporin-3 in the collecting duct (Kwon et al., 2002). The presence of mineralocorticoids such as aldosterone has not been established in non-mammalian vertebrates. It has been proposed that transepithelial salt and fluid secretion mechanisms enable mammalian renal tubules to finely regulate extra-cellular fluid volume and composition always and maintain urine formation during cessation of glomerular filtration (Grantham and Wallace, 2002).

Urine volume is regulated by various factors. Hormonal control of water reabsorption in the distal tubules and collecting ducts by ADH and aldosterone greatly regulates the volume of urine lost from the body. **Atrial natriuretic hormone** also influences the amount of water that is reabsorbed in the kidney since it promotes the loss of sodium ions in urine leading to loss of more water in urine. Atrial natriuretic hormone causes the release of cyclic guanosine monophosphate (cGMP) in renal vessels and tubules while oxytocin acts on its receptors in the macula densa to produce the same compound (cGMP) that closes sodium ion channels in tubular cells (Soares et al., 1999). The effect of atrial natriuretic hormone is opposite to that of ADH and aldosterone. The presence of other solutes in urine other than sodium also determines the volume of urine excreted from the body. A higher concentration of solutes in kidney tubules and ducts is generally associated with a higher volume of urine. In the absence of pharmacological interference in the human being, urinary excretion of sodium ions can vary between less than 0.1% and no more than 3% of the filtered load and that of water can vary between 0.3% and 15% of the same load (Greger, 2000). Neonatal kidneys are not efficient at concentrating urine but dilute urine as efficiently as adult kidneys. The loop of Henle and collecting ducts of neonatal kidneys are greatly similar to those of birds and qualitative changes in the organization of tubules may be responsible for the immature urine-concentrating ability in mammalian neonates (Liu et al., 2001).

URINARY BLADDER AND CLOACA

Urine flows from the kidneys in a posterior or inferior manner through the tube-like ureters to an enlarged structure that varies in different vertebrates (Fig. 13.9). Urine is stored in the structure temporarily before being voided to the outside. In most fishes, the caudal ends of the ureters are slightly enlarged to form structures known as **urigenital sinuses** or **urinary bladders**. Freshwater fish continuously discharge urine into water. Ureters of elasmobranchs lead to a cloaca and open caudal to the opening of the digestive system. Amphibian urinary bladders are large and lead to the cloaca. Some fish, many turtles and lizards have urinary bladders. The bladders store urine temporarily and also act as sites of water absorption. The bladders of many aquatic turtles are used as respiratory organs. Water is normally pumped in and out of such bladders through the cloaca. Other turtles have a pair of **accessory urinary bladders** that evaginate to the outside from the cloaca and are used in gaseous exchange. Most reptiles and birds lack a urinary bladder but have a cloaca into which ureters open. Vertebrates that lack urinary bladders eliminate their wastes with little water mainly since uric acid, the main form of nitrogenous waste in these animals, is in semisolid state

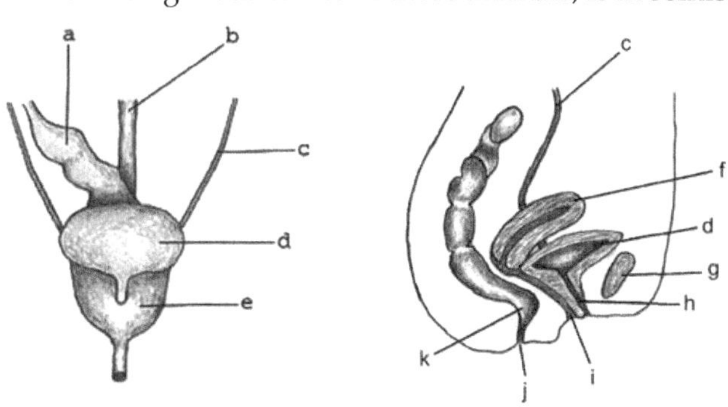

Fig. 13.9 Anterior view of the lower urinary organs of a frog (left) and lateral view showing some of the organs of the lower part of the body trunk in a female human being (right). (a) large intestine, (b) dorsal aorta, (c) left ureter, (d) urinary bladder, (e) cloaca, (f) uterus, (g) pubic symphysis, (h) urethra, (i) vagina, (j) anal opening and (k) rectum.

as a result of its low solubility in water. Monotremes have a cloaca and are the only mammals that lack a urinary bladder. The mammalian urinary bladder can store urine for sometime but water absorption does not take place here.

The lining of the urinary bladder is endodermal in origin and has a **transitional epithelium** that varies in thickness depending on whether the bladder is empty or full. An empty bladder has an epithelium that is several cells thick whereas a distended bladder has about two layers of epithelial cells. The wall of the bladder also contains smooth muscle or **detrusor muscle** that is of mesodermal origin. The mammalian bladder opens to the outside through an **urethra**.

REFERENCES

Beyenbach, K.W. (2004). Kidneys sans glomeruli. Am. J. Physiol. Renal Physiol., 286(5):F811-827.

Grantham, J.J. and Wallace, D.P. (2002). Return of the secretory kidney. Am. J. Physiol. Renal Physiol., 282(1):F1-9.

Greger, R. (2000). Physiology of renal sodium transport. Am. J. Med. Sci., 319(1):51-62.

Hasler, U., Mordasini, D., Bens, M., Bianchi, M., Cluzeaud, F., Rousselot, M., Vandewalle, A., Feraille, E. and Martin, P.Y. (2002). Long term regulation of aquaporin-2 expression in vasopressin-responsive renal collecting duct principal cells. J. Biol. Chem., 277(12):10379-10386.

Hazon, N., Wells, A., Pillans, R.D., Good, J.P., Gary Anderson, W. and Franklin, C.E. (2003). Urea based osmoregulation and endocrine control in elasmobranch fish with special reference to euryhalinity. Comp. Biochem. Physiol. B Biochem. Mol. Biol., 136(4):685-700.

Higashimoto, Y., Nakao, N., Ohkubo, T., Tanaka, M. and Nakashima, K. (2001). Structure and tissue distribution of prolactin receptor mRNA in Japanese flounder (*Paralichthys olivaceus*): conserved and preferential expression in osmoregularory organs. Gen. Comp. Endocrinol., 123(2):170-179.

Hirose, S., Kaneko, T., Naito, N. and Takei, Y. (2003). Molecular biology of major components of chloride cells. Comp. Biochem. Physiol. B Biochem. Mol. Biol., 136(4):593-620.

Honma, Y., Teshigawara, H. and Chiba, A. (1976). Changes in the cells of the adenohypophysis associated with diadromous migration of the three-spine stickleback, *Gasterosteus aculeatus* L. Arch. Histol. Jpn., 39(1):1-14.

Hubschle, T., Kuchenmeister, I. and Gerstberger, R. (1999). Central action of nitric oxide in the saltwater-acclimated duck: modulation of extra-renal sodium excretion and vasotocin release. Brain Res., 825(1-2):22-35.

Hwang, P.P., Lee, T.H., Weng, C.F., Fang, M.J. and Cho, G.Y. (1999). Presence of Na-K-ATPase in mitochondria-rich cells in the yolk-sac epithelium of larvae of the teleost *Oreochromis mossambicus*. Physiol. Biochem. Zool., 72(2):138-144.

Ip, Y.K., Tam, W.L., Wong, W.P. and Chew, S.F. (2005). Marine (*Taeniura lymma*) and freshwater (*Himantura signifer*) elasmobranchs synthesize urea for osmotic water retention. Physiol. Biochem. Zool., 78(4):610-619.

Jouanin, C. (2000). Extrarenal salt secretion in oceanic birds. Ann. Pharm. Fr., 58(5):346-350.

Kidder, G.W. 3rd, Petersen, C.W. and Preston, R.L. (2006). Energetics of osmoregulation: II. Water flux and osmoregulatory work in the euryhaline fish, *Fundulus heteroclitus*. J. Exp. Zoolog. A. Comp. Exp. Biol., 305(4):318-327.

Kikkawa, Y., Virtanen, I. and Miner, J.H. (2003). Mesangial cells organize the glomerular capillaries by adhering to the G domain of laminin α5 in the glomerular basement membrane. J. Cell Biol., 161(1):187-196.

Klussmann, E., Maric, K. and Rosenthal, W. (2000). The mechanisms of aquaporin control in the renal collecting duct. Rev. Physiol. Biochem. Pharmacol., 141:33-95.

Kwon, T.H., Nielsen, J., Masilamani, S., Hager, H., Knepper, M.A., Frokiaer, J. and Nielsen, S. (2002). Regulation of collecting duct AQP3 expression: response to mineralocorticoid. Am. J. Physiol. Renal Physiol., 283(6):F1403-1421.

Lacy, E.R. and Reale, E. (1991). The fine structure of the elasmobranch renal tubule: intermediate, distal and collecting duct segments of the little skate. Am. J. Anat., 192(4):478-497.

Liu, W., Morimoto, T., Kondo, Y., Iinuma, K., Uchida, S. and Imai, M. (2001). 'Avian-type' renal medullary tubule organization causes immaturity of urine-concentrating ability in neonates. Kidney Int., 60(2):680-693.

Loretz, C.A. and Pollina, C. (2000). Natriuretic peptides in fish physiology. Comp. Biochem. Physiol. A Mol. Integr. Physiol., 125(2):169-187.

Manzon, L.A. (2002). The role of prolactin in fish osmoregulation: a review. Gen. Comp. Endocrinol., 125(2):291-310.

Natochin, Y.V. (1996). Evolutionary aspects of renal function. Kidney Int., 49(6):1539-1542.

Natochin, Iuv. (1999). The evolution of osmoregulation in vertebrates (on the centenary of the birth of E.M. Kreps). Ross. Fiziol. Zh. Im. IM Sechenova, 85(4):582-593.

Newbound, D.R. and O'Shea, J.E. (2001). The microanatomy of the rectal salt gland of the Port Jackson Shark, *Heterodontus portusjacksoni* (Meyer) (Heterodontidae): suggestions for a counter-current exchange system. Cells Tissues Organs, 169(2):165-175.

Nishimura, H. (1985). Endocrine control of renal handling of solutes and water in vertebrates. Ren. Physiol., 8(4-5):279-300.

Oren, A. and Ventosa, A. (1999). Benjamin Elazari volcani (1915-1999): Sixty-three years of studies of the microbiology of the Dead Sea. Inter. Microbiol., 2:195-198.

Persson, P.B. (2003). Renin: origin, secretion and synthesis. J. Physiol., 5523:667-671.

Sakamoto, T., Uchida, K. and Yokota, S. (2001). Regulation of the ion-transporting mitochondrion-rich cell during adaptation of teleost fishes to different salinities. Zoolog. Sci., 18(9):1163-1174.

Silva, P., Solomon, R.J. and Epstein, F.H. (1999). Mode of activation of salt secretion by C-type natriuretic peptide in the shark rectal gland. Am. J. Physiol., 277(6Pt. 2):R1725-1732.

Soares, T.J., Coimbra, T.M., Martins, A.R., Pereira, A.G., Carnio, E.C., Branco, L.G., Albuquerque-Araujo, W.I., de Nucci, G., Favaretto, A.L., Gutkowska, J., McCann, S.M. and Antunes-Rodrigues, J. (1999). Atrial natriuretic peptide and oxytocin induce natriuresis by release of cGMP. Proc. Natl. Acad. Sci. U.S.A., 96(1):278-283.

Somero, G.N. (1983). Environmental adaptation of proteins: strategies for the conservation of critical functional and structural traits. Comp. Biochem. Physiol. A, 76(3):621-633.

Takei, Y. (2000). Structural and functional evolution of the natriuretic peptide system in vertebrates. Int. Rev. Cytol., 194:1-66.

Takei, Y. and Hirose, S. (2002). The natriuretic peptide system in eels: a key endocrine system for euryhalinity? Am. J. Physiol. Regul. Integr. Comp. Physiol., 282(4):R940-951.

Treberg, J.R. and Driedzic, W.R. (2002). Elevated levels of trimethylamine oxide in deep-sea fish: evidence for synthesis and inter-tissue physiological importance. J. Exp. Zool., 293(1):39-45.

Tseng, Y.C. and Hwang, P.P. (2008). Some insights into energy metabolism for osmoregulation in fish. Comp. Biochem. Physiol. C. Toxicol. Pharmacol., 148(4):419-429.

Uchiyama, M., Kamijyo, M., Matsuda, K. and Yoshizawa, H. (2000). Terrestrial adaptation and diversity of the kidney functions in the evolution of vertebrates, Amphibia. Biol. Sc. Kidney Int., 49(6):1539-1542.

Weber, G.M., Seale, A.P., Richman III, N.H., Stetson, M.H., Hirano, T. and Grau, E.G. (2004). Hormone release is tied to changes in cell size in osmoreceptive prolactin cell of euryhaline teleost fish, the tilapia *Oreochromis mossambicus*. Gen. Comp. Endocrinol., 138(1):8-13.

Wilson, R.W., Wilson, J.M. and Grosell, M. (2005). Intestinal bicarbonate secretion by marine teleost fish—why and how? Biochim. Biophys. Acta., 1712(2):222.

Withers, P., Hefter, G. and Pang, T.S. (1994). Role of urea and methylamines in buoyancy of elasmobranchs. J. Exp. Biol., 188(1):175-189.

Yokota, S.D., Benyajati, S. and Dantzler, W.H. (1985). Comparative aspects of glomerular filtration in vertebrates. Ren. Physiol., 8(4-5):193-221.

Zeidel, J.D., Mathai, J.C., Campbell, J.D., Ruiz, W.G., Apodaca, G.L., Riordan, J. and Zeidel, M.L. (2005). Selective permeability barrier to urea in shark rectal gland. Am. J. Physiol., Renal Physiol., 289(1):F83-89.

14

Evolution of Vertebrates

Evolution in biological systems is a process that brings about genetic changes in a population that are heritable and are passed on to many generations. Such changes result in an increase in divergence between vertebrates with time. For evolution to occur, alleles have to undergo change in response to agents of evolution. **Alleles** are different variations of the same gene that occupy the same locus on homologous chromosomes. Alleles differ in their DNA sequence and are usually recognized by their phenotypic effects. A single allele from each locus is inherited separately from each parent and alleles code for contrasting forms of a particular trait, for example color.

The rate and direction of evolutionary changes depends on the species line concerned and the time at which the changes occur. Evolution of a species towards adaptability in response to environmental changes often leads to its survival whereas failure to evolve under such circumstances might lead to extinction. Evolution that is continuous over several generations has often resulted in emergence of different species and varieties of vertebrates. Evolution continues to take place in the living vertebrates that also originated from earlier extinct species.

There are various theories that explain evolution based on the agents that bring about changes in the vertebrate genome. There is also evidence to show that evolution has taken place in the past and continues to occur now in vertebrates. Fossil records, anatomical and chemical similarities in some species, geographical distribution of certain vertebrates and genetic changes seen in living species have provided evidence that supports the process of evolution.

FOSSILS

Fossils (L. *fossilis*, dug up) are mineralized remains of plants or animals or other artifacts including footprints (Fig. 14.1) that provide a fossil record in rocks and sedimentary layers or strata. Fossil remains are excavated from sedimentary rock layers where they have been buried and have been compressed into sedimentary rock. A fossil normally has the shape of the original structure that is preserved but is close to the rock in which it is found in chemical composition, color, density and texture. Fossils can be dated according to the age of the rocks in which they are preserved. The age of the rock can be determined using trace amounts of radioactive isotopes such as uranium, thorium, potassium and rubidium whose rate of decay and stable end products are known. Since rocks are stratified, the deeper strata are generally older than shallow ones unless the strata have been deformed or are greatly folded.

Fig. 14.1 Fossils belonging to the extinct aquatic ichthyosaur ('fish-like lizard'). Pyrite (iron-sulfur mineral) also known as 'fool's gold' and phosphate fossils are well preserved and have given a detailed insight into evolution of animal life during the early Cambrian period.

Fossil records have been useful in understanding the nature of extinct vertebrate ancestors as well as linking up living vertebrates to their extinct ancestors that look dissimilar by making use of remains that have transitional traits. Only a small number of vertebrates that lived in the past have been fossilized. Some fossils have been well preserved in one place while others appear incomplete as a result of disintegration of parts or incomplete collection of data. The tree of evolution continues to be refined with time as more evidence is gathered from transitional fossil remains.

Anatomical and chemical similarities are seen in the basic structure of animal cells regardless of whether such cells belong to unicellular organisms or complex multi-cellular vertebrates. All vertebrates that undergo sexual reproduction start life as single fertilized cells known as zygotes. A common structure among many vertebrates is the forelimb that has the same bones which include the scapula, humerus, radius, ulna, carpals and digits as such structures were inherited from a common ancestor or evolved as a result of similar natural processes. Six of the major elements found in living organisms make up most of the material in organisms. All cells of living organisms contain DNA that stores genetic information. Many biological processes are controlled by genes that form parts of the DNA molecule and are responsible for synthesis of thousands of proteins that are found in living organisms and are made up of mainly 20 amino acids that form unique combinations.

The natural geographical distribution of related species shows that isolated areas of the Earth such as islands have evolved mainly distinct vertebrates. The migration of the human being to such areas has introduced new species of vertebrates that have interacted with the indigenous group to change the initial population dynamics. Examples of such areas that were originally isolated include Australia, Hawaii and New Zealand. Genetic changes have occurred over many past generations and continue to occur today in vertebrates. Such changes are brought about by mild or severe environmental changes. Sudden and severe changes result in deaths of many vertebrates leaving the survivors to reproduce and pass on their genetic material. The traits of the survivors will then dominate in subsequent generations. Changes in traits of vertebrates with a longer generation interval take more time to manifest themselves in comparison to animals with shorter intervals. Generations of offspring arising from the surviving population of vertebrates will have many similar traits of their ancestors.

Agents of Evolution

Evolutionary change is brought about by agents or processes that create diversity in the genome and operate either individually or as a group. The agents can also affect the relative rates of survival of alleles. The main processes that influence evolutionary changes form the basis of the overall synthetic theory of evolution and include natural selection, mutation pressure, sexual recombination, genetic drift and gene flow. Natural selection and genetic drift are the most important mechanisms of evolution.

Natural Selection

Environmental conditions favor certain traits in a population and increase chances of adaptation and survival of vertebrates with such characteristics in a natural manner. Natural selection is thus the only process of evolution that adapts a vertebrate to its surrounding environment in a selective manner. There has to be a change in the genotype first before the forces of natural selection can act on the resulting changed phenotype. The vertebrate with a certain favored trait is selected by the existing natural environmental conditions. The theory (mechanism) of natural selection was proposed by the English naturalist Charles Darwin (1809-1882) and Alfred Russel Wallace (1823-1913) (Fig. 14.2). Darwin had previously been on an expedition around the world from 1831 to 1836 during which he served as a naturalist aboard the H.M.S. Beagle on a British science expedition. During the expedition, Darwin collected plant and animal samples wherever he went for further study.

Vertebrates tend to produce more offspring than can be supported by the available resources in the surrounding environment and have to compete for these resources amongst each other. Various vertebrates vary in their phenotypes, some of which are heritable. Those vertebrates with features that are adapted best to the prevailing environmental conditions are most likely to live and reproduce while vertebrates with phenotypes that are not favored by the environment are likely to die early in a concept that has been termed 'survival of fittest'. Heritable adaptive phenotypes will be passed down generations resulting in dominance in a particular environment of vertebrates with favored features. Examples of natural selection include possession of mate attracting traits such as the bright plumage of male birds and strength and presence of large antlers in the male deer that ensure reproductive success.

Fig. 14.2 Photographs of Charles Darwin (left) and Alfred Russel Wallace (right). Russel's independent proposal of evolution by natural selection prompted Darwin to bring forward his detailed theory of evolution earlier than planned and both theories were published in 1859. Darwin's theory of evolution was backed by research findings from an earlier trip around the world.

Mutation Pressure

Mutation is a change that results in genetic variability as it changes the genetic code. The resulting changes are heritable if they involve germ cells and as a result mutations are responsible for evolutionary changes. Without mutations there would be no evolution. Changes in the DNA molecule normally occur during replication and transcription (Fig. 14.3). DNA helicase normally causes separation of the DNA molecule into two strands during replication. DNA polymerase duplicates each of the strands of the existing DNA molecule during transcription resulting in formation of two double stranded DNA molecules that are similar to the original DNA molecule. Changes in the DNA molecule during these processes can result in the wrong type of protein being synthesized. Proteins may also be synthesized at the wrong time or in the wrong type of cell. The changes in the DNA molecule can lead to an excess or deficiency of the protein synthesized. Although mutations can occur without influence of factors external to the DNA molecule, most mutations result from agents known as **mutagens** that damage DNA molecules leading to changes in the genetic code. Small changes in the

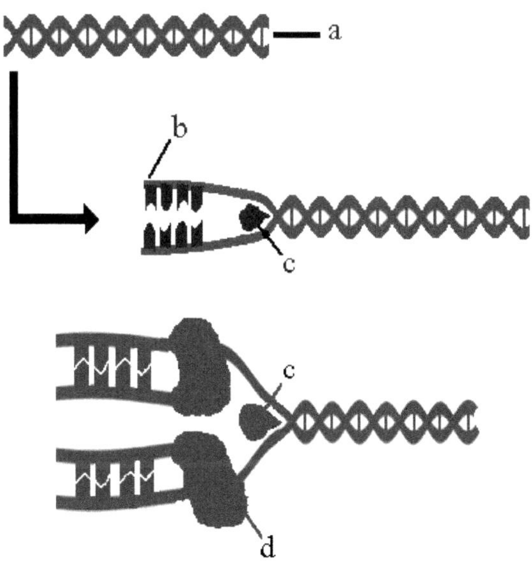

Fig. 14.3 Replication (above) and transcription (below) of a DNA molecule. (a) double stranded DNA, (b) single stranded DNA, (c) DNA helicase and (d) DNA polymerase. In a DNA molecule, the pairing of the four bases is specific. Thymine pairs with adenine while guanine does so with cytosine. Mistakes made by DNA polymerase during formation of new strands result in mutations. Such mistakes are rare and occur approximately once in 10^8 bases.

DNA molecule usually result in noticeable external changes and functions in vertebrates. Some effects of mutations are not observable.

Point mutations are changes that affect one base of the gene sequence in a DNA molecule. Point mutations can be reversible whereby the affected DNA molecule reverts to its original form. Extensive point mutations may involve rearrangement or deletion (loss) of parts of the DNA segment. Deletion mutations may result in production of non-functional proteins since they result in loss of part of the genome. More extensive mutations are the large-scale mutations that involve breakage and rejoining of DNA strands resulting in disruption of large parts of genetic information. Large-scale mutations can be lethal when they involve vital genes since such genes might control enzymes that are involved in many metabolic processes.

The three nucleotides that encode genetic information for a specific amino acid subunit of a protein are known as a **codon**. A nucleotide can be added to or deleted from a codon thus altering the decoding of the entire gene sequence as occurs in **frameshift mutations.** The amino acid sequence of the protein manufactured will be different and the protein is often ineffective. Substitution of one nucleotide by another in a codon results in one amino acid of the protein being different and this could have great effect on the individual. Substitution of an amino acid by another which occurs when the beta hemoglobin gene codon for glutamic acid is substituted by one for valine in the human hemoglobin of red blood cells results in the inherited sickle cell disease, an example of a **missense mutation**. **Nonsense mutations** code for a stop and may truncate a protein molecule resulting in a protein that is not able to catalyze expected reactions normally. In **silent mutations**, a change in a nucleotide ends up coding for the same amino acid as the original codon since most amino acids are encoded by several different codons. Silent mutations cannot be sensed.

Some mutations are beneficial and enable vertebrates to adapt better to their environments. In such cases, the concerned vertebrates will reproduce over several generations to dominate a population of other vertebrates including members of the same species. Most mutations are either neutral or harmful if environmental conditions remain unchanged but could be useful should the conditions change in their favor. Harmful mutations do not normally spread in a population since they kill vertebrates in which they occur and are not passed down to the offspring. Harmful but recessive mutations can persist in a population for a long time but at a low frequency since the dominant normal gene prevents their phenotypic expression. Mutations occur regardless of their effect on survival or reproduction of a vertebrate.

Spontaneous mutations occur during replication of the DNA molecule as a result of the inaccuracy of DNA polymerases, tautomerization of DNA bases and replication slippage of DNA polymerase. A **tautomer** is a structural isomer in dynamic equilibrium. During **duplication slippage** of the DNA polymerase, there is insertion and deletion on short-tandem repeats. Chemical and physical mutagens induce transitional mutations. Several chemicals induce mutations including deaminating agents such as nitrous acid and sodium bisulfite, alkylating agents, intercalating agents and base analogues that comprise purines and pyrimidines that are not found in natural DNA. Base analogues are incorporated into the DNA molecule and are more likely than the natural DNA bases to undergo abnormal pairing. Deaminating and alkylating agents change the structure of the normal DNA bases so that they tend to form abnormal pairing. Physical mutagens include ultraviolet radiation (UV), ionizing radiation and heat. Ionizing and UV radiation alter base pairing or block replication of DNA molecule while heat increases the distance between two bases. Examples of mutations include sickle cell anemia and albinism in the human being and some cases of developmental abnormalities. Scientists are using the tropical zebra fish and the Japanese medaka to understand how genetic mutations occur (Furutani-Seiki and Wittbrodt, 2004) and the role of mutations in the development of cancer. Mutations are relatively easy to obtain and screen in zebra fish when compared to other vertebrates and the fish are also easy to breed.

Sexual Recombination

Sexual recombination occurs in most vertebrates as they reproduce sexually. During meiosis, chromosomes in the developing egg and sperm cells undergo random assortment (recombination) so that each of these cells will contain a haploid number of chromosomes that result form each of the parent's two sets of this genetic material. The possible combinations of offspring that result from sexual reproduction are much more than those which are possible in asexual reproduction. The increased variability of offspring that is associated with sexual reproduction increases chances of survival in vertebrates with this type of reproduction since some will be well adapted to changing environmental conditions. In asexual reproduction, the offspring has a similar genotype to the parent unless a mutation has occurred.

Genetic Drift

Random or genetic drift (Fig. 14.4) entails changes in frequencies of alleles in a vertebrate population that occur by chance and depend on

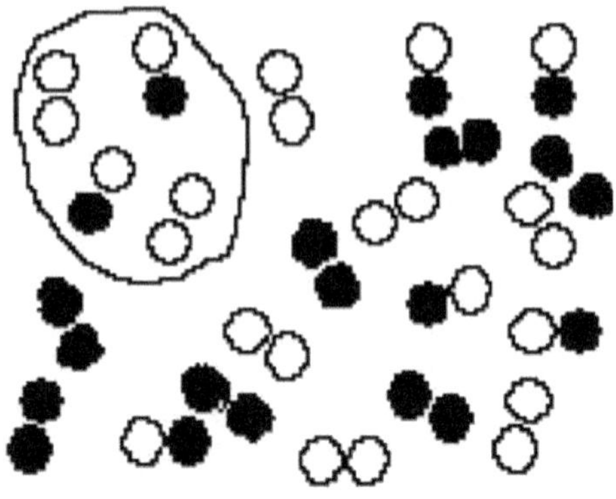

Fig. 14.4 A diagram showing the effect of random drift to the frequency of certain alleles. The larger circle represents a small section of a population or a small and isolated population group. The smaller and empty circles are alleles for blue eyes whereas the smaller and black circles represent brown eyes. Most of the subsequent descendants in the group surrounded by the larger circle will have blue eyes as opposed to the entire population group in the diagram. The effect of a random genetic drift is inversely proportional to population size.

effective population size. The resulting genetic changes result in evolutionary change but not adaptation. Changes in frequency of alleles result from cases where some vertebrates within a species give rise to more offspring than others thus passing on more genes to future generations than the less proliferating group. Such a case could result when there is random variation in the survival and reproduction of various vertebrates. Genetic drift results in one allele replacing the others and becoming part of a population. The rate of allele frequencies varies to some extent depending on population sizes. As allele frequencies may change greatly in small populations, the spread of lethal alleles in genetic drift is seen mainly in such population sizes. A decrease in population size increases cases of inbreeding that result in chances of recessive homozygous alleles increasing. Many recessive alleles are lethal and weaken the survival of a population when in a homozygous state. In large populations, rare alleles may be lost since vertebrates with such genes may die before reaching reproductive maturity or may reproduce less than expected.

Neutral or close to neutral alleles of a gene can be advantageous as more variability can occur in large populations when compared to small populations. Small groups of vertebrates that have undergone random drift may inhabit different environmental conditions from the major population group and such conditions may favor gene combinations that may be advantageous in the new environment.

Gene Flow

A population structure comprises a subdivision of a group of species of vertebrates into units. The whole population group is known as a **metapopulation**. **Subpopulations** are individual components of populations and are also known as local populations or **demes**. Many demes are partly separated from other demes of same species. In a continuous population of vertebrates, there may be no conspicuous subdivisions in structure. A continuous population can be structured whereby certain parts of the population have different gene frequencies but it is not easy to draw exact boundaries between the vertebrates groups. Population groups that are subdivided tend to evolve independently and bring about diversification in a metapopulation.

Gene flow occurs by immigration of vertebrates to other demes and their introduction of new alleles to the new subpopulation. Gene flow can be very low or high and changes allele frequencies. The level of gene flow depends on the number of immigrants arriving at the new deme and the level at which they differ genetically from the resident population. In cases where there is much difference genetically between the two groups of a metapopulation, a small number of immigrants can lead to hybridization that is normally associated with crossing of individuals from genetically different strains, populations or species. Hybridization is associated with the introduction of many new alleles into a population and results in evolutionary changes in the resident population of vertebrates.

HISTORY OF VERTEBRATE EVOLUTION

Evolution of vertebrates has occurred from the ancestral forms of life as a result of gene duplication (Jozefowicz et al., 2003; Taylor and Raes, 2004). Vertebrates originated in the sea from a common ancestor about 500 million years ago. Vertebrates share certain inherited features with the most remarkable being a backbone. An **evolutionary tree** or **cladogram** has been drawn based on cladistics in which organisms are grouped according to shared features to show how the various vertebrates branched off the main trunk and subsequent branches that followed (Fig. 14.5). Species that branch off a common point are closely related in evolutionary terms. There was diversification in vertebrate species with time during evolution and this is represented by an increase in the number of branches in the evolutionary tree towards the present time. The evolutionary tree of vertebrates is undergoing change with time as more information about features of past and present vertebrates is gathered based on the study of fossils and features of vertebrates as

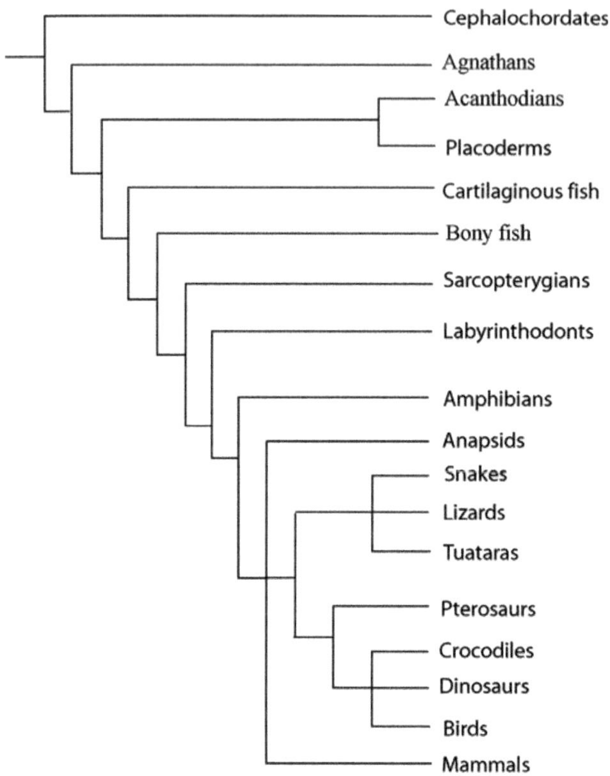

Fig. 14.5 A general cladogram or evolutionary tree of vertebrates. Vertebrates are believed to have shared a common ancestor that resembled the lancelet.

well as DNA analysis. Table 14.1 shows the range and relative abundance of vertebrates with geologic time changes. Many species of vertebrates have become extinct while others have emerged throughout the history of vertebrate evolution.

The earliest known ancestor of chordates is thought to be *Pikaia gracilens* (after Mount Pika; L. *gracilens*, slender or thin) (Fig. 14.6) that was found in the Burgess Shale, near Mt. Pika, in British Columbia, Canada. *Pikaia,* measuring about 40 mm long and with an expanded tail, resembles the lancelet (*Amphioxus*). *Pikaia* is a member of the chordate group that eventually gave rise to vertebrates. There are various theories about the origin of vertebrates. One theory states that vertebrates originated from earlier chordates by paedomorphosis (Gr. *paid*—from *pais,* child; *morphe,* shape) whereby adults retained juvenile characteristics.

Table 14.1 Part of the Geological Time Scale showing the relative abundance of vertebrates in the past to the present time. Mississipian and Pennsylvanian periods make up the Carboniferous period. The first geologic time scale was proposed in 1913 by the British geologist Arthur Holmes.

Era	*Period*	*Time in years (10^6)*	*Events in Vertebrate Evolution*
Paleozoic ('ancient life')	Cambrian	540	Appearance of earliest primitive fish. Mass extinction that wiped out 50% of all animal families.
	Ordovician	510	Radiation of jawless fish. Mass extinction as a result of glaciation at end of period.
	Silurian	438	Diversification of jawless fish and appearance of first jawed fish.
	Devonian	410	Period known as 'age of fishes'. Appearance and diversification of cartilaginous and bony fishes. First amphibians appear at end of period that is also marked by mass extinction which destroyed 30% of all animal families.
	Mississipian	360	Diversification of amphibians.
	Pennsylvanian	325	Appearance of first reptiles.
	Permian	290	Period known as 'age of amphibians'. Amphibians and reptiles dominate. Mammal-like reptiles appear. The continents merge into a single continent known as Pangaea. End of period is marked by the largest mass extinction that claims 95% of mainly marine species.
Mesozoic ('middle life')	Triassic	245	Era known as the 'age of reptiles'. Dinosaurs, crocodyloformes, turtles, ichthyosaurs and mammals appear. Period ends with extinction of 35% of all animal families including labyrinthodont amphibians and all marine reptiles except ichthyosaurs. Dinosaurs radiate into many niches.
	Jurassic	210	Diversification of dinosaurs. Appearance of first birds. Minor extinctions occur at 190 and 160 million years ago.
	Cretaceous	145	Continued radiation of dinosaurs. Feathered dinosaurs and first snakes appear. Mammals diversify. Period ended with mass extinction including dinosaurs and pterosaurs.
Cenozoic ('recent life')	Era divided into various epochs (see Table 13.2)	65	Era known as the 'age of mammals'. Radiation of mammals and birds. First hominids (australopithecines) appeared during the Pliocene epoch (5 to 1.8 million years ago). Mass extinction of large mammals and many birds at end of Pleistocene epoch (11 thousand years ago) followed by human civilization.

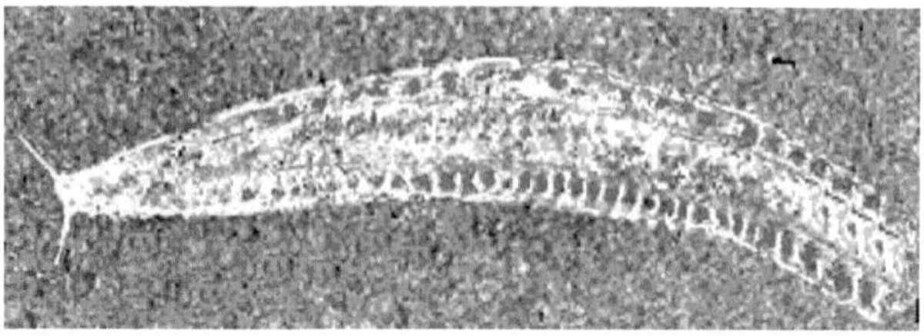

Fig. 14.6 Fossil remains of *Pikaia gracilens*, a relative of the lancelet, from the Burgess Shale collection. *P. gracilens*, a primitive chordate, is believed to be the earliest known ancestor of chordates including vertebrates.

FISH

The earliest craniates (Gr. *kranion*, skull) were the jawless agnathans that are represented by the living hagfish and lampreys. Extinct and armored agnathans known as arandaspids lived about 480 to 440 million years ago and fossil remains of this 5.0 cm long jawless fish (Fig. 14.7) have been found in Australia and South America. Arandaspids are among the oldest craniates in evolutionary terms. Glaciation at the end of the Ordovician period resulted in a massive extinction event that caused extinction of 60% of marine life including arandaspids. With Silurian came sea level rise and a stable climate that resulted in a wide and rapid spread of jawless fish as well as appearance of the first jawed fish and freshwater fishes. Jaws are believed to have evolved from gill arches and have evolved only once in vertebrates history. The evolution of jaws

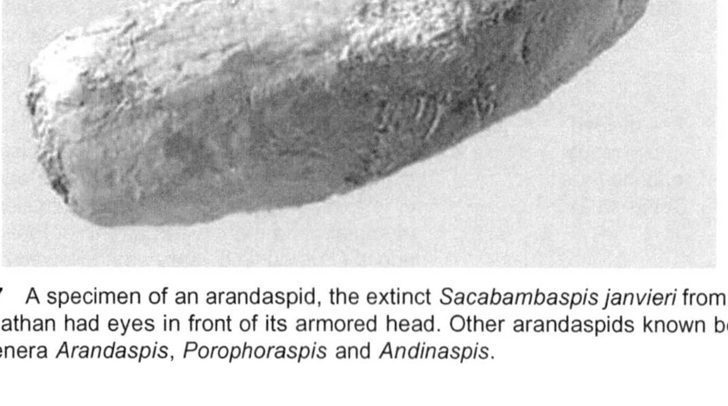

Fig. 14.7 A specimen of an arandaspid, the extinct *Sacabambaspis janvieri* from Bolivia. This agnathan had eyes in front of its armored head. Other arandaspids known belonged to the genera *Arandaspis*, *Porophoraspis* and *Andinaspis*.

enabled fish to prey on smaller organisms unlike their ancestor strainers. There was also a change in the gastrointestinal system with enlargement of the stomach.

Acanthodians (Fig. 14.8) are some of the oldest jawed fish to evolve and appeared during the Silurian period. The fish had large eyes that were located near the front of their blunt heads. Acanthodians became extinct during Permian and their fragmentary fossil remains have been found in North America and China. Acanthodian remains are poorly fossilized. Acanthodians showed little diversity despite having lived for about 160 million years on Earth. The first acanthodians to evolve inhabited marine water but later on became predominantly freshwater fish.

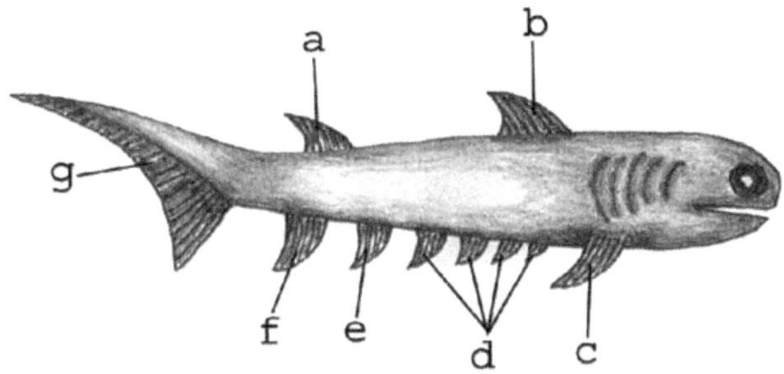

Fig. 14.8 An acanthodian (*Climatius*). (a) posterior dorsal fin, (b) anterior dorsal fin, (c) pectoral fin, (d) intermediate fins, (e) pelvic fin, (f) anal fin and (g) heterocercal tail fin. Bony spines projected from the anterior part of the fins of acanthodians and their caudal fins were heterocercal thus the name 'spiny sharks' in referring to these extinct fish. The bodies of acanthodians were covered by small and diamond shaped scales. Acanthodian fossil remains are rarely preserved as articulated specimens. The presence of large eyes and small nasal capsules indicate that the fish probably depended on vision more than their sense of smell.

Another group of the first jawed fish to evolve were **placoderms** (Fig. 14.9). Most placoderms lacked teeth whose function was played by the bony plates that had sharp cutting edges which resembled meat cleavers and were associated with jaws. Placoderms lived from late Silurian to Permian. During their life on Earth placoderms diversified a lot into various body shapes (ecomorphs) and occupied various ecological niches that were later occupied by other fish after placoderms become extinct. There were more than 250 genera of placoderms making them the most diverse and important group of the earliest vertebrates. The oldest fossil remains of placoderms known are found in China. Some placoderms included the earliest vertebrates to move from seawater to freshwater. The extinction of placoderms seems to have been sudden for reasons that are yet to be known after most survived the great Frasnian-Famennian

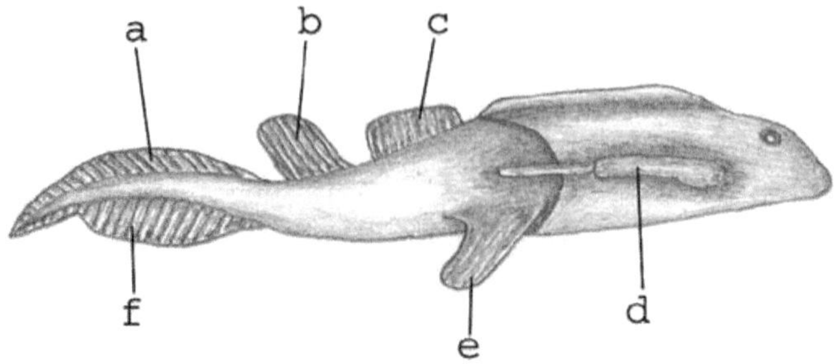

Fig. 14.9 A drawing of the placoderm *Bothriolepis*. (a) epichordal lobe, (b) posterior dorsal fin, (c) anterior dorsal fin, (d) pectoral appendage, (e) pelvic fin and (f) hypochordal lobe. Placoderms had extensive body armor on the head and anterior part of the trunk. Bony scales that sometimes resembled minute denticles embedded in skin covered the rest of the body. A notochord persisted in adult placoderms and their internal skeleton was cartilaginous.

mass extinction at late Devonian that eliminated about 20% of all marine animals. The most affected marine species by the Frasnian-Famennian extinction were those occupying warm bodies of water since there was global cooling.

Cartilaginous fish (see Chapter 2) first appeared between 400 and 450 million years ago and could have shared a common ancestor with placoderms. Cartilaginous fish have showed little change since their first appearance and are sometimes referred to as living fossils (Fig. 14.10). Chondrichthyans have evolved two types of jaws. Most sharks have powerful jaws for biting and crushing food while skates and rays have jaws adapted for feeding on mollusks at the bottom of the sea. Holocephalans first appeared during the Carboniferous period whereas rays evolved in the Jurassic period. The peak of cartilaginous fish

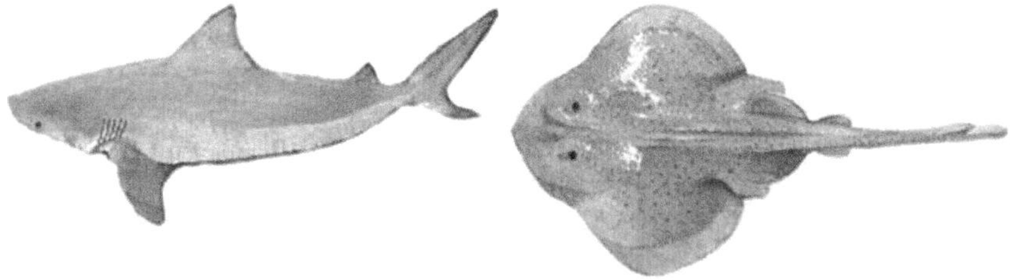

Fig. 14.10 A living shark, the black tip shark (*Carcharhinus limbatus*) (left) and a skate (right). The blacktip has distinctive black markings on its fins and has a maximum reported length of 2.55 m. Hides of the blacktip are used for making leather. Skates and rays are shaped like a kite and are flattened dorsoventrally. The gill slits of skates and rays are located on the ventral part of the head region.

evolution occurred in late Paleozoic with the decline of placoderms. Since cartilage is poorly preserved, the body structures of early sharks are lacking. The teeth of sharks fossilized well and have been used in identifying shark groups. The adaptive immune system started to develop in sharks. Shark lymphocytes are not differentiated into T- and B-types.

Bony fish comprise the most diverse group of vertebrates that have evolved for about 410 million years ago. Fossil remains of the earliest bony fishes appear in freshwater deposits of the Devonian period. The numbers of bony fish increased greatly in the middle of the Devonian period when placoderms and cartilaginous fish began to decline. Ray-finned fish show a variation in the number of Hox clusters that could have resulted from duplications and subsequent lineage-specific gene loss over a period of time (Hoegg and Meyer, 2005). The success of bony fishes can be attributed to the presence of the swim bladder that could have evolved from lungs that were present in some freshwater fish. The three major groups of bony fishes are the ray-fins (Fig. 14.11), lungfishes and the coelacanth (see Chapter 2).

The extinct **rhipidistians** (Fig. 14.12) were similar to living lungfishes though not closely related and had functional lungs. Rhipidistians had bony lobed fins; they used to move in water and on land in search of water. With further evolution, rhipidistians became more terrestrial and

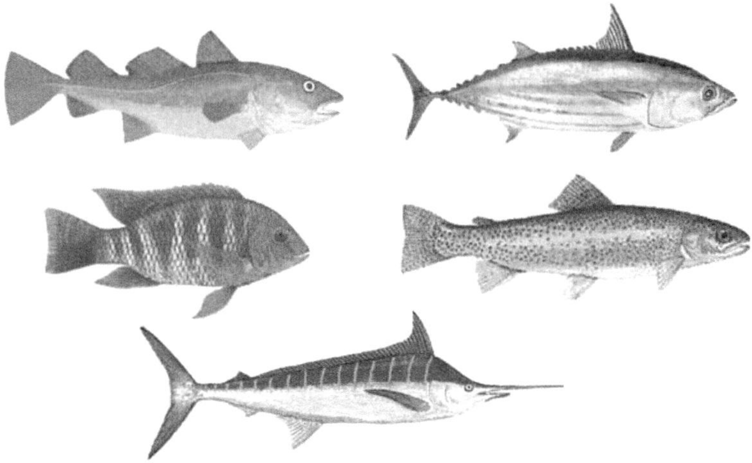

Fig. 14.11 Some ray-finned fishes (actinopterygians). Cod (top left), skipjack tuna (top right), tilapia (middle left), rainbow trout (middle right) and blue marlin (bottom). The fins of actinopterygians are webs of skin that are supported by bony or horny spines. Most fish and aquatic vertebrates are actinopterygians (more than 20,000 species) and have an internal skeleton that has varying amounts of bone.

Fig. 14.12 Fossil of *Eusthenopteron*, an extinct rhipidistian. It is thought that rhipidistians are the ancestors of tetrapods and transition from water to land occurred not later than mid-Devonian. *Eusthenopteron* could breathe air in shallow and muddy water and the internal skeleton of its fins had a distinct humerus, ulna and radius in the fore-fin and femur, tibia and fibula in the pelvic fin.

less aquatic. Rhipidistians became extinct during the Carboniferous period.

AMPHIBIANS

Amphibians were the first vertebrates to make the transition from water to land about 360 million years ago. The tetrapods probably first arose in Euramerica and complete transition occurred over a period of about 25 million years in regions further away (Long and Gordon, 2004). During the period of transition, the land was occupied by plants and arthropods whereas the water had a variety of biological life including trilobites and fishes. For amphibians to be successful on land, the respiratory system had to adapt to breathing oxygen from the air and the problem of losing water through the skin had to be solved. The alternating dry and wet periods at the time of water to land transition made amphibians to criss-cross land in search of other bodies of water when the inhabited bodies of water dried up.

The earliest amphibians are thought to be ichthyostegalians that could have evolved from rhipidistians. Many early amphibians were much larger that the living group and some were as long as 4.6 meters. The early amphibians fed mainly on insects. Since there were no other vertebrates living on land, amphibians were the dominant large terrestrial animals and they rapidly diversified to fill various ecological niches. The largest amphibians were labyrinthodonts (Gr. *labyrinthos*, labyrinth; *odous* or *odont*, tooth) (Fig. 14.13) that possessed sharp and conical teeth with enamel folds that formed a maze-like pattern. There was also a second row of teeth on the roof of the mouth. Labyrinthodonts had a bulky skeleton and short limbs and could have spent most of the time in water since they may have been clumsy walkers on land. Labyrinthodonts dominated terrestrial life for about 100 million years from the end of the Devonian to end of the Permian periods when they became extinct. During Permian, the amphibians were out-competed by

Fig. 14.13 *Ichthyostega* (Gr. 'fish roof'), a labyrinthodont that represents some of the earliest amphibians on earth and lived during Late Devonian. *Ichthyostega* was about 1.5 m long and had seven digits on each hind foot. The caudal fins of labyrinthodonts were supported by lepidotrichia in a similar manner to the fins of actinopterygians (ray-finned fishes).

evolving reptiles that could reproduce on land as a result of their amniotic eggs.

Salamanders and newts (order Caudata) (Fig. 14.14) represent the earliest group of vertebrates on land. Their fossil record is incomplete since their bones do not fossilize well. A 7-inch skeleton that was found in Kazakhstan represents so far the oldest salamander that is about 140 million years old. The specimen might have evolved from an earlier salamander as it is highly specialized. Salamanders and newts could have evolved in North America since most of these amphibians are found on this continent. Most species of salamanders are found in the northern hemisphere and could have come here from North America when the continents were still joined together.

Anurans include frogs and toads (Fig. 14.15) and the fossil record of their extinct ancestors is rare since their bones did not fossilize properly. The earliest fossil remains of anurans were found in Magadascar and

Fig. 14.14 A salamander (left) and the red-spotted newt (right). Salamanders and newts resemble lizards in appearance and the differences between the two amphibians are minor. Newts tend to be generally more aquatic than salamanders and have a rougher skin and flatter tails than salamanders. The skin of salamanders is slimy and their tails are round.

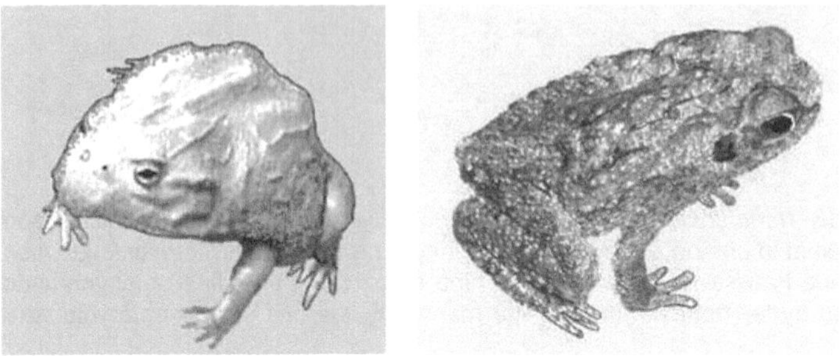

Fig. 14.15 The African Bullfrog *Pyxiecephalus adspersus* (left) and the marine toad *Bufo marinus* (right). Toads are generally similar to frogs though their skin is rougher and drier. Toads tend to be terrestrial and have parotid glands behind their eyes that can produce poisonous substances. The male adult African Bullfrog may be as long as 23 cm and may weigh as much as 2.0 kg making it one of the largest frogs in Africa.

date back to early Triassic. These anurans remains are referred to as *Triadobatrachus* (Fig. 14.16) and measured about four inches in length. *Triadobatrachus* was not a good leaper like the modern anurans since its tibia and fibula were unfused and the rear limbs were proportionally smaller in relation to the body size. The present anuran structure appeared in early Jurassic. Caecilians (Fig. 14.17) too have left a poor fossil record. The earliest fossils date back to 65 million years ago and

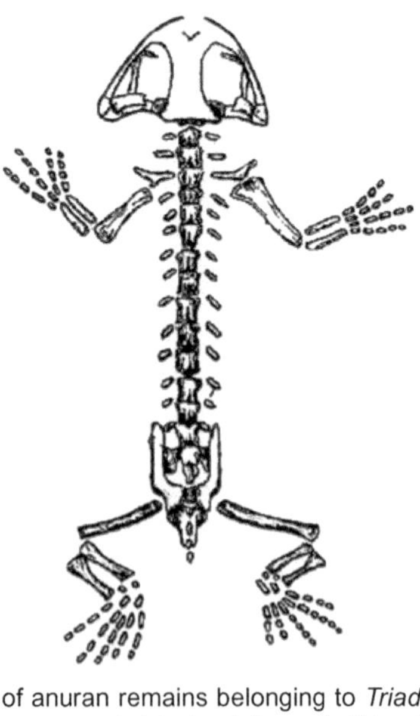

Fig. 14.16 The skeleton of anuran remains belonging to *Triadobatrachus massinoti* that were discovered in Madagascar and date back to early Triassic. *Triadobatrachus* was a poor leaper since the rear limbs were proportionally smaller when compared to body size and the tibia and fibula were unfused.

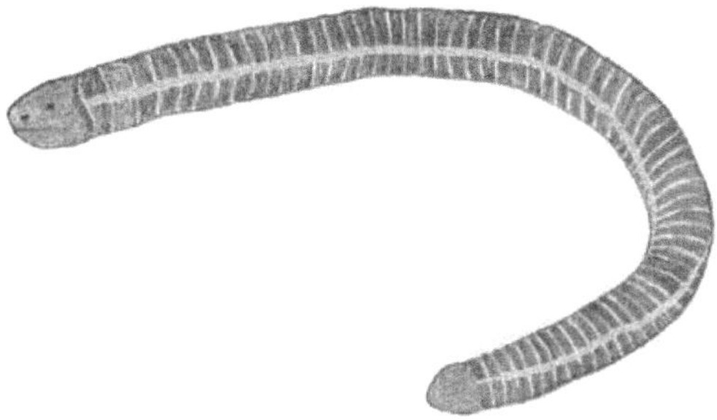

Fig. 14.17 A diagram of a caecilian. Caecilians are burrowing, limbless and worm-like amphibians that are found in the tropical world and lack limbs. The eyes and tails of caecilians are reduced and the skin is segmented with tiny scales. Caecilians have powerful heads for burrowing and generally range in length between 12 and 15 cm though can be as long as 1.5 m. Caecilians are the least studied group of amphibians that belong to the order gymnophiona or apoda.

comprise a single vertebra. Caecilians could have evolved much earlier than this period.

REPTILES

The evolution of reptiles from their ancestors is evident from several fossil remains. Three main types of skulls have evolved in amniotes (Fig. 14.18). **Anapsids** (Gr. *a,* without; *aspid,* loop or bar) had a solid skull with openings for structures including the eyes, nose and spinal cord only. Anapsids are represented by living turtles. **Synapsids** (Gr. *syn,* together) had a pair of holes or fossae behind the eyes (one hole behind each eye and below the temporal bone) in their skulls known as the **infratemporal fossae** that formed the **lateral temporal opening**. The openings increased the surface area for attachment of jaw muscles as well as making the skull relatively lighter. The fossae were surrounded ventrally by a bony loop. Synapsids later on evolved into mammals. **Diapsids** had the pair of openings found in synapsid skulls as well as another pair higher on the

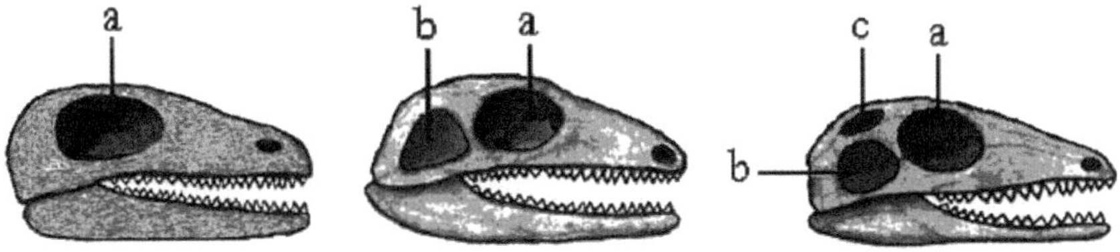

Fig. 14.18 The main types of skulls found in living amniotes. Anapsid (left), synapsid (middle) and diapsid (right). (a) orbit, (b) lateral temporal opening and (c) dorsal temporal opening.

skull known as the **supratemporal fossae** or the **dorsal temporal opening**. Diapsids evolved into lepidosaurs (snakes, lizards and tuataras) and archosaurs (pterosaurs, dinosaurs, crocodiles and birds).

The bones (jugal and quadratojugal) that surround the lower fenestra or infratemporal fossa ventrally (Fig. 14.19) have been lost in squamates leaving the lower jaw unhinged. Squamates are therefore able to open their mouths quite wide. A **euryapsid skull** (Gr. *eurys*, wide) was derived from a diapsid skull and had only a pair of supratemporal fossae since the infratemporal fossae had been lost. The supratemporal fossae were bordered on their lower parts by postorbital and squamosal bones. Euryapsid skulls were found in the extinct plesiosaurs. A minor variation to the euryapsid skull known as a **parapsid skull** was found in the extinct ichthyosaurs. In the parapsid skull, the pair of supratemporal fossae was located high on the skull and was surrounded ventrally by postfrontal and supratemporal bones.

Turtles

Turtles (Chelonia or Testudines) had already evolved by late Triassic when dinosaurs also evolved. By the end of Cretaceous, the turtles had reached their greatest diversity. Several groups of turtles have become extinct in the course of evolution. The first fossil of turtles to be discovered was *Proganochelys* that lived in freshwater and was much larger than living turtles. It is thought that turtle evolution took a relatively short time before *Proganochelys* because of the scanty fossil

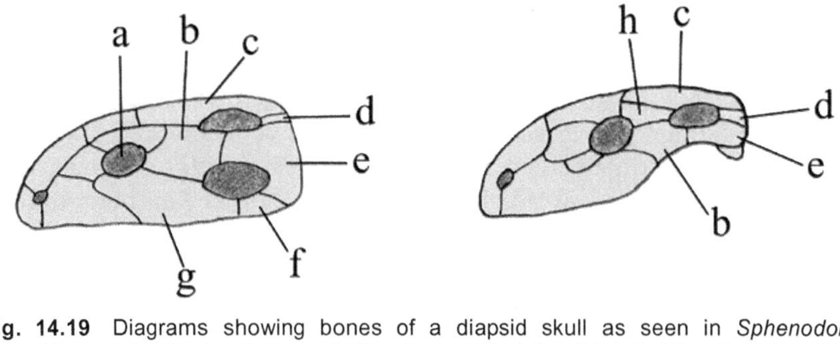

Fig. 14.19 Diagrams showing bones of a diapsid skull as seen in *Sphenodon*, crocodilians and extinct archosaurs (left) and a modified diapsid skull of lizards and snakes (right). (a) orbit, (b) postorbital, (c) parietal, (d) supratemporal, (e) squamosal, (f) quadratojugal, (g) jugal and (h) postfrontal. In the modified diapsid skull, the jugal and quadratojugal that surround the ventral part of the lower temporal fossa have been lost with evolution leaving the lower jaw unhinged and enabling squamates to open their mouths quite widely. In snakes, the squamosal-postorbital arch between the two temporal openings has been lost as well.

record available then. Turtles have changed little in basic structure during evolution. Originally, turtles were terrestrial with limbs that were adapted for life on land. Turtles moved between land and sea many times during the Mesozoic era. The marine turtles evolved flippers or paddles for swimming. After the mass extinction at the end of Mesozoic, four families of sea turtles survived and included Toxochelyidae, Protostegidae, Cheloniidae and Dermochelyidae. Toxochelyidae became extinct during the Eocene epoch and Protostegidae during the Oligocene epoch. As turtles evolved, there was a reduction in the number of bones of the skull and the closing up of the skull resulting in the more solid anapsid state. About 260 species of turtles inhabit the tropics and temperate regions of the world today.

The human being has been a threat to turtles as these reptiles have been fed upon by man for about 2 million years. At maximum risk are land tortoises, some of which have become extinct. Destruction of nesting habitats, removal of eggs and pollution are other threats to turtles.

Synapsids

Synapsids were a group of reptiles with mammal-like features that first appeared during early Carboniferous far before the dinosaurs. The most primitive synapsids were pelycosaurs ('basin lizards') or theromorphs that are also referred to as 'mammal-like reptiles' as they are an intermediate form between reptiles and mammals (Fig. 14.20). The first pelycosaurs were small and resembled large lizards and are thought to have been endothermic. The upper limb bones of pelycosaurs ran in a parallel manner to the ground while the lower radius and ulna or tibia and fibula were placed vertically. With time, pelycosaurs evolved into more different types that were larger in body size. *Dimetrodon* is one of

Fig. 14.20 Some pelycosaurs. *Dimetrodon* (left) and *Titanophoneus* (right). Pelycosaurs (Gr. *pelyx*, bowl; *sauros*, lizard) were small to large (up to 3.0 m long or even longer) early synapsids that first appeared during the late Carboniferous and were the dominant land animals during Early Permian. Other pelycosaurs belonged to the genera *Sphenacodon*, *Edaphosaurus* and *Ophiacodon*.

the most known carnivorous pelycosaurs that had long sail backs which consisted of elongated vertebral spines. Another pelycosaur, the herbivorous moschops, had a massive skull with blunt chisel-like teeth. *Titanophoneus* was a late Permian therapsid that had a massive skull.

The fossil record of synapsids is extensive. The oldest synapsid remains were found in Nova Scotia and are about 320 million years old. By late Carboniferous, synapsids were the most common amniotes. Synapsids and other amniotes occupied the equatorial and subequatorial regions of the Earth. Later synapsids walked on four limbs that were more erect than those of pelycolsaurs and measured about 50 cm to 3.0 m in length. These amniotes had a smaller brain in relation to the mammalian brain and were mainly carnivorous or insectivorous. During their early evolution, the body size of synapsids decreased. The group was mainly active at night and this could be as a result of diversification of the other group of reptiles they had to compete with for resources at daytime. Nearly all synapsids and pelycosaurs became extinct at the end of Permian as a result of a catastrophic event that affected the Earth. A group of pelycosaurs known as Sphenacodontia that included *Dimetrodon* evolved into therapsids that lived from late Early Permian through the Triassic Period. Therapsids were the dominant land animals during the Middle Permian. Among the most successful therapsids were cynodonts that evolved into mammals.

Lepidosaurs

The clade of lepidosaurs comprising snakes, lizards and tuataras (Fig. 14.21) first evolved in Gondwanan territories including Africa and Madagascar, South America and India. This clade of reptiles still maintains lateral undulations of the vertebral column during locomotion. Tuataras (Maori word for 'peaks on the back') first appeared in late Triassic and were a moderately diverse group but most are extinct and are represented by *Sphenodon* that lives off some islands in New Zealand. Squamates include lizards and snakes and are the largest group of modern reptiles. Lizard and snake fossils appear first in early Jurassic and Cretaceous respectively.

The fossil history of snakes is not well known as their skeletons fossilize poorly. Among the oldest snakes were fossils that were discovered in the Sahara Desert of Africa dated about 130 million years ago. A more complete snake skeleton was found in Argentina and was dated to early Cretaceous. Some scientists believe that a group of burrowing monitor-like lizards that lost their external ears and limbs over many years could have been ancestors of snakes. The tiny claw-like

Fig. 14.21 The tuatara *Sphenodon punctatas* (top left), a lizard *Sceloporus undulatus* (top right) and the sharp tailed snake *Contia* (bottom). There are two species of tuataras found on about 30 small but relatively inaccessible islands off the coast of New Zealand.

spurs that are present on either side of the cloaca of boas and pythons (boids) are considered to be remnants of hind limbs of the burrowing ancestors of snakes. These vestiges are usually larger in males. A study using DNA sequence does not support a close relationship between monitor lizards and snakes and supports a terrestrial origin of snakes (Vidal and Hedges, 2004). Boids are the most primitive living snakes. The early ancestors of the modern terrestrial snakes might have been relatives of the present day boids.

When dinosaurs became extinct at the end of Cretaceous, boids became the main snake family on Earth. Approximately 36 million years ago, another group of smaller snakes evolved and coexisted with the boids. This later group of smaller and faster snakes comprised **colubrids** that competed with boids for food and space. The boids remained the dominant snakes until about 20 million years ago after the continental drift. Movement of the Earth's plates further away from the equator was accompanied with climatic cooling, an action that led to the extinction of many boids. The colubrids filled the niches that were previously occupied by boids and diversified to become the dominant snakes of the world. Vipers or solenoglyphs started appearing about 10 million years ago and are considered to be the most evolved snakes. Pit vipers evolved shortly after this period. Pit vipers can detect their warm-blooded prey at night using heat sensitive pits on their heads located between the eyes and nostrils. Snakes considered to be the most specialized are the rattlesnakes and evolved much later than pit vipers. Rattlesnakes have evolved a structure known as the tailrattle that is located at the end of the tail and is made of interlocking modified hollow scales which are rattled possibly as a warning to predators.

Archosaurs

Evolution of archosaurs (Gr. *archon*, ruler; *sauros*, lizard) gave rise to pterosaurs, dinosaurs, crocodiles and birds. Archosaurs evolved during the Triassic period from a more primitive reptile after the Permian mass extinction. At the middle of Triassic, two lines of archosaurs emerged. One branch led to the evolution of crocodiles and the other to dinosaurs and birds.

Dinosaurs

Dinosaurs (Gr. *deinos*, terrible) (Fig. 14.22) evolved from socket-toothed archosaurs during the Triassic period at a time mammals were also evolving. The original ancestors had a sprawling gait similar to that of modern lizards. Socket-toothed archosaurs are thought to have evolved into crocodile-like archosaurs or **thecodonts** that had a semi-erect posture which resembled that of the living crocodile. Thecodonts are the ancestors of dinosaurs, birds, pterosaurs and crocodiles. The earliest dinosaurs were small bipedal carnivores or omnivores that ranged in length from 3.0 m to 4.5 m. The oldest fossil remains of dinosaurs dating

Fig. 14.22 Some of the dinosaurs. Saurischian (left) and ornithischian (right) dinosaurs. Saurischian dinosaurs belonged to the Order Saurischia and were the earliest dinosaurs to evolve about 230 million years ago. Saurischian dinosaurs are believed to be the ancestors of birds and had a hip structure with a pubis that pointed forwards and downwards like that of lizards. The two groups of saurischians were the four-legged herbivores (saropodomorpha) and the two-legged carnivores (theropods). Ornithischian dinosaurs date back to early Jurassic. Some ornithischian dinosaurs were horned and armored while others were 'duck-billed'.

to 230 million years ago were found in Madagascar. Other old fossils of dinosaurs have been found in Argentina.

The early dinosaurs took advantage of habitants left vacant after the mass extinction at end of Permian that wiped out 75% of all tetrapod families. Dinosaurs were then to dominate other land animals for the rest of the Mesozoic era. Dinosaurs evolved into two groups that could be differentiated from each other based on the structure of hipbones.

Saurischian dinosaurs ('reptile-hipped') had the most primitive hip structure (Fig. 14.23) as was present in archosaurs. Saurischian dinosaurs included some to the largest animals that ever lived on land and some weighed as much as 100 tons. Some saurischian dinosaurs were carnivorous while others were herbivorous. These extinct reptiles moved in to occupy territories that had been occupied by synapsids. **Ornthischian dinosaurs** had hipbones that resemble those of birds. These dinosaurs were herbivorous and some had horns.

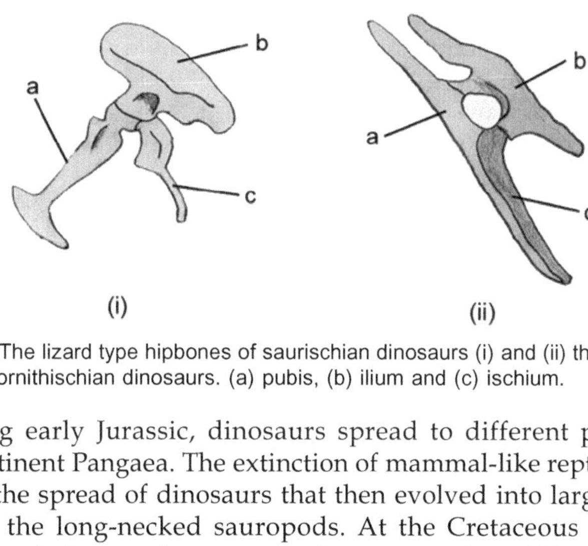

Fig. 14.23 The lizard type hipbones of saurischian dinosaurs (i) and (ii) the bird type hip structure of ornithischian dinosaurs. (a) pubis, (b) ilium and (c) ischium.

During early Jurassic, dinosaurs spread to different parts of the super-continent Pangaea. The extinction of mammal-like reptiles created room for the spread of dinosaurs that then evolved into larger animals, including the long-necked sauropods. At the Cretaceous period, the separation of the continents was almost complete. Dinosaurs were separated from each other at this time by the seas making groups of dinosaurs evolve independently.

At the end of Cretaceous, the population of dinosaurs began to decline and by 65 million years ago had become completely extinct making it the most famous mass extinction on Earth. About 70% of the other vertebrate species on Earth became extinct together with the dinosaurs. It is believed that recovery from the decline in numbers of dinosaurs at end of Cretaceous was prevented by a catastrophic event. There are several theories that attempt to explain the causes of extinction of dinosaurs including climate change and meteoritic impact but none is conclusive. There are many late Cretaceous dinosaur sites in North America.

Pterosaurs

Pterosaurs (Gr. *pteryg*, fin or wings; *sauros*, lizard) or 'winged lizards' (Fig. 14.24) could have evolved from thecodonts. Pterosaurs first appeared during the Triassic period and lived for 150 years before becoming extinct at end of Cretaceous. Pterosaurs were the largest animals to fly with the biggest having a wingspan of about 12 meters (40 feet). The smallest pterosaur was about the size of a sparrow. Fossils representing about 60 genera of pterosaurs have been found. New findings of uncrushed pterosaur skeletal remains with well preserved wing membranes and extensive pterosaur tracks combined with reanalyzing of the relationships of pterosaurs with other reptiles support the original design of the once flying reptile as one with membranes attached to both forelimbs and hindlimbs and with a quadripedal stance and gait when on the ground (Unwin, 1999). The first pterosaurs to evolve had long jaws that bore teeth and long tails but later forms evolved beak-like jaws that lacked teeth and the tail was shortened into a stump. Pterosaurs were adapted to flight and had hollow bones and a highly effective flow-through respiratory system that was capable of sustaining powered flight (Claessens et al., 2009) as they possessed components of a bird-like respiratory system including a series of ventilatory air-sacs (Butler et al., 2009). Their wings were thin membranes of skin that were supported by fibers which attached to the fourth finger that was very long (see Fig. 4.26).

The brain of pterosaurs was more developed than that of dinosaurs of comparable size. Studies have shown that the general neuronal organization of pterosaurs resembles that of birds although pterosaurs

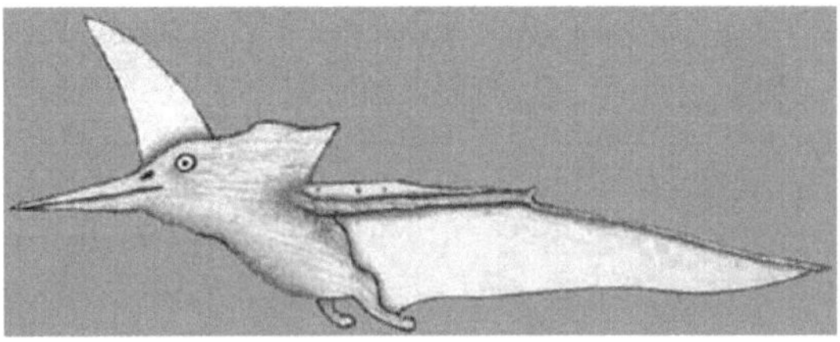

Fig. 14.24 A pterosaur ('winged lizard'). The pterosaur wing was supported mainly by an elongated fourth digit that runs along the forward edge of the wing. The term 'pterodacyl' that refers to the 'wing finger' has sometimes been used to refer to pterosaurs. Pterosaurs appeared in the Triassic period about 230 million years ago and were extinct around 65 million years ago.

had smaller brains relative to body mass than birds and this difference probably had more to do with phylogeny than flight (Witmer et al., 2003). The same study found out that since pterosaurs were visually based aerial predators, their semicircular canals were enlarged reflecting a highly refined organ of equilibrium. The presence of hair on bodies of pterosaurs and the demands of flight have led to the belief that pterosaurs were endothermic. The recent discoveries of a pterosaur egg with an embryonic skeleton and soft tissues in China (Wang and Zhou, 2004), an embryonic fossil of a pterosaur in Argentina (Chiappe et al., 2004) and a pterosaur egg with a structure that was soft and leathery (Ji et al., 2004) has confirmed that pterosaurs were oviparous.

Since extinction of pterosaurs coincided with the appearance of early birds, it is thought that competition with birds could have contributed to the disappearance of the pterosaurs. The small pterosaurs were extinct towards the end of Cretaceous leaving only the large pterosaurs and birds as the only flying vertebrates at the time. Following the mass extinction that ended the Cretaceous, all large pterosaurs died leaving birds, the smaller creatures, as the only flying vertebrates.

Ichthyosaurs

Ichthyosaurs or 'fish lizards' (Fig. 14.25) were fish shaped reptiles that dominated seas for about 245 million years at the time of the dinosaurs and were quite successful. Ichthyosaurs comprised 80 species that ranged in length from about 1.0 m to 15.0 m. The group evolved from a terrestrial ancestor as is evidenced by the presence of two pairs of limbs. Ichthyosaur fossil remains have been found throughout the world. Ichthyosaurs were air-breathing reptiles that evolved a fish-like body, tail fluke that lacked bones and a dorsal fin. Their stocky legs evolved into flippers. There is a missing link between the fossil remains of early ichthyosaurs that would have linked them to their terrestrial ancestors. The fossil remains of some of the oldest ichthyosaurs found belong to

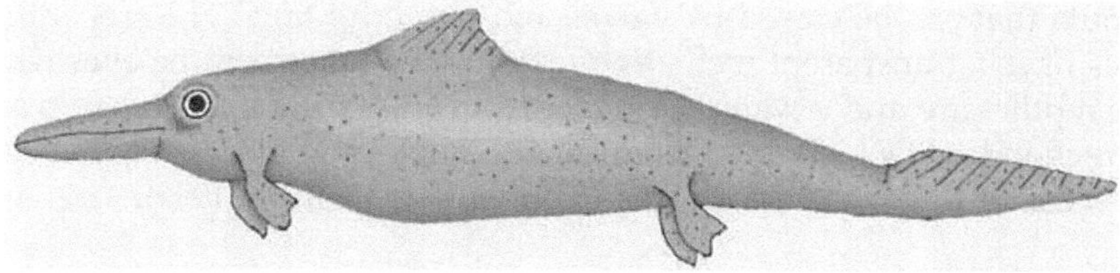

Fig. 14.25 An ichthyosaur ('fish lizard'). Ichthyosaurs lived in water and evolved slightly earlier than dinosaurs around 250 million years ago and were extinct by 90 million years ago. The largest ichthyosaur exceeded 15 m in length while the smallest was about 70 cm long.

Utatsusaurus that was discovered on the northeastern coast of Japan's main island, Honshu, and ***Chaohusaurus*** in China. The outline of *Utatsusaurus* was that of a lizard-like creature with flippers. Scientists believe ichthyosaurs evolved from diapsids at the time lepidosaurs and archosaurs were beginning to evolve.

Crocodiles

Crocodiles evolved from one of the two evolutionary lines of archosaurs about 220 million years ago during the Triassic period. The discovery of well preserved fossil remains of an ancestor to crocodilians in northwestern China known as *Junggarsuchus sloani* that is one meter long shows that the powerful jaws of the crocodile evolved on land much earlier, before crocodiles moved to water. Physiological, anatomical and developmental features of the four chambered crocodilian heart support the paleontological evidence that the ancestors of current crocodilians including stem archosaurs and some dinosaurs were active and endothermic but the crocodilian group reverted to ectothermy when it entered the aquatic environment (Seymour et al., 2004). At that time, the legs and body of crocodiles were still evolving towards walking on land and not swimming. The ancestors of crocodiles walked using their hind limbs that are still longer than the front legs in the living crocodilians that slant forward when they stand.

The earliest crocodiles were known as **protosuchians** and lived at the end of Triassic up to early Jurassic. Their hind legs were highly developed and were placed at right angles to the body. At the beginning of Jurassic, the early crocodiles moved to water and diversified. Protosuchians evolved into **mesosuchians** that lived to beyond Jurassic. The modern crocodile evolved alongside the mesosuchians during Cretaceous and the two groups of reptiles lived together in freshwater though some adapted to seawater. The modern crocodile then displaced mesosuchians during the early Cenozoic era. Crocodiles adapted to an aquatic environment by evolving webbed feet, nostrils on top of their snouts that can be closed by valves and powerful tails. The crocodilian eyes have a transparent nictitating membrane that covers the eyes when the reptiles are under water. Evidence from analysis of nuclear genes and mitochondrial DNA links crocodiles with turtles and places squamates at the base of the evolutionary tree (Hedges and Pauling, 1999).

Plesiosaurs

Plesiosaurs (Gr. *plesios*, near, *sauros*, lizard) (Fig. 14.26) were large and carnivorous aquatic reptiles. The smallest plesiosaurs were about 2.0 m

Fig. 14.26 A drawing (top) and fossil remains (bottom) of a plesiosaur. Plesiosaurs flapped their paddle-like flippers up and down to swim in water in a similar manner to turtles and penguins. The ventral body trunk of plesiosaurs had a series of gastralia (abdominal rib-like bones) that probably stiffened the body to improve on swimming effiency.

long while the largest were 12 to 15 m long. Plesiosaurs first appeared in late Triassic but became extinct at the end of Cretaceous. Plesiosaurs are believed to have resembled turtles with long snake-like necks and small heads. Other species of the aquatic reptile had large heads with strong jaws and short necks. There were two pairs of flippers that were previously limbs in their terrestrial ancestors and a short tail. Plesiosaurs could have evolved from nothosaurs or pistosaurs. The necks and tails of **nothosaurs** ('false lizards') were long whereas their four paddle-like limbs were webbed and possessed fingers and toes. The reptiles measured up to 6 m long and had thin and long heads with many sharp teeth. Nothosaurs flourished during the Triassic period and became extinct at the end of the period. **Pistosaurs** had a nothosaur-like body with a plesiosaur-like head and resembled crocodiles.

BIRDS

Bird fossil records are scanty as their bones are light and hollow making them decompose or be eaten by scavengers before they are entrapped in sediments. The picture of avian evolution is changing and becoming clearer as more fossil remains are discovered. The transition between reptiles and birds is represented by fossils of *Archaeopteryx* (Fig. 14.27) that were discovered in Germany in 1860 and date back to about 150 million years ago. Remains of another bird that resembled *Archaeopteryx* in having wing claws and almost as old were recently discovered in

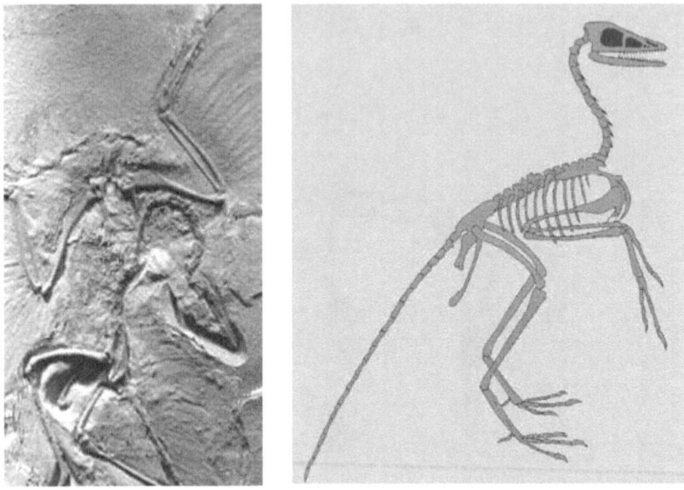

Fig. 14.27 Fossils (left) and skeleton (right) belonging to *Archaeopteryx lithographica* (Gr. *archaio*, ancient; *pteryx*, wing), the Jurassic bird. The remains of this extinct bird that had teeth and long feathers and lived during late Jurassic is thought to be the earliest of all birds and was a dead end evolution as it did not evolve into the modern bird. All fossils belonging to *Archaeopteryx* were dug from lithographic limestone quarries of Bavaria, Germany. The figure shows the Berlin specimen that was discovered in 1877.

China and belonged to *Confuciusornis* that lacked teeth. *Archaeopteryx* weighed about 2.0 kg and had both reptilian and avian features. The reptilian features included pre-maxillae and maxillae that were not covered by horn, non-fused trunk vertebrae, small teeth in jaw sockets, a long tail with 20 vertebrae and an archosaurian pelvic girdle. The neck of *Archaeopteryx* attaches to the skull from the rear and not below as is the case with birds. The avian features of *Archaeopteryx* were feathers, paired clavicles that formed a furcula ('wishbone'), an elongate pubis that is directed backward and a foot with three digits pointing in front and one digit facing backwards. *Archaeopteryx* could be a relative of the ancestors of modern birds and was probably an evolutionary dead-end.

Several non-avian theropod dinosaur fossils with feathers that resemble avian feathers have been discovered in China (Ji et al., 2001; Xu et al., 2001; Norell et al., 2002) and the U.S.A. (Kundrat, 2004) and this shows that feathers evolved in dinosaurs before the emergence of birds. The presence of such feathers and the independent phylogenetic evidence supporting the therapod ancestry of birds strongly corroborate the hypothesis that the integumental appendages in a therapod are homologous with avian feathers (Xu et al., 2001). Lingham-Soliar (2003) urges caution about these theropod feathers since collagen from the hypodermis (blubber) and sub-dermal connective tissue sheath from a dolphin that had been buried for a year was degraded and showed feather-like patterns that were similar to those described as proto-feathers or feathers of the advanced theropod (dromaeosaurid) dinosaurs.

Small theropods were two-legged dinosaurs that had many characteristics in common with modern birds. Theropods had hollow bones, a wishbone, a backward inclining pelvis and three toes per foot. With time, the forelimbs of theropods evolved into longer structures. The wishbone served to support the pectoral muscles that were important in bringing the forelimbs together when catching prey. Theropods might have never been capable of flight. The role played by feathers in theropods is not clear. After *Archaeopterix,* the early diversification of birds was rapid as is evident from the fossil record.

The first living birds to evolve were paleognathae (Fig. 2.14). Many related groups have become extinct. Paleognaths settled into their niches thus slowing down their evolutionary rates. Paleognath fossils of the Mesozoic era are better preserved when compared to those of neornithes whose fossils are even more rare and the reason could be biological rather than geological (Fountaine et al., 2005). Neornithes evolved in Gondwana before the Cretaceous-Tertiary mass extinction event (Cracraft, 2001). The birds then radiated into Laurasia later from Gondwanaland. Phylogenetic studies have shown that the earliest neornithes were heavy-bodied, ground-dwelling, non-marine birds, a fact that could explain the large gap in the early fossil record of birds since fossilization is favored in marine environments (van Tuinen et al., 2000).

FLIGHT

The three main vertebrate groups that have evolved flight include pterosaurs, birds and bats and this shows that flight has evolved three times independently in vertebrate history. The evolution of flight in these major vertebrate groups is a result of convergent evolution. There are other groups of vertebrates that glide using membranes that stretch between the body wall and limbs. Other vertebrates such as the Malay frog flatten their bodies and stretch their limbs to glide. Although evolution of flight has its advantages such as evasion of flightless predators, raising young in an environment that is inaccessible to most other vertebrates and being able to access high niches, there is a cost element involved. Flying is expensive energetically and flying vertebrates have had to evolve certain body shapes and also undergo reduction of body weight. The origin of flight in vertebrates in late Jurassic and Cretaceous happened at a time when levels of atmospheric oxygen were elevated as the high levels of oxygen facilitated aerodynamic force production and enhanced muscle power output (Dudley, 2000).

The three types of flight that have evolved in vertebrates include parachuting, gliding, powered flight and soaring (see Chapter 4). Many theories have been advanced as to how flight evolved. The reasons for evolution of wings can explain why flight came into being. Wings must have evolved for certain reasons before flight occurred. Wings could have evolved from arms that were used in capturing prey, in bipedals that were leaping, for sexual display and in gliding ancestors after they started flapping their forelimbs. Wings could have also evolved for fighting as is seen in birds today. The question of how flight evolved is a difficult one to answer. One theory states that bipedal ancestors to flyers started running fast enough while at the same time flapping the forelimbs until a time they were able to take off the ground. Another theory is the arboreal hypothesis where ancestral vertebrates evolved flight by jumping from one tree to another.

Flightlessness has evolved many times in 13 orders of birds. Flightless birds have evolved from flying ancestors. Evolution of flightless birds has occurred in areas that are geographically isolated and have a relative absence of terrestrial predators. Several structural changes are associated with flightlessness. Flightless birds tend to be big, lack a keeled sternum and have a reduction of muscle and bones of wing and pectoral girdle (see Chapter 2). Such modifications serve the purpose of saving energy. Flightless birds show arrested embryonic development or neoteny whereby there is delayed somatic development while the reproductive organs develop at the normal rate.

HOMEOTHERMY

Since the modern reptiles are ectothermic, extinct reptiles are thought to have had a similar condition. Transition from the ectothermic state to endothermic (homothermic) state of birds and mammals is believed to have occurred in later reptiles. Being endotherms, birds and mammals spend a major portion of their energy resources on maintenance of high body temperatures and metabolic rates. The mechanisms behind the evolution of endothermy remain unclear. Lately, two models that link evolution of high metabolic rates of birds and mammals to evolution of intensive parental care have been proposed (Koteja, 2004). Birds and mammals evolved endothermy independently. Fur and feathers are unlikely to have evolved in direct association with elevated metabolic rates in early birds, mammals or their ancestors (Ruben and Jones, 2000).

Evolution of endothermy will remain obscure since few anatomical and physiological features associated with endothermy are preserved in fossils and the presence of nasal turbinates that are found in extant ectotherms and in some non-mammalian synapsids might provide some insight into the metabolic physiology of extinct vertebrates (Hillenius and Ruben, 2004). Dinosaurs are believed to have had metabolic rates that are intermediate between ectotherms and endotherms and there are several reasons that tend to support this idea.

Dinosaurs were generally big animals with the largest being estimated to have weighed about 100 tones. The small surface to volume ratio in such vertebrates meant that heat could be conserved internally. The development of a secondary palate meant that dinosaurs could chew food and breathe at the same time since endotherms require more food to maintain a stable body temperature and high metabolic rates. Dinosaurs also supported their body weight on their limbs. Such a posture is only seen in endothermic birds and mammals. Supporting body weight on limbs requires muscle tone and control that is believed to be possible by endothermic regulation. Saurischian dinosaurs were bipedal and such a stance is found in endotherms.

In **fibrolamellar bone,** osteocytes are arranged like canals known as the Haversian system. The Haversian system is highly vascularized and is found in birds and mammals. Fibrolamellar bone is thought to have been present in dinosaurs, pterosaurs and advanced synapsids. Fibrolamellar bone is associated with high growth rate that is possible with high metabolic rates. Fibrolamellar bone is associated with endothermy while **lamellar-zonal bone** is found in amphibians and most reptiles. In lamellar-zonal bone, compact bone is laid down by relatively few osteoblasts at the surface. Such bone is poorly vascularized and has a layered appearance that shows incremental growth lines. Fibrolamellar bone is found in some extanct ectotherms including fast growing turtles, crocodilians and lizards. Mammals have left a good fossil record since the appearance of mammal-like reptiles. The fossil record shows the improvement of jaws, teeth and stance as well as appearance of a secondary palate. These features are associated with high metabolic rates and activity.

MAMMALS

Mammals evolved from the reptilian group known as synapsida. Synapsids had a pair of openings or infratemporal fossae in their skulls behind the eyes that increased the surface area for attachment of muscles of mastication. An earlier group of synapsids known as pelycosaurs evolved into therapsids. Therapsids were mammal-like reptiles that dominated the Earth before the arrival of dinosaurs around 225 milllion years ago. Therapsids had their limbs under their bodies and tucked-in elbows that enabled these mammal-like reptiles move on land better than the other reptiles. The metabolic rate of therapsids was higher than that of reptiles since these mammal-like reptiles were endothermic and had to use extra energy to maintain their upright stance. Some therapsids were as large as jackals and co-existed with the dominant dinosaurs.

The early evolution of therapsids gave rise to a group of other mammal-like reptiles known as **cynodonts** ('dog teeth') (Fig. 14.28). The dentition of cynodonts was more specialized and included molars that were used for grinding food in comparison to the earlier reptilian type which comprised of teeth that were designed for catching and holding prey before it was swallowed whole. Later on with time, cynodonts evolved deciduous and permanent teeth. The permanent teeth later evolved into specialized types that served specific functions. There was also a reduction in the number of jawbones resulting in evolution of more middle ear ossicles and improved hearing ability by cynodonts. The presence of a secondary palate in the roof of the mouth enabled cynodonts to eat and breathe at the same time. Cynodonts had whiskers, a sign that they were probably warm-blooded as a result of the presence of hair. At about 200 million years ago, the dominance of terrestrial life by cynodonts was reduced by the diversification of dinosaurs. The size of cynodonts also decreased as they confined themselves to nocturnal activity and underground or rocky habitats. Cynodonts later on became extinct during mid-Jurassic and are believed to have given rise to true mammals. Most of the fairly complete cynodont fossil remains were discovered in South Africa.

Evolution of mammals first occurred during the Triassic at a time when the first dinosaurs appeared and was followed by radiation of the

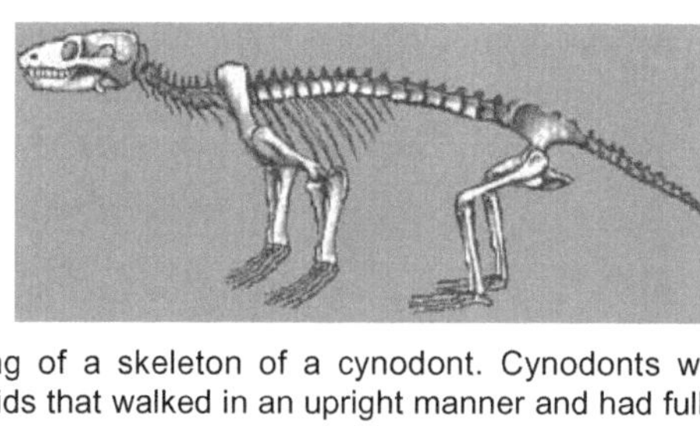

Fig. 14.28 Drawing of a skeleton of a cynodont. Cynodonts were herbivorous and carnivorous therapsids that walked in an upright manner and had fully differentiated teeth. The egg-laying cynodonts were the most successful and one of the most diverse groups of therapsids. A single group of cynodonts evolved into mammals. Some cynodonts were the size of the domestic cat while others were as big as the wolf. The number of bones present in the jaws of cynodonts reduced with some becoming part of the inner ear.

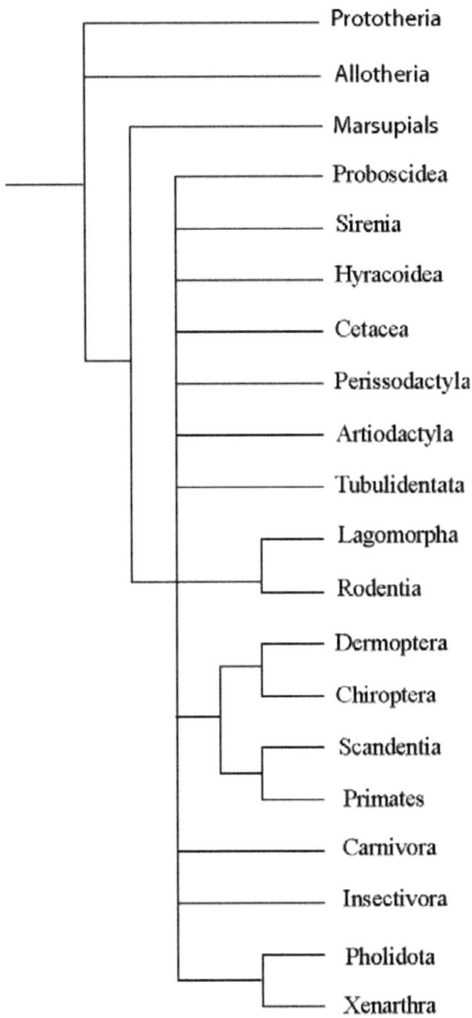

Fig. 14.29 A mammalian evolutionary tree. The evolutionary tree continues to change as more information is gathered about the interrelationships between various vertebrates. Mammal-like reptiles known as cynodonts that evolved from therapsids but became extinct around middle Jurassic are belived to have eventually given rise to mammals.

group into various forms (Fig. 14.29). The earliest mammals were quite small and are thought to have been shrew-like mammals that measured about 2 to 3 cm long. Fossil remains of these early mammals have been found in caves in Wales and around Bristol in the United Kingdom. These early mammals were nocturnal and endothermic.

Later mammals resembled rodents and had long snouts and external ears suggesting that they had a good sense of smell and hearing that was important since these animals were nocturnal and had small eyes. The mode of life of the early mammals could have been similar to that of tenrecs (Fig. 14.30) of Madagascar and the Comoro islands. The nocturnal tenrec maintains a body temperature of about 28°C to 30°C. Since this temperature is much lower than that of most living mammals, it enables the tenrec to maintain its normal body temperature at night without having to spend a lot of energy. The early mammals probably fed at night and avoided the dinosaurs that could not see them or were rendered less active by the low temperatures. The presence of whiskers enhanced the sense of touch of the early mammals. The senses of touch, smell and hearing together with spatial imagination that resulted from a more developed brain enabled the early mammals to know their surrounding environment at night so that they could remain within their territory even though they were not able to see clearly at such a time.

Among the earliest mammals to evolve were ancestors of Allotheria and Prototheria about 190 million years ago. Among **allotherians** (Gr. *allos*, other) were **multituberculates** that first appeared in middle Jurassic and whose molars had many cusps (tubercles) that were arranged in rows. Multituberculates were rodent-like mammals that lived in trees and increased in weight after the extinction of dinosaurs from the size of

Fig. 14.30 The common (greater) tenrec, *Tenrec ecaudatus*. Tenrecs comprise 23 species that are insectivores and are relatives of the African pygmy hedgehog. Tenrecs have a long snout and their backs bear sharp spines. Most tenrecs measure between 140 to 180 mm and weigh from 110 to 250 g. The tailless common tenrec is about 40 cm long and is one of the largest insectivores.

a mouse to that of a beaver. Allotheria evolved independent of other mammals for a long time and were the most diverse and numerous mammals of the Mesozoic era. Extinction of multituberculates in early Oligocene coincided with the evolution of rodents. With no living descendants left, the omnivorous multituberculates are the only branch of mammals to become completely extinct. Many skull fossil remains of multituberculates that lived during the Cretaceous period have been found in the Gobi desert of Mongolia.

The only living representatives of **prototheria** (Gr. *pro*, before) are the monotremes (Fig. 14.31). Montremes branched off the early mammals around 130 million years ago perhaps in Australia. Mammals had by now evolved their distinguishing characteristics. Prototheria had hair over

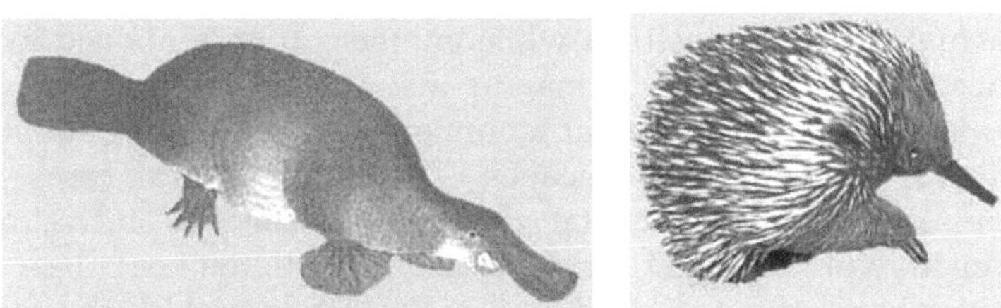

Fig. 14.31 Prototherians are represented by the living duck-billed platypus (left) and two species of echidnas or spiny anteaters (right). Other members of the egg-laying mammal group became extinct in the course of evolution. The ancestors of prototherians first appeared about 190 million years ago and were among the earliest mammals to evolve. The multituberculates and monotremes both have a parafibula bone.

their bodies and fed their young on milk although they lacked nipples.

About 125 million years ago, a line of non-egg laying mammals gave rise to marsupials. Later on, another line branched off the non-egg laying mammalian line as eutherian mammals. The two groups of mammals evolved in two increasingly different directions. The nervous system of mammals evolved at the same time they co-existed with dinosaurs and enabled mammals occupy ecological niches where they could outcompete dinosaurs. Mammals were able to survive conditions that led to the extinction of dinosaurs since they were able to adapt better than dinosaurs. After dinosaurs became extinct about 65 million years ago, mammals moved into niches that were previously occupied by these large reptiles and diversified greatly. The age of mammals perhaps would not have occurred had they relied on a more highly developed brain alone.

Marsupials (Fig. 14.32) evolved in North and South America, Antarctica and Australia. Evolution of marsupials was affected by the drifting of the continents. When the super-continent Pangaea broke apart

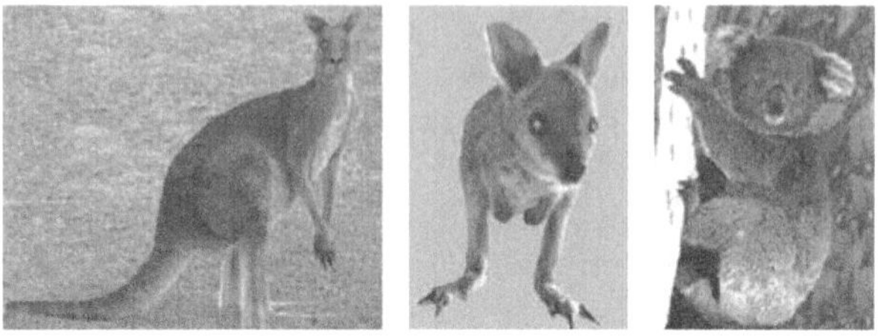

Fig. 14.32 Some marsupials. Kangaroo (left), wallaby (middle) and koala (right). A female marsupial has a pouch (marsupium) in which the relatively undeveloped young, after a short gestation period of four to five weeks, undergoes further development. Kangaroos are the largest marsupials. Wallabies resemble small to medium size kangaroos though stockier in build.

into the current continents, one group of marsupials was isolated on the continental island of Australia while another group remained in South America. The two groups of marsupials evolved in parallel for some time. Marsupials and placental mammals evolved similar adaptations despite forming separate lineages. There are several cases where marsupials resemble placental mammals physically including the Tasmanian 'wolf' or 'tiger' that resembles a wolf, the koala bear which resembles bears and the marsupial mole that resembles the common mole. These animals belonging to major groups of mammals resemble each other as a result of convergent evolution.

Towards the end of Cretaceous, placentals that included herbivores, omnivores and carnivores arrived in South America. Much later, about three million years ago, the lowering of the ocean level created the Panama land bridge. There was migration of mammals between North and South America. The migration of placental carnivores including the saber-toothed cat from North to South America is responsible for the decimation of marsupial carnivores.

Mammalian herbivores first evolved during the late Cretaceous about 100 million years ago. During the Tertiary, medium sized herbivores showed significant radiation. The teeth of herbivores broadened, became extremely hard and high crowned in order to be able to digest plant material that was fibrous. The claws were unnecessary and evolved into hooves as the digestive system adjusted to handling material that was low in nutrients. Later on (about 60 million years ago), the ungulates split into the whale and dolphin and elephant lineages as well as the even-toed artiodactyls and the odd-toed perissodactyls (Fig. 14.33). Some artiodactyls evolved the fore-stomachs, the largest one being the rumen. This stomach stores forage and partially breaks it before sending it back to the mouth for further chewing. Artiodactyls comprise about 210 species that include cattle, antelopes, pigs, hippopotamus,

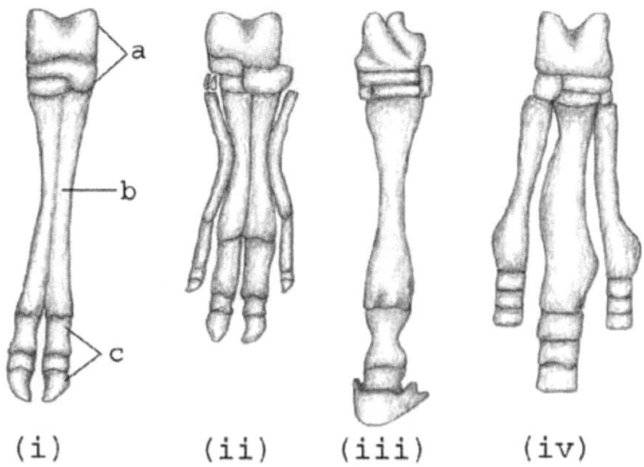

Fig. 14.33 Parts of limbs of an ox (i), pig (ii), horse (iii) and rhino (iv) showing the bones of the distal parts. (a) tarsal bones, (b) fused metatarsal bones and (c) phalanges that make up toes (digits). The ox and pig are artiodactyls (even-toed) whereas the horse and rhino are perissodactyls (odd-toed). The toes in these ungulates each have three (proximal, middle and distal) phalanges. The distal phalanges make contact with the ground and are covered by hooves. The most proximal tarsal bone of each species in the diagram is the talus or astragalus.

camels, giraffes and the deer. The phylogenetic relationship among all 197 species of extanct and recently extinct ruminants has been determined using morphological, ethological and molecular information (Hernandez and Vrba, 2005). Perissodactyls have only 16 species of which the horse is the only exclusive grazer. Other perissodactyls include the rhinoceros, tapirs, zebras and donkeys. The appearance of grazers was followed by evolution of predators for most of the herbivores apart from the large elephants, rhinoceros and their relatives. Herbivores had to evolve long legs for running and elongated skulls that enabled the eyes to see predators while they grazed or browsed.

With about 400 species, **Insectivora** is the third largest mammalian order. Insectivores include moles, shrews, solenodons and the spiny hedgehog (Fig. 14.34). Insectivores feed on insects mainly but also eat other invertebrates and some vertebrates such as fish and lizards. Insectivores are not found in Australia, most of the northern part of South America and the polar region. The teeth of insectivores are unspecialized. Insectivores could have evolved in Europe and Asia before spreading to other parts of the world. The earliest fossils of the group are about 100 million years old and show that insectivores have not changed much in the course of evolution. Since the skeletal remains of insectivores are delicate, their teeth and jaws are the only structures that are preserved and are not easily gathered because of their small size. During Paleocene,

Fig. 14.34 Insectivores. A mole (top left), solenodon (top right), shrew (bottom left) and hedgehog (bottom right). Insectivores are small mammals that resemble rodents. The smooth brain of insectivores is relatively smaller than that of many mammals and the brain case is flat. Insectivores have a long snout and a cloaca. The smallest mammal, the pygmy white-toothed shrew, is an insectivore and weighs 1.2 to 2.7 g. Several species of shrews and solenodon are venomous.

insectivores formed a great proportion of the faunas of North America, Europe and northern Africa. Studies show the existence of a great diversity of sensorimotor specializations of the brain among insectivores adapted to a number of different ecological niches indicating that there has been significant diversification and change in their course of evolution (Catania, 2000).

Rodents (L. *rodere,* to gnaw; *dens* or *dentis,* tooth) (Fig. 14.35) comprise about 2,000 species and are the largest order of placental mammals. The family **Muridae** that comprises rats, mice, voles, hamsters, muskrats and gerbils among other rodents has more than 1100 species. Rodents are native to all continents except Antarctica and are also found on most islands and all habitats on Earth apart from oceans. With the exception of bats, rodents are the only placentals to reach Australia without human introduction.

The common ancestor of rodents separated from other placental orders in the late Cretaceous and during radiation of Rodentia, at least three lineages (Gliridae, Sciuridae and Ctenohystrica) emerged close to the Cretaceous–Tertiary transition time (Huchon et al., 2000). The group Gliridae contains rodents that were indigenous to the Old World with most inhabiting Europe while others are found in Africa and Asia. Gliridae or dormice are small rodents that measure about 70 mm in length and consist of 28 species. Sciurids are a large family of squirrels that are found on all the continents except Australasia (Australia, New Zealand, New Guinea and other neighboring smaller islands). There are

Fig. 14.35 Some members of the order Rodentia. Capybara of South America (top left), squirrel (top right), beaver (bottom left) and porcupine (bottom right). Rodents have two upper and two lower incisors that continue to grow throughout life and have to be worn down by gnawing. The capybara is the world's largest rodent and can weigh up to 40 kg. Beavers weigh 13 to 34 kg and have strong teeth for cutting down trees for making dams and building lodges. A porcupine has quills over its body. Quills are stiff hairs with barbed tips at the end and form part of a porcupine's defense system against predators. Rodents and lagomorphs are also known as glires.

48 genera and 280 species of squirrels. Most rodents of the suborder Ctenohystrica have enlarged infraorbital canals and include the capybara, gundis and porcupines.

Rodents appear in the fossil record towards the end of Paleocene about 54 million years ago and are believed to have originated in Asia. Since rodents share multiple uniquely derived characteristics, they are considered a monophyletic group though the relationship between the 11 families that belong to the suborder **Sciurognathi** has not been fully resolved. The other 18 families of rodents belong to the suborder **Hystricognathi**. Ancestors to rodents could have been the rabbit like anagalids that probably gave rise to lagomorphs. The origin of rodents is not clear since the earliest remains of the group in the fossil record are highly transformed in appearance and earlier intermediate forms have not been found. The uncertainty regarding the position of the rodent root reflects the rapid rodent radiation that occurred in the Paleocene rather than the presence of conflicting phylogenetic and non-phylogenetic signals in the data set (Blanga-Kanfi et al., 2009).

Rodents evolved as plant eaters that relied on bacteria in the gut to break down cellulose. They fed on open roots with their front gnawing teeth (incisors) that grow continuously throughout life to compensate for wear. The incisors have a thick layer of enamel at the front but not at the back part of the tooth and this is responsible for the chisel shape of incisors as they are worn down with gnawing. Rodents lack canine teeth and there is a diastema between incisors and molars.

The greatest diversity of form in rodents is found in South America, a continent that remained isolated for much of the Cenozoic. The major radiation of the South American rodents occurred during the Oligocene and by the Miocene squirrels had evolved. The largest rodent that ever lived so far is *Phoberomys pattersoni* whose fossil remains were discovered in a rich locality of fossil vertebrates in the Upper Miocene of Venezuela (Sanchez-Villagra et al., 2003). *Phoberomys* is estimated to have weighed 700 kg. Another genus with large rodents that became extinct during Pleistocene is *Castoroides* (giant beaver). Giant beavers attained a length of 2.5 m and weighed up to 220 kg making them the largest rodents that ever lived in North America. The extanct rodent with the most archaic characters is the sewellel or mountain beaver of northwestern USA.

Lagomorphs (order Lagomorpha) (Fig. 14.36) include rabbits, hares and pikas and could have originated in northern Asia about 55 million years ago. Lagomorphs have not changed much morphologically in the last 40 million years ago when the fossils of the group became well documented. Rabbits and hares (family Leporidae) moved to North America at end of Eocene and moved to Europe by Pliocene. The pika family (Ochotonidae) spread to Africa, Europe and North America during early Oligocene and are native to the cold climates of mainly Asia, North America and parts of Eastern Europe.

The order **Proboscidea** (animals with trunks or proboscis) is represented by the living two species of the African and one species of Indian elephants (Fig. 14.37). Other members of the group became extinct

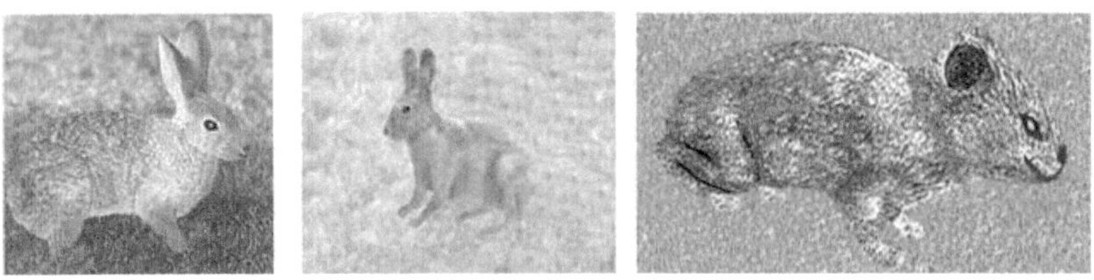

Fig. 14.36 Lagomorphs. Rabbit (left), hare (middle) and pika (right). Lagomorphs have four upper and two lower incisors that grow continuously throughout life thus necessitating constant chewing to stop them from growing too long. Rabbits are sometimes affectionately referred to as bunnies. Many rabbits live underground in burrows and differ from related hares in giving birth to young that are altricial (blind and hairless). Pikas are also known as rock rabbits or coneys and produce high-pitched alarm calls thus the name 'whistling hares'.

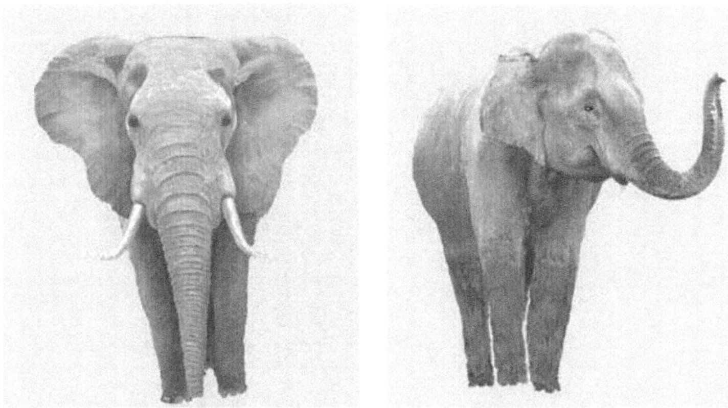

Fig. 14.37 African (left) and Indian (right) elephants. The African elephants are larger than their Indian counterpart making them the largest animals on land. The African elephants can attain weights of 7,500 kg and a height of 3-4 m while the Indian elephant can weigh up to 5,500 kg with a height of 3 m. The upper incisors of elephants grow into tusks that are usually absent in the female Indian elephant.

possibly due their inability to evolve to environmental changes fast enough. Proboscidans evolved about 55 to 60 million years ago possibly from mammals the size of the living pigs known as moeritheres that lived in northern Africa. Proboscidans then occupied all continents of the world with the exception of Australia and Antarctica. According to morphological and biochemical evidence, sirenians (manatees and dugongs) and hyraxes are the closest living relatives of the extanct elephants. Phylogenetic analyses show that the mammoth was more closely related to the Asian rather than the African elephants and the divergence of the three groups of elephants occurred over a short time equivalent to about 7% of the total length of the phylogenetic tree for the three evolutionary lineages (Krause et al., 2006). The three groups of proboscidans belong to the family Elephantidea that originated in Africa.

Hyraxes (Fig. 14.38) belong to the order **Hyracoidea** and most are endemic to Africa with the exception of the bush hyrax that is also found in Sinai and the rock hyrax of Lebanon and Saudi Arabia. The hyrax fossil deposits from the Fayum, Egypt, show that hyraxes were the most important medium-sized ungulates in Africa about 36 million years ago. During this time, hyraxes were quite a diverse group that ranged from animals the size of living hyraxes to a hippopotamus and included semi-aquatic groups. Around 25 million years back, as the radiatiation of bovids was occurring, the diversification of hyraxes was considerably reduced and the group was confined to rocks, trees and bushes. The most notable living hyrax is the rock hyrax, *Procavia capensis*. Living hyraxes still retain primitive features such as imperfect endothermy, cropping of

Fig. 14.38 A hyrax. There are 11 species of hyraxes that belong to the order Hyracoidea. Hyraxes are about the size of the domestic cat with thick fur and are herbivorous in nature. Hyraxes may be the closest living relatives of elephants. Despite the great contrast in the sizes of elephants and hyraxes, the skeletons of these two groups of animals such as skulls and feet are quite similar in structure. Other features that are common between the two groups of animals include the gene sequence, common eye lens proteins, amino acid sequences in hemoglobin and a social structure.

forage using molars instead of incisors and retention of short legs and feet. Hyraxes have plantigrade feet that have hoof-like nails on toes. There are four toes on the fore foot and three toes on the hind foot of hyraxes.

The smallest mammalian order is **Tubulidentata** (L. *tubulus,* small tube or pipe; *dens,* a tooth; *atus,* provided with) and comprises aardvarks or ant bears Fig. 14.39). The earliest aardvark fossils were found in Europe and date back to the Eocene epoch. Aardvarks are found in sub-Saharan Africa and are solitary and nocturnal hoofed mammals that spend the day in a burrow. Aardvarks are considered to be the most primitive living protoungulate. Another small mammalian order is **Pholidota** that has seven species of pangolins (Fig. 14.40) or scaly anteaters. Pangolins are found in tropical and subtropical areas of Africa and Asia. Fossils that resemble the living pangolins have been found in Germany and North America and date back to Eocene and late Oligocene respectively.

The order **Xenarthra** ('strange joints') was formerly known as Edentata ('without teeth') and includes armadillos, two-toed and three-toed sloths and anteaters (Fig. 14.41) that together comprise 29 species of mainly insectivores and herbivores. The earliest fossil remains of xenarthrans date back to about 60 million years ago. Many large xenarthrans such as the giant ground sloths have become extinct. Some

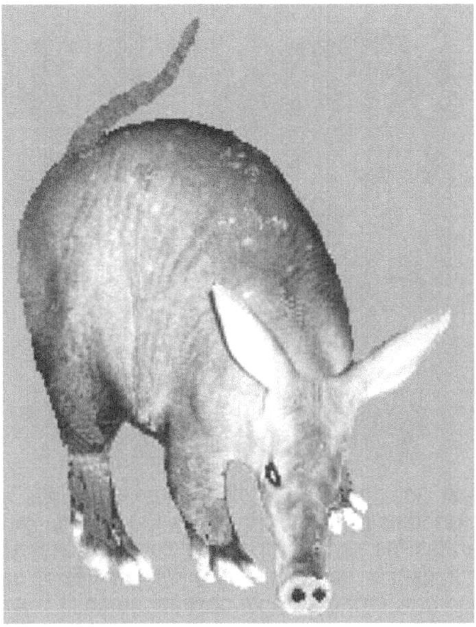

Fig. 14.39 An aardvark (Afrikaner for 'earth hog or pig') or ant bear. The toes of aardvarks are covered by horny material that is an intermediate between claws and hooves. The pulp of an aardvark tooth is tubular and is surrounded by numerous hexagonal prisms of dentine thus the name of the order Tubulidentata. Incisors and canines are lacking in adult aardvarks. Although aardvarks have poor eyesight and are color blind, they have an acute sense of smell. Molecular studies show that the aardvark is a close relative of elephants, hyraxes and sirenians.

Fig. 14.40 A pangolin or scaly anteater. The back and sides of pangolins are covered by large overlapping scales that are a result of agglutinated hairs. Pangolins feed mainly on ants and have muscular tongues that are quite long and attach to the pelvis and last pair of ribs.

extinct xenarthrans such as the giant armored glyptodonts form a group of the largest and most heavily armored vertebrates that ever lived. Xenarthrans were more diverse in the past than at present. Members of the order radiated into South America when the continent had been

Fig. 14.41 Xenarthrans. An armadillo (top left), a three-toed sloth (top right) and the giant anteater (bottom). Xenarthrans lack incisors and canines. Premolars and molars may be present in sloths and armadillos but lack an enamel cover and are cylindrical in shape. The two-toed sloth has two fingers on each fore foot and three toes on each hind foot. In moist conditions, the brownish-gray fur of the sloth appears green in color due to the growth of symbiotic cyanobacteria. The green color camouflages the sloth in its forest environment and the bacteria also provide nutrients to the sloth when licked.

isolated from others during early Cenozoic. At Paleocene, when the land bridge joining North and South America formed, several groups of xenarthrans crossed to the north including armadillos and the extinct giant ground sloths. Within the order, anteaters and sloths are closest to each other phylogenetically. The two genera of sloths belong to two different families that adapted to an arboreal life-style independently (Greenwood et al., 2001). Anteaters totally lack teeth and other xenarthrans lack incisors and may have poorly developed molars. The vertebral joints of xenarthrans differ from those of other mammals since the last thoracic vertebrae and the lumbar vertebrae have extra joints and the pelvic bones articulate with the sacral vertebrae via the ilium and ischium.

Bats and colugos or 'flying lemurs' (Fig. 14.42) of Southeast Asia belong to the orders **Chiroptera** and **Dermoptera** (skin-winged) respectively and branched off the primate line more than 50 million years ago. Bats mainly originated in Laurasia, probably North America (Teeling et al., 2005). A major temperature rise at the time has been responsible for the evolution of the four major lineages of bats and coincided with a rise in plant diversity and the abundance of tertiary insect diversity (Teeling et al., 2005). Lineage diversification rate in bats has not been constant and the largest diversification rate shifts occurred 30 to 50 million years ago (Jones et al., 2005). The earliest bats were insect

Fig. 14.42 A bat (left) and colugo or flying lemur (right). Bats have evolved true flight using wings that result from modified pectoral limbs that support a skin membrane that attaches to the body. Flying lemurs are not lemurs but have lemur-like faces. These dermopterans also do not fly but glide between trees that could be 100 m apart using the patagium. Colugos are slow climbers and do not walk on the ground.

eating and their 50 million year old fossils show that these chiropterans were quite similar to the species of living bats, an indication that they had evolved at the time. During embryological development, cartilages of the third, fourth and fifth digits of the bat wings undergo proliferation and differentiation resulting in great increase in their length due to the presence of bone morphogenic protein 2, an indication that the protein could have played a major role in the evolutionary elongation of bat forelimb digits (Sears et al., 2006).

Bat fossils are not common since their bones are thin and fragile. The larger fruit-eating bats could have had a different origin from other bats and their earliest fossil remains date back to 35 million years ago. Fruit bats differ from insect-eating bats in the structure of their skulls, teeth, cervical vertebrae and pectoral limb skeleton. There are more than 1,000 species of bats today belonging to two sub-orders, the **Megachiroptera** or **Megabats** and the **Microchiroptera** or **Microbats**. Megabats or the Old World fruit bats are mostly larger than microbats and comprise of 170 species that are mostly the flying foxes. Microbats form the remaining species of mainly insect-eating bats that have an echolocating mechanism.

The nocturnal tree dwelling colugos or cobegos that are about the size of the domestic cat could have shared a common ancestor with bats and primates. The order Dermoptera has two species: the Phippine and Malayan colugos. Colugos have an average length of 25 to 40 cm with some being as long as 75 cm. The dermopterans have a broad membrane known as patagium that extends from the neck and body to the tips of fingers, toes and the tail. The patagium is used for gliding.

Tree shrews (Fig. 14.43), previously classified as insectivores, belong to the order **Scandentia** or **Tupaioidea** that has 1 family, 5 genera and

Fig. 14.43 A tree shrew. Although many tree shrews are arboreal, some live on the ground. Tree shrews are territorial and very active. The scandentians are in constant motion in trees, often fighting with each other as they scream. The length of the tail may be slightly shorter or much longer than the rest of the tree shrew body.

about 19 species. The omnivorous tree shrews are found in the deciduous forests of central and southeastern Asia and are related to colugos and primates. Scandentians are squirrel-like in external shape and appearance with long heavily furred tails but do not have the long vibrissae of squirrels. The auditory bullae of tree shrews are complete and the large orbits are bordered posteriorly by a complete postorbital bar. Tree shrews have a good sense of hearing and their forefeet have five toes whereas squirrels have four. The brain to body mass ratio of shrews is quite high when compared with that of other vertebrates including primates.

The ancestors of **marine mammals** (Fig. 14.44) lived on land so these mammals moved to water as a secondary adaptation. Fossil records show that marine mammals returned to water on at least seven separate occasions with five of the major groups still extanct while two are extinct (Uhen, 2007). Although whales and dolphins, seals and sea cows evolved from different terrestrial lineages, they have similar adaptation to living in an aquatic environment. The adaptations these marine mammals share include limbs that have been modified into paddle-like flippers, a thick layer of subcutaneous (hypodermis) fat known as blubber, a torpedo shaped body and various adaptations of internal organs to be able to cope with water depths and demands of diving. Marine mammals have left a poor fossil record making it difficult to know their terrestrial ancestors.

Recent genetic studies show that the closest relatives of whales and dolphins are artiodactyls such as the living ungulates that include cattle, camels and the hippopotamus. Dugongs and manatees are closer to elephants and aardvarks than any other mammal. Analyses of complete mitochondrial genomes that represent all extanct cetacean families show that morphological similarities between *Mesonychia* and Cetacea are a result of evolutionary convergence rather than a common ancestory (Arnason et al., 2004). *Mesonychia* is the presumed ancestor of artiodactyls and cetaceans. The loss of hind limbs in whales has resulted from

Fig. 14.44 Some marine mammals. Dolphin (top left), seal (top right) and walrus (below). The fossil record of marine mammals is poor. Although marine mammals originated from terrestrial mammals, they have evolved similar adaptations to living in water.

selective pressure acting on a wide range of developmental processes and adult traits (Bejder and Hall, 2002). Cladistic analysis also shows that cetaceans are more closely related to artiodactyls than to any mesonychian (Thewissen et al., 2001).

Members of the order **Sirenia** that include dugongs and manatees (Fig. 14.45) as well as the seagrasses they feed on are thought to have evolved in the Tethys Sea area (Fig. 14.58). The fossil record of sirenians can be traced back to about 50 million years ago and comprised about 50 species of the group. Today, only four types of sirenians comprising three manatees and one dugong species exist. On return to the sea, sirenians radiated to the tropical parts of the oceans except the ancestors of the extinct Steller's sea cow that adapted to the temperate waters of the North Pacific. The primitive sirenians that appeared early in Cenozoic were quadripeds that possessed adaptive characteristics to an aquatic environment including absence of paranasal sinuses, retracted nasal openings and dense and thick ribs.

The first **carnivores** evolved about 65 million years ago from the ancestral insectivores or the same ancestors that gave rise to primates and chiropterans and were known as **creodonts** (Fig. 14.46). The creodonts fed on insects away from the dinosaurs. Creodonts were the dominant

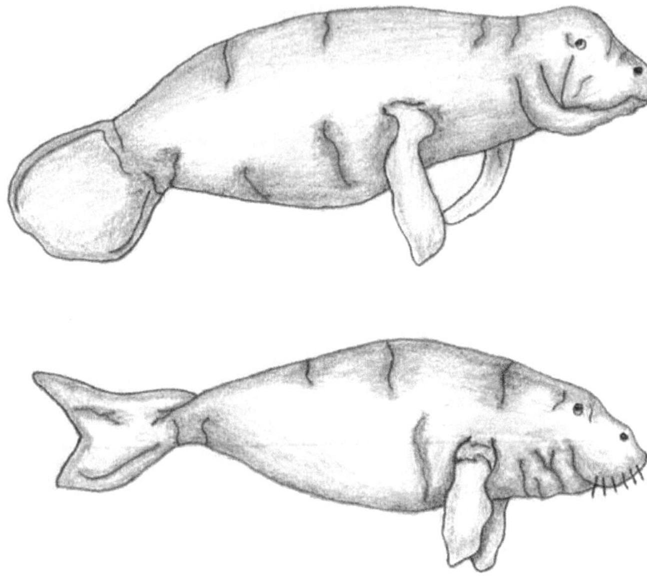

Fig. 14.45 A manatee (above) and dugong (below). Steller's sea cow or great northern manatee was hunted to extinction by 1768 and had adapted to the cold waters of the North Pacific. Sirenians are the only marine mammalian herbivores. Morphologically and molecularly, sirenians are grouped together with elephants and the extinct desmostylians that were hippopotamus-like mammals in a taxon known Tethytheria.

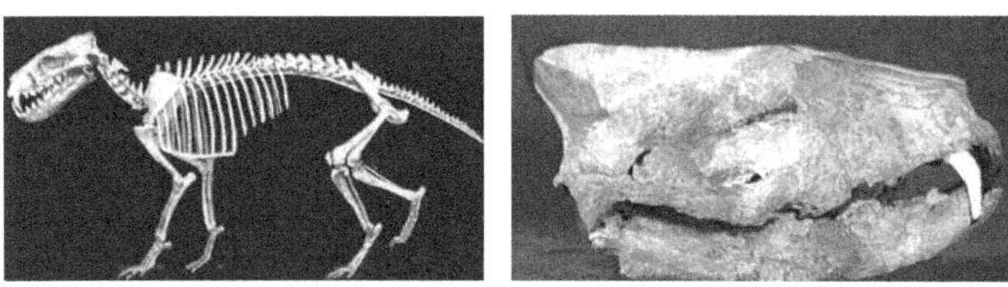

Fig. 14.46 The skeleton of a wolf-like creodont *Hyaenodon* (left) and the skull of a similar creodont (right). Creodonts had a relatively smaller brain than living carnivores and had a plantigrade stance. Living carnivores have a digitigrade stance. Creodonts ranged in size from the size of squirrels to to that of wolves although the largest creodont that ever lived, *Megistotherium*, is believed to have weighed around 800 kg and its fossils were discovered in the Sahara desert of Africa.

creatures in the world and lived in Africa, Europe, Asia and North America from about 55 to 35 million years ago. Living carnivores first appeared about 55 million years ago and eventually dominated the creodonts to become the top predators of the continents to the north from 30 to 20 million years ago and this is attributed to the evolution of the scissor-like carnassial teeth. Creodont carnassials were located far back in the mouth. Creodonts also lacked teeth behind their carnassials for grinding fibrous material. The presence of such teeth in carnivores diversified their feeding habits. Carnivores were able to diversify from their mainly carnivorous feeding habit as a result of such teeth. About 30 million years ago, a change in climate affected creodonts more than carnivores leading to a decline in numbers of the former. Creodonts became extinct about eight million years ago.

As the ungulates evolved their stance by supporting their weight on the tips of their digits that also resulted in an increase in their running speed, carnivores evolved claws for catching their prey and lengthened their leaps by having very flexible spines. The carnivore spine can bend to enable the hind and front limbs to overlap below the body. Some carnivores including dogs and lions learnt to hunt in groups. The cat and dog lines evolved about 55 million years ago from arboreal carnivores. The dog branch, Caniodea, and the cat branch, Feloidae, evolved in the New and Old Worlds respectively. Dog and cat branches evolved independently in these two major regions of the world until the Bering Straight formed a land bridge between America and Eurasia about 30 million years ago. The two branches of carnivores used this land bridge to migrate between America and Europe.

PRIMATES

Primate-like fossil remains first appeared about 60 million years ago after the beginning of the Cenozoic era (last part of Paleocene epoch) (Table 14.2). These remains belonged to members of the genus *Altiatlasius* and were found in geological deposits in Morocco. These primate-like mammals or **protoprimates** (Fig. 14.47) were similar to squirrels and tree shrews in appearance and size. Protoprimates had grasping hands and

Fig. 14.47 Reconstruction of a protoprimate or primate-like mammal, *Purgatorius*, that was the size of a small rat. Protoprimates resembled squirrels and tree shrews and are believed to be a transitional form between insectivores and primates. Protoprimates lived during early Paleocene. *Purgatorius* is thought to have been the size of a mouse and its fossils were discovered in present day Montana.

feet for climbing trees. During the Eocene epoch, as orders of placental mammals were beginning to appear, primates that resemble modern **prosimians** that include lemurs and lorises (Fig. 14.48) began to evolve. Figure 14.49 shows a cladogram of primate evolution. Most prosimian adaptive radiation occurred during the Eocene epoch with the appearance of about 60 genera of these primate ancestors. Remains of these pro-simians are found in Asia, Africa, and Madagascar, North America and Europe.

Table 14.2 The Cenozoia era and its epochs. To the right of the bar is the time scale in years. The current Holocene epoch started around 11 thousand years ago and is thought by some scientists to be one of the interglacial periods of the Pleistocene epoch. The Cenozoic era was formerly divided into Tertiary (66 to 2 million years ago) and Quaternary (2 million years ago to present) periods.

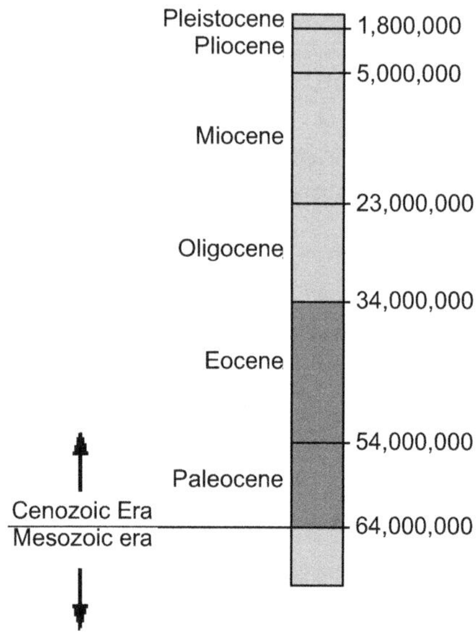

Structural changes occurred with time during the Eocene epoch. The brains and eyes of prosimians became larger as the snouts became shorter. Foramen magnum (L. *foramen*, opening, *magnus*, large) through which the spinal cord leaves the cranium to the spinal canal started to shift from a posterior to ventral position suggesting that the prosimians could partly support themselves upright on two limbs. Most prosimians had become extinct by the end of the Eocene epoch probably due to the appearance of the first monkeys and the cooler temperatures that prevailed.

Fig. 14.48 Some living prosimians. Black lemur (left), slow loris (middle) and tarsier (right). The black lemur is the size of a domestic cat and has a fox-like muzzle. The slow loris is nocturnal and spends the day sleeping up in a tree with its body rolled into a tight ball. Lorises weigh between 85 g and 1.5 kg while tarsiers weigh 80 to 150 g.

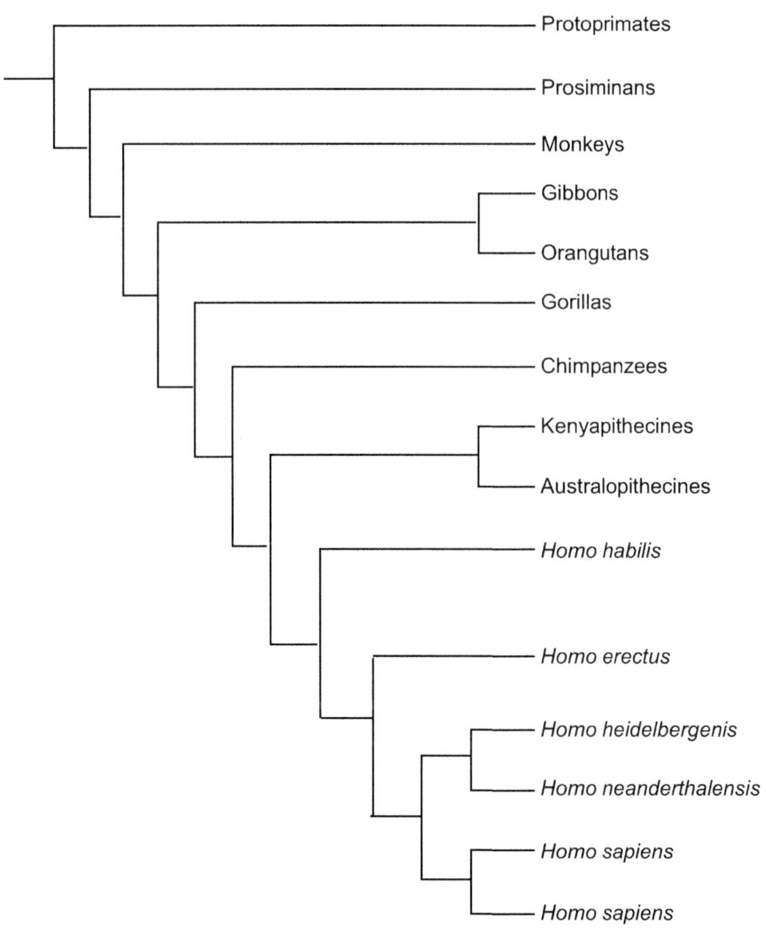

Fig. 14.49 The evolutionary tree of primates. All primates have five inward-closing fingers that have nails instead of claws. Apart from the human being, primates are mainly found in Central and South America, Africa and southern Asia.

Fig. 14.50 Some monkeys. Pygmy marmoset (top left), colobus (top right), baboon (bottom left) and mandrill (bottom right). The mandrill is a type of baboon. There are 264 known species of living monkeys in the world. Monkeys lack any unique traits that are specific to the group and are not present in apes. Most New World monkeys have prehensile tails, a trait that is lacking in Old World monkeys. Monkeys range in size from the pygmy marmoset of South America that is 14 to 16 cm long and weighs 120 to 140 g to the male mandrill of Africa that is about 1.0 m long and weighs 35 kg.

During the Oligocene epoch, few fossil remains of prosimians were present. Most of these remains are from the Fayum deposits in Egypt. The monkeys (Fig. 14.50) appeared during this epoch and had a larger brain, more forward looking eyes, fewer teeth and shorter snouts than prosimians. Monkeys could have out-competed the prosimians to replace them in most environments. The living prosimians such as lemurs, lorises and tarsiers are small creatures that inhabit the tropical forest. Most prosimians are nocturnal or live away from other primates since they may have been marginalized by rodents, monkeys and apes.

Apes (14.51) evolved from monkeys during the Miocene epoch and then continued to dominate the monkeys and the few prosimians present. Apes belong to the superfamily Hominoidea that is divided into the

Fig. 14.51 Apes. Mountain gorilla (left), orangutan (second left), commom chimpanzee (second right) and bonobo (right). Another ape is the small, long armed and very acrobatic tree dwelling gibbon of Southeast Asia. Apes are tailless and have long arms and broad chests. The common chimpanzee and bonobo (pygmy or dwarf chimpanzee) are the two species of chimpanzees and belong to the genus *Pan*. Chimpanzees are the closest living relatives of the human being and their genome is 95 to 98% that of *Homo sapiens*.

families Hylobatidae and Hominidae (hominoids). Hylobatidae comprises 13 species of gibbons that are found in the tropical and subtropical rainforests of Southeast Asia and are also referred to as 'small or lesser apes' whereas gorillas, orangutans, chimpanzees and humans belong to the family Hominidae and are known as the 'great apes'. Apart from gorillas and humans, all apes are agile climbers of trees. The arboreal gibbons are quite acrobatic with arms that are longer than their legs. The wrist of gibbons has a ball and socket joint between the thumb and one of the carpal bones (trapezium) (Fig. 14.52) that permits biaxial movement. Gibbons are able to swing from branch to branch in a movement termed **brachiation** for distances of up to 15 m and speeds of

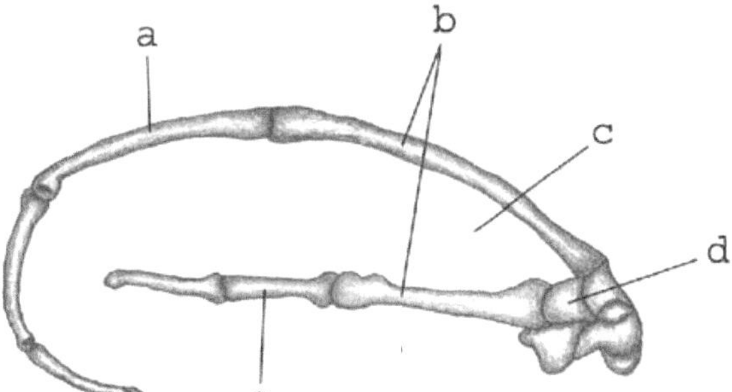

Fig. 14.52 Some of the bones of the distal part of the hand of a gibbon. (a) Second digit, (b) metarcarpal bones, (c) deep cleft between the first and second digits, (d) trapezium (a carpal bone) and (e) first digit or thumb. The metacarpal of the thumb forms a ball and socket joint with the trapezium and such a thumb has a high degree of mobility and can be rotated laterally in an arc giving gibbons an additional 90 degrees of rotational ability within the wrist. Gibbons have an average weight of 5 to 8 kg with arms that are about twice as long as the body trunk. The fingers of gibbons are elongated and curved and their thumbs are relatively some of the longest among primates.

about 55 km/h. The largest gibbons are the siamangs and weigh up to 24 kg.

Among the early apes was *Dryopithecus* (Fig. 14.53), the ancestor to modern apes including man. Close to the end of Miocene, the climate became cooler in the Northern Hemisphere resulting in loss of many primate species with some migrating southwards into Africa and South Asia. *Dryopithecus* was 60 cm long and had a semi-erect posture. The primate was a tree dwelling animal that lived in Eastern Africa in the latter part of Miocene about 12 to 9 million years ago. It later on radiated to the rest of Africa, Asia and Europe. The discovery of a partial skeleton of *Dryopithecus laietanus* in Spain shows that orthograde ('upright') postures and locomotion appeared at least 9.5 million years ago (Moya-sola and Kohler, 1996). *Dryopithecus* evolved into the gorilla line and the chimpanzee/human line in Africa about 9 million year ago. With further evolution, around 6 million years ago, the chimpanzee line separated from the hominid line. The discovery of chimpanzee fossils from the Kapthurin Formation, Kenya shows that representatives of this primate were present in the East African Rift Valley during the middle Pleistocene where they were contemporary with an extinct species of *Homo* (McBrearty and Jablonski, 2005).

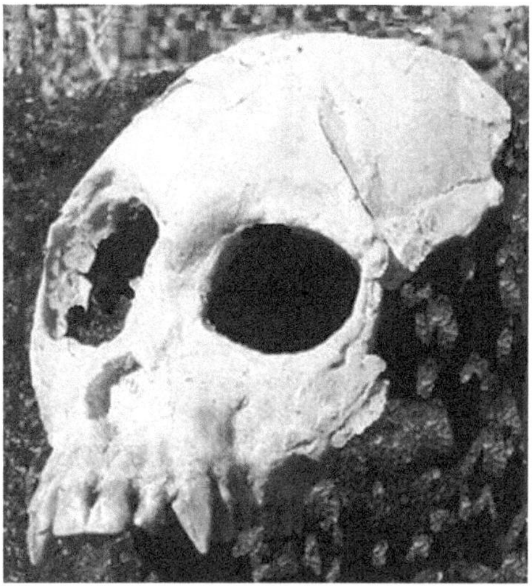

Fig. 14.53 A skull belonging *Dryopithecus africanus* (*Proconsul africanus*) or the 'woodland ape' that was discovered in 1948 in East Africa. The size of a gibbon, *D. africanus* lived during the Miocene epoch. The body weight of *D. africanus* ranged from 4 kg to 70 kg and the primate's fossil record is that of one of the most common hominoids during the Miocene Epoch.

EVOLUTION OF THE HUMAN BEING

Human beings evolved from ape-like humans or hominids (Fig. 14.54). The skull of the earliest hominid dating back to 7 million years ago which had features that place it close to the last common ancestor of chimpanzees and humans was discovered in Chad, Africa (Brunet et al., 2002; 2005). The process that has led to evolution of the modern human being from these ancestors has taken about 6 million years. Bi-pedalism in the human being evolved about 4 million years ago while the large and complex brain as well as ability to communicate using words and use of tools evolved much later. Other traits such as culture, art and other related activities evolved about 100 thousand years ago.

The living humans and great apes share a common ancestor that lived about 6 million years ago. Orangutans and gibbons diverged from the ape lineage about 10 to 8 million years ago. There are various hypotheses about the evolution of living humans. The **multi-regional evolution model** proposes that evolution of living humans occurred throughout the Old World with little population migration and replacement. This hypothesis is supported by the fact that characteristics that are distinct to humans of different geographical regions are also evident in *Homo erectus* remains of the same regions. The **out of Africa model** hypothesizes that evolution of *H. sapiens* took place in East Africa after which there were massive migrations to other parts of the Old World during which there was replacement of the existing *H. erectus*.

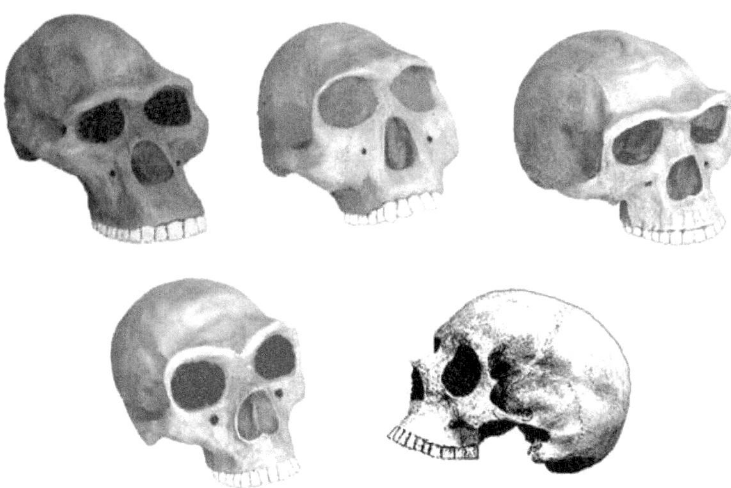

Fig. 14.54 Skulls belonging to various hominid (human family) species. *Australopithecus africanus* (top left), *Homo habilis* (top middle), *H. erectus* (top right), *H. neanderthalensis* (bottom left) and *H. sapiens sapiens* (human being) (bottom right).

The transitional primates between apes and humans lived in eastern and southern Africa between about 6 million years ago. These hominids are now extinct and had bipedal locomotion. The group comprises Kenyapithecines and Australopithecines and had a height of 1.0 to 1.5 m. Australopithecines gave rise to the *Homo* lineage and other Australopithecines about 2.3 to 2.7 million years ago. The two groups co-existed together. *Homo habilis* (skillful or handy person) was the first species of the genus *Homo* to appear around 2.5 million years ago. *H. habilis* had larger brains than *Australopithecus* from which it could have originated and was the least similar in physical features to the human being among the species that belonged to *Homo.* When compared to the living human being, *H. habilis* was short and had disproportionately long arms. The fossil remains of *H. habilis* are associated with stone tools.

Homo erectus (upright man) appeared after *H. habilis* around 1.7 million years ago then spread to tropical areas of the Old World reaching as far as Southeast Asia. About 1.0 to 0.7 million years ago, *H. erectus* spread beyond the tropical Old World to other parts of the Old World. The cranial capacity of *H. erectus* was larger than that of *H. habilis. H. erectus* were hunter gathers and had an average height of 1.79 m (5 feet and 10 inches). The African group continued to evolve into living humans. Other groups of *H. erectus* become extinct. Some *H. erectus* fossils have been found in Java, Indonesia, dating back to 30 to 50 thousand years ago. Studies have shown that there is similar mechanical load-sharing between the lower and upper limbs, and by implication, similar locomotor behavior in early *H. erectus* and modern humans which implies that by the earliest Pleistocene (1.7 million years ago), modern patterns of bipedal behavior were fully established in this early hominid taxon (Ruff, 2008).

Homo heidelbergensis probably originated in Africa or western Eurasia before spreading to other parts of the old world (Rightmire, 2004), A jaw belonging to this species and dated 700 thousand years old was found near Heidelberg making it one of the most ancient fossils of hominids in Europe (Czarnetzki et al., 2003). The brain size of *H. heidelbergensis* was larger than that of *H. erectus* and the faces with large brow ridges in most *H. heidelbergensis* also projected less than that in *H. erectus*. The skeleton and teeth of *H. heidelbergensis* were smaller than those of *H. erectus* but were larger than in the living human being. Sometimes it is difficult to differentiate the fossil remains of *H. erectus* from those of *H. heidelbergensis* since there is no clear dividing line between the two. *H. heidelbergensis* is thought to have evolved into *H. neanderthalensis* although some researchers propose *H. heidelbergensis* as an Afro-European taxon that is ancestral to both Modern humans and Neanderthals (Mounier et al., 2009).

Fossil remains of *H. neanderthalensis* or *H. sapiens neanderthalensis* have been found in Europe and western Asia. The first fossil remains of a Neanderthal skeleton were discovered by quarry workers in 1856 in the Neander valley in Germany. Similar remains were later found in Belgium. Neanderthals are the closest relatives of the living human being and the genomes of the two hominids are at least 99.5% identical (Noonan et al., 2006). The two hominids coexisted together. Studies of mitochondrial DNA show that Neanderthals and the human being last shared a common ancestor about 465,000 years ago (Krings et al., 1999). Neanderthals adapted to the glacial climate of northwestern Europe before disappearing abruptly between 30 and 40 thousand years ago (Mellars, 2004). *H. neanderthalensis* weighed about 80 kg and had muscular short squat physiques for conserving body heat in the ice age environment they inhabited. The primate also had an elongated braincase with a short forehead that had supraorbital buttresses. Other Neanderthal features included a long projecting nose, long skull and long jaw. The mouth was large with robust teeth and the jaw lacked a chinbone. The first hominid to bury the dead was *H. neanderthalensis.* The primate also cultivated land, was a hunter as it relied on a heavy meat and fat diet and used to make a fire for cooking and protection. *H. neanderthalensis* was more intelligent than *H. erectus* and was capable of complex speech.

In Africa, *H. habilis* could have evolved into *H. erectus* that eventually gave rise to *H. sapiens.* The discovery of *H. erectus* remains in Ethiopia that are an intermediary between earlier and later African fossils indicates that the African *H. erectus* was the ancestor of *H. sapiens* (Asfaw et al., 2002). *H. sapiens* (intelligent person) had a greater skull capacity than *H. neanderthalensis. H. sapiens* lived between 300 and 120 thousand years ago and evolved into the living human being. The primate migrated to Eurasia to replace other hominids that occupied niches in this region. The brain had frontal lobes, the jaw had a chin and the size of teeth was smaller than *H. habilis.* Genes that regulate brain size during development such as microcephaly (a congenital defect that is characterized by severely reduced brain size) genes are the main contributors in driving the evolutionary enlargement of the human brain since they are the favored targets of natural selection (Gilbert et al., 2005).

H. sapiens used fire to cook and also shaped stones, wood and bones for making weapons and utensils. The primate also built structures for shelter and also practised painting and sculpturing. The living human being *H. sapiens sapiens* evolved from *H. sapiens.* The earliest known skull remains of the modern human being are about 195 thousand years old and were found in Kibish, Ethiopia (McDougall et al., 2005). The spread of the human being across the Earth is greater than that of any other primate including hominids. Human beings have had far greater influence on the environment and have advanced technologically much more than any previous hominid. The modern human being moved to Europe about 35 thousand years ago. Fossils of the 'Cro-Magnon people' from Les Eysie in France that are about 28 thousand years old belong to the ancestors of Europeans. The first human beings moved to Australia about 60 thousand years ago and to the Americas about 35 thousand years back. The differences between individuals within a group are greater than differences between the averages of each race.

PLEISTOCENE EXTINCTIONS

The Pleistocene epoch (Gr. *pleistos*, most; *kainos*, new or recent) came after Pliocene and before the Holocene epochs. The Pleistocene epoch lasted from 1.8 million years to 11 thousand years ago. Among the ice ages of the Earth's history, Pleistocene is the most known glacial period. At one time during this epoch, ice sheet covered Antarctica, large areas of Europe, the Americas and some areas in Asia. The Pleistocene epoch was characterized by a series of climatic fluctuations. Most of the world's temperate zones were covered with glaciers during the cooling periods that were interrupted by the warmer interglacial seasons during which the glaciers would melt and recede. There were four glacial periods during the Pleistocene epoch. It has been suggested that the current Holocene epoch might be one of the interglacial periods of Pleistocene since the interglacial periods of Pleistocene were of a longer period than the current epoch has lasted. During glacial periods, the beds of some shallow seas were exposed thus connecting previously separated landmasses. The mean annual temperatures during glacial periods were 10° to $13^{\circ}C$ lower then during the Holocene (Callaghan et al., 2004).

Most forms of recent mammals evolved during the Pleistocene. The plants and animals of Pleistocene were similar to those living today in many ways though they differed in spatial distribution depending on climatic changes. Evolutionary changes during this period were not much due to the short period the epoch lasted. Mammals showed the greatest radiation during this period. The large mammals weighing more than 44 kg were known as **megafauna**. The group included mammoths, mastodon, saber-toothed cat, woolly rhinoceros, giant deer, moose-like giraffe, cave bear, bison, sheep, cattle, horses and camels (Fig. 14.55).

At the end of Pleistocene, there were mass extinctions. Close to 70% of the large North American mammalian species were lost. Some of the mammals that became extinct include mammoths, mastodons, saber-toothed cats, native horses and camels, giant ground sloths, stag-moose (a deer) beavers the size of the living black bear and the long-horned

Fig. 14.55 Some of the megafauna that became extinct at the end of Pleistocene. The mammoth (top left), saber-toothed cat or tiger (top right), long-horned bison (bottom left) and giant ground sloth (bottom right). Mammoths and extinct mastodons were relatives of the living elephants. The saber-toothed cat was the size of the lion and had two huge upper canine teeth. The long-horned bison is thought to be a relative of the buffalo. Giant ground sloths were xenarthrans and some weighed about 3 tons and had claws that were as long as 50 cm.

bison. The great teratorn birds with wingspans of 25 feet also became extinct. Europe, Asia and South America experienced some extinctions but to a lower extent. Horses evolved in North America then later on moved across the Bering Strait to the Old World and over the Panama isthmus to South America. The group later evolved into the living horses, donkeys and zebras. After extinction of the ancestral American horses, they were later replaced with descendants of European horses. Anthropogenic causes have been attributed to some of the mammalian Pleistocene extinctions (Barnosky et al., 2004; Diniz-Filho, 2004; Surovell et al., 2005; Pushkina and Raia, 2008).

Before the anthropogenic causes, studies have supported the idea that the late Pleistocene extinctions were environmentally driven by climate changes that triggered habitat fragmentation, species range reduction and population decrease (Pushkina and Raia, 2008).

The **Irish elk** or **giant deer** (*Megaloceros giganteus*) (Fig. 14.56) is the largest deer species that ever lived and was thought to have become extinct around 10 thousand years ago in Europe until remains of the Siberian popuation that lived beyond this period for another 3,000 years were found. Phylogenetic analysis including morphological and DNA

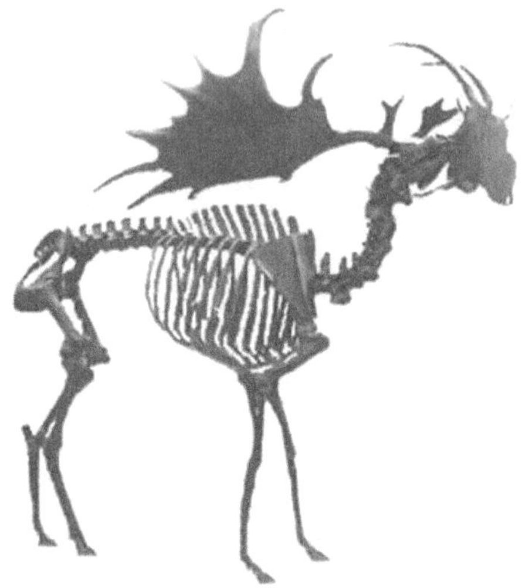

Fig. 14.56 A skeleton of the extinct Irish elk (giant deer or stag-moose). The deer is referred to as 'Irish' because of complete fossils in lake sediments and peat bogs in Ireland. The name 'stag-moose' is also used to refer to the deer since it resembled a cross between an elk and a moose.

sequence evidence support a sister-group relationship of giant deer and fallow deer (Lister et al., 2005). The Irish elk stood at 2.1 m at the shoulder and had antlers that had a span of 3.65 m and weighed 40 kg. It is thought that the Irish elk became extinct as a result of mating preferences over generations. Female Irish elk preferred to mate with males with large antlers and such a trait was passed down generations. The antlers could have been a nutritional burden to the bulls since they were shed annually and contained about 8 kg of calcium and 4 kg of phosphate. The wide span of antlers could have also restricted the movement of the elk in dense forests more than an open environment. The rapid cooling of the climate resulting in the growth of the less productive tundra could have been a contributing factor to the demise of the Irish elk.

ARTIFICIAL SELECTION AND EVOLUTION

Artificial selection is the process of selective breeding for certain traits to alter a certain species through human intervention. The term artificial selection was coined by Charles Darwin to distinguish it from the process of natural selection that occurs without interference or action of the human being. During artificial selection, individuals with the desired traits are encouraged to breed while those with undesirable characteristics are discouraged from perpetrating their genes. With time, down the generation line, only individuals with desirable characteristics will dominate a population. Artificial selection occurs mainly in domesticated vertebrates since individual control of wild species is not easy. Sometimes 'unconscious selection' takes place as artificial selection is being carried out. **'Unconscious selection'** is not intended but occurs in the process of intentionally selecting for a certain trait. A good example of 'unconscious selection' is the reduction of brain size and relatively smaller leg bones in the domestic chicken as there is artificial selection for faster growth rates and larger sizes.

Alternation of the environment consciously or unconsciously by the human being alters the evolution of vertebrates. Vegetation destruction and loss of breeding grounds partly contributed to the extinction of the dodo in Mauritius. Population pressure in many parts of the world together with human activities that have impacted on vertebrates have altered the original natural balance of vertebrate species and also considerably reduced or eliminated certain species. Human beings have practised artificial selection in the livestock and pet industry. In the poultry industry, selection against the brooding gene has led to production of layers that produce many eggs continuously without the desire by the birds to incubate the eggs. New breeds of cattle and pigs have been created based on productivity and desired qualities of products. Dogs have been bred for certain traits including size and aggression. The human being is likely to alter the course of vertebrate evolution to a certain extent by causing the extinction of some species and protecting those species that would have been eliminated under natural conditions.

Eugenics involves social interference through selective breeding to improve human hereditary traits and is the 'self direction of human evolution'. Eugenics has been practised in some parts of the world to reduce hereditary diseases and lessen human suffering and also creation of intelligent people. Various ways have been devised for achieving these goals including selective breeding, birth control and genetic engineering. The reputation of eugenics waned in the 1930s when it was associated with the racial policies of Nazi Germany.

CONTINENTAL DRIFT

The continental drift theory was first proposed by the German meteorologist and geologist Alfred Wegener (1880-1930) (Fig. 14.57) in 1915. Wegener proposed that the Earth's crust drifts slowly on top of a liquid core. In Wegener's hypothesis, there was a large super-continent he named **Pangaea** ('all earth') that started splitting apart during the

Fig. 14.57 A photograph of Alfred Wegener (1880-1930). Although not the only human being to suggest that continents had once been connected, Wegener was the first person to present extensive evidence from various fields to support the view.

Jurassic period about 200 million years ago into the continents **Laurasia** (Eurasia and North America) to the north and **Gondwanaland** (Antarctica, the southern continents and the Indian subcontinent) to the south (Fig. 14.58). Between the two continents was the Tethys Sea. Further plate movements brought about the separation of North America from Eurasia at the same time South America, Africa and India were breaking apart (about 78 million years ago). The last landmasses to separate were Australia and Antarctica. At the end of Cretaceous, the continents had separated into landmasses that resemble the current continents. Wagener's theory was not accepted at the time of its proposal since most geologists believed that the Earth was rigid and continents formed in their present locations and there was no mechanism yet to explain the moving continents.

The theory of **plate tectonics** (Gr. *tekton*, one who constructs) was advanced in the 1960s to explain the movement of the Earth's plates. The outermost part of the interior part of the Earth is made of two layers based on their mechanical properties. The rocky crust that solidified billions of years ago is the outer **lithosphere** while the fluid-like inner layers is **asthenosphere**. The thin but denser oceanic crust in comparison to continental crust lies under the oceans. Crust is dynamic since it is being created and destroyed constantly. Oceanic crust is more dynamic than continental crust. The lithosphere is cooler and more rigid and is broken up into 10 major large plates that float on top of the hotter and softer asthenosphere. These plates including the minor ones move at a speed of about one to 10 cm per year in relation to one another at their boundaries both horizontally and vertically. The direction of movement can also be convergent, divergent or sliding past one another. Other

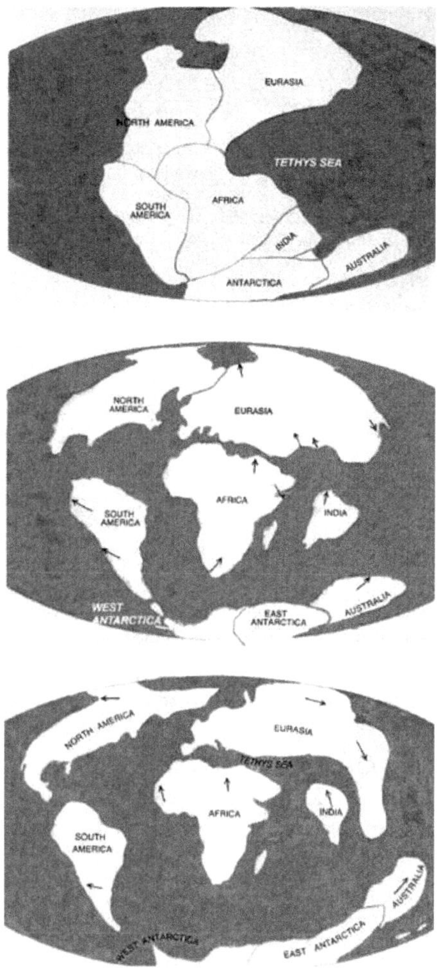

Fig. 14.58 Changes that occurred during the continental drift. The supercontinent Pangaea (top map) started drifting apart around 200 million years ago. Around 78 million years ago (middle map), the separation of the major continents of the world was almost complete. The Indian subcontinent (bottom) map moved northwards and later on converged with Asia leading to the formation of the Himalayas. At 65 million years ago, the continents had assumed a position that is almost similar to the present location of the continents.

activities that are associated with interactions at plate boundaries include earthquakes, volcanic activity, mountain building and oceanic trench formation. Plate tectonic activity is also associated with seafloor spreading.

Continental drift is therefore one of the consequences of plate tectonics. Continents rest on plates and passively move with them. A super-continent blocks the flow of heat from the interior part of the Earth leading to the overheating of the asthenosphere. The lithosphere moves upwards then cracks leading to the rising of magma and pushing of cracked fragments apart. It has been suggested that super-continents form in cycles of about 250 million years as a result of plate tectonics. There were other super-continents that broke up and reassembled before Pangaea. It is not clear how continents reform into super-continents.

There had been evidence to support continental drift before the theory of plate tectonics. Wegener supported his theory based on how the shapes of the continents matched, distribution of fossils, similarity in the sequence of rocks at various locations, past climates and the wandering of the polar-regions. The **continental margins** look like they have common points that were once in contact with each other and these points can fit together like pieces of a jigsaw puzzle. Examples to support such points were the western bulge of Africa that fits into the shape of the eastern coast of North America and the coastal part of Brazil that fits along the western coast of Africa below the bulge.

Fossil remains of plants and animals along the matching coastlines of Africa and South America are quite similar despite the fact that the two coastlines are separated from each other by the wide Atlantic Ocean. This was the strongest reason that convinced Wegener that the two coastlines were once joined together since it was impractical for the common organisms at the two locations to have crossed the vast sea. Wegener also observed that similar plant fossils of the late Paleozoic era were found on various continents, a fact that showed that the current continents were at one time a single landmass.

The **rocks** from the coastlines and mountain belts of Africa and South America had similar structures and were of the same age. Wegener also noted from ancient glacial deposits that ice sheets covered large parts of the Southern hemisphere including India and Australia around 300 million years ago. The glacial striations on rocks showed a pattern that runs from Africa towards the Atlantic Ocean and to South America, an arrangement that would not be possible if the continents were not joined. Such evidence of ice cover in the northern continents is lacking since these continents were located around the equator about 300 million years ago. The coal deposits of the northern hemisphere are evidence of tropical plant deposits when this region was closer to the equator.

REFERENCES

Arnason, U., Gullberg, A. and Janke, A. (2004). Mitogenomic analyses provide new insights into cetacean origin and evolution. Gene, 333:27-34.

Asfaw, B., Gilbert, W.H., Beyene, Y., Hart, W.K., Renne, P.R., WoldeGabriel, G., Vraba, E.S. and White, T.D. (2002). Remains of *Homo erectus* from Bouri, Middle Awash, Ethiopia. Nature, 416(6878):317-320.

Barnosky, A.D., Koch, P.L., Feranec, R.S., Wing, S.L. and Shabel, A.B. (2004). Assessing the causes of late Pleistocene extinctions on the continents. Science, 306(5693):70-75.

Bejder, L. and Hall, B.K. (2002). Limbs in whales and limblessness in other vertebrates: mechanisms of evolutionary and developmental transformation and loss. Evol. Dev., 4(6):445-458.

Blanga-Kanfi, S., Miranda, H., Penn, O., Pupko, T., DeBry, R.W. and Huchon, D. (2009). Rodent phylogeny revised: analysis of six nuclear genes from all major rodent clades. BMC Evol. Biol., 9:71.

Brunet, M., Guy, F., Pilbeam, D., Lieberman, D.E., Likius, A., Mackaye, H.T., Ponce de Leon, M.S., Zollikofer, C.P. and Vignaud, P. (2005). New material of the earliest hominid from the upper Miocene of Chad. Nature, 434(7034):752-755.

Brunet, M., Guy, F., Pilbeam, D., Mackaye, H.T., Likius, A., Ahounta, D., Beauvilain, A., Blondel, C., Bocherens, H., Boisserie, J.R., De Bonis, L., Coppens, Y., Dejax, J., Denys, C., Duringer, P., Eisenmann, V., Fanone, G., Fronty, P., Geraads, D., Lehmann, T., Lihoreau, F., Louchart, A., Mahamat, A., Merceron, G., Mouchelin, G., Otero, O., Pelaez Campomanes, P., Ponce De Leon, M., Rage, J.C., Sapanet, M., Schuster, M., Sudre, J., Tassy, P., Valentin, X., Vignaud, P., Viriot, L., Zazzo, A. and Zollikofer, C. (2002). A new hominid from the Upper Miocene of Chad, Central Africa. Nature, 418(6894):145-151.

Butler, R.J., Barrett, P.M. and Gower, D.J. (2009). Postcranial skeletal pneumaticity and air-sacs in the earliest pterosaurs. Biol. Lett., 5(4):557-560.

Callaghan, T.V., Bjorn, L.O., Chernov, Y., Chapin, T., Christensen, T.R., Huntley, B., Ims, R.A., Johansson, M., Jolly, D., Johansson, S., Matveyeva, N., Panikov, N., Oechel, W. and Shaver, G. (2004). Past changes in Arctic terrestrial ecosystems, climate and UV radiation. Ambio., 33(7):398-403.

Catania, K.C. (2000). Cortical organization in insectivora: the parallel evolution of the sensory periphery and the brain. Brain Behav. Evol., 55(6):311-321.

Chiappe, L.M., Codorniu, L., Grellet-Tinner, G. and Rivarola, D. (2004). Palaeobiology: Argentinian unhatched pterosaur fossil. Nature, 432(7017):571-572.

Claessens, L.P., O'Connor, P.M. and Unwin, D.M. (2009). Respiratory evolution facilitated the origin of pterosaur flight and aerial gigantism. PLOS One, 4(2):e4497.

Cracraft, J. (2001). Avian evolution, Gondwana biogeography and the Cretaceous-Tertiary mass extinction event. Proc. Biol. Sci., 268(1466):459-469.

Czarnetzki, A., Jakob, T. and Pusch, C.M. (2003). Palaeopathological and variant conditions of the *Homo heidelbergensis* type specimen (Mauer, Germany). J. Hum. Evol., 44(4):479-495.

Diniz-Filho, J.A. (2004). Macroecological analyses support an overkill scenario for late Pleistocene extinctions. Braz. J. Biol., 64(3A):407-414.

Dudley, R. (2000). The evolutionary physiology of animal flight: paleobiological and present perspectives. Annu. Rev. Physiol., 62:135-155.

Fountaine, T.M., Benton, M.J., Dyke, G.J. and Nudds, R.L. (2005). The quality of the fossil record of Mesozoic birds. Proc. Biol. Sci., 272(1560):289-294.

Furutani-Seiki, M. and Wittbrodt, J. (2004). Medaka and zebrafish, an evolutionary twin study. Mech. Dev., 121(7-8):629-637.

Gilbert, S.L., Dobyns, W.B. and Lahn, B.T. (2005). Genetic links between brain development and brain evolution. Nat. Rev. Genet., 6(7):581-590.

Greenwood, A.D., Castresana, J., Feldmaier-Fuchs, G. and Paabo, S. (2001). A molecular phylogeny of two extinct sloths. Mol. Phylogenet. Evol., 18(1):94-103.

Hedges, S.B. and Pauling, L.L. (1999). A molecular Phylogeny of Reptiles. Science, 283(5404):998-1001.

Hernandez, F.M. and Vrba, E.S. (2005). A complete estimate of the phylogenetic relationships in Ruminantia: a dated species-level supertree of the extanct ruminants. Biol. Rev. Camb. Philos. Soc., 80(2):269-302.

Hillenius, W.J. and Ruben, J.A. (2004). The evolution of endothermy in terrestrial vertebrates: Who? When? Why? Physiol. Biochem. Zool., 77(6):1019-1042.

Hoegg, S. and Meyer, A. (2005). Hox clusters as models for vertebrate genome evolution. Trends Genet., 21(8):421-424.

Huchon, D., Catzeflis, F.M. and Douzery, E.J.P. (2000). Variance of molecular datings, evolution of rodents and the phylogenetic affinities between Ctenodactylidae and Hystricognathi. Proc. R. Soc. Lond. B 267:393-402.

Ji, Q., Ji, S.A., Cheng, Y.N., You, H.L., Lu, J.C., Liu, Y.Q. and Yuan, C.X. (2004). Palaeontology: pterosaur egg with a leathery shell. Nature, 432(7017):572.

Ji, Q., Norell, M.A., Gao, K.Q., Ji, S.A. and Ren, D. (2001). The distribution of integumentary structures in a feathered dinosaur. Nature, 410(6832):1084-1088.

Jones, K.E., Bininda-Emonds, O.R. and Gittleman, J.L. (2005). Bats, clocks and rocks: diversification patterns in Chiroptera. Evol. Int. J. Org. Evol., 59(10):2243-2255.

Jozefowicz, C., McClintock, J. and Prince, V. (2003). The fates of zebrafish Hox gene duplicates. J. Struct. Funct. Genomics, 3(1-4):185-194.

Koteja, P. (2004). The evolution of concepts on the evolution of endothermy in birds and mammals. Physiol. Biochem. Zool., 77(6):1043-1050.

Krause, J., Dear P.H., Pollack, J.L., Slatkin, M., Spriggs, H., Barnes, I., Lister, A.M., Ebersberger, I., Paabo, S. and Hofreiter, M. (2006). Multiplex amplification of the mammoth mitochondrial genome and the evolution of Elephantidae. Nature, 439(7077):724-727.

Krings, M., Geisert, H., Schmitz, R.W., Krainitzki, H. and Paabo, S. (1999). DNA sequence of the mitochondrial hypervariable region II from the Neandertal type specimen. Proc. Natl. Acad. Sci. USA, 96(10):5581-5585.

Kundrat, M. (2004). When did theropods become feathered?—evidence for pre-*Archaeopterix* feathery appendages. J. Exp. Zoolog. B Mol. Dev. Evol., 302(4):355-364.

Lingham-Soliar, T. (2003). The dinosaurian origin of feathers: perspectives from dolphin (Cetacea) collagen fibers. Naturwissenschaften, 90(12):563-567.

Lister, A.M., Edwards, C.J., Nock, D.A., Bunce, M., van Pijlen, I.A., Bradley, D.G., Thomas, M.G. and Barnes, I. (2005). The phylogenetic position of the 'giant deer' *Megaloceros giganteus*. Nature, 438(7069):850-853.

Long, J.A. and Gordon, M.S. (2004). The greatest step in vertebrate history: a paleobiological review of the fish tetrapod transition. Physiol. Biochem. Zool., 77(5):700-719.

McBrearty, S. and Jablonski, N.G. (2005). First fossil chimpanzee. Nature, 437(7055):105-108.

McDougall, I., Brown, F.H. and Fleagle, J.G. (2005). Stratigraphic placement and age of modern humans from Kibish, Ethiopia. Nature, 433(7027):733-736.

Mellars, P. (2004). Neanderthals and the modern human colonization of Europe. Nature, 432(7016):461-465.

Mounier, A., Marchal, F. and Condemi, S. (2009). Is *Homo heidelbergensis* a distinct species? New insight on the Mauer mandible. J. Hum. Evol., 56(3):219-246.

Moya-Sola, S. and Kohler, M. (1996). A *Dryopithecus* skeleton and the origins of great-ape locomotion. Nature, 379(6561):156-159.

Noonan, J.P., Coop, G., Kudaravalli, S., Smith, D., Krause, J., Alessi, J., Chen, F., Platt, D., Pääbo, S., Pritchard, J.K. and Rubin, E.M. (2006). Sequencing and analysis of Neanderthal genomic DNA. Science, 314(5802):1068-1071.

Norell, M., Ji, Q., Gao, K., Yuan, C., Zhao, Y. and Wang, L. (2002). Palaeontology: 'modern' feathers on a non-avian dinosaur. Nature, 416(6876):36-37.

Pushkina, D. and Raia, P. (2008). Human influence on distribution and extinctions of the late Pleistocene Eurasian Megafauna. J. Hum. Evol., 54(6):769-682.

Rightmire, G.P. (2004). Brain size and encephalization in early to Mid-Pleistocene *Homo*. Am. J. Phys. Anthropol., 124(2):109-123.

Ruben, J.A. and Jones, T.D. (2000). Selective factors associated with the origin of fur and feathers. American Zool., 40:585-596.

Ruff, C. (2008). Femoral/humeral strength in early African *Homo erectus*. J. Hum. Evol., 54(3):383-390.

Sanchez-Villagra, M.R., Aguilema, O. and Horovitz, I. (2003). The anatomy of the world's largest extinct rodent. Science, 301(5640):1708-1710.

Sears, K.E., Behringer, R.R., Rasweiler, J.J. and Niswander, L.A. (2006). Development of bat flight: morphologic and molecular evolution of bat wing digits. Proc. Natl. Acad. Sci. USA, 103(17):6581-6586.

Seymour, R.S., Bennett-Stamper, C.L., Johnston, S.D., Carrier, D.R. and Grigg, G.C. (2004). Evidence for endothermic ancestors of crocodiles at the stem of archosaur evolution. Physiol. Biochem. Zool., 77(6):1051-1067.

Surovell, T., Waguespack, N. and Brantingham, P.J. (2005). Global archaeological evidence for proboscidean overkill. Proc. Natl. Acad. Sci. USA, 102(17):6231-6236.

Taylor, J.S. and Raes, J. (2004). Duplication and divergence: the evolution of new genes and old ideas. Annu. Rev. Genet., 38:615-643.

Teeling, E.C., Springer, M.S., Madsen, O., Bates, P., O'brien, S.J. and Murphy, W.J. (2005). A molecular phylogeny for bats illuminates biogeography and the fossil record. Science, 307(5709):580-584.

Thewissen, J.G., Williams, E.M., Roe, L.J. and Hussain, S.T. (2001). Skeletons of terrestrial cetaceans and the relationship of whales to artiodactyls. Nature, 413(6853):277-281.

Uhen, M.D. (2007). Evolution of marine mammals: back to the sea after 300 million years. Anat. Rec. (Hoboken), 290(6):514-522.

Unwin, D.M. (1999). Pterosaurs: back to the traditional model? Trends Ecol. Evol., 14(7):263-268.

van Tuinen, M., Sibley, C.G. and Hedges, S.B. (2000). The early history of modern birds inferred from DNA sequences of nuclear and mitochondrial ribosomal genes. Mol. Biol. Evol., 17(3):451-457.

Vidal, N. and Hedges, S.B. (2004). Molecular evidence for a terrestrial origin of snakes. Proc. Biol. Sci., 271 suppl. 4:S226-229.

Wang, X. and Zhou, Z. (2004). Palaeontology: pterosaur embryo from the Early Cretaceous. Nature, 429(6992):621.

Witmer, L.M., Chatterjee, S., Franzosa, J. and Rowe, T. (2003). Neuroanatomy of flying reptiles and implications for flight, posture and behavior. Nature, 425(6961):950-953.

Xu, X., Zhou, Z. and Prum, R.O. (2001). Branched integumental structures in *Sinornithosaurus* and the origin of feathers. Nature, 410(6825):200-204.

Index

A
Aardvarks 36, 492, 493, 496
Abducens nerve 135
Abomasum 205, 207
Acanthodians 17, 18, 461
Accelerator muscles 138
Accessory carpal bone 33, 89, 112
Accessory nerve 348
Accommodation 415, 416
Acellular scales 157
Acetabulum 109, 110
Acetylcholine 128, 331, 336, 337
Acoelous vertebrae 90, 91
Acromion process 120
Acrosome 268, 270
Acrosome reaction 270
Actin 73, 78, 79, 119, 129, 130, 133, 141
Actinopterygii 17, 18
Action potential 16, 73, 124, 336, 382-385, 390, 393, 397, 400, 407, 416
Acute pain fibers 14, 314, 363
Adaptive immune system 37
Adaptive radiation of vertebrates 37
Adductor mandibulae 136, 223
Adductor muscle 141, 226
Adenohypophysis 359-362, 364
Adenosine triphosphate 129, 186, 218
Adrenocorticotropic hormone 362
Adipose fins 101
Adipose tissue 54, 55, 166, 170, 175, 188, 215, 297, 324, 373, 387
Adrenal glands 358, 369-371
African tiger fish 22
Afrosoricida 36
Afrotheria 36
Agents of evolution 449, 451
Agnatha 10, 12, 13, 97, 102, 108, 201, 220, 312, 314, 321, 223, 340, 345, 353, 360
Air capillaries 239, 240
Air sacs 28, 29, 230, 237-240, 405, 474
Albumen 223, 261-263, 283
Albumins 322
Aldosterone 325, 330, 356, 369, 378, 439, 444
Alfred Russel 452
Alfred Wegener 511, 512
Alimentary canal 199-202, 211, 312
Allantois 27, 233, 283-286, 435

Alleles 291, 449, 451, 455-457
Allotheria 484, 485
Alpha cells 215, 373
Altiatlasius 499
Altricial chicks 42
Alveolar macrophages 225, 244
Alveoli 175, 176, 227, 241-244, 246
Amacrine cells 411, 419
Amines 337, 426
Amniotes 5, 27, 98, 100, 104, 132, 198, 244, 255, 257, 258, 274, 342, 346, 349, 353, 390, 417, 435, 436, 467, 470
Ammonia 24, 189, 203, 217, 227, 299, 424, 426, 429, 431, 433
Ammonotelic 433
Amnion 27, 283-285, 365
Amphiarthroses 113
Amphibian papilla 401, 402
Amphibians 10, 25-27, 34, 39, 41, 86, 93, 95, 96, 100, 103, 109, 112, 137, 138, 140-143, 145, 151, 156, 162, 163, 192, 194-196, 198, 254, 269, 277, 320, 358, 368, 392, 393, 433, 442, 464, 465, 467, 481
Amphicoelous vertebrae 90
Amphioxus 38, 154, 275, 276, 458
Amphistylic jaw suspension 85
Amphiuma tridactylum 317
Ampulla 253, 254, 385, 386, 394, 405-407
Ampullae of Lorenzini 386
Ampullary organs 385, 386
Anabas 230, 331
Anadromous fish 427
Anal fins 100, 101, 151, 254
Androgens 258, 370, 373, 374
An-estrus 266
Anguilla 231, 260, 427
Annulus 82, 91, 135
Annulus fibrosus 82, 91
Antelope 37, 205, 486
Anterior cavity 409, 414
Anterior cerebral arteries 344
Anterior chamber 409, 414
Anterior communicating artery 344, 345
Antibodies 189, 258, 313, 320, 322
Antidiuretic hormone 325, 364, 433, 443
Antlers 171, 176, 178, 452, 510
Anurans 25, 26, 90, 97, 162, 163, 354, 371, 465, 466
Aortic bodies 385
Apes 97, 112, 210, 346, 502-506
Apocrine secretion 53
Apocrine sweat glands 53, 174
Apodans 25, 61, 102
Aponeurosis 57, 121
Appendicular muscles 132, 141, 143
Aquaporins 443, 444
Aquatic environment 5, 137, 157, 340, 418, 424, 476, 496, 497
Aquatic salamanders 26, 391
Arandaspids 460
Arbor vitae 343
Arboreal locomotion 148
Archaeopteryx 40, 477, 478
Archenteron 132, 274-277, 282, 360
Archinephric duct 254, 255, 259, 431, 434-436
Archosaurs 40, 235, 468, 472, 473, 476
Arctic fox 43
Argentaffin cells 359
Arginine-vasopressin 443
Armadillos 36, 170, 266, 492, 494
Arrector pili 29, 170, 173
Arrectores plumarum 28, 165, 166
Arteries 125, 179, 221, 222, 224, 226, 287, 294, 295, 297, 298, 300-302, 304-308, 310, 312, 327, 328, 330, 344, 345, 438
Arterioles 125, 179, 181, 246, 267, 306-310, 331, 332, 439-441

Arteriovenous anastomoses 179, 181, 306
Artificial selection 510, 511
Artiodactyla 37
Aspartate 337
Aspiration pump 234
Asteriscus 405
Asthenosphere 512, 514
Astrocytes 69-71, 338
Atlantic salmon 160, 427
Atlas 89, 94, 95, 99 116, 117, 146
Atrial natriuretic hormone 330, 378, 444
Atrioventricular bundle 75, 76
Atrioventricular node 75, 76
Atrioventricular valves 303, 326
Atrium 75, 76, 295, 297-306, 330, 378
Auditory ossicles 136, 394, 399
Auditory tube 395, 402
Autonomic ganglion 351
Autonomic nervous system 9, 75, 132, 196, 200, 201, 306, 308, 327, 331, 332, 350-354, 370
Autosomes 253
Autostylic suspension 85
Axis 10, 22, 60, 75, 82, 89, 94, 95, 102, 106, 108, 111, 116, 120, 121, 128, 146, 149, 207, 225, 269, 277, 278, 280, 354
Axolotl 26, 231, 232
Axon 13, 14, 66-72, 336-338, 341, 345-347, 351, 352, 364, 366, 383, 390, 392, 398, 411, 413

B

Baboon 502
Baculum 32, 61
Baleen 171, 176-178, 191
Ball and socket joints 117
Ballistic tongue 138, 192, 193
Barbels 160, 161
Basal cells 50, 182, 390, 430
Basic tissues of vertebrates 47
Basilar artery 344, 345
Basilar membrane 395-397
Basilar papilla 401, 402, 404, 405
Basipterygium 100, 102, 109
Basking shark 11, 191
Basophils 318, 319
Bats 32, 35, 37, 100, 107, 141, 148, 149, 208, 248, 257, 289, 346, 395, 404, 479, 488, 494, 495
Beta-endorphin 337, 363
Beta cells 59, 215, 373
Biaxial joints 117
Biceps femoris muscle 144
Bichirs 18, 39, 158
Bicornuate uterus 266
Bile 212-215, 372, 378
Bipartite uterus 266
Bipedal locomotion 96, 148, 506
Bipolar neurons 67, 413
Birds 3, 4, 9, 10, 27-29, 31, 34, 40-43, 54, 57, 79, 86, 90, 93-97, 100, 104, 106, 107, 109-111, 133, 141, 148-150, 153, 155, 156, 165-168, 173, 180, 189, 208, 237, 240, 261, 304, 363, 404, 429, 433, 472, 475, 478-481, 509, 511
Blastocoel 274
Blastocyst 249, 266, 270, 274, 276, 279, 286, 291
Blastoderm 274
Blastomeres 271-274, 277
Blind spot 409, 414
Blood 5, 10, 13, 17, 24, 34, 41, 42, 49, 50
Blood cell formation 321
Blood cells 34, 79, 83, 213, 214, 225, 245, 246, 309, 315-323, 229, 330, 441, 454
Blood channels 222, 224-226
Blood clotting 188, 320, 323
Blood plasma 322, 324, 328
Blood supply to the mammalian brain 344
Blood supply to muscle 124, 125
Blubber 182, 183, 478, 496

Blue whale 1, 2
Boas 385, 471
Bone 3, 8, 9, 13-15, 17, 18, 22, 28, 30, 32, 33, 35, 53, 57-65, 74, 79, 81-90, 95, 97, 98, 100-102, 104, 106, 107, 109, 111-119, 121, 127, 155, 311, 375, 376, 408, 450, 465, 473, 481, 487, 507, 511
Bone collar 63, 64
Bone homeostasis 64
Bonobo 503
Bony labyrinth 395
Booster hearts 395
Bowfin 20, 21, 202, 260
Bowman's capsule 436-441
Bowman's glands 390
Bowman's membrane 409
Brachiocephalicus muscle 141, 142
Bradykinin 337
Brain 17, 18, 29, 33-35, 59, 66-68, 70, 71, 79, 82, 85, 86, 165, 282, 291, 309, 310, 325, 328, 331, 335, 337-446, 448, 449, 351, 394, 397, 412, 420, 507, 511
Brain natriuretic peptide 378
Brain stem 339, 344, 352
Branchial chambers 222, 230
Branchial grooves 9, 10
Branchiomeric muscles 132, 135, 136, 349
Brockmann bodies 372
Bronchi 49, 57, 234, 238-240, 242, 243, 349, 353
Brood patches 166
Buccal pump 232, 234
Buccinator 136
Bulbourethral gland 256, 257
Bulbus arteriosus 296-298
Bursa 115, 126, 127, 209, 314
Bursa of Fabricius 209, 314

C

C-type natriuretic peptide 378, 432
Calcitonin 59, 65, 356, 367, 368
Calcium ions 77, 128, 129, 131, 134, 323
Cambrian period 459
Camel 41, 43, 315, 487
Canal of Schlemm 414
Capillaries 5, 26, 50, 63, 70, 83, 241, 245, 295, 308, 309, 311, 328-330, 338, 360, 438, 440, 441, 442
Carbohydrates 186-189, 214-217
Carbonic anhydrase 245
Carboniferous period 25, 459, 462, 464
Cardiac cycle 295, 326, 327
Cardiac muscle 3, 72, 75-77, 119, 125, 132, 134, 291, 296, 297, 337, 350
Cardiac sphincter 203
Carnivores 32, 33, 37, 112, 140, 145, 147, 176, 177, 182, 195, 197, 202, 210, 257, 266, 267, 286, 288, 289, 413, 472, 486, 497-499
Carotid bodies 332
Carpal bones 107, 112, 116, 503
Cartilage 3, 14, 17, 18, 32, 53, 58, 59, 62, 63, 79, 81, 83-87, 89, 90, 92, 97, 112-115, 232, 238, 242, 417, 495
Cartilage homeostasis 58, 59
Cartilaginous dorsal plates 92
Cartilaginous joints 113, 114
Caspian Sea 18, 425
Cassowary 29, 31
Catadromous fish 427
Catecholamines 337, 370
Cattle 37, 90, 197, 205-207, 436, 486, 508, 511
Cauda equine 345-346
Caudal fins 13, 20, 39, 101, 102, 461, 465
Caviar 18, 19
Cavum arteriosum 301
Cavum pulmonale 301
Cavum venosum 300, 301
Cell body 66-68, 337, 347
Cellular level 46

Cellulose 187, 206, 489
Cenozoic era 40, 459, 476, 499, 500
Central nervous system 49, 66, 68, 69, 70, 71, 122, 124, 335, 337-339, 342, 351, 354, 366, 378, 382, 383
Ceratotrichia 101
Cerebellum 29, 339-344
Cerebral aqueduct 339
Cerebrum 34, 340-344
Cerebrospinal fluid 338
Ceruminous glands 175, 403
Cetacea 37, 40, 96, 102, 107, 146, 174, 256, 365, 432, 496, 497
Cetartiodactyla 37
Chaenocephalus aceratus 316
Chalazae 262, 263
Chalicotheres 37
Channa 230
Chaohusaurus 476
Charles Darwin 452, 510
Chelonia 27, 468
Chemical level 46
Chestnut 176
Chimaera 14, 16, 202
Chiroptera 37, 404, 494, 495, 497
Chloride cells 226, 430
Cholecalciferol 180, 368
Cholecystokinin-pancreozymin 378
Cholesterol 180, 188, 213, 216, 355, 368
Chondrichthyes 12, 14
Chondroblasts 58, 64
Chondrostei 18
Chordamesoderm 276, 277
Chorioallantoic placenta 264
Chorioamniotic folds 284, 285
Chorion 27, 233, 264, 283-290, 356, 375, 376
Chorionic gonadotropins 356, 375, 376
Choroid layer 409
Choroid plexus 344
Chromaffin cells 370, 371
Chromatids 251, 252
Chromatophores 4, 155, 156, 159, 160, 163, 165
Chromosomes 10, 248, 250-253, 268, 270, 271, 449, 455
Chronic pain fibers 387
Ciliary photoreceptors 412, 420
Circle of Willis 344, 345
Circulatory system 29, 50, 65, 218, 219, 286, 287, 289, 295, 297, 299, 301-311, 313, 315, 317, 319, 321, 323, 325, 327-331, 354, 379, 429, 437, 441
Clades 35, 36, 102
Cladistics 36, 457
Clarias 230
Claspers 16, 108, 254, 255
Clavicle 103-105, 107, 108, 120, 136, 141, 142, 175, 237, 240, 478
Claws 4, 112, 148, 155, 163, 165, 166, 169, 171, 177, 182, 477, 486, 493, 499, 501, 509
Cleavage 271-275, 286
Cleidobrachialis 141
Cleidoic egg 39, 283, 285
Cleidomastoid 136, 141, 142, 349
Cleidomastoid 141
Cleithrum 103, 104
Cloaca 4, 164, 200, 207-209, 235, 254, 256, 257, 259-261, 263, 314, 413, 434, 435, 445, 446, 471, 488
Cochlea 32, 394-397, 400, 404-406
Cochlear duct 395, 396, 404
Cochlear nerve 396, 397
Codon 454
Coelacanth 18, 23, 25, 91, 92, 102, 158, 225, 463
Coelom 221, 224, 276, 282-285, 360, 369, 436
Coelomic cavities 276

Collagen fibers 54, 56-58, 61, 62, 74, 120, 157, 165, 170, 225, 308, 323, 409, 413
Collecting ducts 437, 440, 443-445
Colobus monkey 502
Colored oil droplets 413
Colostrum 290
Colubrids 471
Colugos 37, 149, 150, 494-496
Columella 400, 401, 403
Columnar epithelium 49, 50, 154, 203, 205, 208, 241
Comoro Islands 23
Conducting tissue of the heart 75
Conduction 181, 190, 327
Condyloid joints 117
Cones 412-416, 419
Confuciusornis 478
Conjuctiva 50, 409, 417
Connective tissue 3, 9, 32, 47, 53-58, 61, 62, 64-66, 74, 75, 79, 81, 82, 91, 113, 120, 123, 126, 127, 134, 146, 154, 155, 157, 170, 171, 179, 201, 240, 366, 404, 413, 417, 478
Continental Drift 6, 471, 511, 513, 514
Continuous capillaries 308, 309
Contour feathers 168
Conus arteriosus 297-299
Convection 181, 219
Coprodeum 209
Coprophagy 212
Coracobrachialis 141
Coracoid 103-105, 107, 141, 237
Coracomandibularis 137, 222
Cornea 172, 306, 409, 410, 414, 415, 417-420
Corpus albicans 375, 376
Corpus luteum 266, 267, 363, 365, 375, 376
Corpuscles of Stannius 368
Corticosteroids 188, 369
Corticosterone 369
Cortisol 355, 356, 369, 429
Cortisone 369
Cosmoid scales 158
Crab-eating frog 432
Cranial kinesis 85, 192
Cranial muscles 84, 135
Cranial nerves 335, 340, 348, 349, 352, 354
Creatine 426
Creatinine 422, 426
Cremaster muscle 257
Creodonts 497-499
Cretaceous period 473, 485
Cribriform plate 199, 389
Cristae 405, 407
Crista neglecta 406
Crop 200, 202, 363
Crossopterygians 23, 158
Ctenohystrica 488, 489
Ctenoid scales 159
Cuboidal epithelia 49, 50
Cuchia 230
Cucullaris 136, 349
Cupula 396, 398, 406, 407
Cursorial locomotion 147
Cutaneous muscle 33, 74, 126, 143, 164
Cutaneous receptors 384
Cyclic guanosine monophosphate 444
Cycloid scales 21, 158, 159
Cyclostomes 12, 13, 38, 87, 202, 349, 391, 435
Cynodonts 470, 482, 483

D

Dead Sea 11, 425
Deamination 189, 217, 426, 433
Delamination 273-275
Delta cells 373

Deltoid muscle 142, 143
Deltoideus 136, 142
Demes 457
Dendrites 66-68, 70, 338, 390, 396, 397, 400, 412
Dendritic cells 169
Dendritic knobs 386
Dens 15, 54, 56-58, 65, 74-77, 81, 83, 94, 115, 116, 488, 492
Dense fibrous tissues 54, 56
Dense irregular connective tissue 56
Dense regular connective tissue 56, 81
Depolarization 336, 382, 383
Dermatocranium 87
Dermatome 155, 281, 282
Dermis 28, 154-158, 162-167, 169, 170, 174, 177, 180, 387, 388
Dermoptera 37, 494, 495
Detrusor muscle 446
Developmental biology 4, 22, 248, 249, 251, 253, 255, 257, 259, 261, 285, 287, 289, 291
Devonian period 14, 25, 39, 463
Diabetes mellitus 373
Diaphragm 33, 140, 235, 241-243
Diaphragmatic muscle 235
Diapophyses 92
Diarthroses 113
Diastole 297
Diencephalon 281, 282, 338, 339, 342, 360, 419
Digastricus 136
Digestion 4, 88, 130, 186, 187, 189, 191, 193, 195, 197, 199, 201, 203, 205, 211-213, 215, 217, 310, 339, 367, 373
Digital organs 169, 176, 177
Digital pads 166, 176
Digitigrade foot 112
Digits 89, 96, 106-108, 111-114, 147, 166, 495, 499
Dimetrodon 469, 470
Dinosaurs 28, 40, 42, 106, 164, 167, 459, 468, 469, 471-476, 478, 481, 482, 484, 485, 497
Diphycercal fin 102
Diplospondyly 91
DNA molecule 453
Dodo 29, 141, 511
Dolphins 183, 219, 246, 395, 405, 432, 469
Dopamine 366
Dorsal (adductor) muscles 141
Dorsal intercalary plates 92, 93
Dorsal nerve root 347
Dorsal ramus 347
Dorsal serrate muscles 140
Dorsalis trunci 139
Dorsals 95
Dorsobronchi 240
Down feathers 167, 168
Dryopithecus 504
Duck-billed platypus 485
Ductus arteriosus 228, 298, 299, 304-306
Ductus venosus 305
Dugongs 36, 107, 491, 496, 497
Duplex uterus 265
Dura mater 388
Dynamic equilibrium 407, 455
Dynorphin 363

E

Ear 32, 48, 57, 67, 84-87, 132, 166, 175, 311, 349, 388, 393, 394, 396, 400-403, 405, 407, 482
Ear drum 394, 402
Ebu natriuretic peptide 378
Eccrine sweat glands 53, 174
Echidnas 34, 485
Echolocation 395, 404, 405

Ecoestrogens 374, 375
Ectoderm 9, 70, 132, 154, 158, 194, 196, 275-283, 285, 355, 360, 398
Ectothermy 41, 342, 476
Edaphosaurus 469
Edema 312, 330
Edentata 36, 492
Efferent ductules 256, 257
Elasmobranchs 15, 16, 158, 221, 322, 378, 430, 432, 445
Elastic cartilage 57
Elastic fibers 54, 56-58, 126, 155, 165, 170, 176, 182, 200, 202, 239, 297, 307, 308
Electric catfish 22, 150, 157, 386
Electric eel 22, 150, 151
Electric organs 3, 150, 151, 157
Electrophorus 150, 151, 231
Electroplaques 151
Electroreceptors 16, 22, 151, 383, 385, 386
Elephant birds 31
Elephant shrews 36
Elephants 36, 41, 256, 265, 240, 487, 490-493, 496, 498, 509
Eleutherodactylus 26
Embryo transfer 291
Embryonic disk 271, 272, 174, 275, 277
Emu 29, 31
Encapsulated nerve endings 383, 386, 387
Endocardium 76, 296, 297, 308
Endocrine organs 5, 306, 309, 328, 335
Endocrine reflex 357
Endocrine system 1, 5, 65, 119, 335, 354, 355, 364, 379, 428
Endoderm 84, 200, 275-280, 282-284, 355, 360, 392, 446
Endolymph 311, 395, 396, 399, 402, 405-407
Endomysium 74, 75, 125
Endoneurium 347
Endorphins 363, 364
Endostyle 367, 368
Endothelium 49, 289, 308, 331
Enkephalins 363
Entoglossal process 137
Eocene epoch 469, 492, 499, 500
Eosinophils 318, 319
Epaxial muscles 133, 139
Ependymal cells 69-71
Epiblast 273, 277-279
Epiboly 27
Epibranchial muscles 132, 137
Epibranchial musculature 136
Epicardium 75, 296, 297
Epidermis 154-157, 163, 165, 169-171, 173, 174, 176, 177, 180, 182, 275, 311, 387, 398
Epididymis 48, 254, 256, 257
Epimysium 74, 75, 120
Epinephrine 9, 156, 281, 327, 337, 352, 353, 356, 370
Epineurium 347
Epipubis 35
Epithelial tissue 47, 48, 409
Epithelioid cells 49
Epiural bones 102
Ergot 176
Erythrocytes 5, 55, 312, 315-317, 321
Erythrophores 156
Esophagus 18, 85, 191, 193, 198-204, 206, 207, 220, 221, 224, 229, 230, 232, 349, 368
Estradiol 355, 374
Estrogens 176, 290, 374-376
Estrone 374
Estrous cycle 266, 376
Estrus 266
Euarchontoglires 37
Eugenics 511

Euryhaline species 427
Eusthenopteron 111, 112, 464
Eutherian mammal 35, 265, 269, 274, 283, 286, 485
Evaporation 27, 42, 174, 181
Evolution 2, 6-8, 10, 11-13, 20, 23, 34, 36-38, 40, 71, 97, 121, 183, 207, 246, 338, 378, 395, 424, 434, 453, 468, 471, 476, 479, 483, 486, 491, 498, 505, 511
Evolution of vertebrates 6, 101, 322, 449, 457, 511
Excretion 3, 5, 6, 40, 59, 65, 79, 150, 153, 173, 220
Exocrine glands 49, 40, 51, 52, 154, 358, 359
External auditory meatus 89, 175, 394, 395, 402, 403
External intercostal muscle 243
External oblique muscle 140, 142, 144
Exteroreceptors 383, 384
Extinction 11, 26, 39, 40, 449, 459-462, 469, 471-473, 475, 479, 484, 485, 498, 508, 509, 511
Extra-embryonic coelom 284
Extra-embryonic membranes 27, 278, 283
Extraocular muscles 124, 135, 417
Extrastapes 402
Eyebrows 417, 418
Eyelashes 173, 417, 418
Eyelids 174, 408, 417, 418

F

Facial nerve 392, 393
Facial pit organs 385
Factor X 323
Falciform process 410
Fascicle 74, 75, 120, 347
Fast phasic (white) fibers 125, 131
Fatty acids 188, 197, 206, 207, 212, 216, 357, 373
Feathers 28, 41, 42, 97, 154, 156, 165-169, 173, 181, 403, 478-480
Femur 59, 89, 96, 109-111, 113, 145, 464
Fenestrated capillaries 309, 332
Fetal circulation 304, 305
Fetal hemoglobin 246, 305
Fibrin 320, 321, 323
Fibrinogen 8, 320, 323
Fibrocartilage 57, 58, 91, 113, 114, 121
Fibrolamellar bone 481
Fibrous connective tissues 53, 54
Fibrous joints 113, 407
Fibula 89, 96, 110, 112-114, 464, 466, 469, 485
Filoplumes 168
Fins 10, 11, 13, 15-25, 39, 88, 100-102, 108, 111, 141, 146, 151, 221, 254, 461-465
Fish scales 157, 159
Fishes 5, 9, 10, 14, 16-18, 20-22, 27, 34, 38, 41, 61, 84, 91, 103, 160, 194, 225, 254, 322, 340, 371, 385, 390, 405, 428, 459, 460, 463
Flaps 12, 160, 161, 302, 303, 306
Flehmen reaction 392
Flexor hallucis longus muscle 144
Flight 4, 28, 40, 107, 141, 148-150, 153, 166, 253, 479, 480
Flight feathers 168
Flightlessness 480
Floating ribs 100
Flying lemurs 37, 494, 495
Follicle stimulating hormone 252, 258, 266, 292, 363, 374, 376
Foramen magnum 344, 500
Foramen of panizza 299, 300
Foramen ovale 305
Foregut fermenters 206

Fossils 12, 17, 340, 450, 457, 462, 466, 470, 473, 474, 477-479, 481, 487, 490, 492, 495, 499, 504, 506-508, 510, 514
Fossorials 148
Fourth ventricle 339, 344
Fovea 94, 409, 413, 416
Frameshift mutation 454
Frasnian-Famennian extinction 462
Free nerve endings 180, 383, 386, 387
Freemartin 288
Fur 30, 34, 43, 173, 182, 480, 492, 494
Furcula 105, 107, 478

G

Gamma-aminobutyric acid 337
Ganoid scales 21, 158
Garpikes 20, 21, 158
Gas gland 229
Gastric inhibitory peptide 377, 378
Gastrin 203, 337, 377
Gastrocnemius muscle 144
Gastropore 275
Gastrulation 273-279, 281, 284
Gene flow 451, 457
Genetic drift 451, 455, 456
Genioglossus 137
Geniohyoid 137, 224
Geological time scale 459
Giant anteaters 36
Giant ground sloth 41, 492, 494, 508, 509
Giant panda 111, 268
Gibbons 148, 501, 503-505
Gill filaments 192, 220, 224, 226, 230, 430
Gill slits 9, 10, 15, 38, 220-222, 231, 462
Gills 6, 9, 11, 15-17, 20, 24-26, 39, 136, 151, 219-232, 245, 298, 299, 368, 426, 429-431, 433
Giraffes 37, 487
Gizzards 202
Glands of Moll 175, 417
Glandular epithelia 50
Glaucoma 414
Gliding 116, 117, 126, 148-150, 153, 480, 495
Gliding joints 116, 117
Gliridae 488
Globulins 322
Glomerulus 426, 434, 437, 438, 439-441
Glossopharyngeal 348
Glottis 227, 232, 234, 240, 242
Glucagon 215, 356, 372, 373
Glucocorticoids 369, 370
Gluconeogenesis 187, 216, 217, 373
Glucose 5, 59, 129, 131, 134, 187-189, 206, 211, 215-218, 322, 353, 373, 375, 430, 442
Glutamate 337
Glycerol 187, 188, 212, 216
Glycolysis 187, 215, 216
Goat 37, 142, 161, 205
Golgi tendon organs 122, 123, 384, 388
Gomphoses 113
Gonadocorticoids 369, 370
Gondwanaland 36, 479, 512
Gonopodium 255
Gorillas 501, 503
Gray matter 34, 68, 70, 343, 346, 347, 351
Great Salt Lake 11, 425
Growth hormone 256, 258, 262, 263, 273, 376, 377
Growth rings 159
Guard hair 30, 182
Gustatory cells 393
Gustatory stimuli 392
Glyptodonts 493
Gyri 34, 343

H

Hagfish 12-14, 38, 82, 85, 92, 157, 220, 221, 312, 349, 353, 363, 372, 378, 379, 389, 405, 408, 410, 428, 434, 435, 441, 460

Haikouichthys ercaicunensis 101

Hair 29, 30, 41, 101, 154, 155, 164, 169-183, 185-188, 193, 397, 400, 401, 406, 482, 485, 489, 490, 493

Hair cells 385, 386, 393, 394, 396-400, 403-407

Hair follicle 30, 170, 171, 173-176, 179, 183, 397

Harderian gland 418

Hares 37, 147, 210, 212, 363, 490

Haversian canal 60

Haversian system 60, 82, 481

Head kidney 435

Heart 2, 3, 10, 29, 32, 47, 50, 72, 75, 132, 295-297, 299-306, 308, 311, 326, 327, 330, 353, 365, 437, 441, 476

Hedgehogs 37

Hemal arches 90, 93

Hematopoietic stem cells 134

Hemibranch 221, 223

Hemocytoblast 321

Hemoglobin 8, 213, 319, 229, 245, 246, 305, 317, 322, 454, 492

Hemolymphatic vessels 312

Hemopoiesis 214, 287, 321

Hermaphrodites 249, 250

Heterocercal fin 102

Heterocoelous vertebrae 91

Heterophils 318

Hindgut fermenters 212

Hinge joints 116

Histamine 219, 337

Holobranch 221, 223

Holocene epoch 500, 508

Holocephalans 15, 16, 85, 372, 462

Holocrine secretion 53, 174

Holonephros 434

Holostei 20

Homeotic (Hox) genes 281

Homo erectus 501, 505, 506

Homo habilis 501, 505, 506

Homo heidelbergensis 506

Homo neanderthalensis 501

Homo sapiens 501, 503

Homo sapiens sapiens 507

Homocercal fins 101, 102

Homology 3, 8, 99, 134

Hooves 112, 155, 169, 171, 176, 177, 486, 487, 493

Horizontal septum 98, 132, 136, 139

Hormonal action 5, 6, 356

Hormones 4, 5, 50, 59, 169, 176, 188, 217, 253, 262, 266, 288, 335, 354-359, 361, 363, 367, 379, 428, 437, 438

Horns 4, 154, 155, 169, 178, 265, 266, 473

Horses 37, 41, 112, 288, 289, 508, 509

Humerus 59, 89, 105-107, 108, 111, 116, 120, 147, 239, 450, 464

Hyaline cartilage 57, 58, 62-64, 113, 114

Hyracoidea 491, 492

Hydroxyproline 190

Hyoglossus 137

Hyomandibular cartilage 84, 86

Hyostylic jaw suspension 85

Hypaxial muscles 92, 132, 133, 139, 140

Hypoblast 273-279

Hypobranchial muscles 132, 136, 137, 349

Hypocalcin 368

Hypocentrum 97, 98

Hypodermis 3, 126, 155, 165, 166, 169, 176, 179, 180, 182, 478, 496

Hypoglossal nerve 349

Hypothalamohypophyseal portal system 361, 364
Hypothalamus 181, 250, 310, 325, 339-342, 344, 353, 358, 359, 361, 362, 364, 379, 385, 433, 443
Hypotonic solution 423
Hypural bones 102
Hyraxes 36, 491-493

I

Icefish 316, 317
Ichthyosaurs 40, 459, 468, 475, 476
Ichthyostegalians 464
Iliocostalis 140
Iliopsoas muscle 144
Ilium 35, 89, 96, 105, 109, 110, 209, 412, 473, 494
Immune system 13, 14, 210, 312, 313, 314, 319, 320, 364, 377, 463
Incus 32, 86, 394, 395, 408
Infraspinatus 141
Infratemporal fossa 467, 468, 481
Infundibulum 261, 263-265, 359, 360, 362
Innate immune system 13, 312, 314
Innervation of muscle 112
Insectivora 487
Insulin 59, 215, 356, 357, 372, 373
Integument 3, 28, 47, 50, 83, 146, 153, 167, 275, 478
Integumentary skeleton 83
Interbranchial septae 221, 222
Intercalated disks 75
Interclavicle 104, 105
Intermediate mesoderm 282, 283
Intermediate phasic (pink) fibers 132
Internal carotid arteries 344
Internal intercostals muscle 175
Internal oblique muscle 140
Interneurons 68, 346, 347
Interoreceptors 383, 384
Interrenal cells 370, 371
Interrenal glands 370
Interstitial cells of Leydig 258, 373, 374
Intervertebral disks 57, 82, 91, 113, 147
Intervertebral canal 82, 99
Intestines 49, 52, 73, 201, 202, 205, 207, 208, 210-212, 214, 231, 244, 259, 262, 290, 313, 368, 372, 426, 431
Involution 266, 276-279, 290
Iodopsin 413
Iridophores 156, 160, 163
Iris 73, 132, 409-411, 414, 415, 417, 419, 509, 510
Irish elk 509, 510
Ischiopubic cartilage 109
Ischium 35, 89, 109, 110, 164, 473, 494
Isometric contractions 130, 131
Isotocin 379
Isotonic solution 424

J

Japanese eel 427
Japanese flounder 427, 446
Jawed fish 14, 84, 108, 135, 196, 231, 379, 388, 391, 459-461
Joey 264
Joint capsule 3, 57, 114, 384
Joints 56, 81, 92, 99, 112-117, 119, 126, 127, 143, 145, 166, 240, 407, 492, 494
Junggarsuchus sloani 476
Jurassic period 40, 462, 512
Juxtaglomerular apparatus 438-440

K

Kangaroos 35, 147, 258, 264, 265, 290, 291, 413, 486
Keel 28-31, 100, 237, 480
Keratin 49, 50, 83, 154, 155, 158, 163, 172
Keratinocytes 153, 154, 169, 182
Kidneys 6, 39, 65, 263, 285, 306, 310, 325, 330, 356, 367, 434, 436, 437, 445, 446

Kiwi 29, 31, 390
Koala 186, 264, 265, 486
Krause's end bulbs 387

L

Labyrinthodonts 39, 464, 465
Lacrimal glands 350, 418
Lactiferous duct 175, 176
Lagena 400, 402, 403, 405, 406
Lagomorpha 37, 490
Lamellar zonal bone 481
Lampreys 12, 13, 38, 85, 92, 157, 191, 214, 346, 389, 429, 441, 460
Lapillus 405
Laryngotracheal chamber 232, 233
Larynx 57, 136, 138, 198, 199, 232-237, 241, 242, 349, 362, 366, 404
Lateral line system 16, 349, 393, 396, 398, 404
Lateral rectus muscle 135
Lateral ventricle 339, 344
Laterobronchi 238
Latissimus dorsi muscles 141
Laurasia 37, 479, 494, 512
Laurasiatheria 37
Lemurs 37, 147, 494, 495, 500, 502
Lens 49, 205, 408-411, 414-417, 419, 420
Lepidosaurs 468, 470, 476
Lepidosiren 24, 227, 228
Lepidotrichia 100, 101, 159, 465
Leporidae 490
Lesser panda 111
Leucocytes 5, 55, 134, 318
Levator hyomandibulae 136
Ligaments 3, 56, 57, 81, 114, 115, 117, 143, 145, 147, 306, 410, 411, 414-416
Light organs 161
Lipids 15, 188, 212, 216, 217, 287, 322, 413
Lipotyphyla 36
Lissamphibia 25, 39
Lithosphere 512, 514
Liver 15, 17, 47, 57, 180, 187-199, 202, 211-217, 235, 296, 305, 310, 322, 378, 432, 433, 436
Loaches 231, 292
Long flexor muscles 144
Long-horned bison 509
Longissimus 140
Loop of Henle 429, 432, 433, 437-439, 441-445
Loose connective tissue 54, 126, 127, 170, 182, 200, 258, 360
Loricariidae 231
Lorises 500-502
Lungfish 18, 24, 25, 39, 193, 227, 298, 317
Lungs 15, 17, 23, 25-27, 29, 39, 162, 218, 227, 228, 231-246, 298, 301, 350, 367, 389, 463
Luteinizing hormone 258, 266, 363, 374, 376, 391
Lymph 49, 55, 56, 199-201, 294, 310-314, 321, 395
Lymph follicles 313
Lymph hearts 310, 312
Lymph nodes 312-314, 321
Lymphatic capillaries 310, 311, 329, 330
Lymphatic system 5, 58, 294, 310-312, 338
Lymphocytes 153, 199, 312-314, 319-321, 333, 376, 377, 463
Lymphoid organs 55, 310, 312, 319, 321, 376
Lysozyme 197, 207, 418

M

Macroscelidea 36
Macula densa 440, 444
Maculae 405, 407

Madagascar 23, 29, 31, 42, 466, 470, 473, 484, 500
Malay frog 479
Malleus 32, 86, 394, 395, 408
Mammals 3, 4, 9, 10, 27, 29, 30, 32, 176, 187, 196, 241, 265, 357, 368, 388, 401, 405, 433, 436, 443, 459, 480, 497-500, 508
Mammary glands 29, 33, 52, 53, 175, 265, 267, 290, 363, 365, 376
Mammoth 41, 491, 508, 509
Manatees 36, 107, 491, 497, 498
Mandrill 172, 502
Marine birds 432, 433, 479
Marlins 21, 22
Marsupials 33-35, 148, 149, 264, 265, 267, 290, 485, 486
Marsupium 35, 265, 486
Masseter 136, 142
Massive extinctions 11
Mast cell 54, 313, 318
Mastication 191, 193, 194, 481
Mechanoreceptors 122, 123, 383, 384, 394, 403, 407
Medaka 22, 455
Medial rectus muscle 135
Medulla oblongata 331, 332, 339-344
Megachiroptera 495
Megafauna 508, 509
Meiosis 248, 250-253, 270, 455
Meissner's corpuscles 388
Melanin 156, 160, 165, 168-171, 363, 409, 420
Melanocyte stimulating hormone 356, 363
Melanophore stimulating hormone 156, 363
Melanophores 156, 160, 163, 165, 170
Melatonin 356, 365, 420
Membrane bone 161
Membranous labyrinth 395, 405
Meninges 337, 338
Menstrual cycle 266
Menstruation 266
Merkel's disks 387
Merocrine secretion 53
Mesangial cells 439, 440
Mesencephalon 282, 338, 339, 360
Mesenchymal cells 58, 61, 62, 134, 276, 281
Mesendoderm 275
Mesobronchus 238
Mesoderm 70, 88, 132-134, 155, 275-279, 281-285, 337, 355, 446
Mesonephric duct 436
Mesorchia 253, 254
Mesosuchians 476
Mesotendon 126
Mesothelia 49
Mesotocin 379
Mesovaria 259
Mesozoic era 20, 23, 27, 40, 469, 473, 479, 485, 500
Metabolism 29, 41, 43, 59, 65, 98, 181, 190, 214, 215, 232, 354, 362, 365, 367, 369, 375, 376, 424, 439
Metacarpals 106-108, 147
Metanephric kidneys 436
Metatarsals 110, 113, 147, 148
Metatheria 34, 35
Metencephalon 282, 339, 360
Met-estrus 266
Microchiroptera 404, 495
Microglia 69-71, 79
Middle cerebral arteries 344
Milt 253
Mineralocorticoids 369, 444
Minerals 4, 59, 187, 190, 212, 369
Miocene epoch 502, 504
Missense mutation 454
Mississipian period 39, 459
Moa 8, 9, 31

Moeritheres 491
Mole rat 414
Moles 36, 37, 487
Molting 163, 261, 367
Monkeys 29, 148, 175, 207, 210, 288, 289, 500-502
Monocytes 320
Monotremes 34, 35, 104, 176, 233, 256, 263, 267, 269, 405, 408, 446, 485
Morula 274
Motor end plates 122, 128
Motor nerves 122, 123, 310, 335, 347
Motor unit 124
Mucous glands 157, 162, 164, 198, 202, 241
Mudpuppy 26, 231
Multiaxial joints 115, 117
Multifidi 140
Multipolar neurons 67, 347
Multituberculates 35, 484, 485
Muscle contraction 3, 65, 73, 75, 77, 78, 89, 120, 122, 125, 128-130, 137, 190, 350
Muscle homology 134
Muscle spindles 122, 123, 388
Muscular tissue 47, 72, 132, 198
Muscular system 3, 81, 119, 145, 383
Mutagens 453, 455
Mutation pressure 451, 453
Myelencephalon 282, 339
Myelin sheath 14, 67, 69-71, 128
Mylohyoid 136
Myeloid cells 333
Myocardium 76, 134, 296, 297, 303
Myofibrils 74, 76-78, 133, 134
Myofilaments 75, 77, 78, 131, 133, 134
Myomeres 136, 137, 139, 146
Myoseptae 139
Myosin 73, 78, 79, 119, 129-131, 133
Myotome 132, 133, 135, 140, 230, 281, 282
Myxine glutinosa 428

N

Nails 4, 153, 169, 171, 176, 492, 501
Naming of muscles 127
Nasal gland 433
Nasal turbinates 32, 199, 237, 481
Nasopalatine duct 391, 392
Nasopharynx 234, 241, 242, 395
Natural selection 37, 408, 451, 452, 507, 510
Neoceratodus 24, 227, 228
Neognathae 29, 40
Neopallium 34, 343
Neopterygii 18, 20
Nephric ridge 282, 434-436
Nephrogenic cord 436
Nephron 437, 442
Nervous system 5, 9, 10, 34, 49, 66-71, 74, 75, 122, 124, 132, 166, 173, 181, 196, 200, 201, 276, 281, 306, 327, 331, 335, 338, 350, 358, 378, 379, 382, 383, 485
Neural arch 22, 90-93, 95, 97, 98, 399
Neural crest cells 9, 103, 117, 281, 282, 337
Neural tube 9, 70, 71, 81, 82, 132, 279-282, 285
Neurilemma 70, 71
Neurocranium 85, 89
Neuroglia 66, 69, 70, 281, 346, 361
Neurohypophysis 359-362, 364, 379
Neuromasts 16, 398
Neurons 9, 13, 14, 66-71, 128, 281, 291, 292, 335, 336, 338, 340, 341, 343, 347, 348, 351, 352, 366, 393, 400, 411-413
Neuropeptides 337
Neurotransmitters 335-337, 351, 363, 379

Neurulation 279-281
Neutrophils 71, 154, 318, 319
Nictitating membrane 137, 417, 418, 478
Nidamental gland 259, 260
Nitric oxide 59, 331, 337, 433
Nociceptors 383, 385-387
Nodes of Ranvier 71
Nonsense mutation 454
Nonsteroid hormones 355, 356
Noradrenalin 156, 336
Norepinephrine 156, 327, 337, 352, 353, 356, 370
Notarium 95
Notochord 1, 3, 9, 10, 13, 14, 19, 23, 25, 81-83, 85, 90, 92, 97, 101, 278-282, 285, 462
Nucleus pulposus 82
Nutrition 4, 79, 88, 126, 166, 170, 178, 186, 191, 233, 269, 290, 316, 510
Nutrition and digestion 4, 88, 186

O

Occipitofrontalis 136
Ochotonidae 490
Oculomotor nerve 135, 354
Odontoid process 94, 116
Oilbirds 404
Olfactory bulbs 340
Olfactory epithelium 48, 388-392
Olfactory nerve 349, 389-391, 393
Olfactory neurons 67, 390, 392, 393
Olfactory receptors 388, 390
Oligocene epoch 469, 502
Oligodendrocytes 69-71
Omasum 205-207
Omohyoideus 138
Oogenesis 252, 259
Oogonia 252
Opercularis muscle 401, 402
Operculum 16, 17, 151, 223, 224, 401, 402
Ophiacodon 469
Opisthocoelous vertebrae 90
Opisthonephros 431, 436
Opposum 35, 148, 343
Opsins 412
Optic lobe 339, 341-343
Optic nerve 341-343, 365, 409, 411, 414, 418
Orangutans 501, 503, 505
Orbicularis oculi 136
Ordovician period 460
Organ 3, 20, 21, 46, 51, 56, 57, 151, 153, 156, 162, 180, 183, 193, 198, 366, 376, 386, 391, 396, 408
Organ of Corti 395, 396, 405
Organization of the vertebrate body 46
Ornithischian dinosaurs 40, 472, 473
Oropharynx 224, 242
Os clitoridis 33
Os cordis 32, 61
Os penis 32, 61
Os rostrale 32, 61
Osmoconformers 423, 428
Osmoregulation 3, 6, 153, 220, 363, 366, 378, 423-427, 429, 432, 434
Ossification 20, 23, 32, 58, 61-65, 84, 86, 97, 146, 155
Osteoblasts 59-62, 64, 65, 90, 98, 157, 481
Osteoclasts 62, 64
Osteocytes 59-61, 83, 98, 481
Osteoid 61
Osteons 60
Ostracoderms 12, 13, 25, 38, 83, 87, 155, 420
Ostrich 29-31, 42, 166, 312, 346, 433
Otoconia 406

Otoliths 399-401, 405-407
Outer hair 173
Ova 248, 252, 253, 258
Oval window 394-397, 401-403
Ovary 39, 49, 252, 254, 258-262, 265, 306, 314, 358, 375
Oxytocin 337, 356, 364, 365, 376, 379, 444

P

Pacinian corpuscles 387
Paddlefishes 18, 19, 92, 157
Paleocene epoch 599
Paleognathae 29, 31, 40, 479
Paleozoic era 23, 514
Pancreas 47, 50, 53, 199, 202, 211, 212, 214, 215, 350, 358, 371, 372
Pancreatic islets of Langerhans 215
Pancreatic polypeptide 372, 373
Pangaea 425, 459, 473, 485, 511, 513, 514
Pangolins 37, 492, 493
Parabronchi 239, 240
Parachuting 148, 149, 480
Paranasal sinuses 49, 241, 497
Parapophysis 98, 99
Parasympathetic nervous system 75, 327, 353, 371
Parathyroid hormone 65, 180, 356, 368
Paratympanic organ 404
Paraxial mesoderm 278, 281-283
Parchment-shelled eggs 34
Parthenogenesis 10, 248, 249, 269
Patella 33, 89, 110, 113, 143, 145
Pecten 410
Pectoral fins 15, 16, 18, 25, 221
Pectoral girdle 61, 84, 103-105, 109, 137, 141, 143, 296, 401, 480
Pectoral girdle muscles 141, 143
Pectoral limb 102, 104, 106, 107, 111, 141-143, 495
Pelvic fins 24, 108, 151, 254, 255
Pelvic girdle 27, 96, 102, 103, 107-110, 133, 140, 143, 147, 148, 235, 478
Pelvic girdle muscles 143
Pelycosaurs 469, 470, 481
Pennation 121, 122
Pennsylvanian period 39, 459
Pericardium 49, 57, 295-297
Perichondrium 58, 63, 64, 79
Perimysium 74, 75, 120
Perineurium 347
Periophthalmus 231
Periosteum 60-64, 82, 115, 119, 120, 338
Perissodactyla 37
Perissopterygium 101
Peritoneum 49, 98, 257, 261, 282, 436
Permian period 464
Pesticides 374
Peyer's patches 313
Phalangers 35
Pharyngeal pouches 9, 368, 376
Pharyngula 9, 10
Pharynx 9, 10, 39, 85, 138, 191, 198, 199, 201, 207, 220, 221, 223-225, 227, 232, 233, 243, 366, 392, 393, 400
Pheromones 179, 349, 391, 392, 418
Pholidota 37, 492
Phospholipids 163, 188, 190, 216, 243, 323
Photoreceptors 365, 383, 385, 412, 418, 419
Pigs 29, 37, 112, 173, 288, 289, 413, 486, 491, 511
Pika 458, 490
Pikaia gracilens 458, 460
Pillar cells 225
Pineal gland 339, 343, 344, 358, 365, 366, 418-420
Pinealocytes 365
Pinna 394, 396
Piranhas 22
Pit organs 16, 385

Pit vipers 192, 385, 471, 472
Pituitary gland 169, 266, 281, 310, 325, 339, 341-344, 359-364, 376, 377, 379, 427
Pivot joints 116
Placenta 35, 50, 260, 264, 274, 283, 286-290, 293, 365, 376
Placental lactogens 376
Placoderms 25, 38, 420, 461-463
Placoid scales 14, 158, 159, 194
Plantigrade stance 112, 148, 498
Plasma cells 313, 314, 320, 322
Plate tectonics 512, 514
Platelets 79, 315, 318, 320, 321, 323
Platynereis dumerilii 412
Platysma 136
Pleistocene epoch 459, 500, 508
Plesiosaur 477
Pleurapophyses 96
Pleurocentra 98
Plexuses 200, 201, 248, 348, 353, 387
Pliocene epoch 459
Point mutations 454
Poison glands 52, 157, 162
Polar body 252, 270
Portal lobule 213
Portal system 310, 361, 364, 441
Posterior cavity 409, 414
Posterior cerebral arteries 344
Posterior chamber 409, 414
Posterior communicating arteries 344
Postganglionic neuron 351-353, 370
Post-temporal bone 104
Postzygapophyses 90, 92
Precocial chicks 42
Precocial young 267
Preen gland 166
Preganglionic neuron 351-353
Presbyopia 415
Prezygapophyses 90
Primates 32, 33, 37, 117, 124, 131, 147, 148, 266, 408, 499, 501-503, 506
Primitive meninx 337
Primitive streak 277-279
Proatlas 95
Proboscidea 36, 37, 490
Procoelous vertebrae 90
Proctodeum 209, 210
Proganochelys 468
Progesterone 176, 266, 267, 287, 290, 356, 363, 375, 376
Prolactin 356, 361-363, 376, 427
Prolyl hydroxylase 190
Pronephric kidneys 435
Proprioceptors 123, 383, 384, 407
Prosencephalon 338, 339
Prosimians 500-502
Prostaglandins 6, 188, 216, 357, 358, 440
Prostate gland 256, 367
Protandry 250
Proteins 5, 8, 71-73, 77, 79, 119, 133, 186, 187, 189, 202, 204, 206, 211, 212, 214, 216, 217, 244-246, 252, 269, 287, 292, 322, 324, 355, 357, 442, 443, 451, 453, 454, 492
Prothrombinase 323
Protogyny 250
Protoprimates 499, 501
Protopterus 24, 227, 228, 298
Protosuchians 476
Prototheria 34, 40, 484, 485
Proventriculus 200-205
Pseudostratified epithelia 48, 50
Psoas minor 140, 144
Pterosaurs 100, 107, 149, 150, 459, 468, 472, 474, 479, 481
Pterygoids 136
Pterygophores 101
Pubis 35, 96, 109, 110, 164, 235, 472, 473, 478
Pupil 409, 411, 413-417, 419

Purgatorius 499
Purkinje fibers 76, 327
Pygmy marmoset 502
Pygostyle 96, 97, 166
Pyloric sphincter 201, 203
Pyrite 450

Q

Quadrate bone 30, 32, 84
Quadratus lumborum 140
Quadriceps femoris 143

R

Rabbits 37, 112, 145, 147, 210, 212, 289, 343, 490
Radial glial cells 71
Radiation 27, 37-40, 156, 160, 181, 419, 455, 459, 482, 486, 488-490, 500, 508
Radius 89, 106-108, 112, 113, 116, 120, 147, 316, 450, 464, 469
Raia 151, 509
Ram ventilation 223
Rathke's pouch 359, 360
Ratites 30, 31
Rattlesnakes 186, 198, 472
Rays 14-18, 20, 22, 39, 101, 112, 151, 162, 221, 223, 255, 372, 414, 415, 418, 419, 462
Rectal salt gland 431, 432
Rectus abdominis 121, 133, 140, 142
Reedfish 18, 19, 158
Reflex arc 68, 346, 351, 384
Relaxin 129, 165, 376
Renal corpuscle 437, 439, 440
Renal pelvis 437
Reproduction 4, 5, 10, 33, 153, 217, 249, 252, 260, 263, 290, 349, 354, 365, 391, 392, 450, 454-456
Reptiles 10, 27-29, 34, 39, 41, 86, 90, 95, 143, 155, 194, 203, 253, 278, 286, 299, 339, 363, 389, 402, 403, 433, 459, 476, 481, 485, 489
Respiratory system 4, 48, 198, 209, 218, 219, 223, 232, 239, 464, 474
Rete mirabile 229
Rete testis 256, 257
Reticular cells 56, 83, 207, 312
Reticular formation 339, 340
Reticular tissue 55
Reticuloendothelial system 56, 312
Reticulum 77, 78, 128, 129, 131, 206, 207, 314, 340
Retina 49, 67, 132, 190, 365, 409-416, 419, 420
Retinal 190, 365, 409, 411, 414, 420
Retractor bulbi 135
Retractor lentis 410, 419
Rhabdomeric photoreceptors 412
Rhea 29, 31
Rhino 487
Rhipidistians 23, 25, 463, 464
Rhodopsins 385
Rhombencephalon 338, 339, 360
Rhomboideus muscles 141, 142
Ribs 27, 88, 92, 93, 95, 98-100, 113, 140, 148, 164, 234, 240, 243, 399, 493, 497
Righting reflex 407
Rigor mortis 129, 130
Rodents 32, 37, 112, 145, 148, 202, 212, 257, 265, 286, 292, 343, 373, 488-490, 502
Rods 61, 97, 101, 310, 412-416, 419
Root hair plexuses 387
Round window 394, 395, 397, 401
Ruffini's corpuscle 387
Rumen 175, 205-207, 394, 486

S

Saber-toothed cat or tiger 509
Saccule 395, 399, 406, 407
Sacrum 89, 96, 97, 109, 110, 139, 144, 147

Saddle joints 117
Saliva 193, 194, 196-198, 211, 214, 306, 350, 393
Saltatorial locomotion 147
Sarcolemma 74, 76-78, 112, 128
Sarcomeres 77
Sarcoplasm 74, 77, 78, 128, 129, 131
Sarcoplasmic reticulum 77, 78, 128, 129, 131
Sarcopterygians 22, 23, 108
Saurischian dinosaurs 164, 472, 473, 481
Scala vestibuli 396
Scala tympani 396
Scandentia 37, 495, 496
Scansorial locomotion 148
Scapula 89, 103-108, 116, 120, 136, 142, 143, 402
Scent glands 4, 164, 174, 178, 179
Schwann cells 67, 69, 71
Sciuridae 488
Sclera 409-411, 414
Sclerotome 281, 282
Scrotum 35, 256, 257
Sea cows 36, 496
Sea lizards 433
Sea snakes 433
Seal 107, 497
Sebaceous glands 30, 53, 173-175, 179, 417
Sebum 30, 173, 174
Secondary lamellae 177, 220-226
Secondary meninx 338
Secretin 337, 377, 378
Semen 251, 254, 257, 258, 261
Semicircular canal 394, 395, 399, 400, 402, 407, 475
Seminiferous tubules 251, 254
Semilunar valves 303, 312, 326, 327
Semitendinosus muscle 144
Sense organs 9, 157, 164, 382, 386, 388, 398, 418
Sensory nerves 173, 310, 335, 347, 351, 387
Sensory papillae 406
Sensory receptors 155, 170, 180, 382-384, 386, 387, 392, 407
Septal gills 221
Serotonin 320, 336, 337, 340, 420
Sertoli cells 257, 258
Sesamoid bones 33, 61, 106
Sex reversal 250, 254, 262, 292
Sexual recombination 451, 455
Sharks 11, 13, 14-17, 39, 41, 84, 86, 88, 91, 102, 117, 137, 158, 194, 255, 435, 462, 463
Sheep 37, 173, 205, 207, 508
Shrews 36, 146, 404, 487, 488, 495, 496, 499
Sickle cell disease 454
Silent mutation 454
Silurian period 461
Simple epithelium 48
Simplex uterus 266
Sinoatrial node 75, 76, 327
Sinus venosus 296-300, 327
Sinusoidal capillaries 83, 308, 309, 360
Sirenia 36, 497
Skates 14-16, 151, 221, 223, 462
Skeletal muscle 72-74, 77, 119, 120, 122, 123, 125-131, 133, 134, 141, 203, 353, 384, 398, 417
Skeletal system 2, 3, 8, 25, 65, 72, 81, 83, 84, 88, 190, 235, 281, 338
Skin pigmentation 156, 180
Sloths 36, 41, 492, 494, 508, 509
Slow phasic (red) fibers 125, 131
Smooth muscle 28, 29, 72, 73, 75-77, 130, 132, 134, 201, 207, 285, 308, 330, 365, 384, 410, 411, 417, 439, 440, 446
Soaring 148, 150, 480

Soft palate 198, 199, 242
Solenodon 37, 487, 488
Somatic skeleton 83, 88
Somatolactin 362, 363
Somatomedins 362, 363
Somatopleura 283, 284
Somatostatin 356, 366, 372-374, 377
Somites 90, 132, 133, 281-283, 349
Somitomeres 132, 135, 282
Special sense organs 6, 383, 388
Spermatic cord 256
Spermatids 252
Spermatocytogenesis 251
Spermatogenesis 251, 252, 258, 292, 374
Spermatogonia 251, 256-258
Spermatophores 261
Spermatozoa 248, 251-254, 257, 258, 269, 374
Spermeogenesis 252
Sphenacodon 469, 470
Sphenodon 27, 40, 468, 470, 471
Spinal cord 10, 49, 66-68, 70, 71, 82, 85, 123, 335, 345-347, 351, 354, 366, 467, 500
Spinal nerves 82, 90, 133, 336, 345, 346, 348, 349, 352, 403
Spinalis 140
Spiracle 14, 18, 136, 222, 223, 231
Spiral valve 18, 208, 209, 298
Splanchnopleura 283
Spleen 56, 57, 213, 309, 311, 312, 321, 330, 350, 376
Spongy bone 60-62, 115
Spurs 166, 471
Squamates 40, 193, 302, 310, 468, 470, 476
Stapedius muscle 394, 403
Stapes 32, 85, 87, 394, 396, 400-403, 408
Static equilibrium 407
Steller's sea cow 497, 498
Stenohaline fish 427
Sternal ribs 100
Sternocleidomastoideus 136
Sternomastoid 136
Sternothyroideus 138
Sternum 28-31
Steroid hormones 188, 322, 355-357, 369, 373, 375
Stomach 52, 164, 198, 199, 201-208, 210, 211, 213, 300, 350, 354, 378, 486
Stout infantfish 1, 2, 21
Strap-shaped muscles 121
Stratified epithelium 48, 417
Stratum basale 154, 169, 171, 177
Stratum compactum 155
Stratum corneum 154, 155, 162, 163, 165, 169, 170
Stratum laxum 155
Sturgeons 18, 19, 92, 161, 202, 208, 260, 386, 425
Styloglossus 137
Subcoracoscapularis 143
Subgerminal cavity 271-273
Subserous fascia 126
Sulci 34, 343
Sunfish 22
Superficial digital flexor muscle 143, 145
Superficial fascia 126, 143, 155, 182
Superficial gluteal muscle 144
Supracleithrum 87, 103, 104, 109
Supracoracoideus 141
Suprascapula 103, 401
Supraspinatus 141
Supratemporal fossa 468
Surfactants 243
Suspensory ligaments 409-411, 414-416
Sutures 113
Swallowing 193, 196, 198, 242, 368

Sweat glands 3, 4, 50, 52, 53, 166, 169, 174, 181, 197, 394, 417
Swiftlets 404
Swim bladder 15, 17, 20, 21, 39, 219, 228-230, 243, 246, 259, 399, 463
Swimming 2, 3, 41, 89, 96, 102, 105, 139, 145, 146, 158, 182, 191, 219, 223, 228, 247, 269, 432, 469, 476, 477
Swordfishes 21
Sympathetic nervous system 75, 166, 173, 181, 327, 332, 353, 370, 371
Symphyses 113
Synapsids 39, 407, 408, 467, 469, 470, 473, 481
Synaptic cleft 128, 336, 337
Synaptic knobs 66, 67, 336, 338
Synaptic vesicles 336, 337
Synarthroses 113
Synchondroses 113
Syncytium 74, 271
Syndesmoses 113
Synovial bursa 115
Synovial fluid 114, 115, 126, 127, 324
Synsacrum 96, 97, 109
System level 47
Systole 297, 301, 326, 328

T

Tapetum lucidum 413, 419
Tapirs 37, 487
Tarsal glands 52, 174
Tarsal plate 417
Tarsals 110, 113, 147, 148
Tarsiers 501, 502
Tarsometatarsal bone 111, 166
Taste buds 388, 389, 392, 393
Taste pore 393
Tectorial membrane 396, 397, 401, 406
Teeth 2, 4, 13, 14, 17, 21, 22, 29, 32, 36, 40, 87, 88, 113, 148, 158, 193-195, 199, 234, 281, 464, 474, 482, 490, 506, 507, 509
Telencephalon 282, 338, 339, 341, 360
Teleosts 18, 20-22, 39, 92, 157, 225, 312, 345, 378, 426, 429, 435, 436
Telodendria 67, 336
Temporalis 136
Tendinous cords 120
Tendinous inscriptions 121, 140
Tendon sheaths 119, 126
Tendons 3, 33, 56, 57, 59, 81, 115, 119, 120, 121, 126, 144, 306, 384
Tenrecs 36, 484
Tensor fasciae latea muscle 144
Tensor tympani 394
Teres major muscle 143
Terminal cisternae 78
Terminal nerve 349, 391
Terrestrial locomotion 3, 145, 146
Testis 49, 57, 251, 253-258, 262, 358, 363, 365, 373, 374
Testosterone 188, 250, 258, 355, 356, 363, 373, 374
Tethys sea 425, 497, 512
Thecodonts 472, 474
Therapsids 39, 42, 470, 482, 483
Theria 34
Thermoreceptors 383, 385
Theropods 164, 472, 479
Thick filaments 78
Thin filaments 77, 78, 129, 130
Third ventricle 339, 344
Thymic corpuscles 377
Thymopoietin 377
Thymosin 377
Thymus 56, 312-314, 320, 358, 376, 377
Thymus gland 376, 377
Thyroid gland 49, 65, 362, 366-368
Thyroid stimulating hormone 362
Thyrotropin-releasing hormone 356
Thyroxine 356, 362, 367
Tibia 89, 110, 112-114, 145, 464, 466, 469

Tibiotarsus 96, 110
Tiger salamander 340
Tinamiformes 31
Tissue 3-6, 9, 13, 16, 32, 41, 46-48, 53-56, 62, 66, 70, 75, 79, 81, 121, 127, 187, 198, 215, 245, 267, 324, 331, 357, 366, 370, 379, 409, 428, 437, 446, 475, 478
Titanotheres 37
Tongue 137, 138, 192-194, 196, 198, 206, 230, 232, 242, 349, 391-393, 493
Tonic muscle fibers 130, 131
Tonsils 56, 199, 312
Torpedo 151, 220, 496
Totipotent cells 274
Tragi 173, 175
Transitional epithelium 50, 446
Transverse abdominal muscle 140
Transverse tubules 348
Tree shrews 495, 496, 499
Trematic arteries 222
Triad 77
Triadobatrachus 466
Triassic period 470, 472, 474, 476, 477
Trigeminal nerve 348
Triidothyronine 356, 367
Trimethylamine oxide 426, 428, 431
Trochlear nerve 349
Trophoblast 272, 274, 275, 279, 286, 287
Tropic hormones 355, 379
Tropomyosin 73, 78, 129, 130
Troponin 78, 129
True flight 148, 149, 495
Tuatara 27, 28, 40, 164, 403, 420, 468, 470, 471
Tubulidentata 36, 492, 493
Tunica adventitia 307, 332
Tunica albuginea 256, 257, 374
Tunica intima 307, 308
Tunica media 307, 308
Tunica vaginalis 257
Tupaioidea 495
Turbinates 32, 89, 199, 237, 241, 390, 481
Turtles 27, 28, 40, 90, 163, 164, 194, 196, 235, 300, 301, 433, 445, 468, 469, 476, 477, 481
Tympanic membrane 89, 105-108, 112, 113, 116, 120, 450, 464, 469

U

Ulna 89, 105-108, 112, 116, 120, 450, 464, 469
Ultimobranchial bodies 367, 368
Umbilical arteries 287, 304, 305
Umbilical cord 35, 287, 288, 304, 305
Umbilical vein 287, 305
Uncinate process 100, 105
Unconscious selection 511
Unguligrades 112
Uniaxial joints 116
Unipolar neurons 67, 68
Urbilateria 412
Urea 5, 13, 17, 24, 174, 189, 227, 430, 431, 433, 444
Ureotelic 433
Ureters 209, 263, 264, 434, 445
Urethra 33, 48, 50, 256, 257, 262, 434, 445, 446
Uric acid 5, 40, 150, 189, 217, 286, 322, 424, 426, 433, 445
Uricotelic 434
Urinary bladder 50, 236, 255, 256, 262, 350, 384, 445, 446
Urine formation 437, 441, 444
Urodeles 25, 26, 61, 139, 254, 385, 386
Urodeum 209, 431
Urogenital sinus 259, 263, 264, 431
Uroneurals 22
Urophysis 366
Urostyle 97, 100-102, 147
Urotensin I and II 366
Utatsusaurus 476

Uterus 33, 35, 49, 73, 259, 260, 263, 265, 271, 288, 290, 306, 365, 375, 445
Utricule 405, 406

V

Vaginal sinus 264
Vagus nerve 349, 350, 353, 393
Valves 208, 222, 231, 232, 296, 303, 306, 308, 326, 327, 330, 405, 476
Vas deferens 50, 256, 257
Vasa recta 438, 441-444
Vasa vasorum 306
Vasopressin 364, 379, 443
Veins 125, 179, 284, 287, 300, 308, 311, 312, 326, 330, 332, 366, 441
Ventral (abductor) muscles 141
Ventral intercalary plates 93
Ventral nerve root 68, 346, 347
Ventral rami 133, 348
Ventricle 24, 34, 75, 76, 295-303, 305, 326, 327, 339, 344, 359, 361
Ventricular natriuretic peptide 378
Ventrobronchi 238
Venules 125, 179, 181, 306-309, 332, 441
Vermiform appendix 200, 210
Vertebrae 1, 2, 10, 20, 22, 27, 28, 58, 61, 88-101, 110, 133, 146, 147, 211, 350, 478, 494, 495
Vertebral arteries 344
Vertebral column 10, 23, 27, 62, 81, 82, 89, 92, 93, 97, 102, 106, 143, 147, 148, 298, 345, 351, 470
Vertebral ribs 100
Vesicular gland 257
Vestibular membrane 396
Vestibule 50, 199, 262, 394, 395, 400, 404, 405
Vestibulocochlear nerve 348
Vibrissae 30, 173, 496
Vipers 192, 385, 471, 472
Visceral skeleton 84-86, 135, 136, 220
Vitamin D 59, 65, 180, 190, 368
Vitamin K 180, 190, 323
Vitamins 4, 187, 189, 190, 212, 214
Volkmann's canal 60, 82
Vomeronasal organ 198, 241, 391, 392

W

Wallaby 486
Walrus 288, 497
Weberian ossicles 399
Whale shark 11, 15, 191
Whales 2, 29, 37, 176, 178, 183, 191, 196, 266, 289, 340, 395, 404, 436, 496
White matter 68, 70, 343, 345-347, 351
Whooping crane 237
Wings 28, 29, 105, 107, 109, 148-150, 391, 474, 480, 495, 509

X

Xanthophores 156, 160, 163
Xenarthra 36, 492-494, 509

Y

Yolk sac 27, 34, 35, 206, 260, 263, 264, 274, 283-287, 321, 427

Z

Zebrafish 22, 90
Zona fasciculate 369
Zona glomerulosa 369
Zona pellucida 269, 270, 274, 291
Zona reticularis 370
Zygapophysis 90, 99
Zygomaticus 136
Zygote 5, 252, 264, 270-272, 274, 284, 286, 287, 291, 450

Color Plate Section

Chapter 2

Fig. 2.14 The living representatives of the order Ratitae that comprises flightless birds that belong to the subclass Paleognathae ('old jaw'). Paleognath birds were the first birds to evolve and include the largest of all living birds. A male ostrich (top left), emu (top right), cassowary (bottom left), rhea (bottom middle) and kiwi (bottom right). Some of the extinct ratites include the elephant bird of Madagascar and the moas of New Zealand that are thought to have been eliminated by human beings. Ratites rely on their running speed and strong kicks to defend themselves. The ostrich of Africa is the fastest bipedal runner and can attain speeds of up to 65 km/h while taking strides as long as 4.5 m at times. The other order of paleognath birds is Tinamiformes that includes the tinamous of South and Central America that have a keeled sternum and are weak fliers.

Chapter 4

Fig. 4.32 A giant panda or panda bear (left) and the lesser or red panda (right). Pandas are found mainly in parts of South China, Tibet and Nepal. Male giant pandas weigh up to 115 kg while the lesser panda is slightly larger than the domestic cat and can measure up to 60 cm in length. Pandas are close to both bears and raccoons behaviorally and anatomically. There is much debate among scientists as to where to classify pandas.

Chapter 5

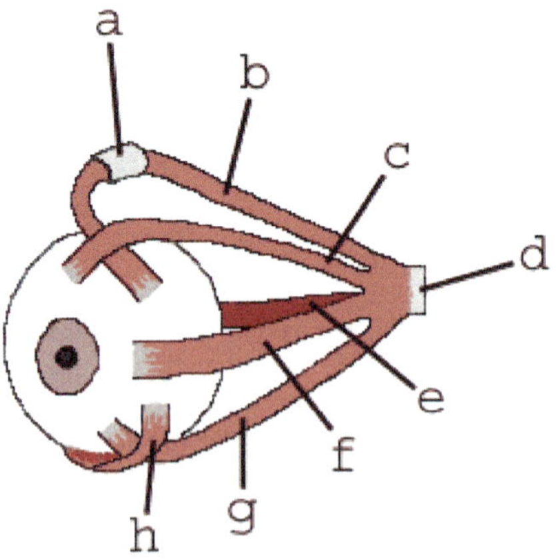

Fig. 5.10 Lateral view of extraocular or extrinsic muscles of the left eye. (a) trochlea, (b) superior oblique, (c) superior rectus, (d) annulus of Zinn, (e) medial rectus, (f) lateral rectus, (g) inferior rectus and (h) inferior oblique.

Chapter 8

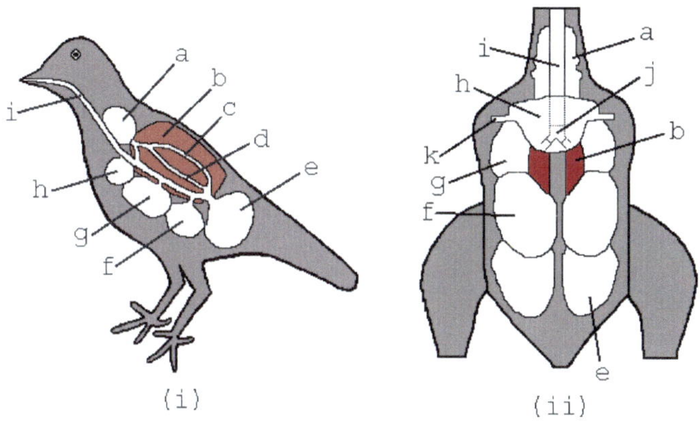

Fig. 8.17 (i) Longitudinal and (ii) frontal views of air sacs of the avian respiratory system. (a) cervical, (e) abdominal, (f) posterior and (g) anterior thoracic and (h) clavicular or interclavicular air sacs; (b) lungs, (c) dorsal and (d) ventral bronchi, (i) trachea (j) syrinx and (k) diverticulum to pneumatic part of humerus.

For Product Safety Concerns and Information please contact our EU representative GPSR@taylorandfrancis.com Taylor & Francis Verlag GmbH, Kaufingerstraße 24, 80331 München, Germany

Batch number: 10725820

Printed by Printforce, the Netherlands